Rock mechanics and engineering

T0292352

Rock mechanics and engineering

CHARLES JAEGER

Dr ès Sc. Tech., M.I.C.E., S.I.A.

Formerly Visiting Professor of Hydro-power Engineering, Imperial College, London and formerly Visiting Professor of Rock Mechanics, Colorado State University

SECOND EDITION

CAMBRIDGE UNIVERSITY PRESS

Cambridge

London   New York   Melbourne

CAMBRIDGE UNIVERSITY PRESS
Cambridge, New York, Melbourne, Madrid, Cape Town, Singapore, São Paulo, Delhi

Cambridge University Press
The Edinburgh Building, Cambridge CB2 8RU, UK

Published in the United States of America by Cambridge University Press, New York

www.cambridge.org
Information on this title: www.cambridge.org/9780521103381

First published 1972
Second edition 1979
This digitally printed version 2009

A catalogue record for this publication is available from the British Library

Library of Congress Cataloguing in Publication data
Jaeger, Charles, 1901–
Rock mechanics and engineering
Bibliography: p. 463   Includes indexes
1. Rock mechanics 2. Civil engineering I. Title
TA706.J3 1978   624'.1513   77–85700

ISBN 978-0-521-21898-6 hardback
ISBN 978-0-521-10338-1 paperback

Contents

Preface   *page* ix
Preface to the first edition   xi

Part 1: Introduction to rock mechanics

1   The historical development of rock mechanics   1
   1.1   The first attempts at rock mechanics   1
   1.2   European and American efforts   2
   1.3   Present trends   4

2   Engineering geology and rock mechanics   5
   2.1   The geologist's approach to rock mechanics   5
   2.2   Typical case histories   11
   2.3   Discussion   25

Part 2: Rock material and rock masses

3   Fundamental concepts and description of fissures   27
   3.1   Definitions   27
   3.2   Structure and anisotropy of rock masses   28
   3.3   Orientation of geological planes   30
   3.4   Statistical density of fissures   31
   3.5   Rock mechanics surveys   32

4   Physical and mechanical properties of rock material   35
   4.1   Physical characteristics   35
   4.2   Anisotropy of rock material   37
   4.3   Modulus of elasticity of rock; Poisson's ratio   40
   4.4   Tensile strength   42
   4.5   Compression tests   45
   4.6   Shear tests   51
   4.7   Dispersion of test results; scale effect and microfissuring   56
   4.8   Correlations of the void index i with some rock characteristics   57
   4.9   Permeability tests   59
   4.10 Correlations between permeability and mechanical properties
          of rock material   68
   4.11 Rock fracture   71

4.12 Classification of rock material   *page* 78
4.13 Filling materials for fractures and faults   82

5 Residual stresses in rock masses *in situ*   85
   5.1 Heim's hypothesis   85
   5.2 Stress relief at the surface of rock masses   86
   5.3 The effective modulus of elasticity and the effective Poisson's
        ratio in rock masses   87
   5.4 Strain and stress about cavities: theory   89
   5.5 *In situ* methods of measuring residual stresses; measuring
        stresses about cavities   94

6 Strains, modulus of deformation and failure in rock masses   102
   6.1 General remarks   102
   6.2 Rock strain and the modulus of deformation; methods of
        measurement   103
   6.3 Other *in situ* tests: shear tests   118
   6.4 Creep of rock masses   123
   6.5 Strain–stress diagrams and interpretation of strain–stress
        curves   128
   6.6 Geophysical methods for testing rock masses   139
   6.7 Engineering classification of jointed rock masses. General
        approach to the problem   154

7 Mathematical approach to strain–stress distribution in rock masses   161
   7.1 Useful formulae   161
   7.2 The half-space of Boussinesq–Cerruti   168
   7.3 Fissured rock masses   178
   7.4 The clastic theory of rock masses   181
   7.5 The finite element method   186

8 Interstitial water in rock material and rock masses   193
   8.1 General remarks   193
   8.2 Some general equations on the flow of water   193
   8.3 Effective stresses in rock masses   196
   8.4 Flow of water in rock masses with large fractures   197
   8.5 Physical and physico-chemical alteration of rock by water   199
   8.6 Aging of rock masses   200

Part 3: Rock mechanics and engineering

9 Rock slopes and rock slides   203
   9.1 Terzaghi's theory; 'the critical slope'; stability of rock faces   204
   9.2 Rock slides; deep-seated lines of rupture   207
   9.3 Effect of interstitial water on slope stability   211
   9.4 External forces which load a slope   216

9.5   The dynamics of rock slides   *page* 218
9.6   Classification of rock falls and rock slides   219
9.7   Supervision of potential rock slides; stabilization of slides   221

10 Galleries, tunnels, mines and underground excavations   226
    10.1   Introduction   226
    10.2   Additional information on stresses about cavities   226
    10.3   Stresses around tunnels and galleries caused by hydrostatic
           pressure inside the conduit   230
    10.4   Minimum overburden above a pressure tunnel   237
    10.5   Overstrained rock about galleries   255
    10.6   Stress and strain measurements in galleries   259
    10.7   New tunnelling techniques in relation to rock mechanics   261
    10.8   Rock bolting   271
    10.9   Classification of jointed rock masses for tunnelling. Esti-
           mate of required rock support based on rock charac-
           teristics   281
    10.10 Estimate of required rock support based on rock deforma-
           tions   287
    10.11 Rock mechanics for underground hydroelectric power-
           stations   308

11 Rock mechanics and dam foundations   325
    11.1   The classical approach to dam foundations   325
    11.2   Shear and horizontal stresses in rock foundations of dams:
           the tensile stresses   325
    11.3   Percolating water through dam foundations: grouting and
           drainage   330
    11.4   Dam foundation design and construction in recent years:
           case histories   347

Part 4: Case histories

12 Dam foundations and tunnelling   359
    12.1   Rock mechanics for Karadj dam   359
    12.2   Analysis of rock properties at the Dez project (Iran)   365
    12.3   Foundation investigations for Morrow Point dam and
           power-plant and Oroville dam and power-plant   367

13 Incidents, accidents, dam disasters   379
    13.1   The Idbar experimental thin-arch dam (Yugoslavia)   380
    13.2   The Frayle arch dam (Peru)   381
    13.3   The Malpasset dam disaster   383
    13.4   Final comments on dam foundations and dam failures   401

14   The Vajont rock slide   *page* 402
     14.1   General information   402
     14.2   The period before 9 October 1963   403
     14.3   Geophysical investigations after the rock slide   406
     14.4   Discussion of the observed facts: *a posteriori* calculations   409
     14.5   The velocity of the slide: the dynamic conditions for an
            accelerated rock slide   411
     14.6   The basic contradictions between 'static' and 'dynamic'
            approaches   413
     14.7   Explanation based on the dynamics of a discontinuous flow
            of masses   414
     14.8   The correlation between rock displacements and the water
            variations in the reservoir   418
     14.9   Summary of the different phases of the Mount Toc rock
            slide   421

15   Two examples of rock slopes supported with cables   424
     15.1   Consolidation of a rock spur on the Simplon Pass road   424
     15.2   Stabilization of a very high rock face for the Tachien Dam
            foundations   427

16   Three examples of large underground hydro-power stations   435
     16.1   The Kariba South Bank power scheme   435
     16.2   The Kariba North Bank power scheme (Kariba Second
            Stage)   443
     16.3   The Waldeck II pumped storage power-station   450

References   463

Appendix 1: Comments on the bibliography   491
Appendix 2: Measurement conversion tables   494
Appendix 3: Table of geological formations and earth history   496
Appendix 4: Some petrographic properties of rocks   497

Author Index   499
Index of geographical names, dam sites, reservoirs, tunnels and caverns   506
Subject Index   509

Preface

At the time the first edition of *Rock Mechanics and Engineering* was being printed, important progress was being made both in theory and practice of rock mechanics. Some new advances were analysed in an 'Appendix' to the book, which is now incorporated, with the necessary additions, in the relevant chapters of the second edition. New developments of the new Austrian tunnelling method (NATM) and similar methods caused the important chapter on underground power-stations to be rewritten, and several new chapters to be added.

The problem of bridging the gap between scientific research in rock mechanics and practical engineering has become more acute. Such bridging has recently been achieved in *Fluid Transients* (Jaeger 1977); it is also vital to applied rock mechanics, as explained in the Preface to the first edition. Many geologists suggest that the rock quality designation (RQD) of Deere is the most reliable parameter for an engineering classification of jointed rock masses. Some geophysicists did not agree and recently introduced their own more complex classification, based on the combination of several parameters describing rock characteristics. Engineers in charge of the construction of large tunnels and underground works were not convinced by these efforts and base their own designs on the rock deformations they expect to occur.

The second edition deals with these problems in several new chapters. There is no better method to deal with them than the close analysis of some case histories. The discussion on the engineering classification of jointed rock masses and the required rock support is illustrated by the description of the second Gotthard Tunnel (16 km long), now under construction and the design of the third, so-called Basis, Tunnel (40 km long).

Many other points require illustration by case histories and two new chapters are introduced. One concerns the stability or instability of rock faces and possible rock slides (chapter 15). The work done for the 300 m high, very steep, rock abutment of Tachien Dam is one of the situations analysed. Underground works is the second subject chosen for extensive new developments (chapter 16). Three very large underground works, Kariba South Bank, Kariba North Bank and Waldeck II are described and analysed, showing the rapid evolution of modern techniques.

Comparing all the case histories on dam foundations, slope stability, rock slides, underground works developed in this second edition brings an answer

x   *Preface*

to the many attempts at classifying jointed rock masses for engineering purposes. There is no universal rule or classification to solve problems of applied
rock mechanics. Any one problem is to be examined in its many aspects
from first principles, using all information available from geology, geophysics,
rock hydrology and engineering.

I should like to express my thanks to Professor F. C. Beavis who pains-
takingly read through my revisions, suggesting a number of corrections and
several most helpful emendations.

C. J.

Pully, July 1978

Preface to the first edition

The first attempts at investigating the mechanical and physical properties of rocks go back to the second half of the last century; but systematic efforts to develop this into a real science of rocks are recent.

One of the more important results of these efforts has been to show the paramount importance of voids, fissures and faults at the level of rock crystals or grains of the rock material and of large masses of rocks *in situ*. Any theory of rock mechanics which considers rocks to be no more than homogeneous masses is just a first approximation. All the properties of rock and rock constants: strength, elasticity, plasticity, perviousness and reactions to sound and to seismic waves depend on the gaps, voids or fissures of the rock as much as on the skeleton of the solid material which builds it up. It is from this point of view that this book has been written.

One of the main purposes of the book is to stress the many links between rock mechanics and engineering. Mining engineers and civil engineers cannot excavate mines or build structures without a knowledge of rock mechanics. This is now accepted by these professions, which include discussions on rock mechanics at their international congresses and symposia. On the other hand, the benefit that rock mechanics gets from the development of modern engineering techniques has seldom been adequately appreciated.

Mining engineering was active at the start of engineering geology and, some decades later, at the birth of the new science of rock mechanics— different from both engineering geology and petrography. It is less well known how great an impulse rock mechanics got from progress and research in civil engineering. Among civil engineers, dam designers were the first to become interested in the progress of rock mechanics. They soon realized that rock foundations were part of the design of any dam. They developed their own techniques for measuring the strength and elasticity of rock masses: *in situ*, at the rock surface, in trenches and in galleries. They recognized the paramount importance of the joints, fissures and faults in the rock masses and contributed to the development of methods for tridimensional representation of families of fissures, testing their shear strength and curing them. The latest tendency of dam designers is to direct their efforts towards a more precise description of the type of rupture occurring at different depths inside the rock masses, and an estimate of the static and dynamic effect of water seeping through the joints and fissures. New chapters of rock mechanics were opened by the joint efforts of dam designers and rock specialists.

Entirely new points of view on the kinematics and dynamics of rock slides, unknown to geophysicists, were developed by engineers responsible for the design of large water reservoirs. More recently, vibrations of the earth's crust and earthquakes were analysed by them.

For many years, tunnel designers worked from empirical rules. At the beginning of the century interest in the strength, elastic and plastic properties of rocks arose. The notion of stress and strain patterns developing about empty galleries or caverns, or around pressure tunnels, is more recent: the sudden undertaking, for many purposes, of large underground works, forced specialists to consider new methods of stress and strain analysis. The finite element method allows the analysis of jointed rock masses, crossed by fissures or faults. More recently still, the effects of rock relaxation about cavities and the drop of the modulus of deformation in relaxed rock were analysed. These entirely new concepts about the behaviour of rock masses are initiating new lines of research in rock mechanics. And progress in engineering is linked with active research in rock mechanics.

Part one of the book (chapters 1 and 2) stresses the importance of the geologist's work, without which the science of rocks would not exist.

Part two (chapters 3 to 8) co-ordinates the knowledge of the physical and mechanical properties of rock acquired from the ample information submitted to the First and Second Congresses and the Sixth Symposium on Rock Mechanics. Chapters 3 to 4 deal with the rock material—laboratory samples—with no major fractures, and chapters 5 to 8 study the properties and behaviour of rock masses *in situ*, and analyse strains and stresses in such masses.

Abstract knowledge of rock properties is of limited use to engineers. It is vital to bridge the gap between the accumulated scientific data on rocks and the requirements of design and field engineers. The second half of the book deals with practical applications. Part three analyses the diverse aspects of rock slope stability (chapter 9) and the strains about cavities excavated in the rock and the modern techniques of underground works (chapter 10). Chapter 11 discusses the very controversial problem of dam rock abutment design.

Part four (chapters 12 to 14) describes typical case histories, which illustrates the more important points developed in parts two and three.

C. J.

Pully, March 1971

Part One
Introduction to rock mechanics

1 The historical development of rock mechanics

1.1 The first attempts at rock mechanics

At the end of last century, geologists studying the formation of the Alps were realizing that tremendous forces were necessary to lift continents and form their mountain chains. Mining engineers and tunnel experts watching rock bursts and rock squeezing in tunnels and galleries, suggested that some 'residual forces' were still at work in rock at great depth. The German tunnel expert Rziha (1874) was probably the first to be concerned with the horizontal component of the forces acting in many tunnels. A few years later Heim (Professor at Zurich University and at Zurich Federal Institute of Technology) suggested that the horizontal force component must be of the same order of magnitude as the vertical component and he forcefully stressed this opinion in several papers (1878–1912). It took many decades for geologists and engineers to realize the importance of the ideas of Heim and Rziha.

In 1920, the Ritom tunnel, which had just been built south of the Alps by the Swiss Federal Railways, was severely damaged. Inspection showed many longitudinal fissures running along the tunnel. The rock strata had a general dip towards the valley and it was feared that water seepage could cause a rock slide. The tunnel was repaired.

At that time, the Swiss Federal Railways were also building the Amsteg tunnel north of the Alps. They decided to start pressure tests in this second tunnel. A dead end of the gallery was sealed off with a concrete plug provided with a manhole and steel cover and was filled with water under pressure. The tunnel diameters were measured by a spider with six branches and the length variations of the six radii versus time were recorded on a rotating disc. The varying water pressure was also recorded and strain-pressure diagrams traced. The bulk modulus of elasticity was estimated as a ratio of stress versus deformation. This was probably the first recording of the elastic deformations of rock masses.

A few years later J. Schmidt (1926a, b) published a thesis in which he

[1]

cleverly combined Heim's ideas about residual stresses in rock, with the newly formulated ideas of rock elasticity to produce the first attempt at a theory of rock mechanics.

It was at this time that steel linings for tunnels and shafts were first introduced, and several authors (Jaeger, 1933), in different countries, produced papers estimating the stresses in the lining as a function of the relative elasticity of the steel and the rock. A few years later the Chilean geologist Fenner (1938) published a thesis which in many respects is similar to that by Schmidt. These two pioneering works were ignored by most engineers until many years later. Some of their theories were confirmed by Terzaghi & Richart (1952).

1.2 European and American efforts

Many engineers were astonished at the wealth of material Talobre was able to produce in 1957 in his excellent treatise *La Mécanique des Roches*. During the preceeding ten years, research in the field of rock mechanics had been slowly gaining momentum, and Talobre's treatise was most timely.

Important research had been going on on both sides of the Atlantic mainly in connection with the mining industry. As early as 1916, Young & Stock listed over a hundred papers dealing with mining problems and the mechanics of subsidence, mostly in relation to coalfields. American mining schools and the U.S. Bureau of Mines were very active and so too were their European counterparts. They were concerned with theoretical problems of stress around rectangular-shaped cavities but were also faced with many practical problems. Techniques were being developed for measuring strains and rock deformations, rock elasticity and convergence of the walls of galleries and cavities. A treatise by Obert & Duvall (1967) gives an extensive bibliography on the early efforts of American and European mining engineers.

Most American experts regard 1950 as the year in which systematic research into rock mechanics began in the USA. American mining schools and universities were increasingly active, mainly on the national level. Nation-wide symposia were organized; the U.S. Bureau of Reclamation at Denver was leading world research on the properties of rock material and rock masses; and an American Society of Engineering Geologists was created, one of its aims being the development of rock mechanics, which for many years was also the concern of the American Geophysical Society, the American Society for Testing and Materials and their research committees. In several American universities the teaching methods in engineering geology were modernized and adapted to the requirements of the petroleum and mining industries.

In Europe during the years 1950 to 1960, the most active centre of research outside the mining schools was probably the University of Vienna, where

Stini created an Austrian Society for Geophysics and Engineering Geology. This expanded rapidly and the 'Austrian School' became well known for its efforts in precisely describing and defining the faults and fissures in rock, far more exactly than is usual in engineering geology. Engineers from many European countries congregated in increasing numbers at the annual congress organized in Salzburg. After deciding against the possibility of linking its efforts with the International Conference on Soil Mechanics, the Salzburg group expanded on its own, forming the core of an independent International Society for Rock Mechanics. This body organized the First International Congress in Lisbon, Portugal, in 1966.

As early as 1951, dam designers started a parallel effort, when a suggestion was submitted by the author to the International Commission on Large Dams (ICOLD) to create a sub-committee on rock mechanics.

Up to this time geologists had been extremely careful in deciding where to build dams. Dams were of relatively moderate size and there were few problems on the stability of rock abutments. Isolated cases of dam rupture were explained by uplift forces or by failing shear strength of the rock. These two points initiated the development of techniques of dam foundation.

The demand for more and more electric power, led to larger dams and to the introduction of bold arch dams. The problems of the strength of rock abutments were becoming more pressing and more difficult. It became imperative to include the elasticity and plasticity of the rock abutments in the mathematical analysis of the arch dams and to consider with greater care the strain and stress distribution in the rock masses. The techniques of testing rock *in situ*, of analysing test results and of exploring rock abutments with galleries attained a high degree of precision. It was felt that dam designers and hydro-power engineers were fully responsible for the structures they were designing and that they could not leave the task of developing methods for rock testing, tunnel construction and dam foundation design to others.

In 1957, a small committee of experts, headed by G. Westerberg (Stockholm), submitted a report recommending the formation, within the organization of ICOLD, of a 'Committee on underground work' whose aim would be to solve the most urgent problems of rock foundation for large dams. The first official meeting of the new committee was at the Sixth Congress of ICOLD in New York, in 1958 (Guthrie-Brown, 1970).

On 3 December 1959, the dam of Malpasset burst, killing about 450 people. The Vajont disaster occurred during the night of 9 October 1963. Members of the 1961 ICOLD Congress (Rome) and the 1964 Edinburgh Congress were deeply shocked by these two disasters. There was no doubt that everybody considered the furthering of rock mechanics as the most urgent task for all dam designers.

1.3 Present trends

The Lisbon Congress (1966) was to show some important increase in the knowledge of physical and mechanical properties of rocks.

The present trends can be summarized as follows:

(1) Research on the structure and microstructure of rocks; causes of weakness and failure. This research mainly concerns universities.

(2) Laboratory testing of rock material; testing methods; standardization of tests, classification of rock material for engineering purposes.

(3) *In situ* tests of rock masses. Physical and mechanical characteristics of rock masses.

(4) The development of new methods of measuring strains, deformations and stresses. Mathematical theories of stresses and strains in homogeneous and non-homogeneous space or half-space and in fissured space. The introduction of new methods and the use of computers.

(5) Rock slopes, stability and safety of structures in rock; design of galleries and cavities, design of rock abutments for dams, consolidation of rock masses.

More recently, the efforts concentrated on:

(6) Research on the statics and dynamics of cleft water and water circulation in rock joints.

(7) The foundation of large dams on rock.

(8) The classification of jointed rock masses for engineering purposes.

(9) Practical applications of the new Austrian tunnelling method and similar methods. The design and construction of underground works.

2 Engineering geology and rock mechanics

2.1 The geologist's approach to rock mechanics

An American geologist, Professor Kiersch (1963), submitted to the annual Congress of the Austrian Society of Rock Mechanics in Salzburg (1962), a most interesting paper on the trends of engineering geology in the United States and on the modern teaching methods introduced at several American universities for training the new generation of engineering geologists. At the same meeting Professor Bjerrum, from Norway, speaking on behalf of A. Casagrande (President of the International Conference on Soil Mechanics) brilliantly explained how rock mechanics should be integrated into soil mechanics as a mere chapter of a wider, more advanced technical science. The Congress reacted by deciding to expand their own efforts and to form the nucleus of an International Society of Rock Mechanics.

Rock mechanics adopts many of the techniques developed in soil mechanics based on the simple law of Coulomb which related shear strength of elastic materials to the friction factor and the normal stress. But the behaviour of rocks is far more complex than the behaviour of soils, and in many cases rock mechanics uses techniques unknown in soil mechanics. The two are parallel but distinct chapters of the science of discontinuous spaces as opposed to the classical theory of strength of materials assuming a continuous space. The links between rock mechanics, engineering geology and classical geology are even more intricate and complex. It is not possible to think about rock mechanics without examining and discussing these links.

Geologic materials possess certain physical, chemical and mechanical characteristics which are a function of their mode of origin and of the subsequent geologic processes that have acted upon them. As stressed by Deere (1968, quoting Miller, 1965) in a lecture given at Swansea University: 'The sum total of these events in the geologic history of a given area lead to a particular lithology, to a particular set of geologic structures and to a particular *in situ* state of stress.' All this geological information is of fundamental importance to rock mechanics.

The geological classification of rocks (lithology) refers to the mineralogy, fabric, chemistry, crystallography and the texture of rocks. Even though rock mechanics has set up a completely different classification, based on strength, deformation and fissuration of rocks, an understanding of the geological classification is of paramount importance.

[5]

There are many obvious similarities between technical methods developed by geologists and methods used by rock mechanicists. One important contribution by Duncan, Dunne & Petty (1968) concerns a detailed analysis of the void index of rock materials classified by their absolute geological age. Duncan writes:

Initially laid down as loose sands or muds, sediments have been subjected to substantial overburden pressures during the course of geological time. As the depositional load increased, lower layers suffered a volume reduction largely attributable to a reduction of the pore volume of the rock. With decreasing pore volume the deposit progressively increased in strength. The fine-grained materials such as clay-shales, shales and marls and mudstones derived their eventual strength largely from this diagenetic process. On the other hand, sands have larger initial pores or spaces between grains, and they may be deposited intermixed with very fine-grained interstitial material which later forms a cement bond around and between grains. At high overburden pressures there is a welding together of individual grains arising largely from high stresses at point contacts between grains and consequently from a pressure-solution effect. We describe as indurated the rock materials with textures created by these more extreme forms of compaction. All through the course of geological time progressive loading results in constant changes in the texture, pore volume and strength of the materials.

2.1.1 The void index *i*

This index *i* is also called the index of alteration. However, there are possible causes of alteration of rock material and of rock masses other than changes in voids and pores and so the term 'void index' *i* will be the only one used here.

The index *i* depends on: the type of rock material: sandstone, shale, mudstone, etc.; on the age of the rock material and on its history, compaction, deformation etc. It is least for indurated sandstone and indurated shales; normally cemented specimens have a lower *i* index than poorly cemented and poorly compacted specimens.

A classification of sandstone according to age and void index for indurated, cemented and poorly cemented material is given in fig. 2.1, and for fine-grained materials, clay-shales, shales, marls and mudstones, in fig. 2.2. A most interesting correlation can be established between the age of rock material and the void index *i*. Geologists and rock mechanicists (Serafim, 1968) use the same technique for establishing this index *i*. The rock material is oven-dried at 105 °C for 12 hours (24 hours by Serafim) and the weight of water absorbed by the dried specimen is determined after immersion in water for 12 or 24 hours. The void index *i* is

$$i = \frac{\text{weight of water absorbed}}{\text{weight of dry rock}}.$$

Indurated sandstones are often 300 to 600 million years old (extending from the beginning of the Cambrian to the end of the Carboniferous periods).

Poorly cemented sandstones are seldom older than Jurassic (about 200 million years); fine-grained indurated rock material is usually over 200 million years old.

Tests on rock materials have shown that swelling strains and swelling stresses can be correlated with the void index. Similarly, the compressive strength, seismic wave velocities and the modulus of elasticity of rock specimens depend on the factor i. There is a direct link between the rock classifications established by geologists, the void index and other rock properties of vital importance to rock mechanics.

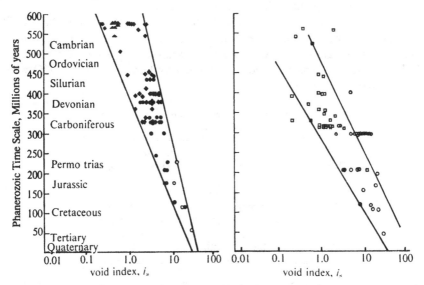

Fig. 2.1 Classification of sandstones: ▲ indurated sandstones; ● cemented sandstones; ○ poorly cemented or compacted sandstones (after Duncan *et al.*, 1968).

Fig. 2.2 Classification of shales, marls, mudstones, etc.: □ indurated; ○ compacted (after Duncan *et al.*, 1968).

Most igneous rocks have a dense interlocking fabric, with only slight directional anisotropy. Some sedimentary rocks are laminated and show considerable anisotropy in mechanical properties; metamorphic rocks may vary over a very large range of properties, most of them important to structural designers.

Some rocks are stable, others are capable of alteration. A massive porphyroblastic gneiss may be compact and competent, but exceptional surface weathering may penetrate several metres deep. Quartz mica gneiss may be poor on the surface, or even microfissured. Amphibolite may have poor weathering characteristics. Wet quartz sand in quartz fissures can create very different conditions in tunnels. Any geological information of this type is of importance to the rock mechanicist and the designer.

Geologists must be able to determine the stability of rock masses, to predict the likelihood of movements along rock joints, slip lines, bedding planes, sheeting, cracks and folliations and be able to warn the engineer of stressed and strained rock masses.

2.1.2 Engineering classification for intact rock

Geologists have introduced other classifications. Deere & Miller (1966) use diagrams relating E_t, modulus of elasticity of rocks at 50% of the ultimate strength to σ_{ult}, the ultimate strength.

Figure 2.3 shows the results for 80 specimens of the granite family, 26 specimens of the diabase family and 70 basalt specimens grouped in a small area of the diagram, whereas values for sedimentary rocks are scattered. Tables 2.1–2.3 are based on tables compiled by Deere & Miller (1966) when summarizing their research.

Table 2.1 *Engineering classification of intact rock on the basis of strength*

Class	Description	Ultimate compressive strength (lb/in²)
A	very high strength	32 000
B	high strength	16 000–32 000
C	medium strength	8 000–16 000
D	low strength	4 000–8 000
E	very low strength	4000

Table 2.2 *Engineering classification of intact rock on the basis of modulus ratio E_t/σ_{ult}*

Class	Description	Modulus ratio
H	high modulus ratio	500
M	medium ratio	200–500
L	low modulus ratio	200

Table 2.3 *Geological classification of intact rock by joint spaces*

Class	Space
very close	2 in (5 cm)
close	2 in to 1 ft (5 cm to 30 cm)
moderately close	1 ft to 3 ft (30 cm to 90 cm)
wide	3 ft to 10 ft (1 m to 3 m)
very wide	10 ft (>3 m)

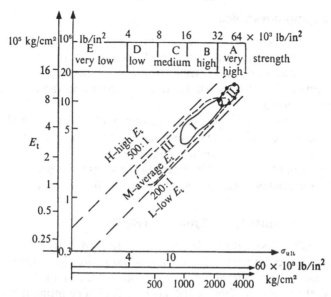

Fig. 2.3   Engineering classification for intact rock: I, granite; II, diabase; III, basalt (after Deere & Miller, 1966).

2.1.3 Classification of joints

Although the word 'joint' is used by geologists for planes of structural weakness apart from faults within the rock mass, it is often vital to differentiate between the various types of joints (Duncan & Sheerman-Chase, 1965, 1966).

Within a sedimentary rock mass, bedding planes may be of considerable lateral extent and relatively uniform throughout any one bed. Shrinkage joints, although often random and discontinuous, are generally present throughout large volumes of the bed. Tectonic joints, produced as a result of regional earth stresses, are usually as prominent at depth as at the surface (Mont Blanc tunnel).

Within an igneous rock mass, cooling joints are more closely spaced near the margins of the intrusion becoming more widely spaced towards the interior of the mass and eventually disappearing. Tension joints and tectonic joints might be expected to be continuous depthwise.

Within metamorphic masses, open cleavage, open schistose and open gneissose planes can be of considerable lateral extent within the one stratum, although these structures are less consistent than the bedding planes of sedimentary masses. Stress relief joints can be expected to show a parallelism to the walls and possibly the floors of the valleys, but to become less frequent in the interior of the mass and eventually to disappear.

This joint classification, suggested by geologists, is very different from the classification adopted by rock mechanicists which will be dealt with later.

2.1.4 Logging boreholes

Geologists obtain a great deal of information about joints and faults by observation of rock outcrops, natural rock cuttings, rock slopes, rock faces and gorges. Additional information is given by inspection of trenches, galleries, adits and shafts; boreholes reaching deep into the rock are often the preferred method.

Logging boreholes is a specialist's responsibility. The cores should be oriented if possible and the main attitudes of the sets of joints should be recorded. Description of the rock types, of the joint spacing, opening and weathering should be described in detail. In some cases the boreholes may be inspected by television cameras specially designed for the purpose.

2.1.5   The rock quality designation (RQD)

The rock quality designation (RQD) classification is based on a modified core-recovery procedure, which, in turn, is based indirectly on the number of fractures and the amount of softening or alteration in the rock mass as observed in the rock cores from a drill hole. Instead of counting the fractures, an indirect measure is obtained by summing the total length of core recovered and counting only those pieces of core which are 4 in (10 cm) or more in length and which are hard and sound. For example, a core run of 60 in length with a total core recovery of 50 in (or 83 %) would with this new definition, be classified as having a recovery of only 34 in, since many small (<4 in) rock pieces would not be counted. This gives an RQD of 57 %.

Following Deere (1968), the relationship between RQD and rock quality is given in table 2.4.

Table 2.4   *Relationship between RQD and rock quality*

RQD (%)	Description of rock quality
0–25	very poor
25–50	poor
50–75	fair
75–90	good
90–100	excellent

The effect of discontinuities in the rock mass can be estimated by comparing the *in situ* compressional wave velocity with the laboratory sonic velocity of an intact core obtained from the same rock mass. The difference in these two dilational velocities is proportional to the structural discontinuities existing in the field. The velocity ratio v/v_L, where v and v_L are the seismic and laboratory sonic compressional wave velocities of the rock mass *in situ* and the intact specimen respectively, was first proposed as a qualitative index by Onodera (1963). The sonic velocity for the core is determined in the laboratory under an axial stress equal to the computed overburden stress at the depth from which the sample was taken.

There is a correlation between rock quality as determined by the velocity ratio and the RQD. For example, for different sources and different rock types where $(v/v_L)^2 = 0.8$, the RQD was 0.70 to 0.90 and where $(v/v_L)^2 = 0.6$, the RQD was 0.50 to 0.80.

There also exists a correlation between the RQD and the fracture frequency (fractures per foot). For example, RQD = 0.75 corresponds to 1 to 3 fractures per foot and RQD = 0.50 corresponds to 2 to 4 fractures per foot.

This is the outline of engineering geology, as pioneered by a few experts. Some of their suggestions will be discussed in greater detail in later chapters.

Rock mechanicists endeavour to fill in the outlines of the geologist with additional detailed information. They try to cover the specific points raised by the design engineer, who is responsible for the reliability of galleries, cavities and dam foundations. The links between engineering geology and rock mechanics have been stressed here, as the techniques of both are intimately related. Another way to see these links is to study some case histories, showing how the work of experts from both sides is interrelated.

2.1.6 The integral sampling method

The previous paragraphs several times refer to the difficulty of obtaining full information from borehole rock samples. The Laboratorio Nacional de Engenharia Civil, Lisbon, has developed the integral sampling method for total recovery of borehole cores (Rocha, 1971a; Jaeger, 1971) including the fine core material lost when using conventional methods. Intact oriented cores may be obtained in which the successive soil layers are kept in their relative positions, and joints, faults and materials filling these defects can be observed. The method consists of drilling a borehole with adequate diameter (> 76 mm) down to the point where the method is to be applied. A narrower coaxial hole is then drilled down from the bottom of the previous one, into which a selected length of steel tube reinforcement (about 32 mm and 26 mm external and internal diameters) is inserted. The reinforcement bar (whose length is usually about 1·5 m) is installed using oriented installing rods and their guiding equipment. The narrow borehole is then filled with an appropriate binder and, after it has set, the larger borehole is drilled so as to overcore the zone of the mass previously reinforced with the bar.

2.2 Typical case histories
2.2.1 Predicting the temperature inside a deep tunnel

British, American and Canadian engineers were the pioneers of tunnelling techniques (Thames tunnel in 1825–8). The 12·8-km-long Mont Cenis tunnel in France was the last to be hand drilled (1870). Engineers gained their first experience of tunnelling under heavy rock-pressure conditions inside the St Gotthard tunnel (1882) and later in the first Simplon tunnel (1906). Geologists were, for many years, trying to help the engineers who were engaged in designing the first long trans-alpine railway tunnels. Their predictions of geological and hydrogeological conditions to be encountered were, at that time, often inaccurate and unreliable. Accurate geological length profiles were usually only established after completion of the excavation.

Rock temperatures are high at great depth, and in saturated air the efficiency of labour falls when the temperature exceeds 25 °C and becomes almost impossible at 35 °C. The death rate among workers in the St Gotthard tunnel was very high. Predictions of temperature is therefore an important part of the preliminary survey for the design of any tunnel, the cover of which reaches 1500 m (5000 ft) or more.

According to Thoma *et al.* (1908) and Andrea (1961) the normal geothermal gradient under plains amounts to 0·031–0·033 °C/m, the reciprocal geothermal gradient being 30–32 m for 1 °C. Where the earth surface is mountainous, the gradient becomes greater under valleys and smaller under the ranges. At great depth the surfaces of equal temperatures (geoisotherms) become parallel planes.

The equation describing the equilibrium condition of heat flow along a vertical section through the mountain containing the centre line of the tunnel is:

$$\frac{\partial^2 \theta}{\partial x^2} + \frac{\partial^2 \theta}{\partial y^2} = 0, \tag{1}$$

where θ is the temperature and x, y are the rectangular co-ordinates (fig. 2.4).

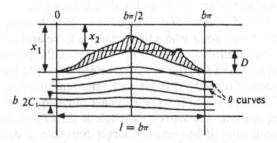

Fig. 2.4   Geothermal gradients (after Andrea, 1961).

The solution of this equation for a regular undulating surface is given by:

$$\theta = C_1 \log_e \{A + \sqrt{(A^2 - 1)}\} + C_2, \tag{2}$$

where

$$A = e^{2x/b} + \sqrt{\{1 - (2\cos 2y/b - e^{2x/b})e^{2x/b}\}}. \tag{3}$$

and C_1 and C_2 are constants. The period of this 'symmetrical' function is $b\pi$ and the θ lines or isotherms are undulating (fig. 2.4).

It can be proved that $2C_1/b$ is equal to the normal gradient $(d\theta/dx)_{x=\infty}$ which determines the value of the constant C_1.

For any value of θ and D (fig. 2.4) the C_2 is given by:

$$C_2 = \theta - C_1 \log_e \left(\frac{e^{D/b} + 1}{e^{D/b} - 1} \right),$$

where D is the height of the isotherm. A curve $\theta = \theta(y, x)$ can be traced from known rock temperatures, measured at the surface of the mountain range and then the isothermic curves inside the rock masses calculated step by step, using $d\theta/dx$ for $y = 0$ and $y = (b\pi/2)$, where $b\pi$ = period of the function. In practice, the mountain range will not be symmetrical and the value of $2C_1/b$ varies with type of rock and fall of strata. It is suggested that equation (1) could, in a general case, be calculated step by step using modern analytical methods.

Andrea (1961) gives a table for the $2C_1/b$ values (table 2.5).

Table 2.5 $2C_1/b$ *values in* °C/m

Fall of strata	Nearly vertical	About 45°	Nearly horizontal
Gneiss	0·027–0·028	0·033	0·034–0·036
Granite	0·027–0·028	0·033	0·034–0·036
Micaschists	0·027–0·028	0·034–0·036	0·037–0·041
Phyllites	0·027–0·028	0·031	0·032–0·033
(if wet: less)		(5·8%)	

The temperature gradient found in one tunnel cannot be assumed in another without further investigation. By applying the gradient measured in the St Gotthard tunnel (1882) the temperature inside the first Simplon tunnel (1906) was predicted to be about 42 °C. In fact it was about 55° C.

Detailed investigations were carried out when designing the Isère–Arc tunnel (1952) to determine the temperature conditions. This tunnel crosses a high mountain ridge. The maximum overburden is about 2000 m and the cross-section of the excavated tunnel bore 43 m² (32 m² after lining). The distance between the extreme headings was 11 700 m, with no intermediary adits. The weir at the tunnel intake on the river Isère was not a major struc-ture; the whole power development depended on the timing of the work inside the long tunnel. The problems of the temperature inside the tunnel and tunnel ventilation were most important to the progress of the work. It was decided to modify two Conway 75 shovels so they could, if necessary, work in parallel; each shovel being able to handle 40 m³/h of crushed rock. The daily progress was about 8 to 12 m. The actual temperature inside the tunnel could be maintained at 34 °C, slightly under the limit of 37 °C predicted by the experts. Two pipes of 1 m diameter and 5 to 6 km in length were used for ventilation. Fresh air was pumped into the tunnel at the rate of 12 m^3/s by 185-h.p. compressors.

2.2.2   Tunnels, galleries and cavities

Predicting geologic formations and the hydrogeology to be encountered along a tunnel is the major task of the geologist.

(1) *Lötschberg* (Switzerland). The 14·6-km-long railway tunnel (1913) was designed to cross the river Kander well beneath the normal water level. Construction ran straight into alluvium which filled the old river bed well below the present river; several workers were killed. The gallery was sealed off and was by-passed by a tunnel about 1500 m long. Lack of systematic geologic prospection before the construction was the direct cause of this disaster.

(2) *Mauvoisin* (Switzerland). The site of the proposed 237-m-high arch dam of Mauvoisin, at that time the highest arch dam in the world, was examined most carefully from the geological point of view. The depth of alluvium filling the narrow gorge at the site was measured seismologically. To enable a direct geological survey to be made, it was decided to drive a gallery deep into the rock abutment on the left-hand side of the gorge. This gallery was planned to pass beneath the gorge and to extend into the right-hand abutment. In addition to a first-hand inspection of the rock, water from several springs was analysed. Later, the geological gallery was used for grouting a deep grout curtain under the dam. When the narrow gorge was reached the gallery unexpectedly penetrated into alluvium, which suddenly gave under the pressure of the water. The gallery was flooded and four engineers were killed.

It is most likely that wrong interpretations were placed on the wave reflections in the narrow triangular-shaped gorge.

(3) *Malgovert* (France) (Pelletier, 1953). Immediately after the Second World War French power-stations were nationalized. Electricité de France was created to administer the whole French electric generating system, with the exception of the power-stations on the river Rhône itself, which were entrusted to the Compagnie Nationale du Rhône, in Lyons. Electricité de France immediately realized that there was an acute shortage of power in post-war France, and possible schemes were hastily examined. A plan proposed for the upper valley of the Isère was found to be too modest. A new scheme was designed for a 180-m-high arch dam at Tignes and the first power-house at Les Brévières (3 × 44 000 h.p., gross head 168·6 m). The

Malgovert tunnel, 9½ miles long, was driven from the portals and through twelve adits. The effective cross-sectional area of the tunnel designed for a discharge of from 47·5 to 50 m³/s was 160 ft², while the excavation ranged from 215 to 290 ft² (20 to 27 m²). The head available between the pool at the Brévières power-station and the end of the Malgovert section was 2463 ft (750 m). The Malgovert power-house is equipped with 4 × 105 000 h.p. Pelton wheels.

On the upstream section, between the intake and adit no. 12, three different methods were used: (*a*) 65-ft² pilot tunnel in invert, (*b*) 65-ft² pilot tunnel in crown, (*c*) full-face (in a few places). The ground behaviour did not give rise to any difficulties, except for the last 1000 ft, where heavy steel supports were needed.

On the downstream section (started in January 1948), between adits 13 and 16, it had been decided to try the full-face method suggested by the geologists. Two large jumbos, both carrying 8 DA-35 Ingersoll-Rand drifters, were available and an American Conway 75 shovel. Forty-eight holes were needed for the full heading. As work proceeded (1949), the quality of the ground steadily deteriorated. Polygonal steel supports were erected at 4-ft centres along the heading. Between adits 13 and 14, it became necessary to revert to the so-called 'Belgian' tunnelling method, using top heading and successive enlargements on part of the tunnel length. Between adits 14 and 16 a 100-ft² invert gallery using drifters for full-face operation was started; the widening of the small pilot gallery was carried out in either one operation or in a number of steps, depending on the type of ground. Heavy longitudinal reinforced concrete beams were used in places as a footing for the polygonal steel arches which maintained the tunnel crown. With these methods it took nine months to drive only 1531 yd.

Difficulties were encountered all the way from adit 13 downwards (Permian–Carboniferous and Carboniferous rock formations). The tunnel caved in with a consequent inflow of sand and water and collapse of the overburden. The ground pressure was so great that the steel ribs used to support the tunnel had to be spaced closer and closer together. When only one foot apart the ribs twisted and the reinforced concrete beams sheared through. Severe damage occurred in one adit where a lateral concrete wall moved inwards by 5 ft and the rail track was lifted by 4 ft in a few hours.

At one point, the tunnel ran into swelling ground, and a stratum of badly crushed shales was located near a fault. These shales were comparatively dry, and driving was easy. A few days after excavation this ground became wet, swelling and developing such a pressure that the steel supports failed, even where they were set flange-to-flange. Work on the heading was stopped and a diversion tunnel driven. But a failure in the roof near a fault caused a heavy inflow of water (nearly 4000 gal/min) which flooded the tunnel. Yet again work was stopped and several pumps brought into action. Progress was held up for several months.

A visitor to the tunnel site could have seen typical signs of a slow rock slide on the mountain slope; not a tree stood upright! Site engineers, commenting on the difficulties encountered, said that because of the pressing demand for electric energy, geological investigations had relied too much on work done before the Second World War. The obvious rock creep was considered to be a superficial movement; rock in the depth was declared to be sound. Facts proved this opinion to be over-optimistic!

Rock slides are still not fully understood. The dynamics of creeping rock masses have yet to be worked out and better methods for determining their depth have still to be devised. (See the Vajont rock slide, chapter 14.)

As a result of this pressurized rock zone, the lining of the tunnel had to be delayed until the creeping rock had stabilized. It was found that the tunnel soffit had been lowered on average by one or two feet over a distance of several kilometres and the cross-sectional area of the tunnel proportionately reduced. Additional pressure losses would have occurred in the water flow with a corresponding loss of electric power. It was decided to restore the tunnel and cut the crown to the planned height before lining the tunnel.

Some poor rock characteristics in the Malgovert tunnel area below adit 13 were not due to creeping movements of the rock and could have been foreseen in a more detailed geological investigation.

In 1946, while driving adit 13, a wet crushed-quartzite seam was encountered at about 426 ft from the portal. The same seam was found again near another portal. In both cases the heading collapsed, the tunnel was filled with wet crushed-quartzite sand and the water pressure found to be 115 lb/in^2.

A specialist contractor was called in to carry out grouting operations to strengthen the crushed rock. A cylinder of rock, about 81 ft diameter and coaxial with the tunnel (19·6 ft diameter bore) to be excavated, was consolidated. A reinforced concrete bulkhead was built to provide protection against fissures in the ground which might be formed as a result of the grouting pressure. After consolidation, the tunnel was excavated, but not to the end of the treated section. Another bulkhead was built and further grouting carried out (2·8-in boreholes were drilled).

The powdery quartzite sand was treated in four stages (Pelletier, 1953) (fig. 2.5).

(a) Sodium silicate solution grouting to prepare the ground.
(b) Cement grouting was continued until the quartzite was compressed at 1·432 lb/in^2.
(c) Sodium silicate and phosphoric acid solution grouting. A grouting pressure of up to 142 lb/in^2 was used for this low-viscosity solution. When the grout sets, after about 15 min, it forms a coherent gel, filling the voids in the consolidated rock. The void ratio is 30%.
(d) Cement grouting keys the ground perfectly by further compression.

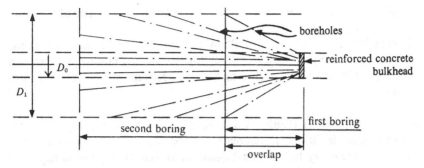

Fig. 2.5 Malgovert tunnel: method of grouting for tunnelling in quartzite. D_0 = tunnel diameter, 19·6 ft; D_1 = consolidated quartzite, 81 ft (after Pelletier, 1953).

The usual methods were used for the excavation, although the consolidated ground was so hard it was necessary to use jack hammers and explosives instead of pneumatic drills. The tunnel was worked in 115-ft stages. No further difficulties or falls occurred. Altogether 3800 tons of cement and 5240 yd³ of gel were used in 16 085 ft of tunnel and 3½ miles of reboring.

(4) *Isère-Arc* (France) (Olivier-Martin & Kobilinsky, 1955). Immediately downstream from the Tignes–Malgovert scheme, the same team of French engineers designed and constructed the Isère–Arc power development. It is interesting to see how experience gained from the Malgovert tunnel was used in the next step of the Isère Valley development.

The Isère–Arc tunnel, from the Isère to the Arc Valley, was designed for a 100 m³/s discharge, double that of Malgovert. A detailed geological survey indicated that the tunnel would cross a weak inclined strata of schists, for which the geologists were able to predict the approximate position and thickness.

The problem facing the engineers and the contractor was how to cross the weak zone of crushed rock. About 60 m were excavated full-face and the tunnel supported with heavy steel arches (I-type differdinger steel weighing 36 to 44 kg/m). The tunnel was lined immediately behind the heading before the rock mass started squeezing inwards. Daily progress, which under ordinary conditions was about 8 to 12 m per day, dropped to 1 m per day.

As the difficulties increased it was decided to change to a bottom pilot tunnel 14 m² in cross-sectional area; a top heading pilot tunnel might have been easier. The bottom heading was chosen because the rail track and the whole equipment was at the level of the tunnel invert. The soffit was excavated laterally as the bottom heading progressed. It took about eight months to cross the 260-m zone.

At the time when the Malgovert and the Isère–Arc tunnels were built, rock bolting was not known to Continental tunnel engineers. Commenting on the Isère–Arc tunnel, the chief resident engineer, Kobilinsky, expressed the opinion that rock bolting would have substantially accelerated the work on this tunnel.

Comparing the two tunnels built by the same team of engineers shows how the more precise geological survey for Isère–Arc allowed the engineers to change from full-face heading to bottom heading with a minimum loss of time.

(5) *Kemano* (Canada) (Libby, Cook & Madill, 1962). Geologists at the Kemano tunnel in British Columbia had noticed a fault crossing the tunnel in hard, medium crystalline quartz–diorite. It ran at approximately right angles to the tunnel and with a nearly vertical dip of 80°. The fault contained mylonite $\frac{1}{8}$ in to 2 in in size (microbrecciated rock) and gouge (a soft clayey material). At some places the wall rock was chloritized and softened for up to several feet on either side of the fault. Two sets of closely spaced fractures, probably related to primary jointing, also occurred on either side of the fault. Fracture members were 4 to 7 ft apart for a distance of 30 ft from the fault. It is known to geologists that faults filled with mylonite and gouge erode deeply to depths of several feet, whereas faults filled with only mylonite or with only gouge erode less.

The engineers in charge of the construction decided not to line it as the tunnel was 16·7 km long and the lining would cost about 12 million dollars.

About two years after completion, pressure losses along the tunnel increased; a rock fall inside the tunnel was suspected. Piezometric measurements made through boreholes from the surface down to the tunnel located the rock fall in the region of the fault. The exact location of the rock fall was confirmed by the reflection time of water hammer waves created by suddenly closing the turbines.

Repairing the damaged tunnel was a major job. The rock fall formed a natural dam inside the tunnel and the water held behind it had to be pumped out. A 500-ft-long, 12-in pipe was used to pump 4000 gal/min (300 litre/s). A timber bulkhead was built upstream of the huge cavern formed in the tunnel roof by the fall. The space between the natural rock fall and the bulkhead was filled with crushed debris to a level just above the original tunnel crown and an attempt was made to shape the surface of the fill correctly. A ventilation duct 4 × 6 ft was provided at the top of the fill. The cavern walls were washed and 1000 yd³ of high-slump concrete was deposited by a continuous operation on top of the fill, over the entire cavern area, to form a protective arch, 6 to 12 ft thick, 35 ft wide and 65 ft long. A layer of $\frac{3}{8}$ to $\frac{3}{4}$ in crushed rock, 3 ft thick, was blown on top of the concrete as a protection against falling rocks. The fill was then excavated and steel ribs positioned to support the protective concrete arch. Finally, the protective arch was blocked and the tunnel lined with concrete. The whole repair operation had been difficult and dangerous because pumping out the water had caused further rock falls.

Minor rock falls along the tunnel were stopped by the conventional method of pneumatic concreting to about 24 in thick. The cost of the repair work was 2·1 million dollars less than that of lining the whole tunnel, but far more than preventative action (local treatment of fractured area) would have been.

(6) *Kandergrund* (Switzerland, 1910) (Jaeger, 1963*b*). Accurate information from the geologists may not be sufficient to trace the real cause of damages. The Kandergrund tunnel (4 km long) was designed and concrete-lined as a free-flow tunnel with a large reservoir chamber excavated in the rock. Later, the tunnel was put under pressure and the reservoir chamber used as a surge tank.

Twice, the concrete lining was severely damaged a few hundred metres from the tunnel inlet. Geologists pointed to very poor rock conditions. The damaged section was relined and the poor rock blocked. The tunnel burst a third time at the same place. Water, seeping through wide fissures in the concrete lining, accumulated over a curved, impervious rock formation and caused a landslide. A forest was destroyed and two people were killed when a house at the foot of the slide collapsed. Geologists again pointed to the obviously poor quality of the rock. For the third time, the rock was grouted and blocked and the tunnel relined.

Three 30-m-long and 5- to 7-cm-wide fissures had formed in the soffit of the tunnel lining and there was evidence that it had been lifted by extremely high hydraulic pressures. Poor quality rock could not explain these conditions.

The rise in pressure inside the tunnel was caused by water hammer waves forming standing resonance waves inside the tunnel. This complex hydraulic phenomenon has since been observed in other hydro-power schemes. The cause of all the trouble was a defective air valve on the pressure pipe line (fig. 2.6). The standing pressure waves caused other minor cracks in the

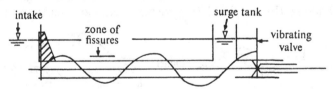

Fig. 2.6 Kandergrund tunnel. Resonance of an odd harmonic (pressure wave) causes severe damage to the tunnel (Jaeger, 1963).

concrete, and where the rock was very weak, the cracks had opened to wide fissures. The geologists had traced the local poor rock conditions but not the real cause, which was a problem of fluid dynamics.

(7) *Mantaro* (Peru). High in the Peruvian Andes at about 2400 m there is a huge bend in the river Mantaro. Consulting engineers designed a scheme utilizing the 1000-m gross head between the arms of the bend along the line I–I (fig. 2.7). Alternatives to this basic scheme were being considered when

news came that aerial surveys had discovered a major error in the mapping of the river course. On official maps, the river penetrated the Peruvian jungle along the course *a* in fig. 2.7. Aerial surveys had shown that there was

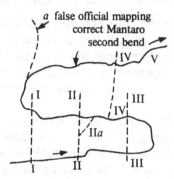

Fig. 2.7 The two bends in the Mantaro river (not to scale) showing the three possible lines for a trans-Andes hydro-power tunnel and the possible power development IV–IV through the second river bend.

another big bend in the Mantaro and that a further fall of 1000 m could be developed. Alternative schemes along the sections II–II or III–III were considered, all equally attractive to the design engineers.

A team of experienced geologists, led by Professor Falconnier (Lausanne), were on the spot for several months. They described the extremely involved geology of the high Andes as a 'mass of granite rock, covered by an older formation of folded and fissured metamorphic rocks'. Springs along the edge of the impervious granite mass indicated the level of the water-table which was inclined from south to north. A cross-section through the tunnel proposed along III–III (fig. 2.8), indicated that both ends would be in

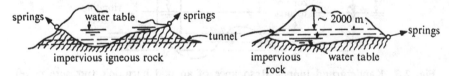

Fig. 2.8 Section III–III. The tunnel would have been under 400 m water pressure.
Fig. 2.9 Section II–II. The tunnel is above the water-table in fissured pressurized rock (overburden about 2000 m).

compact granite but that the central section would be in fissured rock where the water-table was about 400 m above the tunnel. Along the section II–II the tunnel would be almost entirely through fissured pressurized rock but above the water-table (fig. 2.9).

Peruvian engineers had just completed the construction of another, far smaller tunnel to divert water from the upper Mantaro catchment area to the west coast (Pacific) crossing the First Cordillera. They had experienced very severe difficulties due to the number of springs cut by the tunnel. In order to discharge about 2 m³/s of spring-water, the area of this tunnel had to be doubled.

The whole Mantaro problem had to be reconsidered. Route I–I was discarded because development of the second Mantaro bend would have been far more costly, or even impossible with the tunnel in this position. In addition the east end of the tunnel I–I ran into gypsum, and anchorage of the pressure pipe on indifferent rock would have been very difficult. Route III–III was discarded because of the difficulty of driving a tunnel through fissured rock 400 m below the water-table. Route II–II was acceptable to the geologists, in spite of fissuration of the pressurized rock. Route II–IIa was an adaptation for local conditions and towards the possibility of development of the second Mantaro bend.

This is an example of engineering geology because the geologists' findings decided the basic design of the hydroelectric power scheme.

(8) *Santa Giustina.* Italian engineers had similar problems to solve when designing the Santa Giustina power-house, which was to be sited in an underground cavern. Starting from the high-arch Santa Giustina dam, a pressure tunnel crosses hard fissured limestone. The Santa Giustina power-station could have been located in a cavern excavated in this limestone, but the engineers preferred to site it further downstream in pressure-developing clayey schists. The sharply inclined pressure shaft leading to the power-house required an elaborate design, to cross the contact line between the limestone and plastic rocks. The cavern is oval-shaped with a heavy reinforced-concrete lining to withstand the rock pressure. A reinforced-concrete floor at the level of the generator floor is supposed to take the horizontal thrust from the pressurized rock.

(9) *Tunnel lining.* Most rail and road tunnels are lined. Hydro-power engineers are accustomed to balancing the cost of the concrete lining against the cost of energy losses caused by rough unlined tunnels. In some cases the final decision lies with the geologists who have to decide on the stability of the rock or on its capacity to withstand erosion by the water-flow.

French geologists and rock mechanicists (Mayer, 1963) have devised a method by which the resistance of some rock types to erosion is correlated with the porosity of the rock and to its perviousness to air under pressure. The method will be described in detail later.

2.2.3 Storage reservoirs and dam foundations

(1) *Watertightness of storage reservoirs.* Geologists alone are competent to decide on the watertightness of storage reservoirs. Their decision is usually based on a detailed geological survey of large areas to make sure there are

no permeable rock strata through which water could seep out of the proposed reservoir.

In his classical treatise on dams and geology (*Barrages et Géologie*, Paris, 1933) M. Lugeon, one of the great pioneers of engineering geology, mentions two cases where geologists completely overlooked the danger of pervious rock. The two dams were built in Spain in the 1920s, but the reservoirs did not retain any water until expensive repairs were carried out.

The *Camarasa Dam* in Spain (1920) is founded almost entirely on fissured pervious dolomitic limestone (Lugeon, 1933) with impervious marl of the Liassic period several hundred metres below the dam crest. In 1924, Lugeon was called in as an expert geologist. He measured the loss at 11·26 m^3/s and recommended that a grout curtain be driven under the dam itself and from lateral galleries at the level of dam crest. The depth of the curtain varied from 112 to 394 m and was 1029 m in length; 224 boreholes were drilled at a total length of 132 000 m (fig. 2.10).

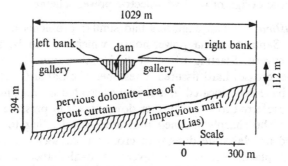

Fig. 2.10   Camarasa dam (Spain). Diagram of the grout curtain (Lugeon, 1933).

The grout curtain absorbed:

cement	40 734 tonnes
ashes	19 675
sand and gravel	129 516
asphalt	790
sawdust	112

Conditions at the *Monte Jaque* dam in Spain on highly fissured limestone were no better (Lugeon, 1933). There the geologist recommended closing all the fissures from the upstream side with mortar. The work was done by hand. The same technique has been used more recently on the abutments of the Chaudanne and the Castillon dams in France.

A very similar problem occurred with the *Dokan Dam* in Iraq (Jones, Lancaster & Gillott, 1958). This high-arch dam (height 111 m, crest length 350 m, radius of upstream face, 120 m) is founded on dolomite, approximately horizontally bedded shale, and higher up the valley sides, limestone. The left bank is a kind of peninsula against which the left abutment thrusts. The shale is rather soft and the ground broken by faults, joints and large caverns.

Several methods for rock consolidation and sealing the dam were discussed at different stages of the geological survey. One similar to that used at Monte Jaque was considered together with a clay blanket covering the reservoir bed upstream of the dam. Finally, the consultants decided on two large grout screens one on either side of the dam. On the left bank (peninsula) a 1348-m-long screen stopped water seeping through the narrow peninsula and the curtain on the right bank (1033 m) stopped leakage through the dolomite.

Some 375 000 m³ of rock were consolidated by injection with 1707 tonnes of cement and 471 tonnes of sand (Jones *et al.*, 1958).

For the curtains (4 528 000 ft²):

Total cement injected	45 000 tonnes
Total sand injected	32 200 tonnes
Length of boreholes	601 000 ft

which is slightly more than required for the Camarasa grout curtain. The cost was £2 400 000 in the 1950s.

In the three examples just given, the major responsibility lay with the geologists, because the final decisions depended on the location and permeability of the rock stratas. More recently, responsibility for the design of grout curtains and rock consolidation has been given to the rock mechanicists. The design of the grout curtain has a direct bearing on the overall stability of the rock abutments; a problem of engineering rather than geology. Examples will be dealt with in detail in subsequent chapters on dam abutment design.

(2) *Reservoir slope stability.* It is well known that a rock slope, which is stable under natural conditions, may become unstable when submerged by an artificial storage reservoir with a varying water level.

As early as 1846, the French engineer Alexandre Collin developed a theory of the stability of clay slopes, which soil mechanics later extended by graphical and analytical methods. The stability of rock slopes is a far more complex problem, as rock masses may not have the relative homogeneity of slopes formed by loose material. The geologist will be called upon to determine the preferred lines of slip, possible faults, along which rock slides may occur.

An example is the tragic case of the *Vajont* rock slide (Jaeger, 1965a). Several geologists, called in at an early stage, either completely ignored the danger of a rock slide or described it as a superficial movement of the rock masses of no danger to the reservoir. Several years later, a geologist (E. Semenza) collaborating with a specialist in rock mechanics (L. Müller) accurately predicted that the rock masses were in slow movement and they correctly located the deep geological surface along which the slide would occur. Here too, a problem which in the past used to be considered only from the geological angle, is moving into the sphere of the rock mechanicists.

(3) *Geology of dam sites.* No major dam should be designed and built without a geologist first submitting a very detailed survey of the area, extending it well beyond the immediate vicinity of the dam foundations. He should be given a detailed brief on which area to investigate and the type of dam which is to be erected.

The site of the *La Jogne* dam (1921) (fig. 2.11), the first high-arch dam to be built in the Swiss Alps, was carefully examined around the bottom and

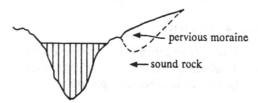

Fig. 2.11   The La Jogne dam (Switzerland). Leakage occurred through the pervious moraine; a trench filled with clay stopped the leak.

lower part of the rock walls of the La Jogne gorge. However, geologists did not discover a former valley, filled with moraine, located at a higher level but below dam crest. When the reservoir filled for the first time considerable losses occurred through this moraine. The reservoir was emptied hastily and a trench was cut through the moraine down to sound rock and filled with compacted clay.

The geologist in charge of investigating the site for the *Malpasset* dam carefully examined the bottom of the valley. He was probably not officially informed that the proposal for a gravity dam had been dropped in favour of a bold, thin-arch dam. Consequently he failed to comment on finding fissured gneiss masses at a higher level and the dam collapsed when they gave way under the thrust.

These are exceptional cases. In thousands of other dams collaboration between geologist and engineer has been excellent. The geologist has been able to produce the required information about the rock strata, their inclination, thickness, contact zones, faults and the stability of the faults, and the mineralogy of the rocks and their chemical stability.

Information about the stability of faults is most important. Many dams have been successfully built across weak strata, so bridging faults (Bort dam in France, Bhakra dam in India), where the geologist has been certain that the area is stable. On the other hand, in an earthquake area to build across a contact zone or any fault likely to produce differential movements would be dangerous for dam foundations. (The Rapel dam in Chile does not cross any dangerous fault.)

The final estimate of the safety of the dam foundations is the responsibility of the designing engineer and the rock mechanicist. Their decision will depend on detailed research on the rock fissures, rock microfissures, shear strength of rock masses, faults, water passages and pressures; on rock strength and on strains and stress distribution in fissured rock masses, in addition to the findings of the geologists.

(4) *Stable rock slopes for pressure pipes.* Often the decision between an underground power-station with pressure shaft and galleries and conventional design with pipeline and power-house above ground depends on the cost, which is determined largely by the local geology. Sound rock inside the mountain and poor rock slopes for anchoring the pipeline would turn the scales in favour of the underground design. Stable rock slopes, providing a good anchorage for pipelines, would favour a conventional design. In such cases the advice of an expert geologist is invaluable to the designing engineer.

2.3 Discussion

The short notes on the historical development of rock mechanics illustrate the slow progress in this field. Mining engineers, petroleum experts and civil engineers designing galleries, tunnels, cavities or dam foundations were faced with increasingly difficult problems. The passage from the general concept of a continuous space or half-space to the concept of fractured, fissured rock masses was often difficult.

It has been suggested that in some cases a problem can be analysed mathematically, assuming the space to be homogeneous and the rock compact and competent, and then, using different mathematical methods, to assume the rock to be fractured and fissured to a large extent. Comparing results, it can often be seen that the real case is somewhere between the two. Circular tunnels were examined on such lines (Jaeger, 1961*b*; Zienkiewicz, Cheung & Stagg 1966), as the axial symmetry of the equations is favourable to their mathematical solution. The Vajont dam abutments were model-tested using a similar approach.

Parallel to the progress of practical techniques the engineering geologists also made constant efforts to master their increasingly difficult problems.

The various case histories summarized in section 2.2 show the breadth of the engineering geologist's work and his responsibilities in connection with rock mechanics. They also show how the work of an engineering geologist in civil engineering or rock mechanics differs from that of a geologist in soil mechanics or the petroleum or mining industries. There are many textbooks dealing with the latter but little has been published on engineering geology, rock mechanics and civil engineering.

Progress in the mechanics of rock masses has not reduced the responsibilities of the geologist. Even though the final responsibility for designs generally lies with rock mechanicists and engineers, the geologists will be asked more and more precise questions. To be able to give accurate answers they will require a knowledge and understanding of the theory and practice of rock mechanics.

Part Two
Rock material and rock masses

---

3 Fundamental concepts and description of fissures

3.1 Definitions

The present practice of rock mechanicists is to consider the rock as it is, a heterogeneous mass. This requires a precise definition of concepts and a very detailed description of the rock masses and their mechanical aspects.

Terzaghi attempted to define 'hard unweathered rock' as chemically intact and with an unconfined strength of more than 5000 lb/in². The majority of shales and slightly decomposed rocks would not fit this definition. He described the *joints* which subdivide the rock mass into individual blocks and assumes that there is no cohesive action across them, they are supposed to be continuous. Other joints are more or less discontinuous. A section following a discontinuous joint cannot enter an adjacent joint without cutting across intact rock. That portion of a section located in intact rock is called a *gap*. If c = cohesion of intact rock, A = total rock section, A_g = total area of the gaps, then the effective cohesion of the rock, c_i, is

$$c_i = cA_g/A.$$

However, the definitions given by Terzaghi were not sufficiently precise, and another set were put forward by the Austrian School, mainly by Müller and summarized in a paper by Klaus W. John (1962). These definitions, with a few modifications introduced by the French experts Farran & Thenoz (1965) will be used throughout this book.

Rock mass refers to any *in situ* rock with all inherent geomechanical anisotropies (John, 1962; Panet, 1967).
Homogeneous zones refer to rock masses with comparable geological and mechanical properties such as type of rock, degree of weathering and decomposition, and rock structure.

The classification of *geological separations* as suggested by Farran & Thenoz (1965) is as follows.

[27]

Microfissures are 1 μm or less in width and about the length of a crystal or two to three molecules of water.

Microfractures are about 0·1 mm or less in width. Their extent is significant despite the fact that they are barely visible to the naked eye. They often depend on the schistosity of the material and have well-defined directions in space.

Macrofractures are wider than 0·1 mm. They may be up to several metres or more in length. The chemical, physical and mechanical properties of the *filling material* are of considerable importance to the overall strength and properties of the rock mass.

Fault usually refers to a large macrofracture with relative displacement of its lips (in French: *faille*).

The classification of fissures proposed by the rock mechanicists is different from that used by the geologists and given in section 2.1. Geologists and rock mechanicists obviously have different views about what is important in the description of rock material and masses.

In the design of a deep tunnel only large geological fractures and faults are important. All the major faults detected on the surface in the Mont Blanc–Aiguille du Midi area were followed deep into the rock masses and were cut by the Mont Blanc tunnel. Macrofractures may be vital to the design of the foundations of the piers of a bridge or the anchorage of dam foundations, Microfissures and microfractures determine the real crushing strength of rock material and masses.

There are many relationships between geological fractures and the microstructure of rocks. This is obvious for sedimentary rocks, but also appears in metamorphic rock masses. For example the geological fractures in the Aiguille du Midi area follow the general microfissuration of the chlorites and feldspars in the rock.

The modern 'engineering classifications of jointed rock masses' discussed in sections 6.7 and 10.9 require a detailed description of all types of joints encasing the rock mass.

Rock or rock material is the smallest element of rock not cut by any fracture; there are always some microfissures in the rock material.

Most rock mechanic specialists make sharp distinctions between the tests *in situ* on rock masses and tests on rock material in the laboratories. Similarly, filling material in macrofractures and faults can be tested *in situ* or in the laboratory.

3.2 Structure and anisotropy of rock masses

According to the Austrian School (Müller, 1963*a*, *b*), the technological properties of a rock mass depend far more on the system of geological separations within the mass than on the strength of the rock material itself.

Therefore, rock mechanics is the mechanics of a discontinuum, that is, a jointed medium. The strength of a rock mass is considered to be a residual strength which, together with its anisotropy, is governed by the interlocking bond of the unit blocks representing the rock mass. The deformability of a rock mass, its anisotropy, its modulus of elasticity and Poisson ratio result predominantly from the internal displacements of the unit blocks within the structure of a rock mass.

Bernaix, in a private report on the rock at Malpasset, mentions that the gneiss was crossed by two families of fissures, the spacing of which was several millimetres for one family and 1 cm for the other. They could hardly be distinguished. He also mentions a Jurassic limestone where the spacing of the microfissures was 10 mm and the spacing of microfractures 36 mm.

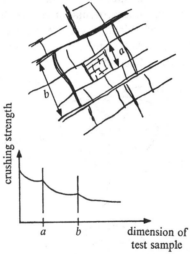

Fig. 3.1 Families of fissures. The crushing strength depends on the size of the sample tested (after Bernaix, private report).

Figure 3.1 shows schematically the structure of the rock mass. According to Bernaix, the crushing strength of the rock samples depends on the spacing of the fissures. This pattern of fissures also explains why the test results depend on the size of the sample.

The pattern of microfissures and microfractures is, on a very small scale, an image of the macrofractures which cross the rock mass. This complexity justifies the remark by Bernaix that microfractures and the macrofractures should be classified separately.

Other properties like wave velocity, elasticity or plasticity, strain and strength also vary with the direction and the stratification. This is true, for example, of the modulus of elasticity which is different in directions parallel and perpendicular to the strata. Even granite shows some anisotropy. For this reason also, the average values and the dispersion factor should be calculated from tests on a large number of samples.

It has recently been shown that the dispersion factor of some characteristic values is one of the most important clues in the classification of rocks. Rock mechanics should try to establish correlations between different rock characteristics; for example, porosity and modulus of elasticity, permeability and rock crushing strength. Similarly, the results of laboratory tests and tests *in situ* should also be correlated. The heterogeneity or lack of homogeneity of rock material and rock masses is another important feature.

3.3 Orientation of geological planes

The type of fissures considered here are macrofractures which are easily distinguished by the naked eye.

The *orientation of geological planes* is defined by the three-dimensional orientation of the line of dip of a particular plane, by the azimuth (between north and the projection of this line on the horizontal plane) and the altazimuth (between the horizontal and the line of dip). Geological planes are graphically represented in two ways.

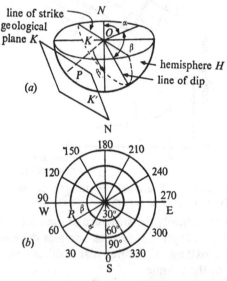

Fig. 3.2 Representation of geological planes. (*a*) Three-dimensional illustration of the method of W. Schmidt; (*b*) hemisphere *H* in equal-area projection.

(1) *Schmidt's method* (fig. 3.2). A joint plane *K* is positioned at the centre of the hemisphere *H*. The line *OP*, normal to plane *K*, will pierce the hemisphere at the pole *P*, this representing the orientation of plane *K*. The poles of all the joints surveyed can be represented on an equal-area projection of the hemisphere, producing a point diagram. The density of points indicates the number of joints with approximately the same direction. Usually there are three main directions for a system of joints.

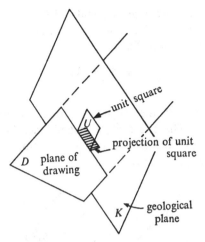

Fig. 3.3   Representation of geological planes (after Müller, 1963*a*).

(2) *Müller* favours another representation, shown in fig. 3.3. A unit square
U, coinciding with the geological plane *K* and located above the horizontal
plane *D* (drawing board) is projected onto *D*, thus defining the orientation
of the plane *K*. For engineering purposes Müller's representation is preferred.
There are other methods to be found (Panet, 1967; Talobre, 1957).

The direction of the major fractures results from the prevailing stress
when the geological fracture occurred. Assuming σ_1 and σ_3 ($\sigma_1 > \sigma_3$) to be
the two plane principal stresses at the time, fractures parallel to σ_1 may have
occurred by brittle fracture or fractures at an angle smaller than $\pi/4$ relative
to σ_3 by shear fracture (visco-plastic failure) (Panet, 1967).

3.4 Statistical density of fissures

The degree of jointing, denoted by κ, indicates the number of intersections
of a particular set of joints per yard or metre measured normal to the geological planes. The average *spacing of joints* is denoted by $d = 1/\kappa$.

The *unit block* refers to the smallest homogeneous rock unit produced by
a system of geological separations intersecting a rock mass. Their representative shape and dimensions can be determined statistically from the orientation and spacing of the joints. The average volume of unit rock blocks is:

$$V_{UB} = (1/\kappa_a)(1/\kappa_b)(1/\kappa_c) = (d_a)(d_b)(d_c),$$

a, *b* and *c* referring to the three familes of joints.

Extent of joints.   In order to evaluate the importance of a particular set of
joints for engineering problems, its extent is determined in both two and
three dimensions.

(1) The two-dimensional extent is calculated (fig. 3.4) as the ratio of the sum of the areas of the set of joints in a section parallel to the joints to the area of the entire rock section. Thus

$$\kappa_2 = \frac{A_1 + A_2 + A_3 + \ldots}{A}.$$

It follows then that $\kappa_2 = 0$ when there are 'no joints' and $\kappa_2 = 1$ when the whole section is joint. There is a close resemblance between the Austrian approach and Terzaghi's description of rock jointing, the Austrian school insisting on more precise definitions.

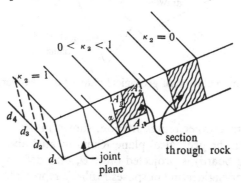

Fig. 3.4   Two-dimensional extent of geological planes. $\kappa_2 = 1$, whole area fissured; $\kappa_2 = 0$, no fissure (after Müller, 1963*a*).

(2) The three-dimensional extent of a particular set of joints indicates the total area of this geological separation intersecting a cubic yard or cubic metre of a rock mass. It is defined by:

$$\kappa_3 = \kappa_2 \kappa,$$

in square yards per cubic yard or the equivalent in metric units.

3.5 Rock mechanics surveys

The engineering geologists are interested in the general description of the area in which a structure is to be built and their approach and the definitions used are given in section 2.1.

The rock mechanicists, usually working in collaboration with geologists, are expected to give a more detailed description of specific areas. The following features of the rock mass are described and measured.

Homogeneous zones and their extent and rock types are identified; samples and drill cores prepared. Geological separations, their geomechanical significance, the orientation of geological planes, jointing and faults are recorded; also width of separations and fillings and the water conditions in pores and faults are measured.

Microfractures and macrofractures are represented in separate diagrams. Microfractures (and to some extent microfissures) determine characteristics of rock material, such as porosity, perviousness, dispersion of crushing strength, etc. The macrofractures determine the gross characteristics of the rock mass such as modulus of elasticity, the Poisson ratio, wave velocity, and the type of strain–stress diagrams.

Contour diagrams representing the orientation of the geological joint system (fig. 3.5) can be traced for fissures and fractures using the system

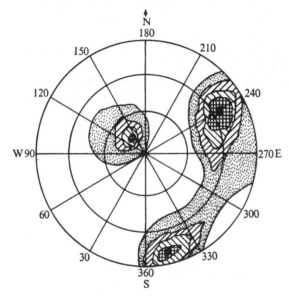

Fig. 3.5 Contour diagram representing the orientation of the geological joint system (after John, 1962).

devised by Schmidt. The lower hemisphere of Schmidt's representation is used to trace lines of equal numbers of joint measurements per unit area on an equal-area projection. Alternatively, a similar representation can be obtained, using Müller's representation of geological planes.

The geochemical properties of the individual joints and macrofractures can be summarized as: the orientation of the joint, angles α and β; the spacing, d and $k = 1/d$; the two-dimensional extent, κ_2, and the three-dimensional extent, $\kappa_3 = \kappa_2\kappa$, of the jointing, width and filling of the joints; and the coefficient of friction, $\tan \phi$, of the filling material.

Another valuable classification of rock gives the compression strength of the rock samples and the spacing of the joints.

The following classification by crushing strength is suggested by some rock mechanicists.

Sound rock	$\sigma = 500$ to 1000 kg/cm^2
Moderately sound, somewhat weathered	$\sigma = 200$ to 500 kg/cm^2
Weak, decomposed and weathered rock	$\sigma = 100$ to 200 kg/cm^2
Completely decomposed	$\sigma = 20$ to 100 kg/cm^2

Others prefer the classification by values of σ_{ult} given by Deere (section 2.1).

The void index i, E, σ_{ult}, the ratio v/v_L and the rock quality designation (RQD) used by some geologists are useful values in rock mechanics.

The information on rock material is usually obtained from laboratory tests on core samples. Information on rock masses requires extensive work on site, such as constructing surface trenches, galleries, bore holes and *in situ* tests. Determining the modulus of elasticity of rock masses and the shear strength of the geological planes are time consuming, and costly. Tests *in situ* may necessitate up to one or two years work by a team of experienced geologists and rock mechanicists.

4 Physical and mechanical properties of rock material

4.1 Physical characteristics

In any rock appraisal, it is important initially to determine the age and nature of the geological environment in which the rock mass exists. Igneous, sedimentary and metamorphic environments differ greatly and the result of rock tests must be related to the appropriate geological situation. The origin, crystalline structure and mineralogical composition of the rock are determined by the geologist or the mineralogist.

Excellent examples of such systematic research on rock material are given in some publications of the Bureau of Reclamation (USA) (Dominy & Bellport, 1965).

4.1.1 Definitions

The following symbols are commonly used:

W_s = weight of solid mineral matter in sample after oven drying to constant weight at 105°C.
W_w = weight of water in the voids.
V_w = volume of water in the voids.
V_v = volume of voids.
V_s = volume of solid mineral matter.
V = total volume of a rock sample, i.e. solid and void.
γ_w = density of water at 4 °C.
γ_s = density of solid mineral matter.
γ_{sat} = saturated density of sample.
γ = bulk density.
γ_d = dry density of sample.

The following physical characteristics can be measured.
Solid mineral grain specific gravity,

$$G_s = \frac{W_s}{V_s \gamma_w} = \frac{\gamma_s}{\gamma_w}.$$

Porosity, $n = V_v/V$.
Void ratio, $e = V_v/V_s = n/(1 - n)$.
Moisture content, $w = W_w/W_s$ expressed as a percentage.
Saturated moisture content (when the voids are fully saturated), $W_w/W_s = i$; (this index is also called the void index, i, and is extensively used as the primary characteristic of rock material, see section 2.1).

[35]

Dry apparent specific gravity,

$$G_b = \frac{W_s}{V\gamma_w} = \frac{\gamma_d}{\gamma_w} = \frac{n}{i}.$$

Saturated apparent specific gravity,

$$G_b^1 = \frac{W_s + W_w}{V\gamma_w} = \gamma_{\text{sat}}/\gamma_w = \frac{G_b}{1-n}.$$

4.1.2 The swelling test

An air- or oven-dried sample is placed in a shallow layer of water and the extent of free swelling is measured on the axis normal to the bedding or major joint system. Water in contact with the lower part of the sample is taken up by the rock by capillary action. The swelling can be measured simultaneously on all three axes to within 0·0001 in. The linear free swelling coefficient is defined as: $\varepsilon_s = \mathrm{d}l/l$, where $\mathrm{d}l$ is the change in the length l of the sample from oven-dried to fully saturated.

This test is used to identify rocks which, for engineering purposes, can be considered as: *cemented*, i.e. they remain as a coherent core or block with no significant swelling on the absorption of water; *compacted*, i.e. they remain as a coherent core or block with significant swelling on the absorption of water.

Some rocks exist as weak compact rock in the dry state and collapse when saturated. Swelling usually reaches its peak within five to ten minutes. Cases have occurred where rock material collapsed after a much longer time, even up to a year later. A test under water should be carried out for a long period on any rock suspected of weakness before it is used for a dam foundation.

The degree of cementation governs the ability of the individual grains to reorientate under the influence of tensile stresses due to the water. If the cement bond is strong, the expansion will be negligible; if the bond is weak considerable swelling will occur.

Table 4.1   *Dilatation in different rocks*

Rock type	Number of different rocks tested	Number of swelling types
Sandstone	23	11
Limestones	13	0
Clays, shales, marls, mudstones	13	11
Granites	12	5

Sixty-one different rocks were tested by Duncan *et al.* (1968). None of the limestones showed swelling within the measurement limits of the apparatus used, although they included a wide range of limestones with the void index varying from 0·2 to 4·3%. This non-dilatational effect results from the fact that calcite will dissolve under pressure at low stress. Table 4.1 summarizes some of the results.

Clays, shales, marls and mudstones displayed the highest values for unit swelling strain. Some disintegrated completely, indicating that the internal tensile stresses exceeded those which the material could sustain. Figure 4.1

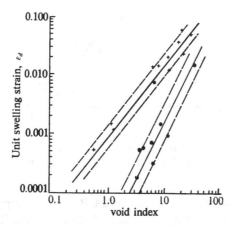

Fig. 4.1 Unit swelling strain versus void index. +, shales and other compacted and indurated fine-grained rock materials; ●, sandstones and medium-grained cemented and indurated rock (after Duncan *et al.*, 1968).

shows how the unit swelling strain ε_d is related logarithmically to the void index i for clayey materials and sandstones.

The saturation swelling stress σ_d is given by the formula:

$$\sigma_d = \varepsilon_d E,$$

where E (Young's modulus) is determined from the ultrasonic wave velocities before and after saturation and ε_d is measured in the laboratory.

More interesting is the relation between the laboratory-determined seismic velocity v (ft/s) and the void index i (fig. 4.2). This important correlation will be examined in more detail later. Correlations have also been established between the unit swelling strain ε_d and the compressive strength (see Serafim, 1968).

4.2 Anisotropy of rock material

Rock mechanicists started testing rocks with the same methods as those used for concrete samples. It soon became obvious that rock materials are far more complex than concrete, and that concrete-testing techniques would

38   *Physical and mechanical properties*

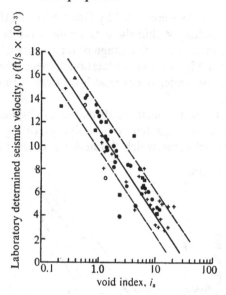

Fig. 4.2   Laboratory-determined seismic velocity versus void index. +, shale;
●, sandstone; ■, limestone; ○, granite; △, basalt (after Duncan *et al.*, 1968).

only give a partial picture of rock properties. Rock anisotropy is one property
which illustrates the difference between rock and concrete. There are several
tests used to measure this.

4.2.1   Point-load test

Cores are drilled in the rock, in three directions at right angles to each other,
so as to cut probable weak planes at a convenient angle. The cores, 2 in in
diameter (±0.025 in tolerance) and 4 to 6 in in length, are cut into discs about
1 in long. The density of the rock is determined and the cores and discs air
dried for at least two weeks prior to testing.

The disc is accurately positioned and compressive load applied through
the central axis by means of a pair of opposing hemispherical indentors.
Assuming a homogeneous isotropic material, the direction of failure would be
expected to be completely random. However, if a line of weakness exists,
there is a tendency for failure to occur in this direction. The number of failures
and their directions are classified by known statistical methods. The average
point-load breaking strength for each plane is determined by Reichmuth's
formula (fig. 4.3):

$$S = KP/d \times t,$$

where S = point-load breaking strength, P = applied load, d = diameter
of specimen and t = its thickness and K = a shape factor = $0.70(t/d)$.

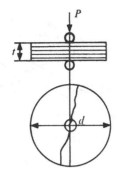

Fig. 4.3 Point-load test.

4.2.2 Line-load tests

The disc to be tested is placed perpendicular to the platens of a testing machine, and a compressive load is applied across the diameter. Specimen failure nearly always occurs along the diameter through which the load is applied; thus, the apparent breaking strength of an anisotropic rock is strongly dependent upon the orientation of the direction of the loading, with respect to pre-existing 'strong' or 'weak' directions. The results are analysed statistically.

4.2.3 The needle test

This test is very similar to the point-load test, but the hemispherical indentors or the spheres are replaced by a needle. The disc rests on a horizontal platen oriented in a 0–0 direction and the needle is loaded. The test is repeated in various directions and the critical loads tabulated against the angles.

A circular diagram can easily be traced which immediately reveals the planes of weakness in the samples.

4.2.4 Correlations

Results of the point-load test and the needle test were found to agree (Paulmann, 1966). Testing granite from Rowan County, N.C. (USA), McWilliams (1966) summarizes his analysis as follows:

There was strong correlation between sonic anisotropy, maximum and minimum tensile strength, preferred direction of failure, and the orientation of certain features of microstructure. These correlations have been extended into a three-dimensional frame of reference through the use of oriented specimens obtained on mutually perpendicular axes from a common source [fig. 4.4]. It was found that a fracture plane was approximately coincident with a prominent plane of foliation. Two other planes were nearly coincident with each other and with microfractures of the quartz, but perpendicular to the plane of prominent foliation.

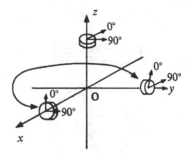

Fig. 4.4 Point-load tests on cores oriented in a three-dimensional frame of reference.

Further correlations have been established between the compressive strength and the modulus of elasticity.

Because of this anisotropy of rock material and the minuteness and density of microfissures and microfractures, most rock samples tested in laboratories are larger than the 'unit volume' of our definitions. This is one reason for the dispersion of test results and it will be discussed further.

4.3 Modulus of elasticity of rock; Poisson's ratio

The *static* modulus of elasticity of rock material, E, and the associated Poisson's ratio, v, are measured on rock samples during uniaxial unconfined compression tests. The methods used are identical to those used for concrete samples. Deformations are measured in the direction of the load and in a direction perpendicular to it.

When tracing a load–strain diagram for the deformation in a direction parallel to the load, it will be seen that the strain seldom varies linearly with the load. A curve instead of a straight line is obtained. The modulus of elasticity can be seen by tracing the secant to this curve between the point of origin of stress and a point corresponding to the test load. Sometimes the value of E_0 at the origin of the strain–stress curve is used to define the modulus. The numerical value of the modulus will depend on the maximum stress applied, on the rate at which the sample is loaded and, when successive loadings and unloadings are applied, on their sequence. Slow loading of sandstone may increase the E value obtained by 'rapid' loading by as much as 30%.

The variation in results can be explained by comparing the small rock samples under examination to a minute 'rock mass' with microfissures or possibly microfractures. When giving figures for the static modulus of elasticity, it is convenient to specify how the sample was loaded. The figures may vary from 10 000 to 500 000 kg/cm^2, the latter figure referring to hard granite samples. Diorite, some limestones and quartzite give values of about 10^6 kg/cm^2; marble may be as high as $1\cdot5 \times 10^6$ kg/cm^2.

Two other parameters E_s, the seismic modulus of elasticity, and G_s, the modulus of rigidity, are used in parallel with the static modulus E. E_s and G_s are obtained by measuring the velocity of longitudinal and transverse waves: seismic shock or ultrasonic waves. The equations relating the moduli to the velocities will be discussed later in sections 6.2 and 6.6 dealing with rock masses.

The most popular method of measuring shear and longitudinal wave parameters in the laboratory is the single pulse method using quartz crystals. Ultrasonic methods have also been used on very small laboratory rock specimens, using procedures similar to those for metals.

Young's modulus can be calculated from the resonance frequency of longitudinal waves or vibrations induced in a specimen (drill core) of rock held by a clamp at its centre. The longitudinal velocity is given by:

$$c_l = 2f_l L,$$

and

$$E_s = c_l^2 \rho.$$

Where:

c_l = longitudinal velocity of sound,
f_l = fundamental longitudinal frequency,
L = length of the specimen in cm or ft,
ρ = rock density.

Similarly, the modulus of rigidity G_s, defined as the shearing stress divided by the shear deformation can be determined from sonic resonance.

Hence, $G_s = c_t^2 \rho$, where $c_t = 2f_t T$ (or transversal velocity). Furthermore, Poisson's ratio, v, can be related to E_s and G_s by

$$v = (E_s/2G_s) - 1 = (f_l^2/2f_t^2) - 1.$$

There is no practical method known for determining the static G for rock samples. The tests also show that when the stresses applied to a sample vary, the velocity of sound changes considerably. In the case of very dense quartzite like that at Tignes (Malgovert Power Development, France), tested under a transverse compression of 250 kg/cm², the velocity of sound changed from 4000 to 6000 m/s when the longitudinal stress was increased from 0 to 1000 kg/cm² (Mayer, 1963).

Comparison of static and sonic measurements of E show that values obtained by static techniques are in general lower than those obtained by sonic measurements (Link, 1966; Clark, 1966). This difference is due to the presence of fractures, cracks or cavities which increase the static yielding by their deformation, the sonic measurements being less affected. Some values of E_s/E_0 for various types of rock (Clark, 1966) are given in table 4.2. Further values have been published by Link (1966).

Table 4.2   *Values of E_s/E_0 for different rocks*

Rock type	E_s/E_0
Chalcedonic limestone	0·85
Oelitic limestone	1·18
Quartzose shale	1·33
Monzonite porphyry	1·36
Quartz diorite	1·42
Biotite schist	1·48
Limestones	1·70–1·86
Quartzose phyllite	2·45
Granite (slightly altered)	2·75
Graphitic phyllite	2·78

The problems discussed here will be reconsidered in more detail when dealing with rock masses. There is often some similarity between the behaviour of fissured rock material and fissured or fractured rock masses.

Similar difficulties arise with Poisson's ratio, which is obtained from the measurement of the strain in a direction perpendicular to the load. According to Link, hitherto published values of Poisson's ratio for rock are almost exclusively determined at stresses below 50% of the fracture strength. During uniaxial compression tests, the deformation perpendicular to the direction of stress increases rather more than the longitudinal strain, and hence the Poisson numbers, $m = 1/v$, decrease considerably. This applies to confining pressures.

The usual values for Poisson's ratio, v, vary between 0·2 and 0·3.

4.4 Tensile strength

4.4.1 Tests

Several methods are used to test the tensile strength of rocks, some of them inspired by the usual methods for testing concrete.

(1) *The 'bending test'*, involves supporting a small rock beam at either end and applying a load in the middle of it. Failure will occur when the stresses under the load are higher than the rock tensile strength.

(2) *A cylindrical rock sample*, 10 mm in diameter and 80 mm in length, for example, can be tested by axial tension. Small metallic rings are glued at the ends of the cylinder. Another method consists of widening the cylinder diameter at both ends in order to fasten it to the testing machine. American laboratories use slightly larger cylinders than indicated here.

(3) *A cylindrical sample*, 36 mm in diameter and 180 mm long is placed in a hollow metal cylinder, slightly longer than the rock sample. The cylinder is rotated about the axis perpendicular to its length. When rotation starts, the rock sample is pressing against the dead end of the cylinder. Knowing the speed of rotation ω, the density ρ of the rock and the section where rupture occurs, the tensile strength of the rock can be calculated. At a distance x from the centre of the sample, the tensile stress,

$$\sigma_x = \tfrac{1}{2}\rho\omega^2 x^2.$$

The maximum stress occurs along the axis of rotation but rupture will occur along a plane of weakness in the sample, which may not coincide with the maximum stress value.

(4) *The Brazilian test or line-load test* (figs. 4.5 and 4.6). The disc to be tested, of thickness l, is placed between the platens of the testing machine, and a

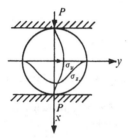 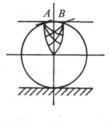

Fig. 4.5 Brazilian test. σ_x, compression stresses; σ_y, tensile stresses.

Fig. 4.6 Brazilian test. Possible rupture by shearing.

compressive load applied across the diameter D. Rupture usually occurs along the main diameter, with a complicated distribution of stress. In the centre of the disc the stresses are:

vertically (compression), $\quad \sigma_x = 6P\pi Dl;$

horizontally (tension), $\quad \sigma_y = -2P\pi Dl = -\dfrac{\sigma_x}{3}.$

When Mohr circles are traced for all the points along the diameter, it is found that rupture may occur at points on the diameter other than the centre and this has been confirmed by tests. It is also found that rupture depends to some extent on the width of the contact area between the platens and the disc. The width can be expressed by an angle of contact, 2α, which is also one of the parameters in the series of tests.

Correlation between results obtained by methods (1) and (2) is good, but the Brazilian test gives widely diverging results about two to four times larger than with other methods (Vouille, 1960). The dispersion of results from tensile tests is usually very large and a number of results is required to obtain acceptable average values. Most laboratories prefer the Brazilian test as it is more reliable than others.

4.4.2 Tensile strength and brittle failure

Rock may fail by brittle fracture or by visco-plastic deformation, the latter case being the more frequent. Brittle failure of rock has been investigated by the Istituto Sperimentale Modelli e Strutture (ISMES) in Bergamo, Italy, in connection with tests on the rupture of rock abutments in dams.

When a cylinder of homogeneous rock is hydrostatically compressed

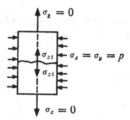

Fig. 4.7 Brittle fracture of rock samples (ISMES laboratory).

along its cylindrical surface only ($\sigma_z = 0$ in fig. 4.7) allowing the vertical axis to deform freely, the following may be observed.

(1) The failure is instantaneous and of a predominantly brittle (bursting) type.
(2) The ruptured surface is not that of a double cone, as might be expected from the shear assumptions, but is always clear-cut, disposed normally to the axis of the cylinder, and in shape which conforms with failure from tension or loss of cohesion.
(3) The 'ideal' principal ultimate stress (σ_z) along the axis (z) of the cylinder is but little removed from the pure tensile strength, that is:

$$R_t = \sigma_{z_i} = -\nu(\sigma_x + \sigma_y) = -2\nu p_i,$$

and hence $p_i = -2$ to $4R_t$ depending on ν, Poisson's ratio. (When a wet clay specimen is subjected to a similar test, the internal shearing creep causes it to fail under large, predominantly plasto-viscous deformations.)

According to Fumagalli (1967) this brittle-type failure seems to occur whenever the classical 'ideal' principal stress equals the limiting value of the uniaxial tensile strength, that is, whenever

$$\sigma_{1_i} = \sigma_1 - \nu(\sigma_2 + \sigma_3) = R_t,$$

where σ_1 is the highest tensile strength or the lowest compressive stress.

The plasto-viscous type failure seems to occur whenever the equation above does not hold or when $\sigma_{l_t} < R_t$. The 'intrinsic curve' which will be discussed later, seems only relevant to plasto-viscous failures.

The ISMES laboratory has published a table (4.3) giving the ultimate biaxial pressure p_i, the Brazilian test strength R_t and the compressive strength R_c for two hard rocks:

Table 4.3 *Biaxial pressure. Brazilian test strength and compressive strength of two hard rock types (kg/cm²)*

Rock type		(1)	(2)	(3)	Average
Mont Blanc granite	p_i	140	125	135	133
	R_t	62	49	48	53
	R_c	1290	1250	1438	1326
Syenite	p_i	200	215	195	203
	R_t	70	64	55	63
	R_c	1570	1027	1210	1269

4.5 Compression tests

Since research into rock behaviour began, compression tests on rock samples, obviously inspired by similar tests on concrete, have been preferred by engineers and geologists. Tests on unconfined rock specimens were found to give an incomplete explanation of rock behaviour *in situ*. The necessity of carrying out triaxial tests was soon recognized. Another aspect of the problem, which has been the subject of considerable investigation, is the effect of interstitial fluid pressure on the strength and deformation characteristics. The results of the following test techniques should therefore be examined: uniaxial compression tests on unconfined rock material, and triaxial tests on confined rock specimens.

This chapter deals with tests where no pore pressure is induced or recorded and tests where pore pressure can be induced and recorded.

Evaluation of the results usually assumes the validity of Coulomb's law for the shear strength τ:

$$\tau = c + \bar{\sigma}' \tan \phi,$$

where
  ϕ = true angle of internal friction,
  c = cohesion or 'no-load' shear strength,
  $\bar{\sigma}'$ = effective stress normal to shear plane $\bar{\sigma}' = \sigma - u$, where σ = external stress, u = pore pressure.

The results are interpreted from Mohr circles, the theory of which is given in sections 4.5 and 7.1.

4.5.1 Uniaxial tests on unconfined specimens

These tests are carried out on small cylinders 1, 2 or 3 in in diameter and usually twice the diameter in length (table 4.4).

The strength of a sample of rock material is characterized by the load under which it collapses, the load being calculated in pounds per square inch or in kilograms per square centimetre (1 kg/cm² equals 14·2 lb/in²).

The dispersion of results is greater for rock specimens than for concrete and the standard deviation for uniaxial tests is greater than the standard deviation for confined triaxial tests. This can be explained as the triaxial compression closing fissures, mainly microfissures, and increasing the compactness of the samples.

It was found that the standard deviation of a series of tests on a large number of samples taken from the same source is a very important characteristic of the rock. This will be examined in conjunction with the scale effect in section 4.7.

Table 4.4 *Strength data for intact rock (after Bieniawski, 1973). (1 MPa = 10·2 kg/cm²)*

Rock type	Uniaxial compressive strength (MPa)		
	Min.	Max.	Mean
Chalk	1·1	1·8	1·5
Rocksalt	15	29	22·0
Coal	13	41	31·6
Siltstone	25	38	32·0
Schist	31	70	43·1
Slate	33	150	70·0
Shale	36	172	95·6
Sandstone	40	179	95·9
Mudstone	52	152	99·3
Marble	60	140	112·5
Limestone	69	180	121·8
Dolomite	83	165	127·3
Andesite	127	138	128·5
Granite	153	233	188·4
Gneiss	159	256	195·0
Basalt	168	359	252·7
Quartzite	200	304	252·0
Dolerite	227	319	280·3
Gabbro	290	326	298·0
Banded ironstone	425	475	450·0
Chert	587	683	635·0

4.5.2 Triaxial tests and equipment

Most laboratories use apparatus capable of testing rock samples with a diameter of 2 to 3 in up to a load of 200 000 to 400 000 lb or about 100 000 to 200 000 kg and a confining pressure of 4000 to 8000 lb/in² or even 12 000 lb/in² (Paris laboratory of the École Polytechnique). The large triaxial shear machine of the Bureau of Reclamation is capable of testing cores 6 in in diameter and 12 in in length under 8 000 000 lb axial load and 125 000 lb/in² lateral pressure.

Initially (up to 1963), triaxial testing of rock material was limited to undrained specimens with no provision for measuring pore pressure. Later observations made apparent the necessity to include pore pressure measurements in order to provide a comprehensive picture of rock strength. Equipment which would accommodate cylindrical specimens of the 2·5-in size (NX cores) was developed in several American laboratories (especially that of the U.S. Army Corps of Engineers). (The standard size used in Paris was 2 cm diameter.)

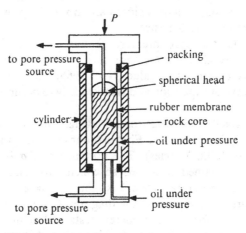

Fig. 4.8 Schematic section through a triaxial compression chamber designed to measure pore pressure.

Figure 4.8 shows a schematic section through a triaxial compression chamber designed to measure pore pressure. This consists of a base, cylindrical triaxial chamber, top, and accompanying hoses, gauges, and accessories. The platens have several holes through which pore water pressure diffuses into the specimen. To achieve high pore-pressures with relatively low-pressure sources a pressure intensifier is used. The confining pressure is held constant by manual operation of a screw piston and compensates for volume changes resulting from strain of the specimen. It can also be regulated automatically by a pressure regulator while setting up the apparatus. In assembly, a small plastic tube is coiled around the specimen to serve as an overflow while the chamber is pumped full of anti-foaming oil. Axial strain during the tests is measured by a transducer. Corrections are made for apparatus deformations during the test in order to determine the true strain and strain rates of the rock specimen.

When starting the tests, the pore water system is filled to the top of the platen and the specimen, dripping wet, is put in position. The top platen and spherical seat are aligned and the specimen and end caps enclosed in an impermeable rubber membrane, and finally the cylindrical chamber and the top are assembled.

A slight load is applied and the confining pressure is gradually brought up to working level. Pore pressure is then induced at both ends of the specimen. It is permitted to back-pressure momentarily, the system is closed from the pressure source, and the specimen progressively loaded at the prescribed rate.

In so-called drained tests, the pore pressure on a saturated specimen is held at zero, and any tendency toward induced pore pressure is permitted to dissipate by drainage through the top and the bottom of the specimen. In an undrained test the system is closed and no drainage is permitted. The induced pore pressures are measured.

During the tests, the pore pressure should be maintained below the confining pressure so that the increase during loading leaves the effective confining pressure $\bar{\sigma}_3$ a positive value (Neff, 1965). An increase in the pore pressure is indicative of expansion and is most likely to occur during advanced stages of failure.

It is suggested (Neff, 1965) that the following diagrams should be traced for a detailed analysis of the tests. First a Mohr circle diagram and intrinsic curve (envelope of failure circles). The normal stress σ is plotted on the horizontal axis for drained saturated specimens, for undrained saturated specimens the effective normal stress $\bar{\sigma}$ is plotted, $\bar{\sigma} = \sigma - u$, and the pore pressure at rupture U, is shown on the diagram. On the vertical axis, shear stress τ is plotted, as for soil or concrete tests and similarly the angle of internal friction and the angle of the shearing plane at rupture.

Further plots should show the deviator stress $\sigma_1 - \sigma_3$ versus the axial strain ε_1, the induced pore pressure $u - u_0$ versus the axial strain, the shear stress on the rupture plane versus the effective normal stress on the same plane (vector curve). It is also possible to express the pore-pressure change Δu which occurs under changes in the principal stresses $\Delta\sigma_1$ and $\Delta\sigma_3$ by Skempton's equation:

$$\Delta u = B[\Delta\sigma_3 + A(\Delta\sigma_1 - \Delta\sigma_3)] \quad \text{with} \quad B \cong 4.$$

The coefficient A is dependent on the relative deviation of the rock behaviour from the elastic theory, which for some specimens is considerable.

Commenting on tests carried out in the Missouri River Division Laboratory, U.S. Army Corps of Engineers, Neff writes that it would seem that actual pore pressure characteristics are best determined from undrained rather than drained tests, in which the pre-pressure build-up is recorded. It is desirable to preload the specimen with the estimated overburden pressure before making observations on the pore-pressure build-up.

4.5.3 Triaxial testing for rock joint strength

In rock masses, failure is likely to occur on a single joint or on a combination of joint surfaces, as shown in fig. 4.9. Lane & Hock (1964) have carried

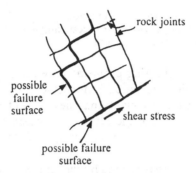

Fig. 4.9 Possible failure surfaces in jointed rock material under triaxial pressure test.

out triaxial tests on rock specimens where an existing joint in the specimen was likely to be the surface along which shear would occur.

The cores selected for testing contained a joint or fresh break which was inclined at 45° to 65° to the horizontal. Tests were also made on particular geological features.

The shear strength in the joint is obtained graphically from Mohr's circles as shown in fig. 4.10. From a circle with diameter $\sigma_1 - \sigma_3$ and a line parallel to the joint where rupture occurred, a point 4 is obtained representing the point of rupture. If this procedure is repeated for several tests, the points representing the rupture fall on a line 1, 2, 3, 4 (fig. 4.10) which is no longer coincident with the conventional intrinsic curve (fig. 4.10 and section 7.1.1).

The Missouri River Division Laboratory has tested granite from the Norad project and quartz monzonite with a machine capable of a 400 000 lb load and confining pressures well over 10 000 lb/in². Intact cores of granite were tested and compared with jointed specimens and with specimens where the wet joints were bonded with epoxy resin. A major improvement was then obtained over the strength of samples with unbonded joints. The tensile strength of the resin (4500 to 7000 lb/in²) is four to five times greater than that of the intact rock, so that in a tension test the core specimens generally broke in the intact rock and not at the glued joints.

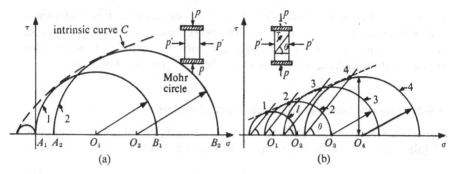

Fig. 4.10   Mohr circles for triaxial compression ($p > p'$): (*a*) typical diagrams for determining the intrinsic curve; (*b*) fissured test sample (angle θ), τ = shear stress along the fissure (after Lane & Hoek, 1964).

Some other aspects of shearing problems were investigated. It was found that the effect of the joint inclination is not large, provided it is kept within 45° to 65°. Some evidence also suggests that the angle of friction ϕ is maximum at low pressures and less when the crystals are sheared off. This is in line with similar experiments by Skempton (1964) and Krsmanović & Langof (1963).

Limited tests with joints well bonded by such minerals as calcite and quartz show quite high strengths, approaching those of the intact cores.

4.5.4   Other tests; additional remarks

For a significant increase in pore pressure as the external stress is increased, the rock should have two basic characteristics: (1) a reasonably interconnected system of pores between the crystals; (2) pores filled with a fluid that is less compressible than the surrounding crystal structure. For these reasons the concept of 'effective confining pressure' is not universally valid (Neff, 1965; Schwartz, 1964; Mazanti & Sowers, 1965).

Tests carried out with conventional equipment assume that the two secondary principal stresses are identical. Investigators were interested in the case where $\sigma_1 > \sigma_2 > \sigma_3$. Instead of a conventional cylindrical core, the test used the hollow cylinder triaxial technique loaded uniformly over the inner, as well as the outer cylindrical faces. The cylinder was also loaded axially. The circumferential stress in the cylinder wall then became larger than the radial pressure on the cylinder.

Summarizing an extensive review of rock failure in the triaxial shear test, Schwartz (1964) mentions that rock fails by either splitting, shearing, or a combination of these (pseudo-shear). Rock failure is ductile or brittle depending upon the degree of confinement. Shear failure will occur in rock if confining pressures are sufficient to prevent splitting. The angle of slip for shear failure is closely predicted by the Mohr criterion.

The importance of compression tests should not be underrated but they do not solve the basic problem of rupture in rock material.

4.6 Shear tests

One of the main conclusions of the preceding paragraphs is that failure of rock samples under axial or triaxial compression usually occurs through shear failure. Many dam designers also believe that the strength of rock abutments for dams depends on the shear strength of the rock. This is why the greatest importance is attached to shear tests. Several testing methods are used as follows.

4.6.1 Punching shear test

The Georgia Institute of Technology utilizes a disc with a thickness of $\frac{1}{4}$ or $\frac{1}{8}$ in cut from a diamond drill core. This is placed in a cylindrical guide and a piston forced through it. The results of such a punching shear tests are comparable to the shear strength with zero confinement (the apparent cohesion intercept of the Mohr diagram). A punching shear test is possibly a representation of punching through hard rock underlain by a soft layer.

4.6.2 Classical shear test

This causes shear failure of a rock sample in a selected plane. The sample is prepared by cutting a rim in it (fig. 4.11). When this is not possible, the rock

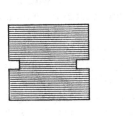

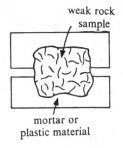

Fig. 4.11 Rock sample prepared for shear test.

Fig. 4.12 Weak rock sample prepared for shear test.

should be cast in a plastic mould (fig. 4.12). The sample is then placed in the testing apparatus and a constant compression load applied in the direction normal to the shearing plane. A shearing force is then applied in the shearing plane which is progressively increased until rupture occurs.

In both the punching and classical tests, the real shear stress distribution along the shearing surfaces is very complicated and only the average shearing strength is obtained. With the classical method it is possible to limit the normal compression strength to small values in order to investigate the apex of the intrinsic curve (fig. 4.10*a*).

The shearing apparatus consists of two stiff boxes catching the prepared sample. Forces are applied to these in such a way that the shearing force acts along the prescribed shearing plane. The upper box is fixed; the lower box slides with a minimum of friction on a horizontal plane. The Paris laboratory uses samples with a shearing surface of about 60 cm² (about 10 in²). The normal compression force may be 500 kg and the horizontal force as high as 3000 kg. The apparatus is equipped with a very stiff dynamometer (80 μm displacement for a force rising from 0 to 3000 kg) and with micrometers measuring the relative displacement of the two boxes and the vertical displacement of the upper box.

The average shear stress τ_{max} is calculated by the formula:

$$\tau_{max} = T_{max}/A,$$

where T_{max} is the force causing rupture and A the area of the sheared plane.

Because of the vertical movement of the upper box, it can be assumed that the shearing surface is lightly inclined and that a certain amount of work is done in the vertical direction. It is suggested that the shear stresses can be corrected and that the real shear is given by:

$$\tau_{real} \cong \tau - \alpha \left(\sigma + \frac{\tau^2}{\sigma} \right),$$

where σ is the normal stress on the shearing surface and α is the angle of inclined shear plane. Skempton (1964) observed that, with increasing strain, ε, the shear strength, τ, of some types of clay reaches a maximum value, τ_{max}, and then decreases to a final value, τ_{ult}. It was found later, that this is due to a realignment of clay particles in the shearing plane. Not all clay types have this characteristic. Krsmanović & Langof (1963) found that the same occurs with some types of shales. Other authors have made similar observations for a variety of rocks. It seems that this property is general for rocks and that it can be explained by the shearing of some crystals with progressive destruction of the rock cohesion. In other cases it seems to be a decrease of the internal friction factor. Figure 4.13 reproduces a typical shearing curve

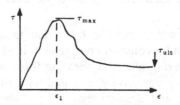

Fig. 4.13 Typical curve showing τ versus the shear deformation ε.

for rocks showing shear strength τ versus strains ε. Assuming that strains of a few millimetres destroy the rock cohesion $c = \tau_0$, the ultimate angle of internal friction is given by:

$$\tan \phi_{\text{ult}} = \tau_{\text{ult}}/\sigma, \qquad \text{for } \tau_0 = 0.$$

Bernaix (1966) has investigated the dispersion of test results, especially for the values of tensile strength, shear strength and cohesion, which are used for tracing the apex of the intrinsic curve. He came to the conclusion that minimum shear strength, main values and maximum values can be used for tracing Mohr circles (see section 7.1.1) and the three corresponding intrinsic curves. As a result the intrinsic curve can be replaced by an 'intrinsic zone' (fig. 4.14).

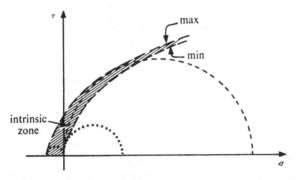

Fig. 4.14 An 'intrinsic zone' replaces the 'intrinsic curve' as a result of the dispersion of test results (after Bernaix, 1966).

This result is important in analysing rock foundations of gravity and arch dams of different types. The rock has usually deteriorated and is weaker near the surface. Stresses transmitted from the dam foundations are limited to about 50 to 80 kg/cm² by the accepted permissible stresses for concrete. These conditions correspond to the apex of the intrinsic curve. Sometimes they are too high for the deteriorated rock strength and the foundations have to be excavated deeper.

4.6.3 Extension of shear tests to fissured bodies

It is imperative to bridge the gap between laboratory shear tests on more or less homogeneous rock material and *in situ* tests on fissured and fractured rock masses. Three workers have been investigating this problem on more or less parallel lines.

Müller & Pacher (1965) tested concrete blocks 70 cm by 70 cm in soft concrete to represent different types of fissures. Hayashi (1966) used models (dimensions not given in the paper) where the cohesionless and deformable properties of joints were simulated by wax paper cast in plaster or by piling up plaster strips.

The variables analysed by Müller were the angle ϕ of the direction of the fissures relative to the main load p_1, the ratio $n = p_1/p_3$ of the main load p_1 to the secondary load p_3 and the frequency of intermittent joints on the same plane $\kappa = e/\bar{e}$; where e and $\bar{e}$ are the spacings (fig. 4.15). Tests were carried

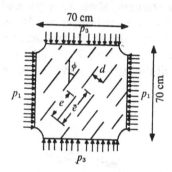

Fig. 4.15   Shear tests on a fissured body (after Müller & Pacher, 1965).

out with one and with several rows of joints. The authors summarized the results in a series of diagrams and compared them with the analytical methods used by Donath (1963). Their conclusions are encouraging and confirm the assumptions on which the circle of Mohr is based.

The testing arrangement used by Hayashi is somewhat different and more like the classical shear box test described, but the results are similar to those of Müller & Pacher. Hayashi's photographs and diagrams (fig. 4.16) show the complexity of the shearing process.

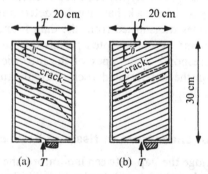

Fig. 4.16   Shear tests by Hayashi (1966) showing the dilatency, direction and shape of bending cracks in a jointed body. (*a*) Negative joint system; (*b*) positive joint system.

Progress is being made on a better interpretation of deformations which develop along an indented fissure, its shear strength to final rupture. Figure

4.17 represents the case of a regularly indented fissure, the force N being constant but small, the shear force T increasing progressively. As long as the resultant force F of these two forces remains within the cone of friction, no

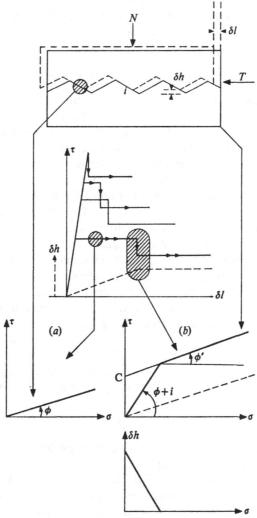

Fig. 4.17 Model of a regularly dented fissure shearing under a constant normal load N. (*a*) local conditions along inclined dent (angle i); (*b*) overall shearing diagram (after Bernaix, 1975).

movement is possible along the face of the dent, inclined at an angle i (see fig. 4.17). When T increases, sliding occurs along a plane i and the fissure opens progressively. The shear strength of the model is given by a Coulomb condition $\tau = \sigma \tan \phi$ along the inclined plane i, causing displacements δh and δl (see Barton, 1971; Bernaix, 1975).

As deformation proceeds, the apices of the dents are broken and a new angle of friction i in the direction of the force T is to be considered. This reasoning has been extended to undulating surfaces and to the case of a variable force N. These investigations are important as the shear strength of fissures is a vital parameter in the engineering classification of jointed rock masses, discussed in sections 6.7 and 10.9.

4.7   Dispersion of test results; scale effect and microfissuring

Compression tests on a large number of concrete blocks show some dispersion no matter what precautions are taken in preparing the samples. It is also known that the average crushing strength of concrete decreases as the size of the sample increases. Theories explain that a large test sample can be thought of as being the sum of several smaller samples, and that the average strength of the larger sample does not depend on the average strength of the smaller samples, but on the lowest strength of any of the small samples (Weibull's Theory). Tests carried out by the U.S. Bureau of Reclamation on a large number of concrete blocks confirm this.

It is to be expected that tests on rock material will show a similar tendency, and they will, to some extent, depend on the microfissuration of the samples. The Paris research laboratory (École Polytechnique) concentrated on these questions when endeavouring to solve some problems connected with the Malpasset dam failure. Assuming that the results of tests in the same series are a_1, a_2, a_3 . . . a_n, then the average value is:

$$M = (a_1 + a_2 + a_3 . . . a_n)/n.$$

Let us write that $v_i = M - a_i$ is the difference between the test result a_i and the mean M, then the standard deviation for this series of test results is given by:

$$S_d = \sqrt{[v_i^2]/(n - 1)}.$$

The standard deviation for tests on rock, whether tensile tests, shear tests or compression tests is often high.

Bernaix tested small cylinders of 10, 36 and 60 mm diameter and height twice the diameter, under uniaxial compression. He tested about 30 to 80 cylinders in each series of experiments and found that the ratio S_d/M is characteristic of some types of rocks. If R_{10} is the mean crushing strengh of the 10-mm cylinders and R_{60} the mean crushing strength of the 60-mm cylinders, the ratio R_{10}/R_{60} is also characteristic of the same rocks. Bernaix's results (1966) are reproduced in table 4.5.

There is an obvious correlation between the values of S_d/M and R_{10}/R_{60} and between the numerical ratios and the degree of fissuration and fracturation. The denser the fissures and fractures, the higher the dispersion of the test results. There is no such correlation between the degree of fissuration

Table 4.5 *Fissures and crushing strength of rock*

Rock type	Fissures	S_d/M	R_{10}/R_{60}
Very poor gneiss	microfissures; microfractures, very intense	0·37	2·90
Poor gneiss	microfissures; microfractures; macrofractures, intense	0·30	1·90
Jurassic limestone	microfissures, very few; macrofractures, intense	0·25	1·40
Biotite gneiss	microfissures, average	0·22	1·25
Compact limestone	no microfissures	0·005	1·0

and the absolute value of the rock strength. A fissured rock, with a high dispersion figure may be hard and show high compression strength; vice versa, a soft compact rock, with few fissures may have a low strength.

Bernaix suggests that the ratios S_d/M and R_{10}/R_{60} can be used for classifying rock masses. They can very easily be obtained from laboratory tests and without expensive tests *in situ*.

4.8 Correlations of the void index *i* with some rock characteristics

Geologists have established a correlation between the void index and the age of the rocks (section 2.1). When some alteration of the rock occurs, the basic void index increases and, for this reason, it is sometimes called 'the alteration index'. That is, the void index is a measure of the compactness of the rock material and of the compression the rock has sustained over millions of years.

Another interesting correlation was established between the void index of some rocks and the unit swelling strain, ε_d. The swelling tests appear to be most important. When carried out over a long period (one year), they may detect some basic weakness of the rock material. It is logical to inquire about other possible correlations, especially those of vital interest to engineering designers.

The Portuguese National Research Laboratory (Lisbon) has investigated such correlations of the void index with rock characteristics, mainly the modulus of elasticity and the rock strength. They worked with granites and gneiss from Portugese dam sites, with rewarding results.

The rock samples are weighed after being soaked in water for 24 hours and again after being dried in an oven at 104 °C. The difference in weight represents the amount of water absorbed by the sample under such standard conditions and is expressed as a percentage of the water absorbed, *i*. This percentage is an indirect measure of the weathering of rocks such as granites, gneiss, shists, etc.

Serafim & Lopez (1961) and Hamrol (1962) have been able to correlate the i values of granite and gneiss with the crushing strength measured in uniaxial and in triaxial compression tests. A similar correlation exists for

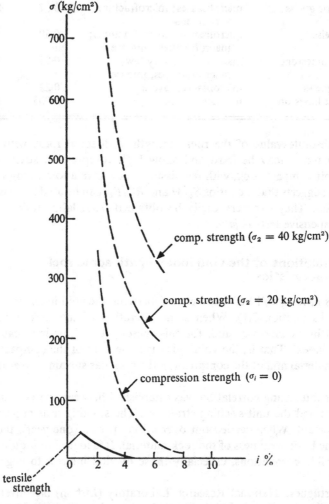

Fig. 4.18   Rock crushing strength, σ, and tensile strength versus void index, i.

the tensile strength and the shear strength. They also proved that the modulus of elasticity of the same rock decreases as the i value increases. Similar research undertaken by other laboratories extended the range of rocks for which these correlations were valid. Figure 4.18 reproduces the main results published by Serafim and others.

These correlations indicate something of the internal structure of rocks and allow a rapid and detailed analysis of large sites.

The definition of the void index used here is that suggested by Serafim (1968), Hamrol (1962) and Mayer (1963), but there are others.

Further research has shown that a distinction must be made between the microfissures and minute cracks, and the more or less spherical voids. This leads to two other definitions of *i* which will be dealt with in sections 4.10 and 4.11.

4.9 Permeability tests

The porosity of rocks is most probably linked to permeability and possibly to some other characteristics. These problems were investigated in several directions by different laboratories.

Farran & Thenoz (1965) at Toulouse have concentrated on micro-fissuration of rock material. Microfissures can be studied under the microscope after impregnation with coloured resins and their specific density calculated in terms of area of fissures per unit of volume. The 'alterability of rocks' depends mainly on the density of the microfissures.

4.9.1 Permeability to air; alteration of rock

The specific density of rocks can be estimated from their permeability to air. To do this air is forced through a rock sample, the pressure on either side of the sample is measured and the gradient calculated. The apparent permeability is calculated from the volume of air seeping through per unit time.

Table 4.6 *Air permeability tests on granites*

Apparent permeability (cm^3/s)	Description of rock sample, rock alterability
1.8×10^{-12}	Very good.
2.5 to 3.2×10^{-12}	Very good.
1×10^{-11}	Very good, no signs of alteration.
1 to 3×10^{-11}	Compact rock, traces of rust along the larger fractures. No traces of colour in the mass of the rock material.
3 to 4×10^{-11}	No alteration in the depth of the sample. Some changes at the surface.
1×10^{-11} to 3×10^{-10}	Quartz crystals getting loose. The rock material coloured within the mass.
1.2 to 3×10^{-10}	Rock is friable, likely to deteriorate into a sandy mass.
1.4 to 3.2×10^{-10}	Granite with few alterations; some voids created by large missing felspars.
1.4 to 5×10^{-9}	Decaying rock, deeply altered.
1.4 to 5×10^{-9}	Decaying rock material, rusty colour within the mass.

Table 4.6, after Mayer (1963), gives the results of air permeability tests on granites.

Further research was undertaken in Toulouse to correlate the progressive formation of microfissures in rock under varying stress conditions, and similar changes in permeability and sonic velocity. Mayer (1966) writes

In this manner the initial effects of the application to rock of stresses capable of causing failures have been discovered before the external appearance of the sample would indicate that cracking had begun. These experiments have shown the existence of a threshold which marks the start of microfissuring and which is clearly shown on permeability–stress logs. In many tests the critical stress is less than one quarter, and sometimes less than a fifth of the rupture stress . . . These same tests showed that in certain granites some of the microfissuring induced by compression was reversible; that microfissuring explained the fatigue phenomena resulting from repeated cycles of compression and decompression in fragile rocks, and finally that in the creeping of rocks at least part of the deformation under constant load corresponded to the occurrence of microfissures with the development of voids . . .

4.9.2 Permeability to water

The same Toulouse laboratory tested the permeability of cylindrical rock samples of 4 cm diameter and 3 cm length under a water pressure of 250 kg/cm^2. The rate of percolation is measured and the effluent analysed.

In sound granite, few calcium ions are leached by the water at about 5 mg/l of water per hour. In the case of granites attacked by water, silica is dissolved and after a certain time redeposited, plugging the smallest voids and decreasing the permeability. This indicates that this rock is likely to deteriorate. In the case of limestones, the opposite effect occurs. When the permeability increases with time limestones are likely to deteriorate.

The French Electricity Board (Electricité de France) used these laboratory tests on selected samples from tunnels. For comparison, tests were also carried out on unlined tunnel sections. After a few months the weaker sections deteriorated under the water action, confirming the laboratory results. On the basis of these tests, a precise lining practice was developed and only safe rock was left unlined.

4.9.3 Permeability tests and techniques

Percolation tests on rock samples are more difficult than corresponding tests on soil, mainly because of the slower rate of percolation. It is difficult to measure rates lower than 0·01 cm^3/h. Though it is desirable to use higher percolation rates, alteration of the rock samples may occur, and they should be watched closely. Gradients up to 1 in 1000 are used in the laboratory to obtain the desired rate of percolation, whereas gradients of only 1 in 10 are usual in nature.

Terzaghi (1962b) distinguishes between 'primary percolation' dependent on microfissures and some of the microfractures, and 'secondary percolation' dependent mainly on microfractures and macrofractures. The latter, although frequently greater than 'primary percolation', cannot be observed under laboratory conditions and can only be estimated accurately from tests *in situ*.

In laboratory examination of primary percolation it is important that the samples be as large as possible. The Paris laboratory of the École Polytechnique started testing samples of 36 mm diameter and 72 mm height on a standard soil test rig but found it necessary to develop a technique better adapted to rock samples. Samples of 60 mm diameter and 150 mm length, using gradients as high as 1 in 1000 were used for the more impervious rock types.

Water percolation is measured as a function of rising pressure gradients and time. Rock samples may, however, deteriorate with the increasing pressure gradient or with time, and it is essential to keep both effects separated. This is determined by observing the percolation rate on a time curve, which shows some horizontal sections of the curve, corresponding to constant percolation flow.

Air enclosed in the pores of the rock samples must be eliminated before starting the tests. It can be done conveniently by saturating the sample with water under vacuum and allowing the water to percolate through it until the effluent water collected from the sample is absolutely free of air bubbles. Complete saturation of dense rock samples may take as long as one week. Water used for percolation tests should be free of dissolved gases. For high-pressure applications it is suggested that the percolation water should be protected with a layer of oil.

The formula to be used for calculating the permeability factor K is:

$$K = QL/pA,$$

where Q is the discharge of water percolating through the sample, L the length of the sample, A the cross-sectional area of the sample and p the pressure differential between the two faces of the sample.

Some research laboratories use apparatus (fig. 4.19) derived from conventional soil-testing equipment, where the rock sample is encapsulated in an epoxy resin, with the object of preventing leakage along the external cylindrical face of the sample. Under practical conditions this is not always possible, and in addition it is difficult to remove the epoxy coating when the sample is required for further tests.

The Paris Laboratory developed an apparatus (fig. 4.20) where the sample is protected with a plastic coating and plunged into pressurized water. The radial component of the water pressure is always superior to the pressure in the sample itself and no water can seep along the cylindrical face.

The upstream and the downstream faces of the cylinder are normal to the axis, and the distance between them is equal to L.

It is possible to classify the rocks in two main categories according to the percolation results in a direction parallel to the axis of the sample. When the

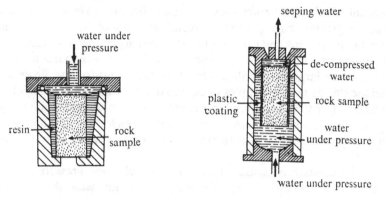

Fig. 4.19 Longitudinal percolation test.   Fig. 4.20 Longitudinal percolation test (Paris Laboratory).

percolation factor is independent of the pressure gradient, it can be assumed that the voids inside the sample are more or less spherical or ellipsoidal. When the rock is microfissured, with fissures much longer in one direction than in the other, the factor K will decrease with the pressure (fig. 4.21).

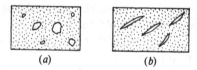

(a)   (b)

Fig. 4.21 Rock samples: (<i>a</i>) with spherical voids; (<i>b</i>) fissured.

Among the rocks tested by Habib & Vouille (1966) were: limestone with 15 to 25% voids ($K = 7.5 \times 10^{-8}$), and hard sandstone with 15 to 21% voids ($K = 2.4 \times 10^{-8}$). In both cases the voids were more or less spherical, and the K value remained constant with varying pressures.

With a microfractured quartz (parallel fractures, 0·1 mm wide) the K value varied from 1.3×10^{-8} to 1.5×10^{-9}, when the pressure rose from 0 to 45 kg/cm^2. The microfractures were then parallel to the axis of the cylindrical sample. When they were at right angles to the axis, percolation through the very compact rock material was practically nil.

Similarly, a stratified hard schist showed a K value that varied from 1.2×10^{-7} to 1.9×10^{-8}, as the pressure rose from 0 to 45 kg/cm^2. This schist was formed from layers of quartz grains alternating with layers of mica. The thickness of the layers was approximately 0·5 mm with many of the micro- and macrofractures parallel to the direction of foliation. Porosity was not more than 5%.

It is probable that increasing pressures tend to close the thin fissures and fractures, and to reduce the free passage to percolating water.

4.9.4 Radial percolation tests

Axially bored samples, 60 mm in diameter and 150 mm long, but otherwise similar to those used in the previous tests were used. The borehole had a diameter of 12 mm and a length of 125 mm. The open end was closed by a tube 25 mm long (fig. 4.22).

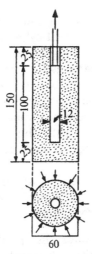

Fig. 4.22 Standard rock sample prepared for radial percolation tests (dimensions in mm) (Ecole Polytechnique Paris Laboratory).

(1) *With radial convergent flow.* The pressure on the periphery of the cylinder is p and greater than atmospheric; the inside of the sample or borehole is at atmospheric and the water percolates in a radial direction inwards. If L is the length of the inside hole, the flow through a cylinder with radius r is:

$$q = K2\pi rL(\mathrm{d}p/\mathrm{d}r).$$

But q must be equal to the flow Q collected inside the inner hole and:

$$\frac{\mathrm{d}p}{\mathrm{d}r} = \frac{1}{r}\frac{Q}{K2\pi L} \quad \text{or} \quad \frac{\mathrm{d}r}{r} = \frac{K2\pi L}{Q}\mathrm{d}p.$$

Integration from $r = R_1$ to R_2 yields:

$$\ln\frac{R_2}{R_1} = K\frac{2\pi L}{Q}p,$$

or,

$$K = \frac{Q}{2\pi Lp}\ln\frac{R_2}{R_1}.$$

These formulae show that Q and therefore dp/dr, the gradient of the percolation flow, are proportional to the pressure p. More detailed analysis shows that all the internal stresses in the cylindrical sample are equally proportional to p. It is therefore not possible with this test series to separate the effect of the field of stresses on the cylinder from the effect of the pressure on the water percolation.

(2) *With radial divergent flow.*   Water under pressure p is introduced in the inner hole of the sample, the external cylindrical face being maintained at atmospheric. The flow of percolating water is reversed, but the pattern of radial flow lines remains the same as for convergent flow. All the internal stresses are again proportional to p, but are now tensile instead of compressive. The gradient of percolation is now reversed and also independent on p. It is now possible to compare results obtained when the sample is under a field of compressive stress to the results obtained for the same sample under tensile stress. Percolation through the ends of the sample is neglected as it can be proved negligible.

The radial percolation test can be carried out under pressures up to 100 kg/cm^2. Discharges as small as $0.01 \text{ cm}^3/\text{s}$ can be measured, which allows the testing of very impervious rock samples.

(3) *Results.*   The Paris Laboratory has produced some most interesting results, comparing tests on rocks with closed pores and rocks with fissures and fractures. These are given in fig. 4.23.

The factor K for rocks with closed pores (St Vaast limestone) is the same for convergent and divergent flow. K is therefore constant and tests are entirely reversible so long as the tensile yield is not reached.

With fissured rock samples the permeability decreases sharply with increase in pressure. The curves show that the test is reversible. When the flow is reversed, permeability increases rapidly until tensile failure occurs. The passage from positive to 'negative' pressures occurs gradually and there is no discontinuity in the value of K near the zero point of the curves. In the region of 'negative' pressures, so long as the point of rupture is not reached, the tests are no longer reversible, as shown by the K versus p curves, but returns to reversibility with increase of positive pressures.

It can be assumed from these tests that permeability does not depend on the gradient of the pressures but only on the absolute value of the pressure. Darcy's Law:

$$Q = KAS',$$

is therefore valid (A = cross-sectional area, S' = slope of the pressure line or gradient of pressures, K = constant).

(4) *Radial percolation under varying strain.*   Figure 4.24 explains the method used for obtaining conditions where the strain on the cylindrical sample is

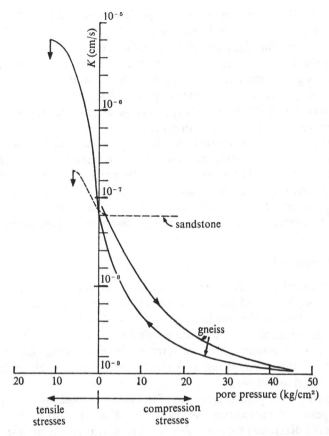

Fig. 4.23 Radial percolation tests. Results for fissured rock (gneiss) and rock with spherical voids (sandstone).

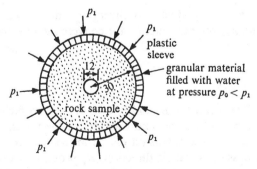

Fig. 4.24 Radial percolation test under a pressure $p_1 > p_0$ (p_0 = water pressure) (dimensions in mm).

different from the test pressure. The sample is surrounded by a thin layer of granular material (for example very small glass spheres 0·1 mm diameter) much more pervious than the sample. This material is surrounded by a plastic sleeve to which a pressure p_1 is applied, which is transmitted to the sample through the granules. If water penetration is at a pressure $p_0 < p_1$, the liquid will percolate through the sample under that pressure. The granular material now transmits the pressure $p_1 - p_0$ to the sample. The stresses and strains are obtained by adding both pressures.

A series of tests can be carried out either by maintaining p_1 constant and varying p_0 or by maintaining p_0 constant and varying p_1. These tests have shown that the permeability coefficient K decreases more rapidly under an increasing strain p_1 than under an increasing water pressure p_0. This test method produces similar strain conditions to those existing, for example, in the rock abutment under a dam where p_1 and p_0 are independent.

4.9.5   Comments

The longitudinal percolation test is the simplest and easiest to interpret, but it has not been used for tests under tensile stress. It cannot be applied to very impervious rocks, the limit being about $K = 10^{-8}$ cm/s.

The radial permeability test under varying strain is the most informative, but it is time consuming. The ordinary radial percolation test should be considered as the standard test and results obtained with this method can be summarized as follows:

A diagram showing the frequencies of the value K_0, obtained for a pressure $p_0 = 0$ (point of intersection of the curve $K = K(p)$ with the ordinate $p = 0$ on the diagrams) for a number of tests, illustrates the homogeneity of the samples tested.

By calculating the parameter:

$$S = \frac{K_{-1}}{K_{50}},$$

where K_{-1} is the permeability factor for diverging radial flow under a pressure of -1 kg/cm² and K_{50} the permeability for converging flow under a pressure of 50 kg/cm².

4.9.6   Some results

There is no correlation between K_0 and S, each of these coefficients depending on differing properties of fissures and fractures of the rocks. The coefficient S probably depends on the extension of the micro- and macrofractures in the sample and possibly on their direction. K_0 probably increases with the thickness of the fissures and fractures demonstrated when coloured water is forced through the sample.

Bernaix (1966) in his thesis describes a highly pervious sandstone with a low crushing strength and with $K_0 = 7 \times 10^{-8}$ to 8×10^{-4} cm/s, but with $S = 1 \cdot 0$. The samples were very homogeneous, without fissures and having only closed pores.

Another series of tests were related to good quality biotite gneiss, where $K_0 = 7 \times 10^{-9}$ to 3×10^{-12} and $S = 4$ to 5. When the gneiss was slightly altered, the perviousness increased to 4×10^{-7} to 9×10^{-8} and $S = 10$ to 25. When the gneiss was completely altered, the coefficient K_0 increased further to 2×10^{-6} or 9×10^{-6} but S dropped to 1 or $1 \cdot 2$, indicating a change in the rock structure.

The most striking example refers to the Malpasset gneiss. The figures obtained by Bernaix were: on the right bank, $K_0 = 3 \times 10^{-11}$ to 3×10^{-5} and $S = 7$ to 200; on the left bank, $K_0 = 3 \times 10^{-11}$ to 3×10^{-5} and $S = 1$ to 200 and more. The K_0 values are quite acceptable, but the S values are

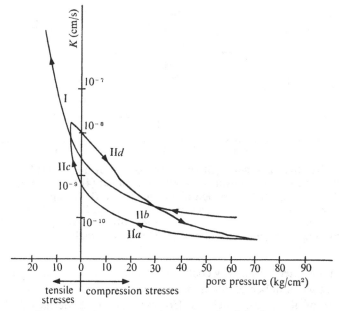

Fig. 4.25   Radial percolation tests. Malpasset gneiss, left abutment (after Bernaix, 1966).

unique to the Malpasset rock. Figure 4.25 reproduces some of the curves obtained by Bernaix for rock on the left bank of the gorge.

Among other rocks tested by the methods described, some ophites and rayolite were so dense that no percolation could be observed ($K_0 = 10^{-13}$). Many granites were characterized by coefficients $K_0 = 10^{-10}$ and $S = 1$ to $1 \cdot 4$, others were more pervious with $K_0 = 3 \times 10^{-9}$ to 5×10^{-8} and $S = 1 \cdot 1$ to 4. Limestones varied from $K_0 = 2 \times 10^{-11}$ and $S = 1$ to $K_0 = 10^{-7}$ and $S = 20$. Gneiss also produced very variable percolation characteristics.

The value $S = 1$ corresponds to rocks with closed pores. A rock sample containing only microfissures will normally have an S value 1 to 5; the S value 10 to 20 indicates the existence of microfractures.

4.10   Correlations between permeability and mechanical properties of rock material

The crushing and permeability tests described in sections 4.7 and 4.9 have established definitions of three major coefficients which accurately characterize rock material: coefficient σ/M which represents the dispersion of material compression test results, coefficient R_{10}/R_{60}, which characterizes the scale effect for uniaxial crushing strength tests on samples 10 mm and

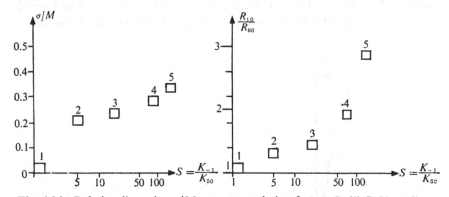

Fig. 4.26   Relative dispersion σ/M versus percolation factor, S: (1) St Vaast limestone; (2) biotite gneiss; (3) fissured Jurassic limestone; (4) Malpasset gneiss (right abutment); (5) Malpasset gneiss (left abutment) Habib & Bernaix.

Fig. 4.27   Scale factor R_{10}/R_{60} versus percolation factor, S: (1) St Vaast limestone; (2) biotite gneiss; (3) fissured Jurassic limestone; (4) Malpasset gneiss (right abutment); (5) Malpasset gneiss (left abutment) Habib & Bernaix.

60 mm in diameter, and finally coefficient S discussed in the previous section. Only a limited number of rocks have been sufficiently tested to allow the calculation of all three coefficients. However, Habib & Bernaix (1966) have been able to publish two diagrams of major interest (figs. 4.26 and 4.27). They reveal a very narrow correlation between different rock characteristics from which it is apparent that the highly fissured Malpasset gneiss is in a class of its own.

Further detailed research on different rock types has shown that most of these properties can be explained by interpreting the theory of rupture of fissured rock by Griffith (section 4.11).

Rock fissuring starts with microfissures at the scale of the crystals. Minute dislocations caused by internal strains may exist in the crystalline mesh. Extension of such microfissures can cause larger microfractures from which rock fracture may develop. There are also some correlations between etch-pits, the minute dislocations and plastic deformations (d'Albissin, 1968; Keith & Gilman, 1960).

The following testing technique has been described by d'Albissin (1968). Small cylindrical samples 36 mm diameter and 72 mm high (or $d = 10$ mm, $h = 20$ mm for single crystals) are tested in a triaxial test rig capable of developing a lateral pressure of 1000 kg/cm and an axial thrust of 8500 kg/cm², causing strain and deformation of the samples.

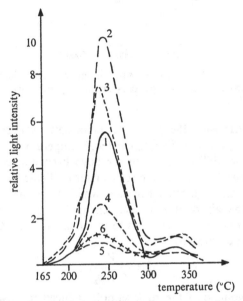

Fig. 4.28   Tests on marble of Mosset. Variations of the natural thermoluminescence with lateral pressure: (1) intact marble; (2–5) later pressures in kg/cm²: (2) 245, (3) 390, (4) 980, (5) 5000; (6) crushed under pressure of 5000 kg/cm² (after d'Albissin, 1968).

These internal dislocations can be detected by thermoluminescence whereby electrons falling from one energy level to a lower one produce light. Crystal dislocations can trap electrons and modify the light emitted. Incident gamma radiation is reflected at another wavelength.

The strained rock samples are reduced to powder, the dimensions of the grains being about 250 to 315 μm (200–315 μm for single crystals) and are then irradiated. The emitted light measured from highly strained and less strained samples at different temperatures (figs. 4.28 and 4.29) gives a measure of the internal dislocations within the rock material. Further correlations were established from other tests involving acid corrosion of polished rock surfaces.

The curves in fig. 4.28 show that a moderate strain causes an increase of the thermoluminescence which then decreases with higher strains (Handin

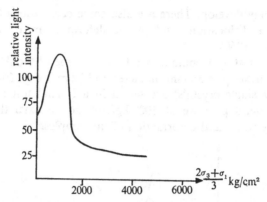

Fig. 4.29 Tests on marble of Mosset. Area under the thermoluminescence curves of Fig. 4.28 plotted against average pressure in kg/cm². Grains of 250–315 μm (after d'Albissin, 1968).

et al., 1957; d'Albissin, 1968). Plastic deformations occurred at lateral pressures of 980 kg/cm² and 5000 kg/cm² and temperatures of about 280 °C.

Additional permeability tests with air have been devised to detect any incidence of chemical change in the rock material (Perami & Thenoz, 1968). Air can penetrate through minute fissures which would not normally accept any water. As the air does not react chemically with the mineral rock materials it does not cause alterations to the rocks. Perami & Thenoz (1968) introduced a void coefficient:

$$i_1 = \frac{V}{V + v_p},$$

where V = dry volume of the rock sample reduced to powder and v_p the volume of the pores. This formula yields:

$$v_p = \frac{V(1 - i_1)}{i_1}.$$

The values V and $V + v_p$ are obtained by measuring the apparent specific weight of the rock sample and the true specific weight when reduced to powder. (The usual technique of Serafim measures the sample wet and after drying for 24 or 48 hours at 105 °C.)

Tests on several granites have shown that when the permeability to air is below a certain limit there is no chemical change in the rock. With higher permeability, granites can be chemically altered by percolating water. The same tests have shown (fig. 4.30) that there are two phases in the permeability curve. During a first phase of increasing uniaxial pressure, the sample is consolidated and the ratio $K/K_0 < 1$ (K_0 = permeability factor for no load). When the pressure increases beyond a 'limit causing microfissuration' the K/K_0 ratio increases rapidly, showing that intensive microfissuration occurs.

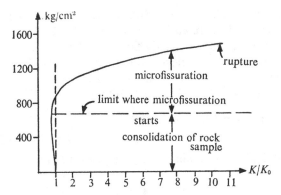

Fig. 4.30 Increasing uniaxial loading causes the granite sample to consolidate at first. Beyond a certain load, microfissuration starts and the relative permeability ratio, K/K_0, increases (after Perami & Thenoz, 1968).

4.11 Rock fracture

Testing rock material under uniaxial or triaxial compression produces several different types of failure. Brittle tensile failure of rock samples may occur under special conditions of uniaxial load (as it also may occur when testing concrete samples). Brittle shear failure is normal under uniaxial and triaxial load tests when the lateral confining pressure is moderate. When the confining pressure is high or very high in the triaxial load test, failure may be attributable to plastic shear. Where a pre-existing weak shear plane exists in the rock, rupture by shear will occur along this particular plane. This aspect has been discussed in detail in preceding sections. The four cases are represented in fig. 4.31.

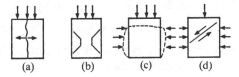

Fig. 4.31 Failure of rock material; (*a*) brittle fracture; (*b*) shear failure; (*c*) visco-plastic deformation; (*d*) shearing along a joint.

The mechanics of brittle fracture of rocks has been analysed by Griffith (1924), who, in his theory, postulates that materials contain flaws or cracks and that in compression large tensile stresses exist near the ends of certain critically orientated cracks. Griffith suggests that the fracture occurs when the most severely stressed crack propagates, the stresses being locally higher than elemental cohesion.

Evidence exists to substantiate the validity of the Griffith model of fracture (as modified by McClintock & Walsh (1962), and by Hoek (1968)). If grain boundaries are considered to be 'Griffith cracks', then the compressive strength of certain rocks is related to crack length, as he suggests.

In fig. 4.32 it is assumed that the Griffith crack has the shape of an ellipse, σ_1 and σ_3 are the principal stresses where $\sigma_1 > \sigma_3$, and $k = \sigma_3/\sigma_1$ with ψ the angle between σ_1 and the crack. σ_n is the stress at the end of the ellipse and σ_N is the maximum of σ_n. Griffith found that:

$$\sigma_N \xi_0 = \tfrac{1}{2}[(\sigma_1 + \sigma_3) - (\sigma_1 - \sigma_3)\cos 2\psi]$$
$$\pm \sqrt{\tfrac{1}{2}[(\sigma_1^2 + \sigma_3^2) - (\sigma_1^2 - \sigma_3^2)\cos 2\psi]},$$

when ξ_0 is a parameter depending on the shape of the crack.

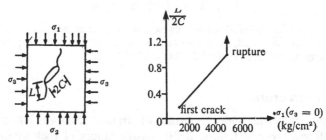

Fig. 4.32 Fracture of test sample. $2C$ = length of initial crack, L = length of secondary fissure (after Bernaix, 1967).

Assuming ψ variable, σ_N is maximum when:

$$\frac{\partial \sigma_N}{\partial \psi} = 0 \quad \text{or} \quad \cos 2\psi = \frac{1-k}{2(1+k)} = \cos 2\psi_c.$$

When $-\infty < k < -0.33$, rupture occurs when σ_3 is equal to the tensile strength of the material.

When $k > -0.33$ the tensile strength σ_t is given by:

$$2\sigma_t = \tfrac{1}{2}[(\sigma_1 + \sigma_3) - (\sigma_1 - \sigma_3)\cos 2\psi] \pm \sqrt{\tfrac{1}{2}[(\sigma_1^2 + \sigma_3^2) - (\sigma_1^2 - \sigma_3^2)\cos 2\psi]}.$$

When the material is isotropic with randomly distributed fissures, some at the critical angle ψ_c, rupture occurs when:

$$\sigma_1 = \frac{-8\sigma_t(1+k)}{(1-k)^2}.$$

Griffith's theory implicitly assumes that the rock material is under tensile stresses. Hoek modified the theory with an assumption that the lips of the crack are under compression. The equations of Griffith–Hoek now become:

$$2\sigma_t = -\tfrac{1}{2}[(\sigma_1 + \sigma_3)\sin 2\psi - \mu\{(\sigma_1 + \sigma_3) - (\sigma_1 - \sigma_3)\cos 2\psi\}]$$

where μ = friction factor, and:

$$\sigma_1 = \frac{-4\sigma_t}{(1-k)\sqrt{[1 + \mu^2]} - \mu(1+k)}.$$

Hoek has examined some schists where the main cracks were parallel with the foliation, with random secondary microfissures depending on the grain. Both formulas were confirmed.

Paulding (1966) disagreed with some of Griffith's conclusions, and used partially broken material to study the growth of cracks during brittle fracture tests. In uniaxial compression this was accomplished simply by placing a supported beam in parallel with the rock specimen to limit the advance of the ram as the load-carrying ability of the specimen decreased. The volumetric strain:

$$\frac{\Delta V}{V} = \epsilon_1 + \epsilon_2 + \epsilon_3, \qquad (\text{with } \epsilon_2 \cong \epsilon_3)$$

was computed from two perpendicular strain gauges, measuring ε_1 and $\epsilon_2 = \epsilon_3$, the principal linear strains.

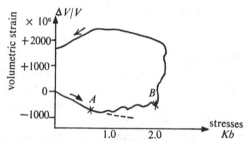

Fig. 4.33 Volumetric strain versus stress curve obtained during uniaxial compression test (after Paulding, 1966).

In fig. 4.33, where $\Delta V/V$ is plotted versus the stress p, the rapid decrease of $\Delta V/V$ from 0 to A is due to the closing of the cracks orientated so as to close under the applied stress p. From A to B the decrease is approximately linear and after B the curve indicates the onset of crack growth (see fig. 4.30).

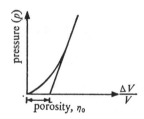

Fig. 4.34 Compressibility test on a jacketed specimen of rock containing narrow cracks, $\Delta V/V$ = volumetric strain (after Walsh, 1965, and Paulding, 1966).

Figure 4.34 shows p versus $\Delta V/V$. According to Walsh the tangent to the end of the curve gives the porosity of the strained specimen. It corresponds to the compressibility of the solid material when cracks are closed.

As a result of a comprehensive series of tests on 28 rock types, Miller (1965) classified the uniaxial stress–strain curves into the six types shown in fig. 4.35. Type I exhibits almost straight-line behaviour until sudden explosive

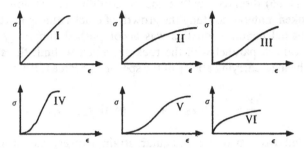

Fig. 4.35 Typical stress–strain curves for rock material in uniaxial compression to failure: (I) elastic; (II) elastic–plastic; (III) plastic–elastic; (IV and (V) plastic–elastic–plastic; (VI) elastic–plastic creep.

failure occurs (basalt, quartzite, diabase, dolomites, very strong limestones). The softer limestones, siltstones, tuff, exhibit a continually increasing inelastic yielding as the failure load point is approached (type II). The type III stress–strain curve is typical of sandstones, granites and schists cored parallel to the foliation. Metamorphic rocks (marbles, gneiss) are represented by curve type IV. Highly compressible rocks (schists cored perpendicular to the foliation) show the S-shaped curve of type V; salt deforms as shown on curve type VI.

Analysis of the $\Delta V/V$ versus p curves (figs. 4.33 and 4.34) permits an indirect estimate of rock fissures and the information thus obtained may be compared with the information obtained from radial permeability tests. Miller's analyses of the stress–strain curves ($\sigma = \sigma(\varepsilon)$) for rock materials make a

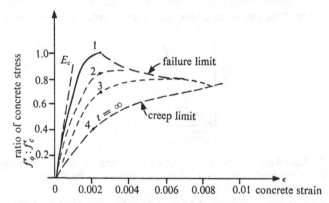

Fig. 4.36 Influence of the rate of loading on strength and modulus of elasticity of concrete. (1) $t = 20$ min; (2) $t = 100$ min; (3) $t = 7$ days; (4) $t = \infty$; f_c' = cylinder strength at 56 days (= 5000 lb/in^2) (after Ruesch, 1960).

useful introduction to the analysis of similar curves obtained by *in situ* tests on rock masses.

The rate of loading is a test variable which affects both the compressive strength and the modulus of elasticity. The effect of loading rate on the behaviour of concrete is well known (fig. 4.36) and a similar reaction might be anticipated when testing rock materials.

Compression tests are normally carried out at a high rate of loading (10 to 100 lb/in²/s) and there is only very limited information on creep of rock material. Serdengecti & Boozer (1961) published figures relating to Berea sandstone and gabbro reproduced in table 4.7. H. R. Hardy (1966)

Table 4.7   *Unconfined compressive strength (lb/in²)*
for sandstone and gabbro

	Time to failure		% strength increase
Rock	30 s	0·03 s	
Berea sandstone	8 000	12 000	50
Gabbro	31 000	40 000	30

published some curves showing the strain increase at constant stress versus time for relatively short time periods of 2000 to 5000 s (fig. 4.37). Hardy utilized a loading jig, incorporating Teflon insert-type loading heads, with

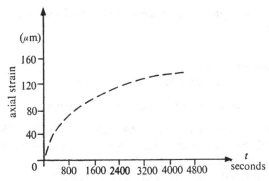

Fig. 4.37   Experimental results from a typical incremental creep test (Wombeyan marble) at high stress ($\sigma = 12\ 445$ lb/in²) (after Hardy, 1966).

which it was possible to uniformly load test specimens. Accurate and uniform loading makes it possible to observe in detail incremental creep behaviour to a total of less than 10 μm in 2200 s.

Information on creep in rock material is still very scarce but it is linked to the void index *i*. Research into this aspect should investigate possible correlations in greater detail.

Knowledge of the inelastic behaviour of geologic material (rocks and minerals) has become increasingly important. Experimental determination of the parameters that may suitably describe it is still very limited. The U.S. Department of Mines has evolved a flexible and programmable load control system designed to carry out a number of different test modes: constant load, constant rate, constant load creep, etc. However, only limited information is available on this programme (Hardy, 1966).

Recent systematic research by a team of French workers on microfissuring of rocks (mainly granite) has confirmed Griffith's theory and shown a relationship between the behaviour of rock samples under stress and strain and their microtexture. Their findings are to some extent supplementary to earlier results (mainly from the English-speaking area).

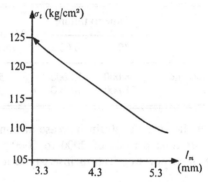

Fig. 4.38   Tensile strength σ_t verus average length of cracks l_m (in mm) (after Montarges, 1968).

Figure 4.38 shows a correlation between tensile strength and the average length of fissures l_m, thus confirming Griffith's law:

$$\sigma_t = A l_m^{-1/2} \quad \text{where } A \text{ is a constant.}$$

Discussion of a diagram similar to fig. 4.34 led Morlier (1968) to a new interpretation of the void index. The porosity index η_0 thus obtained is the porosity due to microfissures whereas i would be the total porosity including the more or less opened spherical voids. This porosity ($i - \eta_0$) has no bearing on the linear elastic deformations of the rock under high load. Figure 4.39 shows how the ratio V/V' varies with η_0, where V' is the limit velocity in rock without fissures and V a wave velocity. The equations of the straight lines obtained are:

$$V_1 = V_1'(1 - 0{\cdot}83 \times 10^3 \eta_0) \text{ for longitudinal waves and}$$

$$V_2 = V_2'(1 - 0{\cdot}5 \times 10^3 \eta_0) \text{ for transversal waves.}$$

The equations were the same for eight different granites.

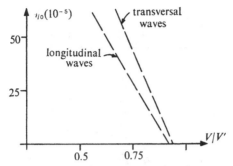

Fig. 4.39   Relative wave velocities, V/V', versus microfissure porosity, η_0, for eight different granites (after Morlier, 1968).

The crushing strength of granites varies with the load increase rate as shown in fig. 4.40 where:

$$\sigma = \sigma_0 + \sigma_1 \log \frac{\sigma'}{\sigma_0},$$

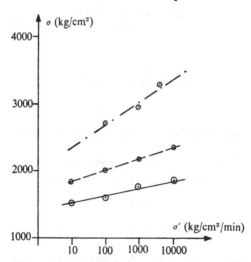

Fig. 4.40   Crushing strength, σ, of three different types of granite versus rate of stress increase, σ' (after Houpert, 1968).

σ' being the rate of load increase in kg/cm^2 per min, and also on the dimensions of the grains (in mm) (fig. 4.41). The curves are given by equations of the type $\sigma = \sigma_i + kd^{-1/2}$.

Gstalder & Marty (1968) sought to establish correlations between the uniaxial crushing strength R_c and the punching shear strength n_r of rocks (fig. 4.42). A distinction was made between the point where the limit of elastic deformation is reached (R_e or n_e) and the point of rupture (R_c or n_r). For very hard rock brittle fracture occurs when $R_c = R_e$ (the two points being the same); for plastic rock $R_e < R_c$ and $n_e < n_r$.

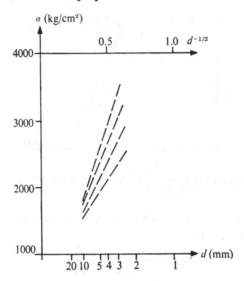

Fig. 4.41   Crushing strength, σ, of granite versus greatest diameter, d (in mm) of mineral components for varying rate of loading (after Houpert, 1968).

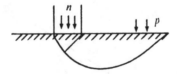

Fig. 4.42   Punching shear test considered for figs. 4.43 and 4.44.

An empirical relationship between n_r and n_e for about 30 different rock types has been established (fig. 4.43). It was found that:

$$n_r \cong 1 \cdot 67 n_e.$$

The correlation between R_c and R_e depends on the shape of the intrinsic curve which, for some limestones, is about a Coulomb straight line and for other hard rocks approximates to a parabola (Torre's rupture hypothesis). This can be seen in fig. 4.44 where (in $100 \, \text{kg/cm}^2$ units): $n_r = 3 \cdot 8 R_c + 194$ for hard granite, quartzite, very hard sandstone, and $n_r = 11 R_c$ for plastic limestones.

The research described in this section has detailed the many correlations between crushing strength, shear fracture, brittle and plastic fractures, wave velocities and microfissures or crystals which confirm Griffith's theory.

4.12   Classification of rock material

The detailed analysis of the physical and mechanical properties of rock material and types of rock failure (chapters 3 and 4) should provide a basis

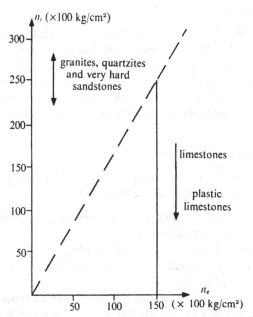

Fig. 4.43 Empirical correlation obtained between n_r and n_e for about thirty different rocks (after Gstalder & Marty, 1968).

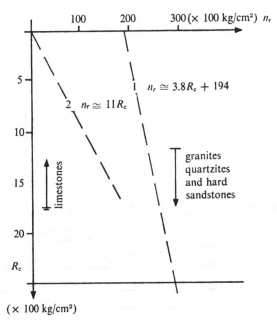

Fig. 4.44 Empirical correlation between n_r and R_c. Curve 1 for granites, hard quartzite and hard sandstone; curve 2 for limestones (in 100 kg/cm² units) (after Gstalder & Marty, 1968).

for rock classification for engineering purposes. In practice, a general classification which would satisfy the needs of the mining and petroleum industries as well as the various branches of civil engineering is no doubt impossible. The rock properties necessary for an unlined pressure tunnel undoubtedly differ from those required for easy rock excavation with no support, etc. Therefore only classifications of limited validity are possible.

4.12.1   Conventional classification

An initial classification of rock materials can be based on some of their mechanical properties and include the following.

(1) Ultimate crushing strength of rock samples under uniaxial loading. Types of rock failure (fig. 4.31).
(2) The behaviour of rock under triaxial loading.
(3) Static and dynamic moduli of elasticity and the ratio E_{stat}/E_{dyn}.
(4) The ratio E/σ_{ult} (fig. 2.3).
(5) The strain–stress curves and their more important types (fig. 4.35).
(6) The tensile strength and brittle fracture (figs. 4.5 and 4.7).
(7) The shear strength: the curves τ = shear strength versus Δs = tangential displacement (fig. 4.45) should be traced and the relevant ratio τ_{max}/τ_{ult} noted.

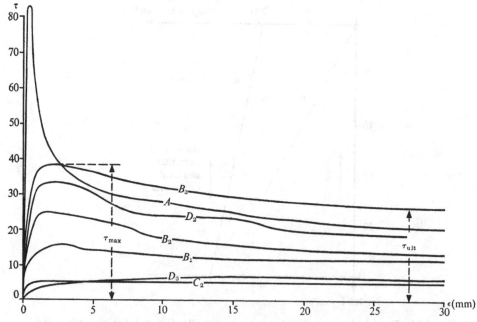

Fig. 4.45   Shear strength of rock, τ, versus shear deformations (after Krsmanović & Langof, 1963).

Besides the ratio τ_{max}/τ_{ult}, the ratio $\tau_{max}/\Delta s_1$ (where Δs_1 is the shear strain necessary to obtain the maximum shear resistance) proved to be of great importance. Krsmanović suggested plotting $\tau_{max}/\Delta s_1$ versus Δs_1; $\tau_{max}/\Delta s_1$ versus $\tau/\sigma_{max} = (C/\sigma) + \tan \phi$ and σ versus τ_{max}/σ, as shown in fig. 4.46. Different types of rocks occupy different areas on this diagram, thus allowing some classification.

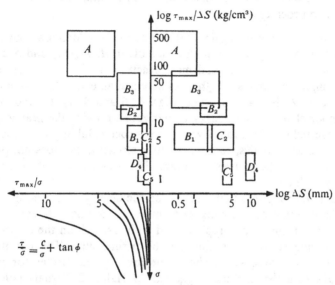

Fig. 4.46   Rock classification suggested by Krsmanović and Langof (1963), based on the shear-strength characteristics of different rock material (ΔS = strain in mm, σ and τ in kg/cm^2).

4.12.2   Classification by void index,

A second classification, by void index, i, is possible and the more important characteristics and correlations are briefly summarized here.

(1) Correlation between age of the rocks and the void index, i. The 'alteration of rocks', possibly measured by the difference between the theoretical value of i and the measured value (figs. 2.1 and 2.2).

(2) Swelling tests and the swelling strain, ε_d, versus the void index, i. Long-term swelling tests (fig. 4.1).

(3) Correlation between the void index, crushing, tensile and shear strength; and the modulus of elasticity (laboratory tests) (Serafim & Lopez, 1961) (fig. 4.18).

(4) Correlation between the void index and the shock wave velocity (fig. 4.2).

(5) Permeability tests with air and water. The permeability factor K. Radial permeability tests. Radial permeability versus strain (fig. 4.23). The $S = K_{-1}/K_{50}$ factor (fig. 4.27).

4.12.3 Classification based on other tests

(1) The dispersion of crushing test results and the scale effect (table 4.5 and fig. 4.27).
(2) The Habib & Bernaix diagrams giving σ/M versus $S = K_{-1}/K_{50}$ and R_{10}/R_{60} versus $S = K_{-1}/K_{60}$ (figs. 4.26 and 4.27).
(3) Tests to detect Griffith cracks, their density and direction.
(4) Tests on rock creep.

It is worth mentioning again that the dangerous Malpasset gneiss was characterized by exceptionally high ratios of: σ/M, R_{10}/R_{60} and S and is a heavily microfissured, micro- and macrofractured rock which in spite of its relatively high crushing strength proved unsuitable for dam foundations. On the Habib & Bernaix diagrams (figs. 4.26 and 4.27) it stands unique, with its poor characteristics easily recognizable, proving the practical value of the suggested classification. Even where a rock exhibits good or excellent mechanical characteristics, such as crushing strength, this does not preclude the possibility of other causes of failure. The proposed classification would reveal any such flaws.

Bernaix (1966) suggests that research on the density of micro- and macrofissures in addition to crushing tests may be essential. Research on the porosity index i should be supplemented with research on the dispersion of test results and scale factor, and on the permeability. The test methods developed by Paulding *et al.* should be investigated further for possible correlations with the methods suggested by Habib & Bernaix (1966). All these tests centre on the density and shape of the voids which weaken rock material.

4.12.4 Rock masses

The problem of the classification of jointed rock masses, as opposed to jointed rock material is a most disputed one. Many geologists believe that the RQD introduced by Deere (see section 2.1.4) is the most reliable parameter for engineering purposes, but several geophysicists disagree, and have developed their own 'engineering classification' (sections 6.7 and 10.9). Tunnel specialists, on their part, base their designs on expected deformations of the rock masses (section 10.10).

4.13 Filling materials for fractures and faults

4.13.1 Laboratory tests

The shear strength of filling material for fractures and faults is very often tested *in situ*. Such tests, if carried out in trenches, on the surface of the rock or deep inside galleries, are always very expensive. Techniques have been developed for testing this in the laboratory, which apart from being cheaper has the advantage of allowing repetitive testing to get an indication of the dispersion of results.

It seems, from tests carried out in Spain by Jimenes-Salas & Uriel (1964), that the scale effect is less pronounced for filling material than for rock material itself.

The thickness of the filling material is not critical. Its shear strength can be higher than that of polished rock faces and the material itself. This can happen when there are geological slips sufficient to cause considerable relative displacement at the lips of a geological fault. In such cases a sample of the polished rock face must be included in the laboratory test.

In general, boreholes are drilled into the rock, penetrating into the fault and the cores obtained. These together with the filling material are then cast in a block of concrete, and tested on shear.

The Paris Laboratory of the École Polytechnique uses a normal testing rig for testing thin fissures, a few millimetres thick. The shearing surface is about 40 cm², maximum 60 cm². When measuring the τ_{max} value, filling material may have a cohesion of a few kg/cm². This cohesion usually disappears when the strain has reached a few millimetres and the angle of friction drops.

For larger fissures, fractures and faults, larger test samples must be used, and the Paris Laboratory has developed special shearing equipment for this purpose. Rock cores of 25 to 30 cm diameter, 50 cm long are extracted, and

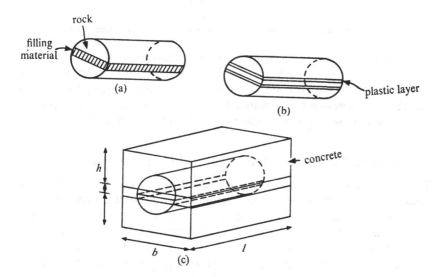

Fig. 4.47 Preparation of a test sample: (*a*) rock core including filling material; (*b*) two layers of plastic material preserve moisture of filling material; (*c*) concrete blocks prepared for shear test.

cast in a concrete block of 60 × 40 cm² and 46 cm in height with a gap of 6 cm for the shearing plane. The shearing surface of the core may be nearly 1000 to 1500 cm². The shear and normal compression forces applied to the shear box are maximum 50 tonnes. These forces are very accurately checked and can be constantly maintained over a period to give strains during creep tests. Horizontal, vertical and angular displacements of the top of the shear box are accurately measured. Pressures on the test core, normal to the box, can reach 700 kg/cm².

Some filling material test results obtained in Paris are summarized in table 4.8.

Table 4.8   *Filling material tests*

Rock	τ_{max}		τ_{ult}	
	C(kg/cm²)	ϕ	C(kg/cm²)	ϕ
Malpasset gneiss, left bank:				
Upstream face	5	60°	0	45°
Downstream face	small	60°	0	40°
Limestone:				
Thin layer	10	42°	0	30°
Thick layer	2	5° for $\sigma < 10$ kg/cm² 30° for $\sigma > 10$ kg/cm²	0	25°–33°30′ average 29°

The figures relating to the upstream and downstream faces of the ruptured Malpasset foundation rock on the left bank are most interesting. The cohesion is obviously low but the friction is reasonably high, even for the τ_{ult} values. Therefore the explanation favoured by many experts that the rupture was caused by a weakness in this clayish seam must be abandoned.

4.13.2 Effect of percolating water in filling material

This is a test which could hardly be carried out *in situ*, but can easily be followed in a laboratory on dry and saturated cores. Results vary, depending on the filling material. The material in the Malpasset seam was not substantially lowered by being saturated. The shear strength of the clayish material filling the joints of some limestones dropped far more. The angle ϕ can drop as low as 10° to 15° when flat surfaces are lubricated with saturated clay interlayerings.

5 Residual stresses in rock masses *in situ*

5.1 Heim's hypothesis

Previous tunnelling experience shows that the rock through which the galleries are excavated is stressed. These stresses, which exist in virgin rock before excavation, are called residual stresses, natural stressing or rock prestressing.

The Swiss geologist Heim (1878), observing the behaviour of rock masses in the big trans-alpine tunnel excavations, was one of the first to conclude that the tunnels were highly stressed in all directions. He assumed that there was a vertical stress component σ_v probably related to the weight of the rock overburden and proportional to it, but added that there was also a horizontal component σ_h to the stresses, most probably of a similar magnitude to the vertical component. At that time his theory, discussed in a series of papers (1878–1912), was extremely bold, although a similar hypothesis had been proposed a few years before by the German tunnel expert Rziha (1874) (fig. 5.1).

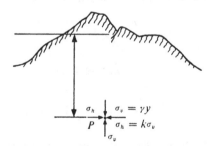

Fig. 5.1   Stressing of a tunnel in all directions ($k = v/(1 - v)$).

Heim related these residual stresses to the gigantic forces which, many million years ago, caused the mountain ridges of the Secondary and Tertiary geological periods to be lifted. According to Heim the residual stresses have a geological origin, a view to some extent confirmed by the fact that very large masses of rock (for example the entire granitic plateau of Sweden) show a horizontal stress component much greater than the vertical stress field. This can only be satisfactorily explained by geological forces.

Terzaghi (1952) also advanced an explanation of the residual stresses based on the elasticity of rock. If rock is strained in the vertical direction by its own weight it should also expand in a lateral direction; the well-known Poisson's coefficient v is a measure of this. This expansion is hindered, at great depth, by the neighbouring masses of rock. Along the vertical plane in the rock mass, lateral displacement is nil in a horizontal direction. This creates horizontal stresses opposing the lateral expansion. If Terzaghi's explanation is accepted, the horizontal stress component would be well below the value predicted by Heim's hypothesis.

A third explanation has been attempted. In a liquid in static equilibrium, at any point P the local pressure is $p = \gamma y$ where γ is the specific weight of the liquid, y the depth under the surface of the liquid, and the pressure p is the same in any direction. This law, known as the hydrostatic distribution of stress, was discovered in the seventeenth century by Pascal, the French physicist and philosopher.

Mathematically, this particular stress distribution is linked to the absence of any shear stresses in static liquid. It has been argued that in static rock masses there is a tendency for shear stresses to be progressivley relieved, giving a hydrostatic stress distribution, with the horizontal stress component equal to the vertical. Since none of these explanations is entirely satisfactory, this problem must be re-examined more thoroughly.

5.2   Stress relief at the surface of rock masses

If a deep gorge is cut through a rock mass (fig. 5.2), natural stresses in three main directions will be found at greater depth, while in a direction normal

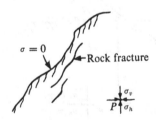

Fig. 5.2   Stress relief at the surface of rock masses.

to the surface there will be no stresses. Geologists have observed that this passage from a three-dimensional stress distribution to a two-dimensional distribution causes rock fissures and fractures in directions parallel to the gorge. These rarely occur at depths in excess of about 50 m. Climatic conditions (temperature gradient at the surface rock) may also contribute to such joints, sometimes being referred to as *sheeting*. Kieslinger (1960), Vienna University, has described many such geological situations, which have occurred mainly in granite. Such fractures, parallel to the gorge, existed on the site of the Vajont dam, causing additional difficulties in designing the dam foundations. The rock had to be consolidated with post-tensioned cables and cement grouting.

Engineering geologists will have to pay great attention to this aspect when examining possible sites for large dams.

5.3   The effective modulus of elasticity and the effective Poisson's ratio in rock masses

5.3.1   Terzaghi's approach

We assume $k = \sigma_h/\sigma_v$ to be the ratio of the horizontal field of stresses σ_h to the vertical field of stresses σ_v in the depth of rock masses. Terzaghi's approach (1952), explained at the beginning of this chapter, related the k ratio to Poisson's ratio, ν, of the rocks as follows.

If the stresses and the displacements are referred to three axes O_x, O_y, O_z, with O_x being vertical and O_y and O_z normal to O_x, then the strains ϵ_x, ϵ_y, and ϵ_z are given by:

$$\epsilon_x = \frac{1}{E}[\sigma_x - \nu(\sigma_y + \sigma_z)],$$

$$\epsilon_y = \frac{1}{E}[\sigma_y - \nu(\sigma_z + \sigma_x)],$$

$$\epsilon_z = \frac{1}{E}[\sigma_z - \nu(\sigma_x + \sigma_y)].$$

In these equations E is the modulus of elasticity and ν the Poisson ratio for rock. Assuming stress symmetry in a horizontal plane we get $\sigma_y = \sigma_z$, and no displacements in a horizontal direction, we have

$$\epsilon_y = \epsilon_z = 0 \quad \text{and} \quad 0 = \sigma_y - \nu\sigma_y - \nu\sigma_x,$$

or

$$\sigma_y = \sigma_z = \frac{\nu}{1-\nu}\sigma_x \quad \text{and} \quad k = \nu/(1-\nu).$$

For $\nu = \frac{1}{5}$ to $\frac{1}{3}$ we get $\sigma_y = \sigma_z = \sigma_x/4$ to $\sigma_x/2$, or $k = 0\cdot25$ to $0\cdot50$ with $k = 0\cdot3$ as the most probable value.

A two-dimensional state of stress ($\sigma_y = 0$) would have produced

$$\epsilon_x = \frac{1}{E}(\sigma_x - \nu\sigma_z) \quad \text{and} \quad \epsilon_z = 0 = \frac{1}{E}(\sigma_z - \nu\sigma_x).$$

With $\sigma_z = \nu\sigma_x$ and $k = \nu$ we would get

$$\epsilon_x = \frac{1}{E}\sigma_x(1 - \nu^2) \quad \text{and} \quad E = \frac{\sigma_x}{\epsilon_x}(1 - \nu^2).$$

The ratio $k = \nu/(1 - \nu)$ (Terzaghi's approach) with the usual values $\nu = 0.3$ to 0.25 or $\nu = 0.2$ accepted for rock material is contradictory to actual measurements made in deep galleries. The most usual results, obtained in several countries, favour Heim's hypothesis: $\sigma_z = \sigma_y = \sigma_z$ and $k = 1$. Working on the Terzaghi formula, $k = 1$ requires $\nu = 0.5$, which is never measured on rock material in practice. This presents an obvious paradox known as 'Heim's paradox'.

5.3.2 Poisson's ratio

The solution assumes that because of fissures and fractures, the effective modulus of rock masses E_{eff} is not equal to the modulus E of rock material. Similarly the effective Poisson's ratio ν_{eff} for rock masses is different from the ν measured on rock material (C. Jaeger, 1966).

Working on a suggestion by Waldorf *et al.* (1963) it is possible to visualize a rock mass cut into parallelepipeds by a series of horizontal fractures, with two other series of vertical fractures at right angles to each other, the vertical direction being called '1'. Assume that δ_1 is the total deformation in the vertical direction '1', with $\delta_1 = \delta_{e_1} + \delta_{p_1}$ where $\delta_{e_1} = \sigma_1 d_1/E$, being the displacement due to the elasticity of the rock with E the modulus of elasticity of the rock material, d_1 the thickness of the rock considered, σ_1 the vertical stress and ν Poisson's ratio. δ_{e_1} obviously represents the elastic deformation of the volume of rock considered. Fractures are also squeezed plastically and we can write (see also 7.2.6.):

$$\delta_{p_1} = \frac{c_1 d_1^2 \sigma_1}{E},$$

where c_1 is a coefficient which depends on ν. Working on Waldorf's assumption it can be shown that the effective modulus E_{eff} measured *in situ* is given by

$$\frac{E_{\text{eff}}}{E} = \frac{\delta_{e_1}}{\delta_{e_1} + \delta_{p_1}} \quad \text{and} \quad \frac{\delta_{p_1}}{\delta_1} = \frac{E - E_{\text{eff}}}{E}.$$

These two equations link the moduli E_{eff} and E to the deformations: the elastic deformation σ_{e_1} depending on the rock material only and the deformation of the fissures depending on the degree of fissuration of the rock masses. (This approach applies when the dimension d_1 is of the same order of magnitude as the loading area.)

A Poisson ratio ν_{eff} can be defined on similar lines (C. Jaeger, 1966). The deformation δ_3 in direction '3', at right angles to direction '1', is

$$\delta_3 = \delta_{e_3} + \delta_{p_3} = \frac{\sigma_3 d_3}{E} + \frac{c_3 d_3^2 \sigma_3}{E},$$

where E is the modulus of elasticity of the rock material, σ_3 the horizontal stress, d_3 a distance and c_3 a coefficient.

Introducing the condition $\epsilon_3 = 0$ for complex total displacements, a fictitious Poisson ratio ν_{eff} must now be introduced instead of the purely physical ν value for rock material, such as

$$\nu_{\text{eff}} = \nu \frac{\delta_{e_3} + \delta_{p_3}}{\delta_{e_3}}.$$

On the other hand $\epsilon = 0$ and $\sigma_2 = \sigma_3$ yields:

$$\sigma_3 - \nu_{\text{eff}}\sigma_1 - \nu_{\text{eff}}\,\sigma_3 = 0 \text{ or } \nu_{\text{eff}} = \frac{\sigma_3}{\sigma_1 + \sigma_3} > \nu,$$

showing how $k = \sigma_3/\sigma_1$ depends on ν_{eff} and, implicitly, $\delta_{e_3}/\delta_{p_3}$. E/E_{eff} and ν/ν_{eff} depend on the density of the fissures normal to the directions '1' and '3'. Assuming for simplicity that

$$\frac{\delta_{p_3}}{\delta_{e_3}} \simeq \frac{\delta_{p_1}}{\delta_{e_1}},$$

we obtain

$$\frac{E_{\text{eff}}}{E} = \frac{\nu}{\nu_{\text{eff}}} = \nu \frac{(\sigma_1 + \sigma_3)}{\sigma_3}.$$

In cases where the basic assumption of Heim is correct and $\sigma_1 = \sigma_v = \sigma_h = \sigma_3$ (as is usually found at great depth) then $E_{\text{eff}} = 2\nu E$.

With $\nu = 0.25$, $E_{\text{eff}} = E/2$, as found by Waldorf on site measurements. Similarly, $\nu_{\text{eff}} = 0.5$, as required by Heim's hypothesis for $\sigma_v = \sigma_h$ and $\epsilon_h = \epsilon_3 = 0$.

It is important to realize that all the values E, E_{eff}, d_1, d_3, σ_1, σ_3, ν_1, δ_{e_1}, δ_{e_3}, can be measured or calculated.

5.4   Strain and stress about cavities

5.4.1   The elastic field theory

Mechanical engineering textbooks solve the classical stress problem of a steel plate perforated by a circular hole assuming the steel plate to be uniformly stressed by a unidirectional field of tensile stresses σ. The stresses around the circular hole are calculated using Airy functions.

It is a simple matter to reverse this problem and calculate the stresses about a circular gallery excavated in rock. When loaded by its own weight only, the compression stress field in the rock is a uniform potential vertical field. Assuming a horizontal component of forces also, the two potential fields can be added.

This problem of rock mechanics was first solved by Schmidt (1926), and in 1938 the Chilean geologist, Fenner, published a similar solution. Terzaghi

& Richart (1952) have explored the problem in more detail in a classic paper, assuming different ratios for horizontal and vertical residual stress conditions and different cavity shapes; circular or elliptical galleries, spherical cavities, etc. Mining engineers tackled the problem of rectangular mining galleries.

An Airy function F of two variables (x, y) must be established to meet a general condition:

$$\frac{\partial^4 F(x, y)}{\partial x^4} + \frac{2 \, \partial^4 F(x, y)}{\partial x^2 \partial y^2} + \frac{\partial^4 F(x, y)}{\partial y^4} = 0,$$

and additionally to satisfy the boundary conditions of the problem. The stresses σ_x, σ_y, and τ_{xy} in the (x, y) directions at the depth y are then functions of x and y (fig. 5.3):

$$\sigma_x = \frac{\partial^2 F}{\partial y^2}, \qquad \sigma_y = \frac{\partial^2 F}{\partial x^2} - \gamma y \quad \text{and} \quad \tau_{xy} = -\frac{\partial^2 F}{\partial x \, \partial y}.$$

In these equations the field of external forces is accepted as the vertical weight component of the rock in the y direction and it explains the term γy, where γ is the specific weight of the rock.

Fig. 5.3   Airy function of two variables.

In the case of a problem with axial symmetry like that of a circular hole in a steel plate or a circular gallery in rock, the variables to be considered are obviously r/x and the angle θ (fig. 5.4), where x is now measured in the radial direction and the stresses to be observed at any point P inside the rock are σ_r in the radial direction σ_t in a circumferential direction and the shear stresses τ_{rt} in the same direction as the gallery.

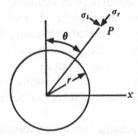

Fig. 5.4   Stresses about a circular gallery.

Assuming the field of external forces to be a potential field of vertical uniform loads p with no horizontal field of forces, the detailed calculation (Terzaghi & Richart, 1952) shows that

$$\sigma_r = \frac{p}{2}\left(1 - \frac{r^2}{x^2}\right) + \frac{p}{2}\left(1 + \frac{3r^4}{x^4} - \frac{4r^2}{x^2}\right)\cos 2\theta,$$

$$\sigma_t = \frac{p}{2}\left(1 + \frac{r^2}{x^2}\right) - \frac{p}{2}\left(1 + \frac{3r^4}{x^4}\right)\cos 2\theta,$$

$$\tau_{rt} = \frac{p}{2}\left(1 - \frac{3r^4}{x^4} + \frac{2r^2}{x^2}\right)\sin 2\theta,$$

where the radius of the circular gallery is r and x is the distance from the centre O of the gallery to point P in the rock $(x > r)$. The angle θ is zero in the vertical direction, upwards directed (fig. 5.4). The boundary conditions are obviously: for $x = r$, $\sigma_r = 0$ and $\tau = 0$ all along the gallery. For $x = \infty$, $\sigma_r = p$ in the vertical direction $(\theta = 0)$. For $x = \infty$, $\sigma_r = 0$ in the horizontal direction $(\theta = 90°)$. Therefore, along the vertical axis for $\theta = 0$ and $\cos 2\theta = 1$

$$\sigma_r = \frac{P}{2}\left(2 - \frac{5r^2}{x^2} + \frac{3r^4}{x^4}\right),$$

which yields, as expected:

for $x = r$ $\sigma_r = 0$,   for $x = \infty$ $\sigma_r = p$.

Along the horizontal axis, for $\theta = 90°$ and $\cos 2\theta = -1$:

$$\sigma_r = \frac{3p}{2}\frac{r^2}{x^2}\left(1 - \frac{r^2}{x^2}\right),$$

and so

for $x = r$ $\sigma_r = 0$,   for $x = \infty$, $\sigma_r = 0$,

as required.

The tangential or circumferential stresses σ_t are as follows:

$$\theta = 0,\quad \sigma_t = \frac{p}{2}\left(\frac{r^2}{x^2} - \frac{3r^4}{x^4}\right),$$

and for

$$x = r,\quad \sigma_t = \frac{p}{2}(1 - 3) = -p.$$

For

$$\theta = 90°,\quad \sigma_t = \frac{p}{2}\left(2 + \frac{r^2}{x^2} + \frac{3r^4}{x^4}\right),$$

which yields for $x = r$, $\sigma_t = 3p$. The σ_r and σ_t vary with x (figs 5.5a and 5.5b). The circumferential stress σ_t along a horizontal diameter varies from $\sigma_t = 3p$ in immediate vicinity of the tunnel bore to $\sigma_t = p$ at great distance; at a distance $x = 2r$, σ_t is still $\sigma_t = 1\cdot22p$. The high circumferential stress for $x = r$ may cause an overstraining of the tunnel wall and subsequent rock failure.

The stresses along a vertical diameter are even more critical. Tensile stress is $\sigma_t = -p$ in a circumferential direction near the tunnel bore. These stresses σ_t remain negative until the point $1 - 3r^2/x^2 = 0$ or $x = r\sqrt{3} = 1\cdot73r$ is reached when the σ_t become positive. Along a horizontal diameter, the radial stresses σ_r start from $\sigma_r = 0$ along the tunnel wall and increase to

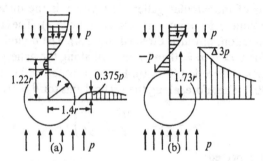

Fig. 5.5   (a) Diagram of radial stresses, σ_r; (b) diagram of circumferential stresses, σ_t.

a maximum $\sigma_r = 0\cdot375p$ for $x = r\sqrt{2}$ and then decrease. Along a vertical diameter, the σ_r start from $\sigma_r = 0$ to pass through slightly negative values. There is a value $\sigma_r = 0$ for $x = 1\cdot22r$. These negative values are low: for example, for $x = 1\cdot1$, $\sigma_r = -0\cdot04p$ only. There is an obvious danger that these negative values may overstrain the tunnel soffit where tensile strength may be low after rock blasting.

The solution suggested by Heim's hypothesis, where horizontal residual stresses σ_h are equal to the vertical residual stresses σ_v ($\sigma_h = \sigma_v = p$), is obtained by superimposing two solutions as before, but at right angles. Because of the symmetry of the solutions ($\sigma_h = \sigma_v$), σ_r and σ_t do not depend any more on the angle ϕ and for any value θ.

$$\sigma_r = p(1 - r^2/x^2) \qquad \text{with } \sigma_r = 0 \quad \text{for } x = r,$$

and

$$\sigma_t = p(1 + r^2/x^2) \qquad \text{with } \sigma_t = 2p \quad \text{for } x = r.$$

A cylinder of uniformly strained rock with maximum circumferential stresses $\sigma_t = 2p$ is formed round the excavated gallery.

If the unlined tunnel were filled with water at a pressure p, equal to the weight of the rock overburden, the pressures in the rock masses would revert to a uniform residual stress value p and be balanced, provided that $\sigma_v = \sigma_h$. If the unlined tunnel were bored in a rock mass, subjected only to a field of vertical forces $\sigma_v = p$ with no horizontal field $\sigma_h = 0$, and then filled with water at pressure p, severe stresses near the tunnel soffit would reach $\sigma_t = -2p$ and fissures would most probably develop there. Tunnel designers must know the residual rock stresses in the rock mass through which a tunnel has to be excavated, as the final stressed state depends on the combination of hydrostatic pressure and residual stresses.

In many cases the horizontal component of the residual stresses is $\sigma_h = k\sigma_v$ with $0 < k < 1$ (fig. 5.6). Tables have been calculated and stress diagrams published for characteristic values of k (Terzaghi & Richart, 1952).

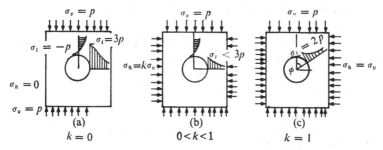

Fig. 5.6 (a) $\sigma_v = p$, $\sigma_h = 0$, $k = 0$; for $\theta = 0$, $\sigma_t = -p$ and for $\theta = 90°$, $\sigma_t = 3p$. (b) $\sigma_v = p$, $\sigma_h = k\sigma_v$, $k < 1$; for $\theta = 0$, $\sigma_t > -p$; and for $\theta = 90°$, $\sigma_t < 3p$. (c) $\sigma_v = p$, $\sigma_h = k\sigma_v$, $k = 1$; for any value of θ, $\sigma_t = 2p$.

The theories developed in this chapter may be used for estimating borehole diameter deformations under radial pressure.

According to the method of Merril & Peterson, the radial deformation of a circular hole in ideal rock is used to calculate the plane stress and strain around the hole. For uniaxial load, plane stress:

$$u = \frac{d\sigma_x}{E}(1 + 2\cos 2\theta);$$

for uniaxial load, plane strain:

$$u = \frac{d\sigma_x}{E}(1 - \nu^2)(1 + 2\cos 2\theta);$$

for biaxial load, plane stress:

$$u = \frac{d}{E}[(\sigma_x + \sigma_y) + 2(\sigma_x - \sigma_y)\cos 2\theta];$$

and for biaxial load, plane strain:

$$u = d\left(\frac{1 - \nu^2}{E}\right)[(\sigma_x + \sigma_y) + 2(\sigma_x - \sigma_y)\cos 2\theta];$$

where u is radial displacement, d is diameter of hole, σ_x, σ_y are mutually perpendicular applied stresses, θ is angle clockwise from σ_x and ν is Poisson's ratio. Furthermore when u_1, u_2, u_3 = borehole deformations at 60° angles,

$$\sigma_x + \sigma_y = \frac{E}{3d}(u_1 + u_2 + u_3).$$

5.4.2 Extensions of the theory

(*a*) When rock is being overstressed and overstrained around the cavity the theory of elastic deformation no longer applies. Plastic deformations caused by overstraining and loosening of the rock penetrates into the rock mass to a depth R_L which can be estimated with the method of Kastner (1962, see section 10.10).

(*b*) Overstrained rock should be supported. A simplified theory of rock support is developed in sections 10.3.3 and 10.3.4.

(*c*) Problems concerning some special boundary conditions are often complicated. These are summarized in section 10.3.5.

5.5 *In situ* methods of measuring residual stresses; measuring stresses about cavities

5.5.1 Measurements of residual stresses

The conventional method for measuring residual stresses in rock is to bond a strain meter to the rock (fig. 5.7). The indicated value is taken as the zero

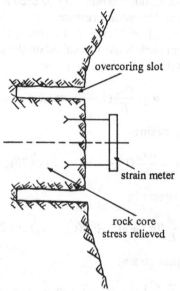

Fig. 5.7   Measurement of residual strains by stress relieving a rock core.

point. When an annular slot is carefully driven around the strain meter the rock core to which the strain meter is attached is progressively relieved. Assuming elastic rock deformations, the value indicated by the strain meter is equal to the residual strain in the rock in the direction of the meter. After the test the core is extracted and its modulus of elasticity determined in the laboratory. In order to obtain full information on rock strain a rosette of three strain meters is used. From the resulting measurements the principal strains ε_1 and ε_2 can be deduced. An additional strain meter may be used for eliminating temperature strains.

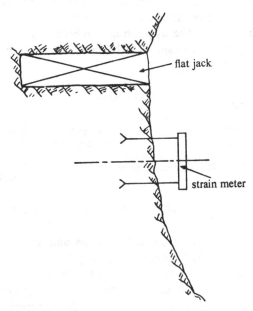

Fig. 5.8 Measurement of residual stresses by using a flat jack (Freyssinet type).

In order to obtain stress values, the modulus of elasticity E of the rock has to be assumed or measured. An average value E_m is taken to be the same in all directions. The arrangement shown in fig. 5.8 allows direct measurement of the stresses. A strain meter is fixed to the rock wall and the zero point observed. A slot is then excavated, which relieves the strains and stresses in that area of the tunnel wall. A flat jack (also called a Freyssinet jack) is introduced in the slot and the pressure observed for which the strain meter again reads 'zero point'. This pressure p is assumed to be equal to the rock residual stress in a direction perpendicular to the flat jack. A rosette of three strain meters in combination with three flat jacks is used to obtain the principal rock stresses σ_1, σ_2 and σ_3. This method was introduced in the USA by the Bureau of Reclamation in 1947. Professor G. Oberti from ISMES, and others began to use it in Europe at about the same time.

The use of flat jacks is also advocated by Rocha (1971a), head of the Laboratorio Nacional de Enghenaria Civil (LNEC), Lisbon, who has developed a machine which cuts one or more coplanar contiguous slots by means of a disc saw. The large flat jack is inserted in the slot and deformeters at different points inside the flat jack measure the opening of the slot.

Kujundzic and others (1966, 1974) checked simultaneously, *in situ*, the static elastic modulus, E_{stat}, with flat jacks introduced into a slot cut in the rock mass and the dynamic elastic modulus, E_{dyn}, by means of seismic waves crossing the same mass of rock (Jaeger, 1971).

A commercially available alternative is a triaxial borehole deformation gauge which is introduced into a small diameter hole (usually 1·5 in), of maximum depth of 40 ft. The overcoring borehole is 6 in in diameter. The deformation gauge has six fingers which measure the borehole diameter in three directions 60° apart (fig. 5.9).

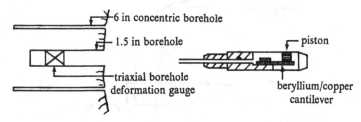

Fig. 5.9   Borehole deformation gauge (after Obert & Duvall, 1967).

The first method described measures strains and stresses on the tunnel walls, or preferably at the tunnel face, usually to a depth of about 5 ft in areas where blasting has shattered the rock. Further disturbances of the natural residual stress distribution is caused by the gallery itself. Some research into this, working with models, has been carried out. Terzaghi's opinion is that, for at least a few feet from the exposed surfaces in the rock excavations caused by blasting, the stress conditions in the rock vary erratically over very short distances.

Alternative methods are advocated where measurements are carried out at the bottom end of boreholes. These methods were developed on both sides of the Atlantic, mainly by mining engineers, who are less interested in the real value of residual stresses in rock, than in minute changes in strains and stresses which affect rock stability; but their methods can be used for direct measurement of residual stresses.

One method developed by the National Coal Board in Great Britain (fig. 5.10) comprises a high-modulus brass plug with embedded electrical resistance strain gauges. A tapered prestressed plug is bonded with epoxide cement into a borehole with a matching taper at the measurement point. The resistance changes that occur in the strain gauges before and after trepanning the slot around the borehole are measured using a strain bridge capable of

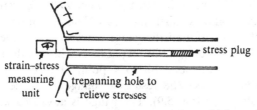

Fig. 5.10 Measurement of *in situ* rock stresses using a high-modulus stress plug set in a long borehole (National Coal Board and Sheffield University).

measuring to 1 microstrain. Since the plug is unidirectional, these measurements have to be repeated at three different angles, say 45° or 60° to one another in the plane measurement.

Hast (1958) drilled 20-m deep, 26-mm diameter boreholes by increments of between 50 cm and 100 cm. After each increment a gauging device comprising a pressure cell 10 mm high (6 mm diameter) with wedges was installed. The annular groove drilled round each borehole has a diameter of 87 mm. Measurements were made in succession in three directions at 60° apart, the cell being moved by 10 cm stages within the rock.

A team at Sheffield University (Great Britain) (Roberts *et al.*, 1962) inserted a cylinder of very birefringent glass into a borehole. The cylinder was prestressed by jacking platens at two diametrically opposed points. The magnitude of the prestressing is given directly in terms of an interference fringe pattern which appears in the plug when viewed by polarized light, and which changes when the surrounding rock is relieved by trepanning. The sensitivity of the meter is determined by the length of the plug and the elastic constants of the rock masses. Three readings have to be taken at different angles. This device can detect changes in borehole diameters of 10^{-5} in. The disc method has also been used to measure changes in strains on the rock surface by bonding directly to the surface.

American mining engineers have developed similar methods mainly to detect changes in the rock stresses. According to Terzaghi, prerequisites for reliable stress computation are rock masses with only moderate anisotropy and joint spacing greater than a few feet.

E. R. Leeman (1970) developed a method which involves cementing an electrical strain gauge rosette, or 'doorstopper' (so called because of its shape), at the bottom of a flat-ended, polished borehole. At least three boreholes, drilled in different directions, are required for this method. Difficulties were encountered in water-bearing rock masses (Ruacana underground power station, Cunene River; Chunnett, 1976; Van Heerden, 1974).

5.5.2 *In situ* stresses in rock mass

The method used by the U.S. Bureau of Mines (Merrill & Peterson, 1961) to determine *in situ* stresses beneath the rock surface is to drill EX (1·50 in) holes to a depth of about 20 ft into the tunnel wall. The gauge, developed by the Bureau and slightly smaller than the EX hole, is inserted into the hole. At this point, the rock around the gauge is under stress due to the weight of overburden and its geological history. The EX hole is then concentrically overcored with a 6-in diamond bit to a point just beyond the gauge to relieve the *in situ* stresses. The resulting change in diameter is measured with the borehole gauge (fig. 5.9). The gauge is advanced at 6-in intervals and oriented in various directions 45° apart, successively overcoring. From these observations, and from laboratory tests to determine the modulus of elasticity, the existing two-dimensional stress field is computed in a plane perpendicular to the drill hole.

The 6-in 'doughnut-shaped' cores are placed in a pressure sleeve. The borehole guage is again installed in the EX hole and changes in hole diameter measured as pressure is applied radially to the exterior surface. A thick-wall cylinder formula is used to calculate the modulus of elasticity.

5.5.3   Some results of rock measurements; residual stresses and strains about cavities

Results of measurements on residual rock stresses published in many countries, mainly by the U.S. Bureau of Reclamation, show a wide range of values. In many cases, if a rough average were taken, Heim's bold assumption ($k = 1$) would seem to be confirmed. In a few cases $k > 1$ was found (Portugal, surge tank chamber excavation). In Sweden there is a general overstressing of granite in the horizontal direction, with an average value $k = 3$ (Hast & Nilsson, 1964).

To supplement the direct measurement of residual stresses, strains around cavities have been measured in some tunnels. Some information is available on Straight Creek tunnel which was excavated in granite (Hartmann, 1966; C. Jaeger, 1966). Rock displacements have been measured in a radial direction in boreholes drilled in the tunnel soffit at different angles. From published diagrams it appears that areas of slight tensile stress exist almost everywhere near the tunnel soffit. It is remarkable that in several instances the radial displacements indicate radial tensile stresses in existence beyond the point $x_1 = 1 \cdot 22r$, which theoretically is the limit for tensile stresses even in the worst case, $k = 0$. Tensile stresses appear up to $x_1 = 1 \cdot 8r$. The higher the average density of fractures, the larger the ratio x_1/r.

Theories cannot be developed whilst the available information is so limited, but it is defined that tensile stresses tend to extend in fissured rock beyond the accepted normal, and that the weaker the rock the more they spread. This behaviour of rock under moderate tensile strain and stress is quite different from that under moderate compression stresses and it can possibly be explained by Griffith's theory of rupture of rocks.

5.5.4 Special instruments

The modern trend towards lighter rock supports in tunnels and galleries requires precise knowledge of the actual strains and stresses developing in the rock, together with methods for registering minute deformations in the rock masses.

Some of the specialized instruments already available are an indication of how interest has recently shifted from measurements made on the surface of the rock or at shallow depth to borehole measurements at some distance from the rock surface.

(1) *Rod type extensometer for boreholes.* This is a rod which is fixed to the bottom of the borehole and protected by a pipe or hollow rock-bolt which (if it is attached to the sides of the drill by grouting or anchors) may also provide the reference point for collar measurements. The relative distance between the end of the centre rod and the end of the protective pipe is measured by a dial gauge, strain meter or transducer. The comparative rigidity of the protective pipe and centre rod makes installation of this instrument difficult or impossible in boreholes exceeding one tunnel diameter in length. The usual zone influence around a tunnel opening is more than two diameters, so that both fixed points may be displaced and total displacement is not measured.

(2) *Wire type extensometers for boreholes.* Wire type extensometers may be tensioned with weights or with springs. They can be either single position which provide information on two separate fixed points contained within a borehole, or the more recently developed multiple position which makes it possible to obtain much more useful and detailed data from a single borehole (fig. 5.11).

Fig. 5.11 Schematic arrangement of a three-point borehole extensometer (Terrametrics).

$$\Delta l = M \left(l + \frac{kl}{Ef} \right)$$

where Δl = length variation in mm; M = dial gauge reading in mm; k = spring constant in kg/mm; f = cross-sectional area of measuring wire in mm² (2 mm steel wire). In the figure a = tension head; b = anchors; c = grouted protection pipe; d = measuring wires.

With the wire coil-spring type, an invar wire is fixed to the anchors placed inside the borehole. The 'air-end' of the measuring wire is clamped to a tensioning pipe incorporating a spring to apply tension to the wire. The instrument head is enclosed in a watertight housing. This extensometer is best suited to long-term measurements.

The four-point coil extensometer consists of three independent spring-tensioned wires, combined in a single housing with anchorages at three different points inside a borehole. Completely waterproof, the thermically insulated sensor head is fixed outside the borehole. The anchor assemblies may be grouted individually, by the use of inflatable air pockets. The annular space between the borehole walls and the plastic pipe protecting the wires is filled with injected grout. The spacer pipes are coated with grease, so the grout does not bond to them except at the anchor assembly points. The measuring accuracy is ±0·01 in (±0·25 mm), and the measuring range is ±0·4 in (±10 mm) but this can be increased to ±1 in (±25 mm). An improved version with eight fixed points and a sensing head capable of remotely measuring their axial displacements is also on the market.

Such measurements, taken successively with time and tunnel face advance, provide the basis for plotting rock mass strain gradients. It is possible to outline stable zones, and those which are being compacted, compressed, loosened, or placed in tension. Active major fracture zones penetrated by the instrument's borehole can also be located.

(3) *Borehole deflectometer.* The borehole deflectometer (e.g. System Terrametrics, fig. 5.12) moves in a drill hole of size 5 in, inside a 4-in plastic

Fig. 5.12   Borehole deflectometer (Terrametrics) a = deflection arm; b = deflection arm head; c = ball bearing supported ball rollers; D = drill hole diameter = 4 in or larger; α = deflection angle (measuring sensitivity ±0·001 inch).

tube. A high degree of protection against water and extreme temperatures must be provided. The borehole deflectometer measures rock deformation normal to the axis of the borehole in any desired measuring plane.

It consists of two flexibly connected parts: the guide housing and the deflection arm. A pivoted joint between the deflection arm head and the guide housing allows the arm to vary its position as the instrument is moved up or down the cased borehole. Ball bearing supported ball rollers are mounted on both housings in a triaxial arrangement. One of each of the three rollers is spring loaded. Two sets of rollers provide longitudinal stability for the guide housing while the deflection arm has only one set. An anchored and tensioned steel wire passes through a knife-edge orifice plate in the guide housing. Deflection of the borehole results in a corresponding bending of the wire and the angular deviation of the wire is measured by electrical transducers. Semifixed multiple-point borehole instruments are also available. Precise monitoring of rock mass movements normal to the borehole axis at a number of points is thus possible.

More information on measurements of rock deformations *in situ* will be given in section 10.11 on underground hydro-electric power stations, and in sections 16.1 on Kariba South Bank and 16.3 on Waldeck II pumped storage station.

5.5.5 Static equilibrium method

Consider a gallery of width D driven in rock. The vertical field of natural stresses before the opening is driven is unknown. Consider a plane xy perpendicular to the tunnel. Before the gallery is driven, the total force on the rock mass between the points x_1 and x_2 is

$$\int_{+x_1}^{+x_2} \sigma_v \, dx = \sigma_v l, \quad \text{with} \int_{+x_1}^{+x_2} dx = x_2 - x_1 = l.$$

After excavation of the gallery, this vertical force remains unchanged. A number of stress gauges $G_1 \ldots G_n$ are installed on both sides of the tunnel and stress increases caused by excavation of the gallery $\sigma_1, \sigma_2 \ldots \sigma_n$ are measured (fig. 5.13).

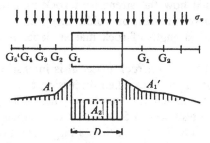

Fig. 5.13 Static equilibrium method for determining the vertical component of the field of stresses σ_v.

To satisfy the condition of static equilibrium along $l = (x_2 - x_1)$,

$$\sigma_v D = A_2 = (A_1 + A_1'),$$

which yields the vertical stress field σ_v before driving the gallery (see Morgan & Panek, 1963).

6 Strains, modulus of deformation and failure in rock masses

6.1 General remarks

Tests carried out on samples have shown the complexity of rock behaviour when under strains and stresses. It has been proved that deformations, strength and rupture of rock material, depend on the degree of fissuration of the sample, the porosity and water content and the dispersion of the test results. Griffith's theory is at the root of the problem of rupture of rock material.

Rock material can be used as small-scale models of rock masses. But microfractures and geological faults, bedding and stratification in the mass intensify the defects observed in it. Measurements of the modulus of deformation largely depend on the fractures and faults observed in the mass to such an extent that in addition to a modulus of elasticity E, a modulus of total deformation E_{total} including non-elastic, non-reversible deformations of the mass must be defined, and differ from the modulus of elasticity of the rock.

Chapter 5 described how the strength of rock material depended upon microfissures and microfractures; on voids, their density and shape. It will now be seen that the strength of rock masses depends on microfractures, joints and faults.

The scale effect observed on rock material is intensified in tests on rock masses. It is of great importance when discussing test methods and results to be fully aware of the actual volume of rock mass under examination. For example, the Talobre tests with a 50-ton jack on a small bearing plate may involve a far smaller volume of rock than a large size test with 4 × 300-ton jacks. Similarly, a deformation test with a 20-ft- or 30-ft-long cable anchored in rock will concern a very much larger volume of rock than a test with a cylindrical jack in a borehole.

Tests to determine the modulus of deformation of rock masses usually cause some progressive internal fracturing of the rock, which can be observed either on a strain–stress diagram or on a strain–time diagram. The degree to which the rock mass is stressed during the test is important. For example, the Austrian test method keeps the stresses well below the rock strength, whereas the method devised by Talobre systematically overstresses the rock in order to reach its final crushing strength.

Failing rock cohesion, rock shear strength, rock tensile strength or rock brittleness are all possible causes of rock failure. For this reason we will include shear and other tests in this chapter.

Correct interpretation of test results is often difficult, especially when the rock mass is not homogeneous but stratified. The usual approach is to establish mathematical relationships between the strains and stresses and to check them by *in situ* tests. Many analytical models have attempted to simulate *in situ* conditions and these will be discussed in this and succeeding chapters. Ideally, any method should be checked by direct calibration and more attention should be given to this particular aspect of the problem.

6.2 Rock strain and the modulus of deformation; methods of measurement

Measuring the modulus of elasticity of rock samples in the laboratory is a simple and routine part of checking the physical properties of rock materials. To do the same with rock masses *in situ* is, however, a far more difficult task.

One of the first tests on rock masses *in situ* was the measurement of rock deformations during construction of the Amsteg tunnel (1920–1). A limited length of the tunnel was plugged with concrete, and the space behind filled with water under pressure p; the deformations of the tunnel diameters D were measured in different directions. Curves showing the unit rock strain ε versus p could be drawn, giving an average value for the modulus of elasticity E of the rock mass *in situ* (see fig. 6.1). Engineers realized that the value E could be used in estimating the required thickness of the concrete or steel

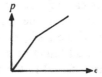

Fig. 6.1 Hydrostatic pressure, p, in a gallery versus rock strain, ε.

lining for a pressure tunnel. Several theories were developed during the late nineteen twenties and early thirties.

A few years later, when large arch dams were built it was realized that elasticity of the rock abutments could not be ignored when calculating the elastic arches or shells forming the dam. Measuring the elasticity of the rock for dam abutments became a most important feature of rock mechanics.

Experience gained in testing rock masses established that the curves relating the rock deformations ϵ to the rock stresses σ were yielding important additional information. The $\epsilon = \epsilon(\sigma)$ curves can be used for classifying rock masses for engineering purposes exactly as $\epsilon = \epsilon(\sigma)$ curves are used to classify rock material. There are correlations between alteration of rock masses and their E-values and between the E-values and the rock crushing strength.

6.2.1 The pressurized gallery (water loading test)

The oldest technique is the one used inside the Amsteg tunnel. Similar tests have since been carried out in many countries (fig. 6.2).

The principle of this method is always the same: water is pumped inside the cavity and the pressure is maintained at a required level. Before any measurement can begin the temperatures of the water and the rock mass have to be equalized. Accurate measurement of radial displacements is the most

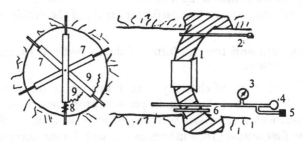

Fig. 6.2 Water loading test on an unlined gallery in rock. (1) Manhole; (2) air outlet; (3) pressure gauge; (4) water meter; (5) water outlet; (6) cable tube; (7) invar rods; (8) vibrating wire meter; (9) cable.

important part of the test. The method used at Amsteg was rather primitive. The displacements at six points of the tunnel's circumference were directly relayed through six stiff radial rods to a disc of metal located in the centre. The disc was rotated during the test to record displacements versus time.

More sophisticated tests were carried out for the Eichen pressure shaft in Switzerland. In addition to measuring the modulus of elasticity and the total modulus of deformation of the rock, designers were concerned with stresses developing in the thin steel lining of a steep pressure shaft and the proportion of the load being transferred to the rock. Detailed theories have been developed for estimating this transfer, taking into account the modulus of elasticity of both rock and steel. Another aspect of the tests concerned the inward buckling strength of the thin steel shell when stressed by outside pressure from water filling the rock joints. New designs to reduce this danger were tested.

The tests were carried out on a short gallery excavated in the rock in a direction closely parallel to the proposed shaft. A great number of stresses and strains were measured simultaneously and transmitted to an electronic recorder. Because of the relative shortness of the length ΔL on which the uniform radial pressure is aplied, the deformation curves along ΔL show a maximum deflection in the middle of the length (Gilg & Dietlicher, 1965). The results confirmed the basic theories of elasticity used for the design of steel-lined pressure shafts.

Hydro-power tunnels are often built at relatively shallow depths where the rock is anisotropic. Radial measurements in different directions are required to account for the actual E values in each direction. It is essential for all measurements to be related to a stable tunnel axis.

As will be seen later, when developing the theory of tunnel linings and borehole tests, the hydrostatic pressure in the test gallery causes circumferential tensile stresses. The state of the rock is very complex and fissuration is likely to be caused by tensile failure rather than by visco-plastic deformation. This type of test provides information vital to tunnel designers, particularly as natural residual rock stresses are also included. It will nevertheless be appreciated that the modulus of elasticity E and the modulus E_{total} thus obtained may differ from values obtained by other methods. The E value is calculated from (Jaeger, 1933)

$$E = \frac{Pa^2(1 + \nu)}{r\Delta r},$$

where P = internal pressure, a = radius to rock face (assuming circular chamber), r = radius to point where deflection is measured, Δr = change of radius r, ν = Poisson's ratio.

6.2.2 The Austrian method

This method, developed by Austrian engineers (see Seeber, 1961) and also used in Switzerland, is an alternative to that just described. The radial load is not applied by water pressure but by a cylinder which radially compresses a limited length ΔL of the tunnel (fig. 6.3). Yugoslav engineers (Kujundzic, 1965) used similar techniques, the main difference being that they applied the radial load with flat jacks.

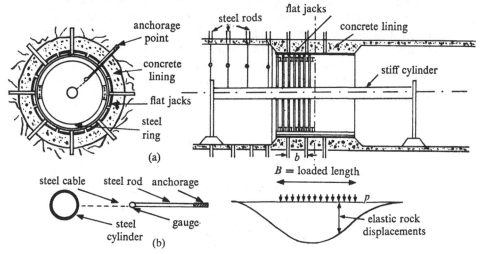

Fig. 6.3 (*a*) Austrian loading test (after Seeber, 1961); $B = 1·75$ to $2·00$ m, $b \cong 0·40$ m. (*b*) Detail of the radial measuring equipment.

Tests of this type were carried out in the 2-m-diameter Kaunertalkraftwerk tunnel in Austria. The rock quality varies considerably throughout the tunnel and a great number of sections had to be tested to make sure the overall lining was strong enough. The rock quality was tested with seismic waves between measuring points. The Austrians were careful not to overstress the rock and to keep deformations elastic.

As shown in fig. 6.3 the testing rig consisted of a 2-m-long cylindrical steel frame which was put into compression with radial wedges (the radial pressure on the rock can reach 65 kg/cm²). Radial displacements are not measured on the contact surface of the rock and concrete lining but about 15 cm inside them.

Deformations caused by a cylindrical jack such as the one described here are not alike even if each steel ring forming the jack is uniformly loaded. Theoretically, the deformations vary along the cylinder, describing what looks like a gaussian probability curve. The cylinder can easily be displaced along the tunnel. It can be seen from fig. 6.3 that all the deformations stem from a stable axis.

This method can determine the required thickness of the lining, but it is less informative about rock elasticity or plasticity.

6.2.3 Plate-bearing test

With this method a load is applied to a flat surface of the rock and the resulting surface deformations measured. The test can be carried out in an open surface trench, inside a tunnel or gallery, or in a specially excavated rock cavity. Inside galleries are often preferred because it is easier to support the test rig on the roof or opposing wall. Loading tests can be horizontal or vertical and the plate transmitting the load to the rock can be elastic or rigid; the stress distribution under the plate depends on its shape and elasticity. Rock settlements are measured under the plate or on an axis at some distance from it. The applied force may vary from about 50 tonnes (Talobre jacks) to 1200 tonnes (4 × 300 tonnes) for the tests carried out inside the Bort tunnel (France) (fig. 6.4).

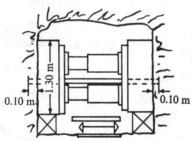

Fig. 6.4   4 × 300-tonne jacks, mobile on rails, used inside the Bort tunnel (France) (after Talobre, 1962).

The theoretical basis is the well known Boussinesq solution for the normal displacement of the surface of a semi-infinite elastic solid under the action of a point normal load. This expression relates the surface displacements with the applied loads as:

$$S_0 = \frac{P(1 - \nu^2)}{\pi Er},$$

in which S_0 represents the normal displacement of the surface at radius r from a concentrated normal load P; ν is the Poisson ratio and E is Young's modulus.

Surface displacements can be expressed in the form:

$$S_0^* = \overline{m}P\frac{(1 - \nu^2)}{E\sqrt{A}}$$

in which S_0^* is the average displacement of the loaded surface, A is the area of the loaded surface, and $\overline{m}$ represents a coefficient dependent on the shape of the loaded surface and the distribution of the load and stiffness of the plate.

At Morrow Point dam the Bureau of Reclamation used 2 × 200-ton capacity jacks 3 ft 10 in apart (fig. 6.5); load was transmitted through steel

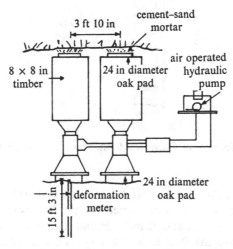

Fig. 6.5 Jacks employed by U.S. Bureau of Reclamation for jacking tests (Morrow Point dam).

jack shoes to hydraulic cushions and concrete bearing pads (24-in sides). The bearing pads were cast directly against the rock surfaces which had been trimmed by hand tools to a depth of about 2 ft beyond the original tunnel wall. The hydraulic cushions were connected to a pressure-control switch, which accurately maintained the desired load. AX (2-in) holes were drilled in the centre of each jack to a depth of about 16 ft. Each hole was provided with an extensometer and changes in tunnel diameter were measured between the two jacks with a micrometer.

As an alternative, Clark drilled a $1\frac{15}{16}$-in borehole, 16 ft deep (fig. 6.6). A tubing of outside diameter 1 in was grouted in the hole and the tubing provided with a special joint-meter. In addition cold rolled steel anchors were imbedded in line with the vertical jack in the ceiling and in the floor, to which

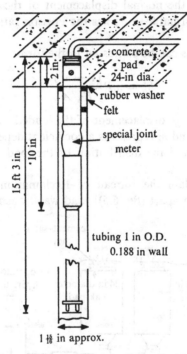

Fig. 6.6 Rock deformation gauge for foundation jacking tests (U.S. Bureau of Reclamation).

invar rods were attached. The tunnel deformation was measured by observing the increase of the gap with a micrometer as the load was applied. The displacements are usually about 0·01 to 0·1 in and can be conveniently measured by dial gauge reading to 0·0001 in with a range of 0·5 in. The deformation at a depth z is calculated with Habib's formula:

$$w_z = \frac{P}{E}\frac{(1 + \nu)}{2\pi a}\left[\frac{az}{a^2 + z^2} + 2(1 - \nu)\tan^{-1}(a/z)\right],$$

where

a = radius of the bearing plate,

z = vertical depth under the plate.

Similar techniques were used for the Dez Project (Iran), the Karadj arch dam, and many others.

Some experts believe that a bearing pad of 24 in (61 cm) square is too narrow and that the volume of rock strained by the load and the depth reached by the strain are too small. Rocha *et al.* (1955) mentions loads of 300 tonnes over an area of 1 m² and Stucky (1953) 720 tonnes over 1·2 m².

Interpretation of the results of plate-bearing tests is complicated by the presence of the loose zone having a much greater deformability than that of the undisturbed rock mass. Manfredini *et al.* (1975) have developed a scientific method of interpretation of the test results, based on the finite element method. With this method it is possible to obtain the deformability of both the loose zone, and the undisturbed rock, through surface measurements of displacements of points located at different distances from the loading plate, and on the walls of a borehole drilled at the centre of the loading area.

In situ *triaxial tests*.   Rock failure under a loaded plate occurs by visco-plastic deformation of the rock (see Prandtl–Terzaghi's formula). Swiss

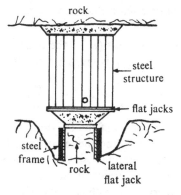

Fig. 6.7   Triaxial compression tests on rock *in situ*. The crushing strength of rock is substantially reduced when compared to the standard compression test.

engineers (Gilg & Dietlicher, 1965) developed a method (fig. 6.7) for testing rock under a triaxial stress field similar to that obtained in a laboratory. A block of rock is cut out of the rock mass *in situ* and restrained laterally by a steel frame. When vertically loaded, lateral flat jacks located inside the rigid frame develop horizontal pressures. Failure occurs at far lower loads than those applied in the classical plate-bearing test.

6.2.4   Flat jacks

Flat jacks – sometimes called Freyssinet jacks – used for measuring residual stresses, may also be used for measuring strains, deformations and stresses. The E values will be derived from the strain–stress curves $\epsilon = \epsilon(\sigma)$.

A slot is cut into the rock by drilling a line of overlapping holes and the flat jack is grouted into the slot so that each face is in uniform contact with the rock. Pressure is applied to the rock and the field of stresses perpendicular to the jack are measured. Then the average displacements of the rock can be deduced from the volumes of liquid pumped into the jack. The modulus of elasticity is evaluated again using the Boussinesq solution. The test is repeated in two directions to obtain the vertical stress field σ_v, the horizontal field σ_h and the modulus E of the rock masses. Rock anisotropy can be detected.

Another method for measuring the modulus E (fig. 6.8) demonstrates that a reference point R will move with a component of displacement u in a

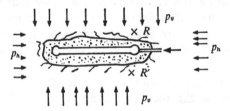

Fig. 6.8   Reference point method for measuring modulus, E.

direction parallel to the main stress (p_v in the figure, assuming the flat jack to be in the horizontal position). u is given by the function:

$$u = A\frac{p}{E} + B\frac{p_h}{E} - C\frac{p_v}{E},$$

in which A, B and C are functions of the Poisson's ratio, of the geometry of the elliptical hole, of the flat jack and the location of R. The measurement is accomplished by placing two reference pins R and R' symmetrically opposite R. With an extensometer measurements are accurate to about 0·0005 in. Two measurements are carried out with two flat jacks at right angles yielding p_v and p_h for $u = 0$. The displacement u measured prior to installing the flat jack (when $p = 0$) when stress relieving the rock, may now be used to calculate E for known values of p_v and p_h, putting $p = 0$ in the previous equation.

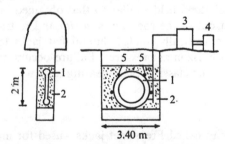

Fig. 6.9   Hydraulic flat-jack test (Yugoslavia). (1) Hydraulic flat-jack; (2) packing; (3) reservoir; (4) pump; (5) deflection guages.

This method sometimes yields higher values for the modulus of elasticity E than that determined by the plate-bearing method, rock-bolt method or convergence tests. At Oroville dam, in sound rock in the area of the underground power-house, the flat-jack method gave E values about five times higher than those obtained with other methods. Other authors have stated however, that convergence tests give higher figures.

The Yugoslavs have published information describing an arrangement where a large flat jack is introduced in a slot excavated in the invert or in the walls of a gallery, at right angles to the axis of the gallery (fig. 6.9).

6.2.5 Miscellaneous tests

Cylindrical jacks. These were introduced in boreholes to test soils. Mayer, giving the Third Rankine Lecture in London in 1963, demonstrated a powerful form of cylindrical jack for rock testing, developed by the Centre d'Études du Bâtiment (Paris). It is possible to measure the modulus of deformation in boreholes of standard diameter from 76 mm to 86 mm (3 in to $3\frac{1}{2}$ in). Electricité de France has designed another apparatus which can be used for boreholes of 160 mm and higher pressures (fig. 6.10).

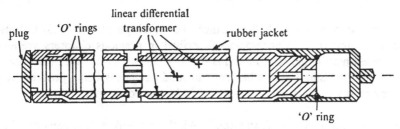

Fig. 6.10 Cylindrical jack for rock testing in boreholes.

The theory of a radially strained borehole is easy to develop from the special case of pressure galleries. The radial stress, σ_r, and the circumferential stress, σ_t, on the inner surface of the borehole are related to the pressure transmitted by the jack, p, as

$$\sigma_r = -\sigma_t = p.$$

The radial deformation u of the radius r of the borehole is related to E, the modulus of deformation of the rock, through the relation

$$u = \frac{r}{E}\left(\sigma_t - \frac{1}{m}\sigma_r\right) = -\frac{r}{E}\left(1 + \frac{1}{m}\right)p,$$

where $1/m = \nu =$ Poisson's ratio.

According to Mayer the increase ΔD of the borehole diameter D_0 is given by

$$\frac{\Delta D}{D_0} = \frac{1 + \nu}{2E} [\sigma_v + \sigma_h + (3 - 4\nu)(\sigma_v - \sigma_h) \cos \theta],$$

where θ is the inclination of the borehole to the vertical.

Mayer did not rely on this theory and had the cylindrical jack calibrated in blocks of marble or concrete against which the modulus of elasticity could easily be directly checked. A French report to the Eighth Congress on Large Dams (1964) indicates how calibration of the jack was carried out in a block of concrete 1·70 m square, solidly cemented to sound rock inside a gallery (La Bathie). The surface of the concrete was flush with the rock surface and the measurements were carried out at a depth of 0·70 m. Load was increased by steps of 10 kg/cm².

Instruments of this type can be used in the exploratory stage. Used in boreholes slanted in various directions, they will show the extent of fissurations or cracking in the rock at different depths and reveal any anisotropy. Initial tests in a considerably fissured gneiss revealed decompression of the rock towards the slope of a valley as well as a progressive improvement of the rock mass with increasing depth. Results obtained on various other sites prove that this type of test is of great use.

Convergence tests (*rock bolts*). Since the development of rock bolting, rock bolts and rock extensometers described in previous chapters are also used for estimating the rock modulus E. Rock bolts have an additional use in measuring the rising strain when rock around cavities is being progressively relieved of stress. This rising strain should be recorded against time.

Cable method of in situ *rock stressing.* All the methods previously described can be criticized because they can only test a small rock mass. For example, in borehole tests the stresses decrease rapidly, starting from the edge of the borehole. At a distance of only one diameter the stresses in competent, unfissured, rock decrease to one-ninth of the test pressure in the borehole. Similarly, with plate-bearing tests on competent rock the stresses do not penetrate deep into the rock mass, but strains of deeper penetration do occur when the rock is fissured. For this reason several authors suggested the use of tensioned cables for *in situ* tests to both the Seventh and Eighth Congress on Large Dams, Rome, 1961, and Edinburgh, 1964.

For such a test (Zienkiewicz & Stagg, 1966) the load is applied through a steel cable anchored at depth in a small borehole: a similar arrangement is shown in fig. 6.11. In order that the reaction at anchorage point will not appreciably affect the displacements at rock surface, a minimum anchorage depth of eight to ten times that of the bearing-pad diameter is tentatively recommended. The loaded area can easily be made large enough in relation

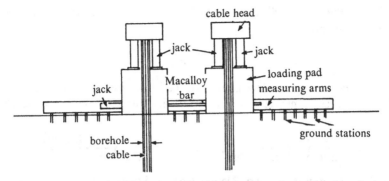

Fig. 6.11   Cable tests with four vertical cables, each tensioned by three jacks and horizontal jacks and tie bars (after Zienkiewicz & Stagg, 1967).

to the diameter of the borehole cutting through it, to justify neglecting the presence of the borehole in the stress analysis. Loads of up to 1000 tons can be applied in this way using a single cable, thus allowing a large volume of rock to be influenced. Several cables could be used to apply even greater loads if needed.

The cable approach has distinct advantages. In particular, the rock can be tested at the exact foundation location and in the directions in which the actual loads of the structure (dam abutment) will be exerted. The test could be repeated at various levels of excavation, using the same borehole and cable to obtain information about how rock characteristics vary with depth. With two adjacent cables (Zienkiewicz & Stagg, 1966, 1967), loads tangential to the surface can be applied, and information obtained about the variation of the elastic modulus with direction of load (fig. 6.11). This is an important point since the majority of rock masses have more or less anisotropic deformability.

The problem of displacements due to point tangential loads was theoretically solved by Cerruti in 1887, giving a solution analogous to Boussinesq (1885) and which can be integrated in the same manner to yield the displacements due to loads distributed over prescribed areas (Jaeger, 1950). Mitchell established expressions corresponding to the Boussinesq and Cerruti solutions for the case of transversely isotropic material (i.e. material in which there is a plane of isotropy and possessing differing properties in a direction normal to this plane). This system is reasonably typical of many stratified rocks (it possesses five independent elastic constants: two elastic moduli, two Poisson ratios and an independent shear modulus).

It can be shown (Mitchell, 1900) that if the elastic modulus in the direction perpendicular to the surface is E, and if nE is the elastic modulus in the plane parallel to the surface, then the formulae for the average displacements of a single square pad with a side a are:

$$S^* = 2 \cdot 97 A_1 \frac{p}{a},$$

where S^* is the average displacement of the pad in a direction perpendicular to the surface and

$$\bar{u} = 2\!\cdot\!97B_1 \frac{Q}{a},$$

where $\bar{u}$ is the average displacement of the pad in a direction tangential to the surface. In these formulae $v_1 = v_2 = 0$, and A_1 and B_1 are functions of E and n:

$$A_1 = \frac{1}{2\pi E}\left(1 + \frac{1}{n} + \frac{2}{\sqrt{n}}\right), \quad \text{and} \quad B_1 = \frac{1}{2\pi E}\left(\frac{\sqrt{[2(1+n)]}}{n}\right).$$

Field tests carried out by Zienkiewicz & Stagg yielded $E_v = 3\!\cdot\!60$ to $3\!\cdot\!95 \times 10^5$ lb/in² and $E_h = 1\!\cdot\!98$ to $1\!\cdot\!78 \times 10^5$ lb/in².

In similar lines Jaeger (Seventh Congress on Large Dams, 1961) suggested a combination of the cylindrical jack test and the cable test with the cable

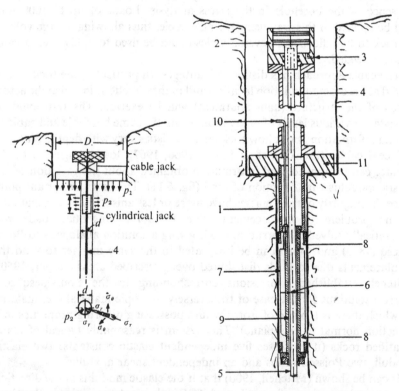

Fig. 6.12 Combined cylindrical jack test and cable test. (1) Rock under test; (2) cylinder for axial thrust; (3) oil pressure admission to (2); (4) cable; (5) lower end of cable enclosed in rock; (6) borehole in rock for radial loading; (7) rubber hose inflated for radial loading of rock; (8) canvas-reinforced ends of (7) vulcanized to serrated ends of tube (9); (9) tube; (10) oil pressure admission to rubber hose (7); (11) steel block for axial loading of rock.

passing through the cylindrical jack (fig. 6.12). The cylindrical jack causes tensile stresses to develop in a circumferential direction and the test would yield valuable information on the tensile strength of rock *in situ* when under triaxial strain and on the *E* modulus under such *in situ* conditions. Very little is known about tensile strength of rock masses *in situ*, despite the fact that local tensile failure and brittle fractures of rock masses may cause the final collapse of a concrete dam.

6.2.6 Residual stresses and modulus of elasticity. The Pertusillo dam case

The foundations of the Pertusillo dam (Italy) are an excellent example of the importance of rock compactness and its effect on the modulus of elasticity (Fumagalli, 1966*a*).

At depth, under the project dam (100 m high), the rock is a marl-clay sandstone gres; the overburden is a conglomerate, the smallest thickness of which is about 20 m (fig. 6.13). This conglomerate is characterized by the

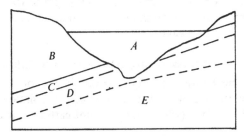

Fig. 6.13 Geology of the Pertusillo dam (after Fumagalli, 1966*a*)

A = concrete, $E = 250\,000$ kg/cm², $\gamma = 2.5$.
B = soft sandstone with clay,
$\qquad E{\downarrow} = 25\,000$ kg/cm², $\overrightarrow{E} = 40\,000$ kg/cm², $\gamma = 2.4$, $\phi = 30°$.
C = stony sandstone, $E{\downarrow} = 35\,000$ kg/cm², $\overrightarrow{E} = 50\,000$ kg/cm², $\gamma = 2.4$, $\phi = 30°$.
D = conglomerates, $E{\downarrow} = 50\,000$ kg/cm² mean modulus of the whole mass, $\gamma = 2.4$, $\phi = 30°$.
E = Marl-clay sandstone mass, $E = 10\,000$ kg/cm, compressibility modulus, $\gamma = 2.5$, $\phi = 25°$, $c = 1$ kg/cm².

following figures: $E = 50\,000$ kg/cm², $\lambda = 2.4$ tonnes/m³, $\phi = 30°$ (angle of internal friction).

The underlying gres was tested in deep prospecting galleries: rock cores of 60 to 70 cm diameter were isolated and pressure was applied inside the cut by means of cylindrical rubber pockets. Specific radial deformations, measured with extensometers, were very high, $\epsilon = 5$ to 7×10^{-3} and an $E = 10\,000$ to $12\,000$ kg/cm² was found, with $\gamma = 2.5$ tonnes/m³ and $\phi = 25°$. Rock cohesion was very low ($c = 1$ kg/cm²), for the foundations of a high-arch dam (the figures are lower than those of the gneiss foundations of Malpasset).

Seismic measurements of the *in situ* rock elasticity, however, gave E values as high as $E = 500\,000$ kg/cm^2, pointing to the fact that the *in situ* rock was highly precompressed and had been compacted during a very long geological period. High rock compaction was maintained by the weight of the overburden conglomerates. Excavation of prospecting galleries and drilling rock cores had relieved these residual stresses and the E value obtained by the tests was correspondingly low.

Rock samples were systematically tested in the laboratory with a triaxial equipment capable of developing 2000 tonnes. The tests proved that the deformation modulus for low loads was low, but that the material could be compacted after several loading cycles; the final modulus of elasticity was high.

The design of the dam was checked on a model up to rupture by overloading. The arch dam ruptured, but not the rock abutment correctly represented to scale. It was concluded that the equilibrium conditions of the dam and rock were perfectly safe within ample elastic limits. Although the material was of low cohesion and with a small angle of friction there was a high degree of compactness maintained by a considerable overburden load due to the overlying rock strata.

6.2.7   Correlation of measurements of the modulus *E*

In situ tests. Some figures have been given in section 4.3 correlating E values obtained by the seismic and static methods. The figures are taken from a table published by Clark (1966), and it is not certain that they all refer to laboratory tests.

It is not to be expected that figures concerning the E values obtained by different methods of *in situ* strain measurements should correlate; they depend on: (1) the rock fissuration and the residual stresses in the rock, (2) the degree of chemical and physical alteration of rock material, of fissures and joints, (3) the degree of wetting of the rock and water percolating through the fissures, (4) on the method used for rock testing: some methods cause fissuration by tensile stresses whereas plate-bearing tests usually cause rupture by shear. The state of residual stresses in the rock mass is very important.

Test results may depend on the degree of stress relieving of the rock near the surface or in the vicinity of cavities and galleries. They also depend on the ratio of the horizontal field of residual stresses σ_h to the vertical field σ_v. Assuming $\sigma_h \cong \sigma_v = p$ during the driving of the tunnel, circumferential compression stresses σ_t will develop, opposing the tensile stresses σ_t caused by pressure chamber tests. Radial stresses, however, would diminish as the tunnel is driven, becoming zero at the edge of the cavity. In the event that $\sigma_h \cong 0$, vertical fissures may develop, lowering the E figures obtained by the pressurized gallery method. Plate-bearing tests are similarly influenced by the state of residual stresses.

Recent tests carried out by French engineers show the static E values measured in radially stressed boreholes to be 1·5 to 5 or 8 times higher than simultaneous tests with the usual plate-bearing test method (see section 6.6).

If flat jacks are installed in a direction perpendicular to the tunnel axis it will be found that the longitudinal stress, σ_l, has not been diminished during the excavation of the tunnel. The rock under such conditions remains nearer to its original condition of soundness than does rock fissured on the radial direction.

Convergence tests are likely to yield a different modulus than plate-bearing tests. Because of physical limitations, it is often necessary to install rock bolt extensometers a short distance back from the tunnel heading and some small amount of unmeasured radial deformation may already have occurred. Secondly, a scale effect: rock bolt extensometers 10 ft to 30 ft in length will span the disturbed fractured zone near the surface, as well as a sizeable thickness of sound, undisturbed rock beyond; the measured modulus of this mass of rock may be a so-called 'effective' or average modulus. Plate-bearing tests are usually carried out on rock already stress relieved.

At the Oroville dam the average modulus measured at five different sites was: with the tunnel convergence method, $2·6 \times 10^6$ lb/in²; with flat jacks method, $7·8 \times 10^6$ lb/in².

Comparison of field and laboratory results. It is generally agreed that the main reason for differences between laboratory and field values is due to the jointing of the rock masses which differs in size and direction with the micro-fissures of the rock material.

Other reasons are as follows. *Alteration of the rock*: this term can mean alteration of the joints or alteration of the rock itself. As mentioned in section 4.8, it has been Portuguese research workers who have been mainly responsible for studying the alteration of rock and its effect on rock compressive and tensile strengths. They have also shown that the modulus of elasticity of the rock, E, depends on the factor of 'rock alterability' or 'void index' i. Shuk (1964*b*) confirms the Portuguese findings. He tested a phyllite and found the following average results:

i	E
0·3	85 000 kg/cm²
2·0	55 000 kg/cm²

Effect of saturation: it has been found that saturation or wetting of the intact rock can lower the value of the modulus of deformation. Tests conducted by Shuk on samples of phyllite–quartzite indicate that the reduction can be as much as 70% of the dry rock values after three days immersion in water. Bernaix found only a 30% drop in strength of some filling material in joints of fissured limestone. *Rate of strain*: many authors (Talobre, Seeber, Hardy, etc.) mention the decrease in strength properties and of the E value with the lessening rate of strain. Laboratory tests are usually performed more rapidly than field tests. The reduction of the E value may be as much as 50%. Seeber finds similar reduction for rock masses tested *in situ* at different rates.

Velocity of seismic waves.   The velocity of seismic waves in rock material and rock masses varies with the void index, i. (See section 6.6.1.)

6.3   Other *in situ* tests: shear tests. Tensile strength of rock masses

6.3.1   Shear tests

In situ shear tests are next in importance to *in situ* tests on rock deformability. Detailed information on such tests is available either on rock shear strength or on tests of concrete blocks adhered to rock surfaces.

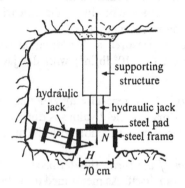

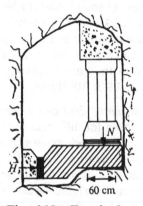

Fig. 6.14   Shear test (Portugese arrangement, Rocha). The force, P, has a horizontal component, H, causing shear failure.

Fig. 6.15   French shear test (Electricité de France).

Portuguese tests.   Serafim & Lopez at the Fifth International Conference on Soil Mechanics, Paris, July 1961, discussed the technique they used in Portugal (fig. 6.14). They carried out field shear tests on rock blocks of reasonable size, attached to the base rock, and on concrete blocks moulded against the rock surface. Such blocks include some fissures, seams and local alterations of the foundation rock (fig. 6.15).

The blocks (70×70 cm and 30 cm high) were surrounded by very rigid metallic frames. The blocks and the foundation were kept saturated during the tests. Tangential displacements were measured at two points and the normal ones at four points of the blocks, up to failure. Normal forces were applied first, and only when stabilization of the deformations was reached, were inclined forces applied. These were gradually increased until stabilization of the displacements occurred at each step.

The results were interpreted by tracing Coulomb's lines on a (σ, τ) diagram. Two criteria decided the 'tangential stress at failure.' One considered that failure occurred at maximum tangential stress and, the second, that failure occurred when the direction of the vertical displacement of the downstream side of the block was inverted (fig. 6.16).

Forty-four tests were carried out, each parallel and perpendicular to the schistosity planes of analogous rocks, showing approximately the same index of porosity. Shear tests parallel to the schistosity planes gave slightly lower τ_{max} values than other tests. (See also section 12.3 on Morrow Point dam.)

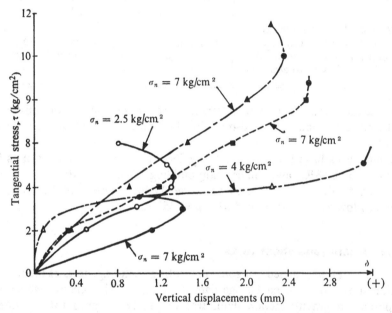

Fig. 6.16 Vertical displacement curves showing the inversion criteria of failure of field shear tests (after Serafim & Lopez, 1961).

Failure took place in the rock in almost every rupture test on concrete blocks moulded against rock. Some of the results are shown in table 6.1.

U.S. Bureau of Reclamation field tests. A 15 × 15 in by 8 in high block was prepared. The block was made to project above the floor level of a tunnel and was surrounded by a structural steel jacket. Shear and normal loads were applied. The shear load was inclined so that its line of thrust passed through the centre of the shear zone, between the projecting test block and the rock mass, thereby eliminating overturning moments. Instrumentation consisted of dial gauges which measured both vertical and horizontal deformations of the block. The results were correlated with laboratory tests on large blocks (approximately 1 yd³) and with triaxial tests to give a better interpretation of the shearing resistance of the block. Tests were also conducted, utilizing

Table 6.1 *Shear strength of rocks tested* in situ (*after Serafim & Lopez, 1961*).

Rock	Direction of test	Criteria for failure	Index of porosity (i)	Shearing strength kg/m²	φ°
Weathered granites	upstream	maximum stress τ_{max}	3	13·4	62·5
			7	3·5	52
			10	2·2	46·5
	downstream	inversion of displacement	3	6·5	63·5
			7	2·2	49
			10	1·5	42
Weathered schists and concrete on schists	parallel to cleavage perpendicular to cleavage	maximum stress τ_{max}	—	0 to 4	59 to 64
			—	3·5 to 9	60

concrete blocks cast on rock to determine the cohesion and friction resistance at the place of contact. (See also section 12.3 on the tests at Morrow Point dam.)

Similar tests have been conducted by Electricité de France, as well as in Spain, the USSR, and other dam-building countries. Bollo (France) reporting to the 1964 Congress on Large Dams, writes that he found φ values as low as 12°–17°, others with φ = 45°.

6.3.2 Anchorage shear tests

J. A. Banks (1957) carried out some interesting anchorage tests at the power station site of the Allt na Lairige dam in Scotland. The scheme, which incorporated a gravity section dam, was originally designed for a limited storage (8·3% of the annual run-off). A prestressed concrete design was investigated (83 ft high) and it was estimated that storage could be increased by 10% at practically the same capital expenditure by building a much higher dam. The anchorage method first used at the Cheurfas dam in Algeria was adopted, except that anchorage bars and plates were used instead of cables. The bedrock at Allt na Lairige is granite, a far more compact foundation than that of the Algerian dam.

The method used to test the strength of such an anchorage is clearly shown in fig. 6.17. Rock surface displacements are measured with precision levels. Rock displacements and concrete displacements relative to the bottom of the pit well are measured with high-tensile steel wires. Figure 6.18 shows the vertical displacement of the rock surface (line *A*) and of the concrete plug (line *B*) relative to the rock at base of the pit. Line *C* gives the absolute displacement of the rock surface, measured by precision level (accuracy ± $\frac{1}{100}$ in). For all practical purposes the test results were reassuring. With a depth of only 18 ft for the test anchorage as against 26 ft for the anchorage under the dam, a pull of 4400 tons did not damage an anchorage designed for normal $P = 1176$ tons. The tests were carried out with a bare rock surface, whereas under the dam the full weight of the dam plus the opposite

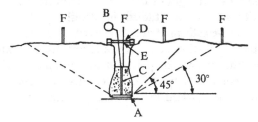

Fig. 6.17 Detail of test anchorage at Allt na Lairige dam: (*A*) Freyssinet flat jack; (*B*) hand pump; (*C*) deflectometer; (*D*) crossbeam; (*E*) steel tube duct; (*F*) staffs for vertical displacement measurement (after Banks, 1957).

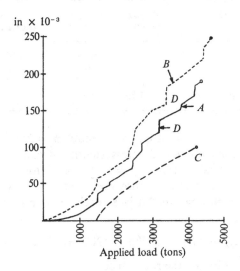

Fig. 6.18 Deformations recorded at rock anchorage test at Allt na Lairige dam: (*A*) vertical displacement of rock surface relative to rock at base of pit under flat jack; (*B*) vertical displacement of concrete plug relative to rock at base of pit under flat jack; (*C*) approximate absolute displacement of rock surface measured by precision engineering level; (*D*) relaxation periods (after Banks, 1957).

force $-P$ would hold the rock surface. There is no danger to the foundation as long as weights and anchoring forces balance the overturning moment from the horizontal hydrostatic thrust.

The analysis of stress conditions during the test shows that the anchorage forces were transmitted by shear. The weight of the inverted rock cone of 18 ft height and 18 ft radius at the base is about 435 tons. This is quite inadequate to hold down an uplift thrust of 4400 tons and obviously the uplift force was balanced not by the weight of the rock but mainly by shear stresses. Figure 6.18 shows that deformations were more or less elastic for $P = 1000$ tons. Beyond that plastic deformations and small ruptures occurred, but the shear strength of the rock was not finally destroyed.

This complexity should be studied in detail by direct mathematical analysis or with photo-elastic methods. It is likely that stresses in the immediate vicinity of the application point of the forces are high and that strains will locally reach the domain of plastic deformations, which substantially complicates the theoretical approach. A tentative calculation using Boussinesq's formulae is summarized in table 6.2. The maximum shear stress, τ,

Table 6.2   *Allt na Lairige dam, stresses about the anchorage*

Depth above anchorage plate	$P = 1000$ tonnes		$P = 4400$ tonnes	
	$\sigma_{v\ max}$ (kg/cm²)	$\sigma_{h\ max}$ (kg/cm²)	$\sigma_{v\ max}$ (kg/cm²)	$\sigma_{h\ max}$ (kg/cm²)
$x = 0.5$ m	192	10·6	840	46·5
1·0 m	48	2·6	210	11·6
5·5 m	1·6	0·08	7	3·8
7·95 m	0·75	0·04	—	—

is less than 30% of $\sigma_{v\,max}$ in a direction about $\phi = 25°$ from the vertical. The most dangerous horizontal stress (tensile stress) occurs for $\phi = 0$. The maximum vertical stress $\sigma_{max} = 840$ kg/cm² for $P = 4400$ tons as tabulated, is lessened locally because the force P was transmitted to the concrete anchorage through six flat jacks each 870 mm in diameter bringing the stress down to

$$\sigma_{max} = 4400 \times 10^3/(6 \times 5900) = 120 \text{ kg/cm}^2.$$

The flat jacks burst before the rock failed.

6.3.3 Tensile strength of rock masses

Up to now attention has been concentrated on rock crushing strength. Dam designers are very informative about vertical and shear stresses transmitted from the concrete foundation to the rock foundation at the base of gravity, buttress or arch dams, but they rarely give any information on how these stresses are absorbed within the rock mass.

Some investigations going more deeply into this vital aspect have shown that shear stresses parallel to the surface of a half-space cause tensile stresses to develop inside the half-space. Zienkiewicz has calculated the tensile stresses below prestressed dams in the deeper rock mass between the dam heel and the anchorage of the cable. At the Seventh Congress on Large Dams, Rome, 1961, the problems caused by such stresses were examined and it was suggested that rock may not be able to stand tensile strains. In the case of rock rupture along the line of highest tensile stresses, the dam may be in danger of overturning or sliding. Zienkiewicz also suggested that a similar area of tensile stresses existed in the rock foundation near the heel of buttresses in a buttress dam. Jimense-Salas & Uriel (1964) and others examined the problem for an ordinary gravity dam and found that the zones under tensile stresses were more extensive than the areas mentioned by Zienkiewicz for anchored dams. This is not improbable, considering how the shape of the dam may influence stress distributions in the rock.

Measurements made inside Straight Creek tunnel (USA) showed that definite areas were under tensile stresses along the soffit which was being excavated in reasonably sound rock. These strained areas could have been investigated with differential rock-bolt extensometers. It is interesting to note that the areas under tension extend beyond those forecast by the theory which assumes sound homogeneous rock and the worst possible case of $K = \sigma_h/\sigma_v = 0$ (see section 5.4). The area under tension extends even more deeply into the rock masses with increasing density of the rock jointing. This shows the importance of tensile stresses which develop in rock masses and their obvious danger to the stability of large engineering structures.

There is no known method for the direct measurement of tensile strength of rock *in situ*. An indirect method is to trace the so-called intrinsic curve (using circles of Mohr) in the area near the point $\sigma = 0$. Tensile strength is expected to be about the same magnitude as the rock cohesion, but there is no proof that this applies for rock masses.

The whole problem of tensile strength of rock masses will be discussed from another aspect in the sections dealing with tunnel and dam engineering. Model tests of dam abutment carried out by Fumagalli (1966b) at the ISMES Laboratory in Bergamo, show that there is a danger of rock failure by brittle fracture when the dam foundations do not penetrate deep enough into the rock. At greater depths failure occurs by shear fracture of the rock.

6.4 Creep of rock masses

Rock creep can be observed *in situ* and in laboratories under conditions of constant load versus time. These slow rock deformations do not usually cause rock rupture and they can be seen in salt mines and in the foundation of some large dams. Slow rock deformations of another type can be observed inside a tunnel shortly after it has been driven, when overstressed rock reaches its limit of elastic deformation or even the limits of its crushing strength.

The two types of rock creep occur under constant load. Repeated loading and unloading causes rock deformation curves versus time, which although similar to rock creep curves result in internal ruptures of rock masses, even when final failure does not occur. Such deformations cannot be classified under 'creep' and they will be discussed separately.

6.4.1   Creep in salt mines

Rock salt at depth is under hydrostatic stress, and a cavity opened in it has a tendency to close. Stress distribution around a cylindrical or spherical cavity is such that the rock is triaxially strained, the radial component being the minor principal stress in the vicinity of the cavity. Time displacements in a radial direction are therefore dependent upon the creep characteristics of salt in triaxial compression. In figs 6.19 and 6.20 some results of creep tests

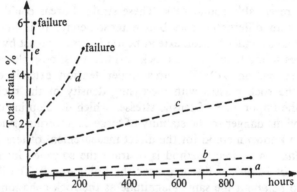

Fig. 6.19   Triaxial creep tests on rock salt. Stress difference: (*a*) 1000 lb/in²; (*b*) 1500 lb/in²; (*c*) 2500 lb/in²; (*d*) 3125 lb/in²; (*e*) 3650 lb/in² (after U.S. Army W.E.S., 1963).

conducted in triaxial and uniaxial stress state are given. It can be seen that at higher temperatures, the total rock salt strain is substantially increased.

According to these tests the radial strain ϵ_r in salt mines can be calculated by using the formula:

$$\epsilon_r = 1.87 \times 10^{-13}(\sigma_1 - \sigma_3)^{2.98}t^{0.36},$$

where $(\sigma_1 - \sigma_3)$ is the stress difference in lb/in² and t the time in hours.

Griggs (1936) performed a series of experiments on rocks to determine their creep characteristics. According to his findings, where rocks are subjected to stress for long periods of time, the terms 'elastic limit', 'set point', and 'strength' lose their ordinary meaning because they usually define properties for short period tests. (Typical creep tests using field jacks maintain constant pressure for six days and drop back to zero for one day before raising the pressure again to a higher level for a further six days.)

In an application of creep analysis to the deformation of a circular shaft in salt the displacements were expressed by the formula:

$$U_r = -pB(a^2/r)\log(1 + bt),$$

where U_r = radial displacement, p = pre-excavation pressure in the rock around the shaft, t = time in days, a = shaft radius, r = radial distance to the point in the solid, B, b = constants. For salt Barron & Toews (1963) found $B = 76.6$ to 90.8 and $b = 1/2.7$. More information on rock creep will be given in section 6.5.

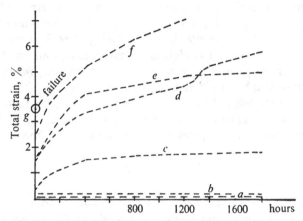

Fig. 6.20   Uniaxial compression tests on rock salt (after U.S. Army, W.E.S., 1963).

	compressive stress (lb/in²)	temperature (°C)
a	525	23
b	750	65
c	1750	23
d	1750	65
e	2250	23
f	2250	65
g	3000	23

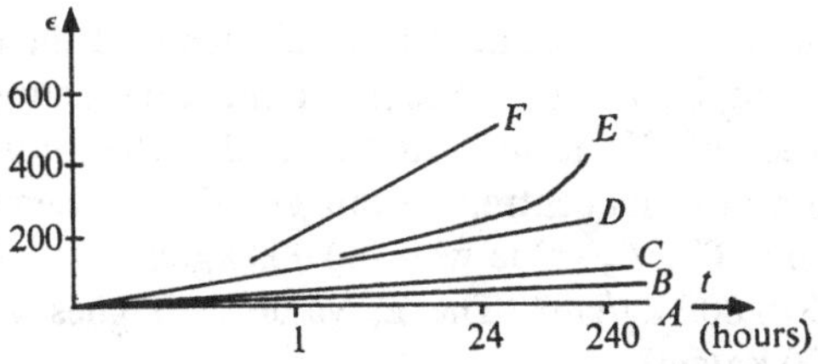

Fig. 6.21   Creep curves: (A) compact quartzite $\sigma = 56$ kg/cm²; (B) schist $\sigma = 48$ kg/cm²; (C) argilolith $\sigma = 56$ kg/cm²; (D) weathered granite $\sigma = 19$ kg/cm²; (E) very weathered schist $\sigma = 30$ kg/cm²; (F) schist $\sigma = 48$ kg/cm²; t = time in hours, ϵ = strain in microns (by French National Committee on Large Dams, 1964).

A research committee of Electricité de France (1964) similarly observed creep curves after several loadings and unloadings. The creep deformations caused by a steady load following variable loadings were always far smaller than those previously noticed (about 5 to 20%). The greater part of creep deformations was found to be reversible. In one case a loaded schist failed after five days (fig. 6.21).

6.4.2   Checking rock deformability against settlement of large dams

The deformations and displacements of large dams are systematically measured over long periods, and detailed material is sometimes available which could be used to calculate the deformability of rock masses in reverse order. Deformations caused by weight and pressure of the dam and hydrostatic load involve very large masses of rock, far larger than any *in situ* test. The scale effect must also be considered in the comparative analysis of the results.

Let E_c be the modulus of elasticity of the concrete, E_r the modulus of the rock and $\eta = E_c/E_r$. The radial deflection of the cantilevers, δ, depends on η and it can be assumed (Gicot, 1964) that this function δ is linear and that:

$$\delta = \frac{1}{E_c}(a + b\eta),$$

where a and b are constants and E_c is usually known approximately. The problem is in calculating η and E_r from a series of tests. If possible, the method should yield values for the elastic modulus, for the modulus of total deformation of the rock and at least some information on the creep of large rock masses.

French engineers have buried related sonor extensometers (Coyne type) both in rock and concrete. Assuming the field stress to be about the same on both sides of the foundation line, we find that the ratio of measured deformations yields the ratio η of the two moduli. For example, for the La Palisse dam the E_c value was $E_c = 240\,000$ kg/cm² and $E_r = 85\,000$ to $130\,000$ kg/cm². The curves obtained indicate that at the beginning of filling, the rock settlement was far greater than the settlement of the concrete until a stable η value was reached.

A second technique is to measure the displacement of the abutments of an arch dam. For example, the upper arch of the Chaudanne dam (France) transferred a thrust of 275 tonnes to the rock. The radial displacement of the arch was 8 mm at its centre, the tangential displacement of the abutments was 0·85 mm. The E_c value was $200\,000$ kg/cm² and the calculated E_r value was then $320\,000$ kg/cm². The E_r value at Tignes was found to be $110\,000$ to $170\,000$ kg/cm².

At the Eighth Congress on Large Dams (1964) the French reported that various testing methods gave widely differing results and that there seemed to be no definite correlation between them. The French authors believe that E_r values obtained from measurements of dam deformations are higher than those obtained by direct measurements of rock deformations.

Similar reports were published about the Lanoux and Grandval dams. (France). At Lanoux, the reversible deformations were 40% of the total during *in situ* rock tests and 50% when analysing the deformations of the dam. Detailed information is also available on the Rossens dam, which was built on sandstone (Gicot, 1964). Jacking tests, as well as laboratory tests, have shown a horizontal modulus of deformability ranging from 20 000 to 40 000 kg/cm² with a minimum of 8000 to 10 000 kg/cm² and a maximum of 80 000 to 120 000 kg/cm². The so-called plastic deformations formed 20 to 50% of the total.

Vertically the test showed on an average deformability about 50% higher than the horizontal factor. The E_c value for concrete was 400 000 kg/cm². The dam was designed for $\eta = 10$ for the arches and $\eta = 15$ for the cantilevers and checked for $\eta = 20$, and 30, taking into account a possible creep of the sandstone.

The deflections of the dam were carefully and systematically measured over the periods 1949 to 1951 and again from 1952 to 1962. Gicot analysed these deformations and expressed them as

$$\frac{1}{E_c} = \alpha, \quad \frac{1}{E_r} = \beta, \quad \eta = \frac{E_c}{E_r} = \frac{\beta}{\alpha},$$

and $\delta =$ the calculated radial deflection of the cantilevers which becomes

$$\delta = \frac{1}{E_c}(a + b\eta) = a\alpha + b\beta.$$

The observed deflection is δ_0 and the difference between computed and measured deflections is

$$\epsilon = a\alpha + b\beta - \delta_0.$$

The conditions for $\sum\epsilon^2 =$ minimum, require:

$$\sum\epsilon\frac{\partial\epsilon}{\partial\alpha} = 0 \quad \text{and} \quad \sum\epsilon\frac{\partial\epsilon}{\partial\beta} = 0;$$

Therefore,

$$\sum a(a\alpha + b\beta - \delta_0) = 0,$$

and

$$\sum b(a\alpha + b\beta - \delta_0) = 0.$$

These equations yield:

$$\alpha = \frac{\sum ab \sum b\delta_0 - \sum b^2 \sum a\delta_0}{[\sum ab]^2 - \sum a^2 \sum b^2}$$

and

$$\beta = \frac{\sum ab \sum a\delta_0 - \sum a^2 \sum b\delta_0}{[\sum ab]^2 - \sum a^2 \sum b^2}$$

By using this method it has been found over the period from 1952 to 1962 that: $E_c = 380\,000$ kg/cm^2 and $\eta = 4\cdot2$ as against $E_c = 400\,000$ kg/cm^2 and $\eta = 3$ to $4\cdot5$ for the period from 1949 to 1952. The corresponding E_r values are higher than those measured direct from the rock. Gicot tried to separate elastic and inelastic deformations. He found that during the first period (1949–51) inelastic deformations closely followed the water levels.

An attempt was made to compute the possible effect of concrete creep. A 'final specific creep' of $0\cdot002\%$ for $\sigma = 1$ kg/cm^2 was assumed. With such a creep the irreversible deformations due to rock correspond to a value $\Delta\eta = 4\cdot0$ and the total value $\eta_{\text{total}} = 4\cdot2 + 4\cdot0 = 8\cdot2$. If the creep of the concrete is assumed to be only $0\cdot001\%$, the value of $\Delta\eta = 5\cdot6$ and $\eta_{\text{total}} = 9\cdot8$. These values correspond to an apparent total modulus of the rock (including elasticity and plasticity) in the region of $45\,000$ kg/cm^2 while the real modulus of elasticity of the rock is about $90\,000$ kg/cm^2.

6.5   Strain–stress diagrams and interpretation of strain–stress curves

6.5.1   Definitions

These diagrams basically record the deformations, ϵ, of rock masses caused by a stress, σ, applied to the surface. The diagram $\sigma = \sigma(\epsilon)$ or strain–stress diagram is designed to give basic information on the behaviour of fissured rock masses *in situ* under uniform loading. Results depend on the method used for obtaining rock deformations, on the point of measurement (e.g. on the surface of the rock, and the depth of the rock mass), or at some distance laterally to the applied load. They also depend on the rate at which the load is applied and on its direction. Loading a borehole with a cylindrical jack may take a few minutes whereas the Austrian technique for loading tunnels and galleries spreads over weeks.

How certain stress, σ, is obtained is also important. Figure 6.22 is a typical stress–strain diagram where ϵ are the strains of deformations and σ the average uniform compression stress. On such a diagram the curves OA correspond to a loading of the rock from $\sigma = 0$ to $\sigma = \sigma_1$. Usually the curve is concave downwards, corresponding to a non-linear stress–strain deformation. When unloading the rock mass and reducing the stresses σ, the deformation does not revert to the O point ($\epsilon = 0$) but to a point D_1. Reloading progressively to a stress $\sigma_2 > \sigma_1$ and unloading again yields a curve $D_1A_2D_2$. It is assumed that the deformations OD_1, OD_2 are non-reversible and are probably caused by the closing of fissures or small fractures. The lines A_1D_1 and A_2D_2 would correspond to the average elastic deformation of the mass of fissured rock. The lines A_1D_1 and A_2D_2 are sometimes parallel; it may require several loadings and unloadings to obtain a constant angle α for these lines. Elastic rock deformation is not always reached.

Several different definitions of Young's modulus can be seen from fig. 6.22. The line A_1D_1 represents the modulus of elasticity $E_{\sigma_1} = \tan \alpha_1$ for the first

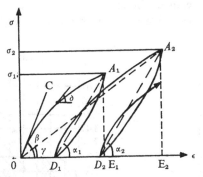

Fig. 6.22   Typical stress–strain curve for *in situ* rock compression tests. Definition of the modulus of elasticity $E = \tan \alpha$, of the modulus of total deformation $E_{\text{total}} = \tan \gamma$ and of $k = OD/OE$.

loading to an average rock stress σ_1; the line A_2D_2 the modulus of elasticity $E_{\sigma_2} = \tan \alpha_2$ for the second loading up to σ_2. The E_σ values correspond to the more or less elastic rock deformations. The line OA_2 defines a $E_{\text{total}} = \tan \gamma$ which corresponds to the total deformation when loading and unloading the fissured rock mass up to σ_1 then to σ_2. E_{total} is the 'modulus of total deformation'. Tracing a tangent OC to the origin O of the curve OA_1 defines a modulus $E_0 = \tan \beta$. Theoretically E_0 should be the same as the static modulus when measured on rock samples in the laboratory; but there is not enough evidence to support this. Finally, at any point of the curve $\sigma = \sigma(\epsilon)$ a local modulus $E = \mathrm{d}p/\mathrm{d}\epsilon = \tan \delta$ could be defined. This indicates rock response to load increases at a certain point of the curve $\sigma(\epsilon)$. It may also indicate some internal rock failure or changes in rock characteristics.

The curve OA_1A_2 is obtained by slow loading. Austrian experts have introduced a similar curve OBB_1 obtained by rapid loading of the rock (fig. 6.23). They call it OA_1A_2 or OB_1B_2 'work curve' (*Arbeitslinie*) as it is

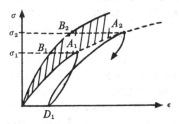

Fig. 6.23   Slow and rapid loading of rock masses shown on the curves OA_1A_2 and OB_1B_2. The shaded area represents the work of internal rock deformation caused by slow loading.

obviously related to the internal work of deformation in the rock. The shaded area within the two curves OB_1B_2 and OA_1A_2 is called *Arbeitsbereich* and is a measure of the work of internal rock deformation under long duration load tests when the load rises slowly. This is not true creep.

The E_e and E_{total} values obtained from the load deformation curves $\sigma = \sigma(\varepsilon)$ are of primary value as is the ratio $k = OD_1/OE_1$ of the plastic

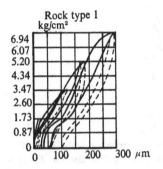

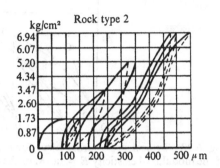

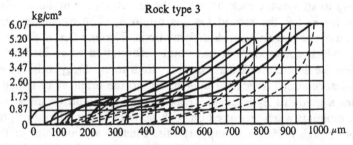

Fig. 6.24 Strain–stress curves for different rock types (after Oberti, 1960).

deformation OD_1 which is irreversible to the total deformation OE_1. American geologists rightly consider this ratio to be an intensive characteristic of rock masses. Additionally, the shape of the curve $\sigma = \sigma(\varepsilon)$ yields important information on the behaviour of rock masses, and possibly on their structure. The E_0, E_e, E_{total} and k values indicate the behaviour of rock mass under strain and stress. These figures should always be accompanied by additional information giving the value σ for which the E_e, E_{total} and k values were measured and defining the σ value determination method.

Most of the strain–stress curves obtained in tunnels and galleries concerned elastic and plastic deformations that did not reach the final crushing strength of the rock masses. Out of 300 deformation curves recently recorded by Electricité de France only three concerned rock masses where rupture had actually occurred. Pressures of 320 kg/cm² were used in the Roselend and in the Gittaz galleries (France) without causing rupture of the rock.

6.5.2 Interpretating strain and stress curves

Oberti, in a paper read in 1960 to the Salzburg meeting of the Austrian Society of Rock Mechanics, submitted several typical strain and stress curves characteristic of good, indifferent and poor rock (fig. 6.24). In the case of good rock (type 1), the k value is small, and elastic deformation is reached rapidly. It is obvious that the rock shown in the type 2 diagram is less reliable and that type 3 has very poor characteristics.

Experts try to analyse the curves $\sigma = \sigma(\epsilon)$ in detail. If the deformation curve is a straight line, it indicates an elastic behaviour of the rock (fig. 6.25*a*). A point of discontinuity (point *A* on fig. 6.25*b*) may indicate some internal rupture in the rock mass, possibly a local shear failure. A departure from a straight line of deformation may suggest internal plastic deformation (fig. 6.25*c*).

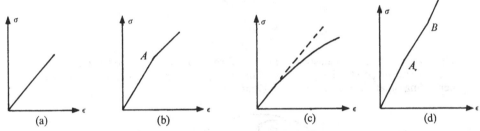

(a)   (b)   (c)   (d)

Fig. 6.25   Types of deformation curves.

More information about the behaviour of rock masses can be obtained from systematic analysis of the successive loops of the strain–stress curves. Mazenot has carried out such an analysis using a large number of strain and stress measurements supplied by Electricité de France. His interpretation of the curve is based on a theory developed by Talobre (see section 6.5.3).

Some curves clearly show not one, but two points of discontinuity. Shuk (1963) produced a curve (Fig. 6.25*d*) on metamorphic rock described as a phyllite–quartzite. There are two points where the elastic limit had been reached: *A*, for the phyllitic phase (small crystals), and *B*, a higher limit for the quartzite phase with larger crystals. The general curvature and direction of the line should also be observed.

The shape of the curve obtained when unloading the rock also has to be analysed for possible discontinuities or indications of internal failures. When the load decreases sharply, without a corresponding reversal of deformations, the curve may confirm that some internal areas of rupture are no longer behaving elastically (Talobre). It is normal practice to load the rock in at least two directions, in order to detect possible directions of weakness. These exist in any stratified rock, and may even exist in most crystalline or metamorphic rocks. In stratified rock masses, the modulus of elasticity parallel to the strata is often two or three times greater than the modulus measured in a direction normal to the strata. Assuming the values $E_1 > E_2$ to be known in the direction parallel and perpendicular to the main rock stratification, the value E_α at any angle α to E_1 can be calculated assuming the deformation by analogy to the sketch in fig. 6.25a. The E values are given by an ellipse with E_1 and E_2 as the main axis (Jaecklin, 1965b); fig. 6.26a. In this respect the circular jacking of a gallery allows a far closer examination

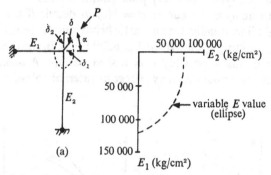

Fig. 6.26a   Modulus of elasticity in stratified rocks. E_1 and E_2 have been measured parallel and perpendicular to the rock main stratification. E varies from E_2 to E_1 depending on angle α (after Jaecklin, 1965b).

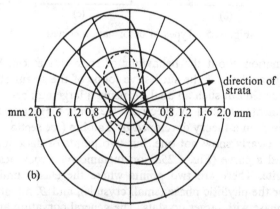

Fig. 6.26b   Typical radial deformation diagram obtained in the Kaunertal pressure tunnel. Broken line, elastic deformations; solid line, total deformations (parallel layers of marl, 0.30 m thick, 2.00 m distant) (after Seeber, 1964).

of rock-masses. It is possible to record the deformations, ϵ, in any direction for any load, σ, in a circular diagram (fig. 6.26b). The contour lines corresponding to elastic deformations can be traced. And, in addition, the characteristic directions of maximum deformations (normal to stratification) and minimum deformations (parallel to stratification) $\sigma = \sigma(\epsilon)$ can be found and analysed. These are needed for planning the lining of a tunnel or gallery.

The strain–stress diagram can also be used to detect possible creep of the rock masses. In fig. 6.27 rock deformation increases by $\Delta\epsilon$ as rock stress decreases, which obviously corresponds to rock creep as a function of time. Constant load tests were carried out at Electricité de France over a period of 240 hours, with loads varying between 19 kg/cm² for decomposed granite, 30 kg/cm² for a weak schist (curve *E*) and 80 to 160 kg/cm² for very hard limestone and gneiss. Curve *E* in fig. 6.21 is interesting because it shows the accelerated creep of a weak schist after about 200 hours.

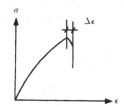

Fig. 6.27   Strain–stress diagram.

These French tests show that creep deformations can be reversible or non-reversible, instantaneous or delayed, but that it is difficult on a strain-stress diagram to separate instantaneous and delayed non-reversible deformations. It is believed that delayed deformations may represent about 5% to 20% of apparently instantaneous deformations. Furthermore it is difficult to expect test results obtained on a few cubic metres of rock over a period of ten days to represent exactly the deformation of a dam rock abutment after several years under varying hydrostatic loads. In some cases the creep

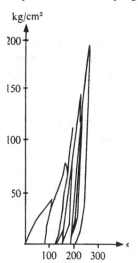

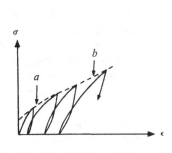

Fig. 6.28   Concave strain–stress curve (after French National Committee on Large Dams).

Fig. 6.29   The strain–stress curve is first convex (*a*), then straight (*b*).

deformation is practically reversible. In one case, it was found that the deformation ϵ had returned to $\epsilon = 0$ three months after suppression of the load ($\sigma = 0$).

The general trend of the strain–stress curve may give some further information on the rock structure. Figure 6.28 shows that after several loading and unloading cycles the envelope to the strain curve is concave, which indicates progressive compression of voids and fissures and consolidation of the rock. In fig. 6.29 the general trend of total deformations is a straight line, which probably means there was some initial residual stress on the rock. Diagrams (*a*) and (*b*) of fig. 6.30 published by a research team of Electricité

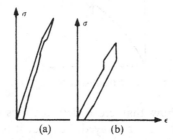

Fig. 6.30   Strain–stress curves: (*a*) dry rock; (*b*) impregnated rock (after French National Committee on Large Dams).

de France refers to dry and to impregnated gneiss showing the effect of impregnation of rock masses.

6.5.3   The Talobre diagram

J. Talobre (1957) uses a movable 50-tonne ram during pressure tests in galleries. The load, P, is transmitted to rock masses through a relatively small rigid steel plate and pressures are increased so that the rock is stressed well beyond the yield point. Internal ruptures occur in the overstrained rock causing so-called 'plastic' deformations.

To interpret the corresponding test results, Talobre traces simultaneously the strain and stress curves on a $\sigma = \sigma(\epsilon)$ diagram and the Mohr circles on the $\tau = \tau(\sigma)$ plane (fig. 6.31). Mohr's theory is that when material ruptures the circle of Mohr representing that particular state of stresses is a tangent to an 'envelope curve' or 'instrinsic curve, C'. (Curve C is the envelope to all circles of Mohr for which rupture occurs.) He uses this theory to interpret *in situ* test results.

For example, starting from a point 1 on a curve $\sigma = \sigma(\epsilon)$, the main stress σ_1 is assumed to be known ($\sigma_1 = p$); the other two stresses σ_2 and σ_3 can be measured directly with flat jacks. This allows the relative circle to be known on the (σ, τ) diagram. When decreasing the load from point 1 to 2, deformations remain elastic, as shown by the gradient of the curve 1 to 2. With further pressure decreases, deformations become inelastic. According to Talobre this means that lateral pressures (σ_3) take over. The circles from 2 to 3 and also from 3 to 4, obtained when pressures are increased again, are a tangent to the 'intrinsic curve, C'. From point 4 to point 6, deformations are again elastic and the Mohr circles should no longer touch the curve C. By

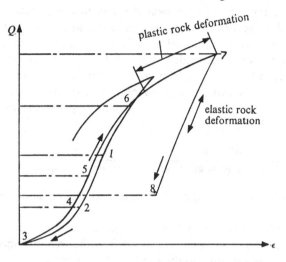

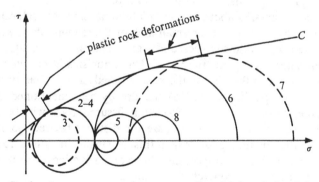

Fig. 6.31 Strain–stress curves and corresponding Mohr circles (after Talobre, 1957).

increasing the pressure in the jack beyond point 6 the rock again yields plastically to high pressures, and deformations become inelastic. From 6 to 7, circles of Mohr will again touch the curve C.

According to the theory and technique developed by Talobre, segments of the intrinsic curve C can be traced when overloading the rock and causing internal rupture of some areas in the rock mass. Other segments of the curve C are obtained when unloading the rock. This causes stresses in the rock to become unbalanced, and further internal ruptures occur. When the second principal stress is either known or measured Talobre uses the formula:

$$\frac{\sigma_3 - \sigma_1}{2c \cotan \phi + \sigma_3 + \sigma_1} = \sin \phi,$$

which will be demonstrated in the next chapter. In this formula σ_3 and σ_1 are the principal stress. For example, starting from some results obtained on mica schists inside the Bort tunnel, plastic deformation started when σ_3 reached 20 kg/cm². Tests proceeded with increasing pressure to 40 kg/cm² when the second principal stress was $\sigma_1 = x$. Pressure was then dropped, causing elastic rock deformations until plastic deformation recommenced for $\sigma_3 = 6$ kg/cm². If we assume that the second principal stress $\sigma_1 = x$ remains unchanged, the equation then becomes

$$\frac{40 - x}{2c \cotan \phi + 40 + x} = \frac{x - 6}{2c \cotan \phi + x + 6}.$$

Assuming c cotan $\phi = 10$ kg/cm² we find that the equation yields $x = 18$ kg/cm².

Some of Talobre's critics say his method is guesswork for either the second principal stress $\sigma = x$ or the value c cotan ϕ. But the second principal stress at the start of the tests is equal to the residual stress and should be measured. A second criticism involves the changes in direction of the second principal stress during loading caused by Poisson's ratio, v.

More general remarks about loading-plate dimensions are relevant to any plate-bearing test. Because of the small area of the loading-plates stresses do not penetrate deeply into the rock and weak areas are not located, even at shallow depth under the plate. In spite of these criticisms Electricité de France now accept Talobre's theories as the basis of analysis for their strain–stress curves. Mazenot's publication (1965) has already been mentioned in section 6.5.2. Further information on jack-testing fissured rock will be given in section 12.1. It is important to compare the distance between joints to the diameter of the loading pad. When this is small the joints reduce the measured E value. Scale is important when comparing local test results with conditions under a large concrete foundation.

6.5.4 The distribution of stresses in rock masses

Two main types of failure, brittle fracture and visco-plastic deformation, have been observed on rock material in addition to shear fracture. Visco-plastic has been seen on rock masses during shear tests or uniaxial compression tests (Gilg, 1965) but real rupture of the rock masses under a plate-bearing test, (which usually causes the collapse of soils) is rarely observed. Actual rock failure has been seen in several major dam ruptures, sometimes from excessive pressures. The conditions prevailing at the periphery of a concrete arch or gravity dam are such that compression stresses, shear stresses and moments are transmitted from the concrete abutment to the rock foundations, whereupon rock strains and stress develop, causing the surface to yield and deform. This deformation modifies the stress at the concrete boundary. The rock mass may be considered as a homogeneous isotropic half-space or, alternatively, as a fissured mass of rock or a clastic accumulation of independent blocks (fig. 6.32).

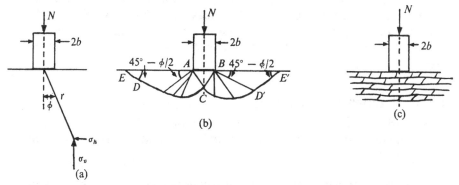

Fig. 6.32 Loading of rock: (a) homogenous rock (formulas of Boussinesq); (b) rock with little or no cohesion, conditions for rupture of rock; (c) stratified fissured rock.

6.5.5 Brittle fracture and visco-plastic failure

Compression and shear tests on rock masses *in situ*, triaxial and shear laboratory tests on rock material may be supplemented by additional information obtained in the laboratories on rock mass models where the stratifications, faults and cracks are correctly represented.

Brittle fracture occurs whenever the classical 'ideal' principal stress equals the limiting value of uniaxial tensile strength. The plasto-viscous type seems to occur in all other cases and the higher the average isotropic triaxial compressive stress $\sigma_0 = (\sigma_1 + \sigma_2 + \sigma_3)/3$ with respect to the diverging stresses $(\sigma_1 - \sigma_0)$, $(\sigma_2 - \sigma_0)$ and $(\sigma_3 - \sigma_0)$ the more important the microcracking and the greater the deformations. The theory of the intrinsic curve to be developed later may be satisfactorily applied to plasto-viscous failures.

In rock mass, the brittle type of failure normally occurs at the crystal layers where the isotropic triaxial compression stress is usually less intense. Owing to their stiffness, cohesion bonds are not assisted by numerous hyperstatic connections within the rock mass and they fail abruptly as soon as the elastic deformation limits are exceeded.

In a paper on the stability of arch dam rock abutments Fumagalli (1967) describes the progressive failure of rock masses by plasto-viscous deformation as follows:

As the load increases the deformation processes affect the resisting structure by successive frontiers located ever deeper within the rock and farther from the force application plane. As the frontiers move deeper, the cohesion bonds in the receding area gradually decrease while resisting stresses due to internal friction of the material are called upon to mutually cooperate through a gradual redistribution of the stresses.

At constant loading the deformation speed decreases and finally becomes nil as long as possible state of static equilibrium is possible.

When collapse occurs it is usually a global phenomenon.

He describes these gradual plasto-viscous deformations after close observation of the collapse of an abutment mass, on laboratory dam models. This type of collapse cannot be satisfactorily observed during *in situ* jacking tests under compression, and it is seen even less clearly during *in situ* shear tests.

6.5.6   Theories and methods

It is clearly appreciated that there are limitations in the different theories and methods used to estimate stress distribution inside a strained–stressed mass of rock.

Classical homogeneous isotropic elastic half-space equations.   Rock has been simulated to a first approximation of homogeneous, isotropic space. The calculation of stresses about a cavity located deep in the rock mass is a typical example of how a problem can initially be solved by the classical equations of strength of material. Another example is the case of a half-space loaded by forces at its surface. The equations of Boussinesq (1885) and Cerruti solve most of the problems (fig. 6.32a). The Prandtl–Terzaghi theory (fig. 6.32b) is an extension of Boussinesq's approach.

Fissured rock.   Müller & Pacher developed a method for estimating the stress distribution in a mass of fissured rock. It assumes that a detailed survey of the rock mass has been made, and that the direction, continuity and spacing of the main families of fissures are known. The tensile strength and the shear strength of the rock at any point of the mass and in any direction depend on these fissures.

Clastic rock.   The previous method assumes rock masses to be capable of shear and tensile strength and thus to be able to transmit stresses across a fissure: at least to some extent.

The clastic mass consists of blocks of rock, or other materials, piled on top of each other with no possibility of shear or tensile stresses being transmitted from one block to the next. Only compression stresses are transmitted across a contact face or point between blocks. This theory has recently been developed by several authors, and Trollope (1968) applied it to rock mechanics. Krsmanović (1967b) built and tested some models representing this.

Modern stress–strain analysis. Modern methods of stress–strain analysis may be used for solving particular problems. These will be dealt with in chapter 7. The mathematical approach based on Boussinesq's equations yields over-optimistic results, but the alternative methods developed by Müller, Trollope, Zienkiewicz and others are more realistic.

6.6 Geophysical methods for testing rock masses

Direct *in situ* measurement of the static modulus of elasticity of rock masses requires cumbersome and costly equipment. The information obtained is extremely valuable, but the method is not adaptable to surveying large areas. Economic geophysical methods have been developed for surveying large areas of alluvium or bedrock. The *seismic method* directly measures the dynamic modulus of elasticity of rock masses for comparison with the static modulus. In addition it gives valuable information on the degree of fissuration. *Resistivity tests,* used in soil mechanics can also be used in rock mechanics.

6.6.1 The seismic method

(1) *Sources.* The principle of the seismic method is to send an elastic wave or vibration through alluvium or rock and to pick up this wave at a distance with receivers or geophones. The signals should give the velocity of the waves, the length and trace of the path they have been following and the modulus of elasticity of the medium inside which they have been propagating. This wave may be produced by various methods: by a heavy blow with a hammer; by an explosion on the surface or slightly below the surface of the ground; by a vibrator or a source of ultrasonic waves. For example, Evison (1953) used a high-frequency vibrator in shattered rock, resting on a thin layer of mud. Electric sparks exploded underwater at a frequency of 2 to 4 per second

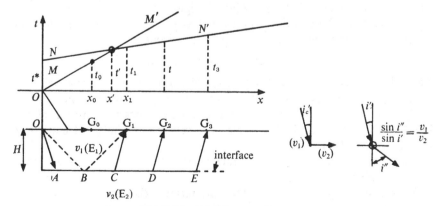

Fig. 6.33 Seismic wave theory.

were successful in a survey of the English Channel bedrock. Prospecting at greater depths, an explosion of a mixture of propane and oxygen in a small underwater chamber allowed the shock wave to penetrate the bedrock.

Explosions or acoustic waves have been used between boreholes for exploring rock layers at different depths. This method has been used in tunnels and galleries to determine the depth at which rock is deteriorating through blasting.

The principle of the seismic path is best explained by fig. 6.33. A wave starts from point O at the surface of the soil or rock to be tested. It is known that harder rock is located at an unknown depth, and assumed that the modulus of elasticity of the upper layer, E_1, is less than the modulus of the lower layer, E_2. Wave velocities v_1 and v_2 in the two media depend on E_1 and E_2 and v_1 will be smaller than v_2 ($v_1 < v_2$).

A complex system of waves emanate from point O and an even more complex system is reflected back or deflected and then reflected at points A, B, C, D, of the interface. These are registered by receivers located at the points G_1, G_2, G_3, along a straight line at known distances from O.

(2) *The wave path.*   The general wave equation is

$$\frac{\partial^2 u}{\partial t^2} = C^2 \nabla u,$$

where C is a constant and u the disturbance which propagates with time. In the case of cylindrical symmetry it becomes

$$\frac{\partial^2 u}{\partial t^2} = C^2 \frac{1}{r} \frac{\partial}{\partial t} \left(r \frac{\partial u}{\partial r} \right),$$

and for spherical symmetry:

$$\frac{\partial^2 (ru)}{\partial t^2} = C^2 \frac{\partial^2 (ru)}{\partial r^2},$$

where r is the distance from the point of origin.

In many cases the waves can be treated as plane waves (Jaeger, 1933) in which case the integral of the wave equation becomes

$$u = g(x - vt) + f(x + vt),$$

g and f being functions of the distance, x, and time t, v being the wave velocity.

In geophysics, the path of waves and their velocity are of most interest. The analysis of the wave path and its reflection determines the position of the interface of different geological strata. The velocity v permits calculation of the dynamic rock modulus.

Time can be measured very accurately and a time–distance curve is constructed with axis x and t.

A wave starting from O may reach the receiver G_0 directly. A wave can reach the detector G_1 by one of three paths: The direct path OG_1, the reflected path OBG_1, or the refracted path $OACG_1$. If work is carried out on first arrivals only the reflected path, which is always greater than the direct path, and is covered at the same velocity v_1, is automatically eliminated. For the direct path the time–distance curve is a straight line OMM' . . . passing through the origin. Its slope is by definition equal to the velocity v_1. For the refracted path, the pulse travels down to the interface with the velocity v_1 and its incident is at the so-called 'critical angle', i'_C. After refraction, the impulse travels along the boundary with the velocity v_2 (supposedly higher than v_1) and leaves the interface at the critical angle to reach G_1. As the distances AC, AD, AE, are covered along the interface at the higher velocity v_2, there will be a point when the wave starting from O and following the refracted path will arrive at one of the detectors G_1, G_2, or G_3, before the direct waves. The points of arrival are now on another time–distance curve, the straight line NN, which does not go through the origin O. Let $ON = t^*$; it can be shown that the depth H of the upper layer to the interface is given by

$$H = \frac{v_1 v_2 t^*}{2\sqrt{(v_2^2 - v_1^2)}} = \frac{x^2}{2}\frac{1 - \sin i'}{\cos i'} = \frac{x'}{2}\sqrt{\left(\frac{v_2 - v_1}{v_2 + v_1}\right)}$$

In a few cases the method does not work well: when the interface is a weathered rock (weathered granite); when the two layers are similar and E_1 only slightly less than E_2, or when the upper layer is covered by a thin layer with high modulus of elasticity $E_0 > E_1$. Seismic tests in general concern rather low stress levels by means of high-frequency vibration of very small amplitude. They are not as precise as jacking tests, but they can economically indicate variations in rock quality over a greater portion of the foundation.

(3) *Wave velocity and the* in situ *dynamic rock modulus.* When a shock is produced on the surface or inside a body with modulus of elasticity E_d and density ρ, shock waves progress in all directions. Various types of waves can be detected: the dilatational, longitudinal or compressional waves with higher velocity v_e, and the transversal or shear waves with velocity v_s inferior to v_e.

According to the elastic theory the following equations relate longitudinal and transversal velocities v_e and v_s, the Poisson ratio of the rock mass, ν, and the dynamic modulus, E_d:

$$E_d = v_e^2\rho\,\frac{(1 + \nu)(1 - 2\nu)}{(1 - \nu)}, \tag{1}$$

$$E_d = 2v_s^2\rho(1 + \nu), \tag{2}$$

$$\nu = \left[\frac{1}{2}\left(\frac{v_e}{v_s}\right)^2 - 1\right] \Big/ \left[\left(\frac{v_e}{v_s}\right)^2 - 1\right]. \tag{3}$$

It is usually difficult to register correctly the transversal waves and very often only the first longitudinal wave is registered for the calculation of E_d. The Poisson ratio, ν, is estimated unless it can be measured on rock samples.

Numerical example (Evison, 1953):

Concrete
Longitudinal wave velocity, $v_e = 11\ 700$ ft/s
Transversal wave velocity, $v_s = 7100$ ft/s
Specific weight, $\rho = 137$ lb/ft^3
Poisson ratio, $\nu = 0.21$
Modulus, $E_d = 3\ 600\ 000$ lb/in^2 (252 000 kg/cm^2)

Fissured rock (*indifferent quality*)

Longitudinal wave velocity, $v_e = 6500$ ft/s
Transversal wave velocity, $v_s = 2500$ ft/s
Specific weight, $\rho = 122$ lb/ft^3
Poisson ratio, $\nu = 0.41$
Modulus, $E_d = 460\ 000$ lb/in^2 (32 200 kg/cm^2)

The Poisson ratio of this rock is rather high, probably due to fissuration.

One of the most interesting examples of systematic site investigations with seismic waves was at the rock abutment of the high Vajont arch dam. The rock was tested before and after grouting, which was stopped when the overall wave velocity reached at least 3000 m/s. Other seismic tests were carried out on the slopes of the Vajont gorge and Mount Toc; they illustrated the progressive deterioration of the rock along the slope, before final collapse occurred.

Ultrasonic waves are used inside tunnels. The time taken by longitudinal waves travelling between two boreholes is recorded so that the depth at which rock blasting causes damage to rock masses may be measured. The wave velocity is less near the surface of damaged rock than at depth in sound rock. Usually it becomes normal at depths of from 1·50 to 2·00 m corresponding to the damaged zone.

Oliveira (1974/75) reports on a 25-km-long hydro-power tunnel, 3 m in diameter, excavated in Portugal. Two solutions were considered: a free-flow tunnel at level 90 m and a pressure tunnel at level 75 m, the top of the hills reaching 120 m. The seismic method proved that both solutions were to be located in dubious decomposed phyllites ($v_e = 2000$ m/s); the better rock was located too low. The free-flow solution at level 90 m was adopted.

(4) *The velocity ratio* v/v_L. The effect of discontinuities in the rock mass can be estimated by comparing the *in situ* longitudinal wave velocity, $v_e = v$, to the laboratory sonic velocity, v_L, of intact core obtained from the same rock mass. The difference in these two longitudinal velocities is caused by

the structural discontinuities which exist in the field. Figure 6.34 shows how the ratio v/v_L is correlated to the rock quality designation (RQD) (Deere, 1968). Duncan (1967) and Morlier (1968) have proved the correlation between

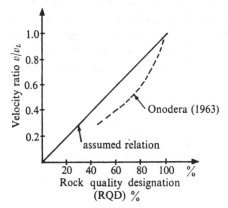

Fig. 6.34 Correlation of rock quality as determined by velocity ratio and RQD (after Deere *et al.*, 1966).

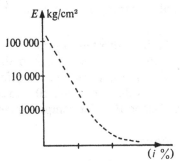

Fig. 6.35 Correlation between E and i measured in laboratory (after Hamrol, 1962).

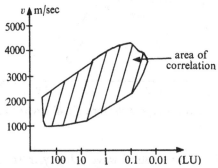

Fig. 6.36 Correlation between velocity (in m/s) and permeability in Lugeon units for two sites on granitic foundations. One Lugeon unit corresponds to loss of 1 litre per minute per metre borehole under the pressure of 10 kg/cm (after the French National Committee on Large Dams, 1964).

the v_L figures and the void index. Hamrol (1962) established similar relations between E_{static} (laboratory) and the same void index i (fig. 6.35). Finally, a French research team (1964) proved the correlation between the longitudinal velocity v_L and the rock permeability factor measured in Lugeon units (LU) (fig. 6.36). Morlier (1968) has extended research to a similar ratio v/v_L for very dense rocks without voids but with slight fissuration. Instead of the 'porosity index,' i, introduced by Duncan, Serafim, Hamrol and others, he developed the 'fissuration porosity', η'_0 (fig. 6.37). When a compact rock is

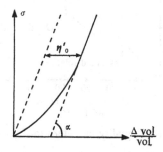

Fig. 6.37   Definition of the fissuration porosity, η_0, (after Morlier, 1968).

compressed equally on all sides hydrostatically the relative decrease, Δ vol/vol, of the volume versus σ follows a curve similar to that in fig. 6.37; tan α represents the elastic modulus for purely elastic deformations and η'_0 the closing of the fissures. The η'_0 value varies from 10^{-5} to 50×10^{-5} for granites and from 100×10^{-5} to 300×10^{-5} for compact quartzites.

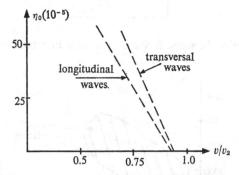

Fig. 6.38   Correlation between η_0 and v/v_L for eight different granites (after Morlier, 1968).

Assuming v'_L to be the theoretical wave velocity in sound, compact rock material, Morlier traces the lines v/v'_L versus η'_0 for different granites, quartzites and similar rocks. He finds that the ratio v/v_L is approximately a linear function of η'_0 (fig. 6.38). This confirms that any void, or fissure reduces the velocity of sound and shock waves; this applies equally to the longitudinal and transversal waves.

There is, therefore, a closely interrelated group of factors or figures (v, v_L, v/v_L, η'_o, i, E, Lugeon units, etc.) which all directly or indirectly measure the porosity and fissuration of rock material and rock masses. These correlations require more detailed investigation as they are the basis of other rock characteristics.

(5) *The rebound number R.* Schmidt designed a hammer for testing the strength of concrete. When the plunger is pressed against the surface of the concrete the mass of the hammer is released. After impact, the mass rebounds to a height indicated by a pointer against a scale. This height is the rebound number R. It has been proved that there is a definite connection between the E values of the concrete, the concrete uniaxial crushing strength and R.

These types of hammers have been used on concrete for the past twenty years and many authors have suggested that they be applied to *in situ* rock tests. Duncan (1967) has published diagrams showing interesting correlations between R and v_L and the crushing strength of rock (fig. 6.39a, b). The measure of the rebound number R is a rapid *in situ* test method requiring additional detailed and precise investigation. Correlation of R with the void index i depends on the type of rock and it is not very coherent (fig. 6.39b). It does not give any information on the fissuration of the whole rock mass when used *in situ*. It does, however, detect local weak spots.

Research by Habib, Vouille & Audibert (1965) has shown that uniaxial or triaxial pressures of up to 500 or 1000 kg/cm² considerably increase the velocity of some longitudinal and transversal waves. Figure 6.40 clearly shows that these velocity increases are due to the closing of pores and fissures by the uniaxial load. Triaxial loading causes similar velocity changes.

(6) *Further research on correlations between E and v.* Roussel (1968), referring to an unpublished thesis by Schneider, reproduces a very interesting diagram where a curve, $k = E_d$ (dynamic)$/E_s$ (static) is traced versus λ_T (in metres), λ_T being the wavelength and E_d the dynamic modulus of transverse waves developing in rocks. The curve confirms the wide range of variations of E_d/E_s from about 1 to 13 (fig. 6.41), (Clark had found $E_d/E_s = 0.85$ to 29). The correlation with λ_T is astonishingly good. (Note the location of the 'Malpasset gneiss' number 13 in the diagram well at the upper end of the curve). Figure 6.42 shows a similar correlation between E_s and the frequency of the translatory waves. The wave-velocity measurements were carried out by the falling hammer method on a base of about 40 m (fig. 6.43).

Roussel (1968) published a detailed theory explaining this correlation. His approach is based on the 'Kelvin–Voigt' model of visco-elastic vibrations, represented in fig. 6.44. Such a system uses the equation

$$m\frac{\mathrm{d}^2 Z}{\mathrm{d}t^2} + \eta\frac{\mathrm{d}Z}{\mathrm{d}t} + EZ = 0,$$

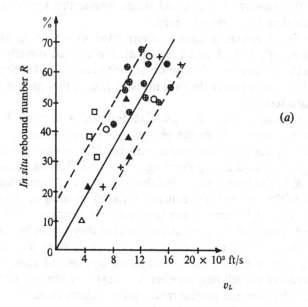

(a)

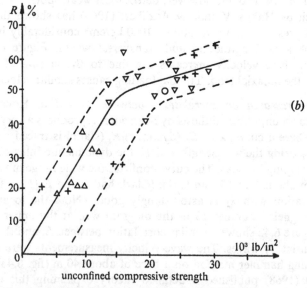

(b)

Fig. 6.39 (a) Correlation between the *in situ* rebound number R (as %) and the laboratory-determined seismic velocity v_L; ▲, shale; □, coal; ●, basalt; ○, granite; +, mudstone, ⊕, sandstone. (b) Correlation between R and crushing strength: ○, granite; △, shale; +, mudstone; ▽ sandstone. (c) Correlation between rebound number R and void index (after Duncan, 1966).

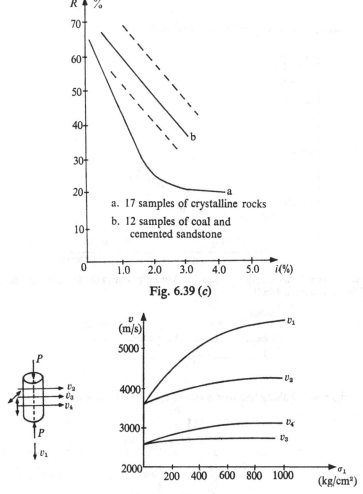

Fig. 6.39 (c)

Fig. 6.40 Increase of wave velocities under uniaxial compression of rock sample (after Habib, Vouille & Audibert, 1965).

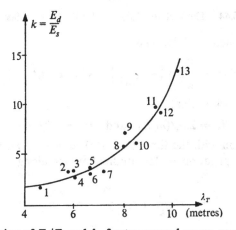

Fig. 6.41 Correlation of E_d/E_s and λ_T for transversal waves, measured on thirteen dam sites (after Roussel, 1968).

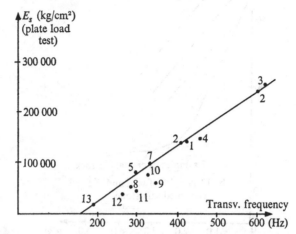

Fig. 6.42   Correlation of E_s (plate load tests) in kg/cm^2 and the transversal frequency (in Hz) (after Roussel, 1968).

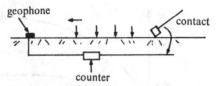

Fig. 6.43   Falling hammer method used on a base of about 40 m.

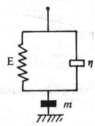

Fig. 6.44   The Kelvin–Voigt method for visco-elasticity.

the solution of which is of the type

$$Z = Z_0 \left[\exp \left(-\frac{\eta t}{2m} \right) \right] \cos \left[\sqrt{(E/m)}t + \phi \right].$$

The period of rock oscillation is

$$T_0 = 2\pi \sqrt{(m/E)} \quad \text{and} \quad \omega_0 = \sqrt{(E/m)}.$$

The amortization with the time is (in s^{-1}): $a = -\eta/2m$. Extension of the theory to the propagation of longitudinal waves along an axis yields the equation

$$(\lambda + 2G) \frac{\partial^2 u}{\partial x^2} + \rho\eta \frac{\partial^3 u}{\partial^2 x \partial_t} - \rho \frac{\partial^2 u}{\partial t^2} = 0.$$

A possible solution is

$$u = u_0 \left[\exp\left(-\alpha x\right)\right] \left[\exp\left\{i\omega(t - x/c)\right\}\right].$$

(The λ coefficient of Lamé is obviously different from the wavelength $\lambda = 2\pi c/\omega$ where ω is the wave frequency.

In this equation u is the displacement of an oscillating point in the direction of Ox; α is the coefficient of amortization along Ox (in m^{-1}) and c the velocity of the sound or shock waves. Furthermore, $\lambda + 2G$ are Lamé's coefficients. The values of α, ω and c can be measured and related to the rock characteristics $\lambda + 2G, \rho$ and η. It is therefore possible to show that:

$$\eta = \frac{2\alpha c^3 \omega^2}{(\omega^2 + \alpha^3 c^2)^2}$$

and $\eta = 0$ when $\alpha = 0$ (elastic vibration). When examining the rock from the purely static point of view,

$$\lambda + 2G = E_s \frac{1 - \nu_s}{(1 + \nu_s)(1 - 2\nu_s)} = E_s f(\nu_s),$$

where ν is the Poisson ratio and the suffix refers to static values only (excluding viscosity effects).

On the other hand, when discussing the wave propagation it could be proved that:

$$\lambda + 2G = \rho c^2 \omega^2 \frac{\omega^2 - \alpha^2 c^2}{(\omega^2 + \alpha^2 c^2)^2}$$

and

$$\rho c^2 = E_d \frac{1 - \nu_d}{(1 + \nu_d)(1 - 2\nu_d)} = E_d f(\nu_d),$$

where d refers to dynamic values of E and ν.

Finally:

$$\frac{E_s f(\nu_s)}{E_d f(\nu_d)} = \frac{\omega^2 - \alpha^2 c^2}{(\omega^2 + \alpha^2 c^2)^2} \omega^2.$$

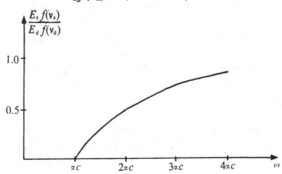

Fig. 6.45 Theoretical curve $E_s f(\nu_s)/E_d f(\nu_d)$ versus ω for longitudinal waves (after Roussel, 1968).

This curve is represented in fig. 6.45 which shows how $E_s f(v_s)/E_d f(v_d)$ varies with ω. It is very difficult to use this curve as the ratio $f(v_s)/f(v_d)$ varies considerably.

A similar theoretical development can be achieved assuming transversal instead of longitudinal waves. The displacements u parallel to Ox are replaced by the displacements v perpendicular to Ox and the function $f(v)$ by the function: $g(v) = 1/2(1 + v)$, which now varies very little with v. It is possible to assume that $g(v_s)/g(v_d) = 1$ and therefore that

$$\frac{E_s g(v_s)}{E_d g(v_d)} = \frac{E_s}{E_d} = \frac{\omega^2 - \alpha^2 c^2}{(\omega^2 + \alpha^2 c^2)^2} \omega^2.$$

This curve is traced in fig. 6.46 where some of the points measured *in situ* on dam sites are also marked with transversal waves replacing longitudinal waves. The trend of the theoretical curve and of the measured empirical

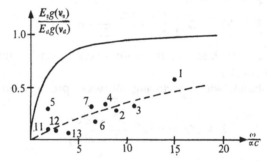

Fig. 6.46 Correlation of $E_s g(v_s)/E_d g v_d$ and $\omega/\alpha c$ for transversal waves (after Roussel, 1968).

curve are the same, but there is still a gap between them. Roussel thinks that this gap would be smaller if the E_s values were based not on plate-loading tests (giving low E_s values) but on borehole tests which yield higher E_s values. Rock fissuration too is another factor which may explain the gap.

Results of this theoretical analysis are most satisfactory because they show that E_s and E_d must be different but related through a series of physical coefficients, some of them being capable of direct measurement. The analysis also emphasizes the often-neglected transversal waves.

The efforts by Roussel, Morlier, Duncan, Serafim, Bernaix and others show how the whole structure of rock material and rock masses depends on voids and fissures.

(7) There are situations in which it is important to study vibration levels. This was done at Cabora Bassa (Oliveira 1974, 1975), where the maximum particle velocities (v_r = radial, v_v = vertical, v_t = transversal) were determined at a concreting site near the blasting site of a nearby cavity. Some types of vibrations are detrimental to setting concrete.

(8) *The proposed Channel Tunnel* (Reynolds, 1961). In 1802, a French engineer, Albert Mathieu, put forward a plan for constructing a tunnel under the English Channel. For many years (1837–67) the French geologist, Gamond, investigated the Channel bottom. Later, another prominent French geologist, de Lapparant (1875–6) made 7700 underwater soundings and recovered 3276 samples from the sea bed. Test galleries were excavated on both sides of the Channel.

The project was more strongly backed in France than in Great Britain. A French company was formed, financed partly by the Compagnie Internationale du Canal de Suez and the French railways (Compagnie des Chemins de Fer du Nord). On the British side, the idea was energetically pursued by a British financial group.

After the First World War, while technical and economical reports favoured the scheme, it was turned down for general political and military reasons. After the Second World War, the political outlook had changed sufficiently to remove any major opposition.

Between 1850 and 1960 many alternative ideas were submitted. They included two parallel dams for rail and road transport, between which barges and small ships could safely cross from Calais and Dover; a dam crossing the Channel; several large bridge projects; and finally, a tube laid on the Channel bottom.

After the Second World War an international study group began checking the geological information obtained by de Lapparant and others. The main risks were heavily fissured or broken rocks and buried valleys filled with water-bearing sand or gravel which might be encountered at excavation level because the hydrostatic water pressure would be such that crossing these would be both difficult and expensive.

It has been established since 1867 (Gamond) that the basal half of the lower chalk formation has all the features desirable for satisfactory tunnelling. The lower chalk is about one quarter of the total thickness of chalk formation. It is a massive grey rock composed almost entirely of finely divided calcium carbonate of organic origin, free of flints and without cracks or fissures. De Lapparant had mapped this lower chalk formation but it was essential to check his findings in detail.

The Study Group first undertook some land trials. Evidence from boreholes did not show a sufficient velocity difference in the middle and lower chalk to provide a distinguishing boundary to the structure of these formations. According to H. R. Reynolds (1961), a seismic reflection method called 'Sonar System' was adopted in 1958. This utilized a high-frequency sound source, which was installed beneath a motor launch. Results were good, but the energy source was insufficiently powerful to overcome the rather absorbant nature of the chalk strata. A 'Sparker' method was then adapted which used an energy source from a 12 000-volt spark discharged under water. The spark is produced at an electrode gap on the end of a cable towed by a survey launch. The gap can be regulated from $\frac{1}{4}$- to $\frac{1}{2}$-s intervals. The energy waves reflected are picked up by a hydrophone towed on a parallel line. The signals are amplified, filtered and printed on a chart, thus producing a continuous record of the underwater geological formation.

Reynolds also adds that another source of energy which could have been used was the RASS (Repeatable Acoustic Seismic Source) which utilizes the explosion from a mixture of propane and oxygen. These gases are detonated by means of a sparking plug in a small chamber installed in a torpedo-shaped container towed behind the launch. The pulses from this have higher energy and lower frequency than spark pulses and they are used for deep penetration below the sea bed. Reflections have been obtained in the Gulf of Mexico at depths of as much as 4000 ft. The two methods can be used at the same time to provide a simultaneous record of both shallow and deep strata.

6.6.2 Electric resistivity method

The electric resistivity of rock, ρ', depends on the type of rock and on the moisture content of pores and fissures. Table 6.3 gives an idea of the range of values which have been measured.

Table 6.3 *Rock resistivity, ρ', in ohms/cm*

Crystalline rocks, low porosity,	50 000–1 000 000
Consolidated sediments	5 000–100 000
Sand and gravel	8 000–150 000
Silt and clay	100–15 000
Sand and salt water	100–1 000
Sand and slightly salt water	1 000–10 000

Resistivity measurements of granite, for example, will detect weak degraded moist rocks.

The main problem is in detecting geological strata with different electric resistivity. For example, wet rock may have a resistivity level of one-fifth or less than that of dry rock. In fig. 6.47 a borehole is put under tension between the levels E_1 and E_2. The voltage difference is measured between two mobile points S_1 and S_2. Varying rock resistivity of the strata versus depth is recorded.

In fig. 6.48a the equipotential lines in a homgeneous mass of rock between the points E_1 and E_2 are circles. If the electric current comes in contact with low-resistivity strata (fig. 6.48b) the flow lines are deformed. The electric flux concentrates in the low-resistivity strata. Assuming that the distance $E_1E_2 = L$ in fig. 6.49 is progressively increased, and the distance between S_1 and S_2 is maintained: for short distances $E_1E_2 = L$, the flow lines are spherical, corresponding to a high rock resistivity ρ'. Assuming an increase

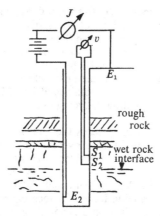

Fig. 6.47   Electric resistivity test in borehole.

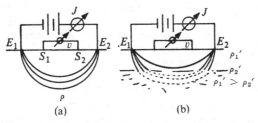

Fig. 6.48   Resistivity test. (*a*) Constant resistivity, ρ'; (*b*) varying resistivity, $\rho' > \rho_2'$.

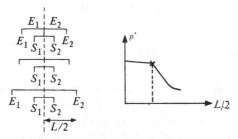

Fig. 6.49   Point determining depth of wet rock.

in the distance $E_1E_2 = L$ resistivity will drop as soon as the flow lines enter in contact with low resistivity (wet rock). In fig. 6.49, the sharp drop indicates the depth point of wet rock.

Moore (1961) comments on shock waves and resistivity as follows:

The seismic test will normally have good application to foundation studies for bridges, buildings and dams, and can be used to good advantage in determining the character of the materials present in roadway cuts and at the portals of proposed tunnel sites. This test has little value for use in locating materials of construction such as sand and gravel and it may prove rather ineffectual in a detailed study of landslide areas. Also in areas where thin layers of relatively hard material, such as sandstone, limestone or comparatively fresh and dense volcanic rock are underlain by weathered materials or formations of lesser density, the seismic test will be ineffective since the higher velocity wave in the dense layer will reach the detectors first and tend to mask the effect of the arrival of the waves through the lower layers lying below . . . At the present time the Bureau of Public Roads, [Washington] makes use of the seismic test to supplement subsurface data obtained by use of the resistivity tests and to prove the results obtained at 1 or 2 locations where resistivity tests were made when some question exists regarding the analysis of the resistivity data.

The resistivity test is equally as good as the seismic test for foundation studies when proper calibration of the test is made over known materials and it has obtained excellent results on many slope design problems. This test is not greatly hampered by thin layers of hard material overlying less dense formations, it being possible to locate a shale bed or other less-resistant material beneath sandstone or other materials possessing high resistivity. Its use for locating construction materials has been well demonstrated in the field and reported in the literature. In landslide studies, since the water associated with many landslide areas is likely to affect the values of the measured resistivity, the resistivity test shows promise as being a method for obtaining more accurate and detailed information on the formations or conditions associated with the slip surface. Also, the possibility that the water that causes the sliding being responsible for measurable electric potentials that can be associated with a particular slip surface or contact between soils of differing moisture content or plasticity must not be ignored.

These comments are based on experience in road construction and they should be supplemented by seismic tests in rock mechanics for solving foundation problems, rock elasticity and efficiency of grouting. Seismic tests are vital in detecting incipient rock slides in progressively deteriorating rock masses (see chapter 14, the Vajont rock slide).

6.7   Engineering classification of jointed rock masses. General approach to the problem

Denkhaus (1970) pointed out the existence of a gap between the acquisition of rock mechanics data and the final decisions of engineers. How can engineers best use the information given to them by geologists, geophysicists and specialists in rock mechanics? What type of information is required and by whom? The consulting engineer wants to know how rock will behave over a long period of time, what strains it will develop, what stresses it will withstand and what load will be transmitted to the structures. The contractor is mainly concerned with short-term problems, with rock stability and rock behaviour during the different phases of the construction work. The planning and the progress of the construction will depend on rock properties in which the consulting engineer may be less interested. Major problems concerning legal responsibilities and price levels may depend on different interpretations of rock characteristics.

(*a*) Before discussing rock classifications established by geophysicists and rock mechanicists, it is necessary to mention the approach to such a problem by geologists. The useful work of Duncan, Deere and others has been dealt with in sections 2.1.1, 4.1.2 and 6.6.1. Dearman (1974) has recently published an interesting paper summarizing a geologist's approach to engineering rock classifications, based on his experience with geology in the British Isles. This author produced a table detailing a classification of igneous rocks on the basis of mineralogy and grain size (a classical approach to be found in many textbooks). Other tables give similar classifications of sedimentary rocks. He then summarizes the work of Duncan, Jones, Deere, etc., which has

Rock mass classification

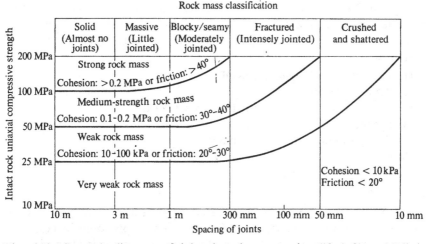

Fig. 6.50 Strength diagram of jointed rock masses (modified from Müller).

already been amply discussed in the preceding chapters. Such a geologist's classification could be consulted by engineers, since it is based on modern geophysical information.

(*b*) One of the first attempts by a geophysicist to systematically describe characteristics in order to establish an engineering classification of jointed rock masses is a well-known diagram prepared by Müller-Salzburg, reproduced as fig. 6.50.

Müller groups the rock masses in four classes: strong, medium strength, weak and very weak. He uses two main rock characteristics, combining the uniaxial compressive strength of intact rock and the joint spacing. Additionally there is mention on his diagram of the rock cohesion or of the angle of friction which can be expected for each class of rock.

(*c*) Geophysicists have been impressed with the advantages of the rock quality designation (RQD) proposed by Deere (1963–68) (see section 2.1.4). In recent times some experts have felt that it would be useful to supplement the information given by the usual description of general area and of local site geology and the RQD index with additional information on joints and faults and other rock properties.

A considerable amount of work has been done in analysing hundreds of cases, by a comparison of rock characteristics and the behaviour of the engineering structures (Coates, 1964; Cecil, 1970*a*, *b*; Wickham, 1972; Tiedmann; Skinner; etc.) – the International Society for Rock Mechanics and the International Association of Engineering Geology have both appointed Commissions to study rock classifications.

Bieniawski (1973) states that while a rock classification of jointed rock masses, based on the inherent properties of the rock mass itself, should be capable of application to practical engineering problems, it should be general enough so that the same rock would always be classified in the same way, regardless of how it is being used. The classification he suggests was developed from an analysis of many earlier classifications. It incorporates the following parameters:

(1) Rock quality designation (RQD).
(2) State of weathering.
(3) Uniaxial compression strength of intact rock.
(4) Spacing of joints or bedding.
(5) Strike and dip of joints, bedding.
(6) Openness of joints.
(7) Continuity of joints.
(8) Ground water inflow.

Based on these parameters, an engineering classification of jointed rock masses, termed the 'Geomechanics Classification' is proposed in table 6.4. For the purpose of this classification, it is necessary to divide the rock mass into a number of domains, each having similar structural characteristics; for example the same rock type or the same joint spacing. The boundaries of a domain may coincide with such geological features as faults or dykes. Experience shows how rapidly rock characteristics change when crossing contact zones.

For simple classification, rock masses are grouped into only five classes. The various parameters need not necessarily conform to the same rock class: for example, a rock with an RQD of 70% (class 3) may have a joint spacing of 0·2 m (class 4) and display no ground-water inflow (class 2).

(1) It is easy to establish five classes for the first parameter, the RQD index varying from 0 to 25%, 25 to 50%, etc.

(2) For the second parameter, rock weathering, Bieniawski suggests the following classes:

Unweathered: no visible sign of weathering.
Slightly weathered: penetrating weathering developed on open discontinuity surfaces, slight weathering of rock material.
Moderately weathered: slight discoloration extends through the greater part of the rock mass.

Table 6.4 *Geomechanics classification of jointed rock masses (after Bieniawski)*

Item	Class No and its description	1	2	3	4	5
		Very good	Good	Fair	Poor	Very poor
1	Rock quality RQD (%)	90–100	75–90	50–75	25–50	<25
2	Weathering	Unweathered	Slightly weathered	Moderately weathered	Highly weathered	Completely weathered
3	Intact rock strength, MPa	>200	100–200	50–100	25–50	<25
4	Spacing of joints	>3 m	1–3 m	0·3–1 m	50–300 mm	<50 mm
5	Separation of joints	<0·1 mm	<0·1 mm	0·1–1 mm	1–5 mm	>5 mm
6	Continuity of joints	Not continuous	Not continuous	Continuous, no gouge	Continuous, with gouge	Continuous, with gouge
7	Ground-water inflow (per 10 m of adit)	None	None	Slight <25 litres/min	Moderate 25–125 litres/min	Heavy >125 litres/min
8	Strike and dip orientations	Very favourable	Favourable	Fair	Unfavourable	Very unfavourable

Highly weathered. weathering extends throughout rock mass, rock material partly friable.

Completely weathered: rock totally discoloured and decomposed.

(3) The third parameter concerns the uniaxial compressive strength of intact rock. The rocks can be grouped in five classes (1 MPa = 10·2 kg/cm²).

Very low strength,   1–25 MPa: chalk, rock salt.
Low strength,   25–50 MPa: coal, saltstone, schist.
Medium strength,   50–100 MPa: sandstone, slate, shale.
High strength,   100–200 MPa: marble, granite, gneiss.
Very high strength > 200 MPa quartzite, dolorite, gabbro, basalt.

(4) and (5)   Five classes are chosen for the spacing of joints, varying from class 5, for distances smaller than 50 mm to class 1, when the distance is larger than 3 m (as can be seen in table 6.4). Similarly the separation of joints is assumed to vary between larger than 5 mm (class 5) to smaller than 0·1 mm (class 1).

(6) to (8)   These parameters are classified as shown in table 6.4.

It is advantageous to assign a rating to each parameter by a weighted numerical value. The final rock class rating will be the sum of weighted values determined for the individual parameters, higher numbers reflecting better conditions and hence lesser support in the case of tunnels. Based on a study of Wickham *et al.* the importance of rating as shown in table 6.5 is proposed by Bieniawski.

It should be noted, for example, that items 1 and 2, representing data obtainable from cores, receive 25% rating, while the intact rock strength (item 3) is worth 10%. Items 4 and 5, representing field data on joints, receive as much as 45%.

Bieniawski has developed some further aspects of this 'Geomechanics Classification' concerning tunnel design and construction. These will be discussed in section 10.9, together with further research work on similar problems.

The eight parameters introduced by Bieniawski allow a fair description of jointed rock masses, but there are some limitations. According to Bieniawski 'special caution should be exercised in the case of shales and other swelling materials. These rocks are characterized by a wide variation in their engineering properties, particularly their durability (resistance to weathering) under conditions of wetting and drying' (see Franklin, 1972 and Olivier, 1973). A similar remark could be made for rock masses with a tendency to slow creep.

A Norwegian team (Barton *et al.*, 1974) objects that Bieniawski has almost ignored three important properties of rock masses, namely the roughness of

Table 6.5 *Importance ratings*

(a) Individual ratings for classification parameters

Item	Parameter		Class			
		1	2	3	4	5
1	Rock quality RQD	16	14	12	7	3
2	Weathering	9	7	5	3	1
3	Intact rock strength	10	5	2	1	0
4	Spacing of joints	30	25	20	10	5
5	Separation of joints	5	5	4	3	1
6	Continuity of joints	5	5	3	0	0
7	Ground water	10	10	8	5	2
8	Strike and dip orientations { Tunnels	15	13	10	5	3
	{ Foundations	15	13	10	0	−10

(b) Total ratings for rock mass classes

Class no.	1	2	3	4	5
Description of Class	Very good rock	Good rock	Fair rock	Poor rock	Very poor rock
Total rating	90–100	70–90	50–70	25–50	<25

joints, the frictional strength of joint fillings and the rock load. The importance of the natural residual stresses in the rock mass will be demonstrated at some length in section 10.9.

Important cases exist where a precise, systematic description of the rock mass characteristics, along the lines suggested by Bieniawski or others, fails to detect vital causes of potential failure of the structure. In part four, several case histories will be discussed where a combination of factors, some of them unforeseen at the design stage, caused major troubles or even the collapse of structures, the description of the rock masses having failed to disclose inherent weaknesses. The cases of the collapse of the Malpasset dam, of the rock falls at Kariba North Bank Machine Hall excavation and the great Vajont rock slide will illustrate this point. Similar cases were mentioned by Barbier (1974); final decisions on where to locate dams were taken on the basis of geological exploration at large, rather than local geology, almost disregarding rock characteristics.

7 Mathematical approach to strain–stress distribution in rock masses

7.1 Useful formulae

The complexity of a fissured strained rock mass has already been demonstrated. Many theories and methods of approach have been suggested to deal with this problem. In section 6.5.4 four analytical methods were proposed and we will now examine these in greater detail. They are: classical homogeneous isotopic elastic space (the equation of Boussinesq and Cerruti for the half-space); fissured rock; clastic mass of rock; and, modern stress–strain analysis.

7.1.1 The circle of Mohr; shear stresses; the intrinsic curve

The theory of Mohr's circle is given in detail in most textbooks on strength of materials, and in many textbooks on soil mechanics. It is, however, an essential part of any theory of rock mechanics, and as such is summarized here very briefly.

The basic problem can be stated as: stresses σ_x, σ_y and τ_{xy}, acting on two orthogonal surfaces x and y at point O, are known; the stresses σ, τ acting on a surface at an angle α to Ox have to be calculated. Projection of all forces on the directions of σ and τ yields (fig. 7.1a):

$$\sigma = \frac{\sigma_y + \sigma_x}{2} + \frac{\sigma_y - \sigma_x}{2} \cos 2\alpha + \tau \sin 2\alpha, \tag{1}$$

$$\tau = \frac{\sigma_y - \sigma_x}{2} \sin 2\alpha + \tau \cos 2\alpha. \tag{2}$$

The maximum and minimum values σ_{max} or σ_{min} are obtained for $\sigma/\partial\alpha = 0$, this yields:

$$\tan 2\alpha = \frac{2\tau}{\sigma_x - \sigma_y}, \tag{3a}$$

$$\sin 2\alpha = \pm \frac{2\tau}{\sqrt{[(\sigma_x - \sigma_y)^2 + 4\tau^2]}}, \tag{3b}$$

$$\cos 2\alpha = \pm \frac{\sigma_y - \sigma_x}{\sqrt{[(\sigma_x - \sigma_y)^2 + 4\tau^2]}}. \tag{3c}$$

[161]

The corresponding values of σ_{max} and σ_{min} are

$$\frac{\sigma_{max}}{\sigma_{min}} = \frac{\sigma_z + \sigma_y}{2} \pm \tfrac{1}{2}\sqrt{[(\sigma_z - \sigma_y)^2 + 4\tau^2]}. \tag{4}$$

Similarly, the maximum of the shear stress τ_{max} is obtained, writing that $\partial\tau/\partial\alpha = 0$.

This condition yields:

$$\tan 2\alpha' = \frac{\sigma_z - \sigma_y}{2\tau} = \frac{1}{\tan 2\alpha}. \tag{5}$$

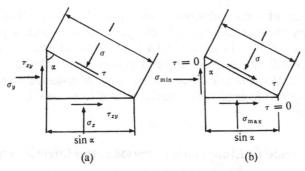

Fig. 7.1   Stresses on orthogonal surfaces.

The angles α' and α are thus related and

$$\alpha' = \alpha \pm \pi/4, \tag{6}$$

which means that the maximum shear stress τ_{max} is at an angle of 45° to the principal stresses σ_{max} and σ_{min}. Furthermore the condition $\partial\tau/\partial\alpha = 0$ yields:

$$\tau_{max} = \frac{\sigma_{max} - \sigma_{min}}{2} = \pm\tfrac{1}{2}\sqrt{[(\sigma_z - \sigma_y)^2 + 4\tau^2]}. \tag{7}$$

The compression or tensile stress normal to τ_{max} is

$$\sigma_{\tau_{min}} = \frac{\sigma_z + \sigma_y}{2} = \frac{\sigma_{max} + \sigma_{min}}{2}. \tag{8}$$

Let us now refer the σ_z and σ_y values and the angles α to the principal stresses σ_{max} and σ_{min}. The shear stress normal to the principal stresses is $\tau = 0$. Therefore (fig. 7.1b):

$$\sigma = \frac{\sigma_{max} + \sigma_{min}}{2} + \frac{\sigma_{min} - \sigma_{max}}{2}\cos 2\alpha, \tag{9}$$

$$\tau = \frac{\sigma_{max} - \sigma_{min}}{2}\sin 2\alpha. \tag{10}$$

The equations just developed can be represented on Mohr's circle which is defined as: on a system of axis (σ, τ) assuming $OA = \sigma_{min}$ and $OB = \sigma_{max}$ (fig. 7.2). The 'circle of Mohr' is a circle with radius $MA = MB = \frac{1}{2}$ $(\sigma_{max} - \sigma_{min})$ and centre in M so that $OM = \frac{1}{2}(\sigma_{max} + \sigma_{min})$.

When we trace a radius MC at an angle 2α to the axis we see that

$$OD = \sigma = \frac{\sigma_{max} + \sigma_{min}}{2} + \frac{\sigma_{min} - \sigma_{max}}{2} \cos 2\alpha,$$

$$DC = \tau = \frac{\sigma_{max} - \sigma_{min}}{2} \sin 2\alpha.$$

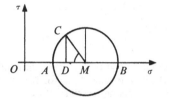 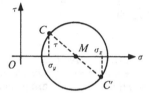

Fig. 7.2 The circle of Mohr.   Fig. 7.3 Co-ordinates of C, C'.

Conversely, assuming σ_x, σ_y and τ_{xy} to be known, they are the co-ordinates of points C and C' in (σ,τ) fig. 7.3. The centre, M, of the circle of Mohr is obviously on the straight line CC' and its radius is $MC = MC'$.

We have

$$OM = \frac{\sigma_x + \sigma_y}{2} = \frac{\sigma_{max} + \sigma_{min}}{2}.$$

These are the properties of the circle of Mohr which are an essential tool for further analysis of stresses in rock.

Definition of the intrinsic curve (fig. 7.4). Rupture of rock may occur under uniaxial compression (circle 1 in fig. 7.4), or under uniaxial tensile stress (circle 2) or by visco–plastic deformation and shear fracture. The theory of Mohr explains that, when all the circles corresponding to failures have been traced they have a common envelope 'C' called the 'intrinsic curve'. In most cases the curve C is parabolic.

7.1.2 The law of Coulomb–Mohr

Coulomb's law when used in soil mechanics is:

$$\tau = c + \sigma \tan \phi \tag{11}$$

(pore pressure on the rock is neglected in this equation). This law determines the maximum shear stress τ at which rupture will occur along a plane, c is the internal cohesion of rock masses and ϕ the angle of internal friction, σ is the compression stress normal to the plane of rupture.

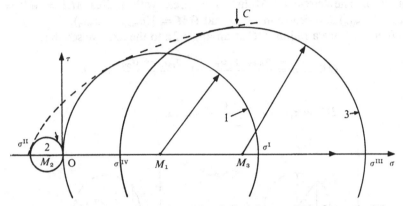

Fig. 7.4   Intrinsic curve C for rock. Mohr circle 1 (centre M_1) σ^{I} = uniaxial compression strength. Mohr circle 2 (centre M_2) σ^{II} = tensile strength. Mohr circle 3 (centre M_3) the σ^{III} and σ^{IV} are principal stresses causing triaxial failure.

The law can be illustrated on circles of Mohr for the case $c = 0$ and for $c > 0$:

(1) *Cohesion is negligible.* $c = 0$ (fig. 7.5). The condition of Coulomb for shear rupture is represented on the (σ,τ) plane by a straight line through the

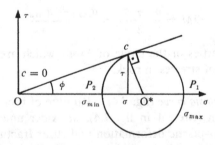

Fig. 7.5   The law of Coulomb for $c = 0$.

origin O and $\tau = \sigma \tan \phi$. Rupture occurs when the circle of Mohr is tangential to this line (fig. 7.5), and yields the condition:

$$\frac{\sigma_{\min}}{\sigma_{\max}} \geqslant \frac{OO^* - O^*P_2}{OO^* + O^*P_1} = \frac{OO^* - R}{OO^* + R} = \frac{1 - \sin \phi}{1 + \sin \phi},$$

or

$$\frac{\sigma_{\min}}{\sigma_{\max}} \geqslant \tan^2 \left(\frac{\pi}{4} - \frac{\phi}{2} \right). \tag{12}$$

(2) *Cohesion is not negligible, $c > 0$.* Figure 7.6 yields:

$$\frac{\sigma_{\min} + c \cot \phi}{\sigma_{\max} + c \cot \phi} \geqslant \frac{1 - \sin \phi}{1 + \sin \phi},$$

(13)

or

$$\sin \phi \geqslant \frac{\sigma_{\max} - \sigma_{\min}}{\sigma_{\max} + \sigma_{\min} + 2c \cot \phi}.$$

(14)

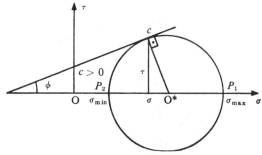

Fig. 7.6 The law of Coulomb for $c > 0$.

If Coulomb's law is accepted as the determining condition for rupture by shear, the intrinsic curve C is replaced by the Coulomb straight line.

7.1.3 Plane of rupture in jointed rock

The rock strata to be examined for safety against shear rupture has planes of weakness inclined at an angle to the principal stresses. The Coulomb condition is now assumed to be

$$\tau \leqslant c + (\sigma - u) \tan \phi$$

(15)

where u is the pore pressure or uplift pressure caused by water in the joints

The projection of the forces $Z = \sigma_z \cos \beta$ and $Y = \sigma_y \sin \beta$ on the directions of σ and τ as indicated in fig. 7.7 yields:

$$\sigma = + \sigma_z \cos \beta \cos \beta + \sigma_y \sin \beta \sin \beta,$$

$$\tau = + \sigma_z \cos \beta \sin \beta - \sigma_y \sin \beta \cos \beta.$$

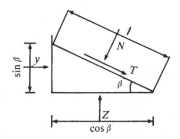

Fig. 7.7 Plane of rupture in jointed rock.

Introducing σ and τ in the Coulomb equation it becomes

$$\sigma_z \cos \beta \sin (\phi - \beta) + \sigma_y \sin \beta \cos (\phi - \beta)$$
$$+ c \cos \phi - u \sin \phi \geqslant 0. \tag{16}$$

For $c = 0$ and $u = 0$ the simplified equation is

$$\frac{\sigma_y}{\sigma_z} = -\frac{\cos \beta \sin (\phi - \beta)}{\sin \beta \cos (\phi - \beta)} = -\frac{\tan (\phi - \beta)}{\tan \beta} = \frac{\tan (\beta - \phi)}{\tan \beta} \tag{17}$$

which is the condition for limiting equilibrium.

Talobre (1957) has given a very good example showing how this formula can be used (fig. 7.8).

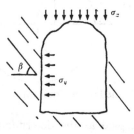

Fig. 7.8   Stability of a gallery excavated in inclined strata (after Talobre, 1957).

An unlined tunnel is excavated in stratified rock. The angle of the stratification planes to the horizontal is β. The vertical field of residual stresses due to the overburden is $\sigma_z = 20$ kg/cm².

The density of rock bolting required to stabilize the rock is unknown. When $\beta > \phi$

$$\beta = 50°, \qquad \phi = 40°, \qquad \beta - \phi = 10°.$$

In addition $u = 0$ and $c = 0$. The simplified formula (17) yields:

$$\frac{\sigma_y}{\sigma_z} = \frac{\tan (\beta - \phi)}{\tan \beta} = \frac{\tan 10°}{\tan 50°} = \frac{0 \cdot 1763}{1 \cdot 734} \simeq 0 \cdot 10.$$

The pressure exerted by the rock bolts in a direction normal to the vertical tunnel walls is therefore:

$$\sigma_y = 0 \cdot 10 \sigma_z = 0 \cdot 10 \times 20 \text{ kg/cm}^2 = 2 \text{ kg/cm}^2,$$

or

$$\sigma_y = 20 \text{ tonne/m}^2.$$

This simplified example, suggested by Talobre (1957), assumes conditions occurring when the rock mass is being ruptured, neglecting locked-in stresses. As long as the mass remains elastic, its behaviour depends on residual stresses (see sections 5.4 and 10.2).

7.1.4 The effective shear stress: the Prandtl–Terzaghi equation

The general law of Coulomb:

$$\tau \leq c + \sigma \tan \phi, \qquad (11)$$

shows that to avoid rupture by shearing, the shear stress should be less than the right member of the inequation. Rupture occurs for $\tau = \tau_{\text{lim}}$ ($\tau_{\text{lim}} =$ shear strength).

An effective shear stress is introduced as

$$\tau^* = \tau - \sigma \tan \phi, \qquad (18)$$

τ^* is the part of τ which is taken by shearing only and not by friction. According to the previous law rupture will occur when $\tau^* = c$ or rock cohesion. This situation is in many cases more important than the condition of Coulomb. It is therefore important to find the maximum of τ^* or τ^*_{max}. Let

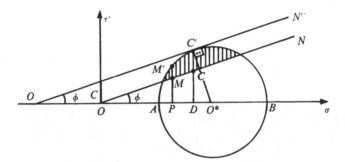

Fig. 7.9 The maximum effective shear strength $\tau^*_{\text{max}} = CC'$.

us consider any point within a stressed body (fig. 7.9), σ_{max} and σ_{min} are the principal stresses, ϕ the angle of internal friction.

The corresponding circle of Mohr has a diameter $AB = \sigma^*_{\text{max}} - \sigma_{\text{min}}$, O^* being the centre of the circle, with:

$$OO^* = \frac{\sigma_{\text{max}} + \sigma_{\text{min}}}{2} = \frac{\sigma_x + \sigma_y}{2}.$$

A straight line ON is traced through the origin at an angle ϕ with axis σ and a parallel $O'N'$ to ON, tangential to the circle of Mohr at the point C'. Consideration of the values $\sigma = OP$ and $\tau = PM'$, reveals that

$$PM = \sigma \tan \phi$$

and that according to the previous definition

$$MM' = \tau^* = \tau - \sigma \tan \phi.$$

The maximum of τ^* is obviously

$$\tau_{\max} = C'C = C'D - CD,$$

$$\tau_{\max} = \frac{AB}{2} = O^*C = \tfrac{1}{2}(\sigma_{\max} - \sigma_{\min}),$$

$$C'D = \tau_{\max} \cos \phi,$$

$$CD = (OO^* - DO^*) \tan \phi = \frac{\sigma_{\max} + \sigma_{\min}}{2} \tan \phi$$

$$- \tau_{\max} \sin \phi \tan \phi.$$

or

$$\tau_{\max}^* = \frac{\tau_{\max}}{\cos \phi} - \frac{\sigma_x + \sigma_y}{2} \times \frac{\sin \phi}{\cos \phi},$$

and

$$\tau_{\max}^* = \frac{1}{2 \cos y} [\sqrt{\{(\sigma_x - \sigma_y)^2 + 4\tau^2\}} - (\sigma_x + \sigma_y) \sin \phi]. \qquad (19)$$

When $\tau_{\max}^*$ reaches the critical value of internal cohesion, there is danger of failure. $\tau_{\max}^*$ occurs at an angle $\alpha = \tfrac{1}{4}\pi - \phi/2$ to the principal stress. According to this theory the dangerous τ value is therefore not $\tau_{\max}$ at an angle of 45° to the principal stresses but $\tau_{\max}^*$.

Equation (19) is well known as a means of estimating the maximum effective shear stress. Found in old French textbooks on the strength of materials, it was introduced into soil mechanics by Prandtl and Terzaghi.

7.2 The half-space of Boussinesq–Cerruti

In 1855, Boussinesq published a most important paper dealing with the strain–stress distribution in a homogeneous half-space loaded with a single force normal to the free surface. This was followed in 1888 by a paper by Cerruti in which he analysed a similar problem for an isolated force acting in a direction parallel to the surface of the half-space. Both were based on the theory that the addition of several forces to groups of forces or loads is permissible. A whole range cf similar problems can be solved in an extension of this basic analysis. When more elaborate methods are used for analysing stress–strain distributions in the non-homogeneous fissured half-space, the results are normally compared to a Boussinesq–Cerruti solution. These theories are well known and only the main results will be given here

7.2.1   The isolated concentrated load normal to a half-space

A force P is supposed to act at point O in a direction normal to the plane surface of the half-space. At a point M at a distance r from O, at an angle θ to the normal, the stresses (fig. 7.10):

σ_x = horizontal radial,

σ_y = horizontal tangential,

σ_z = vertical,

and the shear stress τ_{xz} is given by:

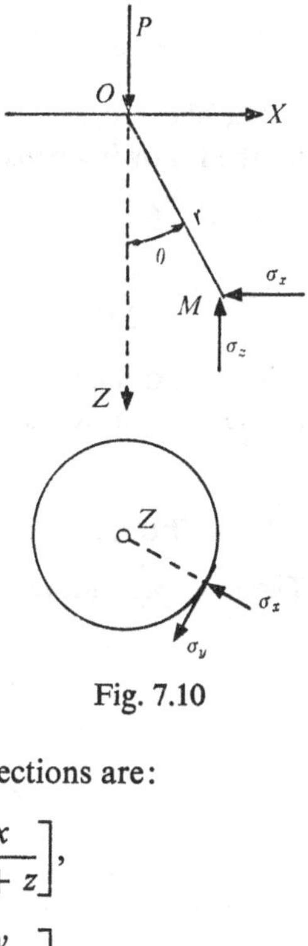

$$\sigma_x = \frac{P}{2\pi}\left[\frac{3x^2 z}{r^5} - (1 - 2\nu)\,\frac{1}{r(r + z)}\right]$$

$$= \frac{P}{2\pi r^2}\left(3\cos\theta\sin^2\theta - (1 - 2\nu)\,\frac{1}{1 + \cos\theta}\right),$$

$$\sigma_y = -\frac{P}{2\pi}\left[(1 - 2\nu)\,\frac{z}{r^3} - (1 - 2\nu)\,\frac{1}{r(r + z)}\right]$$

$$= -(1 - 2\nu)\frac{P}{2\pi r^2}\left(\cos\theta - \frac{1}{1 + \cos\theta}\right),$$

$$\sigma_z = \frac{3Pz^3}{2\pi r^5} = \frac{3P}{2\pi}\times\frac{\cos^3\theta}{r^2},$$

$$\tau_{xz} = \frac{3P}{2\pi}\times\frac{z^2 x}{r^5} = \frac{3P}{2\pi r^2}\cos^2\theta\sin\theta.$$

Fig. 7.10

The displacements of the point M in the x, y and z directions are:

$$u = \frac{P}{2\pi r}\times\frac{1 + \nu}{E}\left[\frac{x^2}{r^2} - (1 - 2\nu)\,\frac{x}{r + z}\right],$$

$$v = \frac{P}{2\pi r}\times\frac{1 + \nu}{E}\left[\frac{yz}{r^2} - (1 - 2\nu)\,\frac{y}{r + z}\right],$$

$$w = \frac{P}{2\pi r}\times\frac{1 + \nu}{E}\left[\frac{z^2}{r^2} + 2(1 - \nu)\right].$$

The vertical stresses σ_z which do not depend on ν, are obviously maximum when $\theta = 0$, $r = z$ and

$$\sigma_{\max} = \frac{3P}{2\pi z^2}.$$

Some results for $P = 1000$ tonnes are summarized in table 7.1.

The most dangerous horizontal stress σ_z occurs for $\theta = 0$ when the stresses are tensile as given in table 7.2.

<table>
<tr><td colspan="2" align="center">Table 7.1</td><td colspan="2" align="center">Table 7.2</td></tr>
<tr><td>z</td><td>σ_{max}</td><td>z</td><td>σ_{max}</td></tr>
<tr><td>0·5 m</td><td>192 kg/cm²</td><td>0·5 m</td><td>10·6 kg/cm²</td></tr>
<tr><td>1·0</td><td>48</td><td>1·0</td><td>2·6</td></tr>
<tr><td>5·5</td><td>1·6</td><td>5·5</td><td>0·08</td></tr>
<tr><td>7·95</td><td>0·75</td><td>7·95</td><td>0·04</td></tr>
</table>

The shear stresses are obtained by multiplying the σ_{max} values (table 7.1) by the following constants, C_1 depending on the angle θ:

$\theta = 0°$	$C_1 = 0$	$\theta = 45°$	$C_1 = 0\cdot177$
25°	0·284	60°	0·054
30°	0·282	90°	0

The maximum shear stress is about 30% of σ_{max} in a direction about 25° to the vertical. Rock failure will probably occur by shear fracture.

7.2.2 The half-space loaded by a line load

The load per unit length of the line is p. The stresses are:

$$\sigma_z = \frac{2p}{\pi} \frac{zx^2}{(x^2 + z^2)^2} = \frac{2p}{\pi r} \sin^2 \theta \cos \theta,$$

$$\sigma_z = \frac{2p}{\pi} \frac{z^3}{(x^2 + z^2)^2} = \frac{2p}{\pi r} \cos^3 \theta,$$

$$\tau_{zz} = \frac{2p}{\pi} \frac{xz^2}{(x^2 + z^2)^2} = \frac{2p}{\pi r} \sin \theta \cos^2 \theta.$$

7.2.3 Strains under the loaded rectangle, the circular plate

Rock deformations occurring under a loaded rectangular area are most important for the analysis of gravity and of arch dam foundations. Similarly formulae for the deflections occurring under loaded plates are required for evaluating the strains and stresses occurring during plate load tests. It should be realized that a loading test on rock masses *in situ* is a tridimensional problem where the Poisson ratio $v = 1/m$ is an important factor. Results will also depend on the test rig used. Small misalignments of the loads may cause serious errors in estimating the Young's modulus.

The analysis of these problems starts with two formulae developed by Boussinesq (1885) and two developed by Cerruti (1888).

Isolated loads.   An isolated vertical force P is assumed to load the half-space at the point O^*. The deformations δ_n vertical and δ_r radial at a point O are given by the formulae of Boussinesq (1885), (fig. 7.11) (with $m = 1/\nu$):

$$\delta_n = \frac{P}{r\pi E}\left(1 - \frac{1}{m^2}\right);$$ (1)

$$\delta_M = \delta_r = -\frac{P}{2\pi r}\frac{(m-2)(m+1)}{Em^2}; \qquad \delta_x = \frac{x}{r}\delta_r.$$ (2)

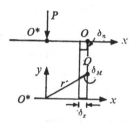

Fig. 7.11

δ_n is the downward deflection, parallel to the force P and δ_r the radial displacement (fig. 7.11).

Similarly, Cerruti established two formulae for the deformation caused by a load Q parallel to the half-space (fig. 7.12).

$$\delta_n = \frac{Qx}{Er^2}\frac{(m-2)(m+1)}{2\pi m^2},$$ (3)

$$\delta_x = \frac{m+1}{m}\frac{Q}{\pi E}\left(\frac{1}{r} - \frac{1}{m}\frac{y^2}{r^3}\right).$$ (4)

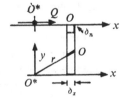

Fig. 7.12

Loaded areas.   Figure 7.13 shows a rectangle $A = ab$ under the uniformly distributed load $N = \sigma A$. An element of area $dA = dx\,dy$ is therefore loaded with load $\sigma\,dA$; which causes a vertical displacement of point O equal to:

$$\Delta\delta_n = \frac{\sigma\,dx\,dy}{\pi Er}\left(1 - \frac{1}{m^2}\right) = \frac{N}{EA}\frac{(1-1/m^2)}{\pi}\frac{dx\,dy}{\sqrt{(x^2 + y^2)}}.$$ (5)

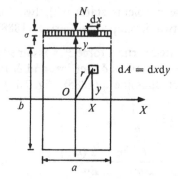

Fig. 7.13

Under the whole load $N = \sigma A$, the vertical displacement of the point O, at the centre of the rectangle is (Jaeger, 1950)

$$\delta_n = \frac{N}{EA} \frac{(1 - 1/m^2)}{\pi} \int \frac{dx\, dy}{\sqrt{(x^2 + y^2)}} = \frac{cNa}{EA}. \qquad (6)$$

The factor c depends on the ratio b/a. For:

$b/a =$	1·0	2·0	4·0	10·0	20·0
$c =$	1·08	1·47	1·88	2·44	2·86

Waldorf *et al.* (1963) suggest the use of the following formula, with square root A instead of A below the line:

$$\bar{u} = \frac{\bar{m}(\Sigma P)(1 - v^2)}{E\sqrt{A}}$$

where $\bar{u}$ is the average deflection under the plate, A its area and $\bar{m}$ a coefficient.

For uniform loading on circular and rectangular areas, the values of $\bar{m}$ are given in table 7.3 (after Waldorf, Veltrop & Curtis, 1963).

Table 7.3

Circle	Rectangle with ratio of sides						
	1:1	1:1·5	1:2	1:3	1:5	1:10	1:100
0·96	0·95	0·94	0·92	0·88	0·82	0·71	0·37

Under circular plates for various load distributions, table 7.4 has been given for the value $\bar{m}$.

Under a rigid die high pressures occur at the edges but displacements are the same everywhere. Under a rigid sphere high pressures develop at the centre.

Table 7.4

Displacement	Type of load		
	Rigid die	Uniform	Rigid sphere
Centre	0·89	1·13	1·33
Average	0·89	0·96	1·00

Alternatively, the following formulae can be found in the literature (where r = radius of the plate):

$$\text{Stiff circular plate:} \quad \bar{u} = \frac{P}{2E}\frac{1 - v^2}{r}. \tag{7}$$

Elastic circular plate deflections.

$$\text{Average deflection:} \quad \bar{u} = 0\cdot54\frac{P}{E}\frac{(1 - v^2)}{r}. \tag{8}$$

$$\text{Centre of the plate:} \quad u_c = \frac{2}{\pi}\frac{P}{E}\frac{(1 - v^2)}{r}. \tag{9}$$

$$\text{Along edge of plate:} \quad u_e = \frac{4}{\pi^2}\frac{P}{E}\frac{(1 - v^2)}{r}. \tag{10}$$

All these formulae are based on the theory developed by Boussinesq for load normal to the half-space.

7.2.4 Displacements caused by a moment M or by a shear force Q

A triangular load (fig. 7.14) is equivalent to a moment M acting on the surface of the half-space. Boussinesq's theory can be used to estimate the surface displacements under such a moment. The moment M causes a unit angular deflection, α obtained by integration over a rectangle $A = ab$ (fig. 7.14):

$$\alpha M = \frac{0\cdot6M}{EI}, \tag{11}$$

and a horizontal displacement of point O equal to

$$\delta_m M = \frac{0\cdot08a^2}{EI}M, \tag{12}$$

where E is the modulus of the rock and $I = a^3b/12$ is the moment of inertia of the area on which the triangular load is applied (fig. 7.15).

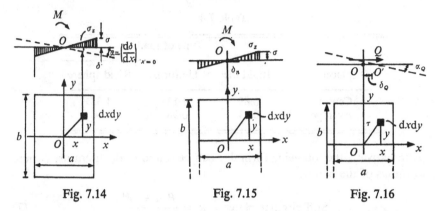

Fig. 7.14   Fig. 7.15   Fig. 7.16

Similarly, a force Q parallel to the surface of the half-space causes a horizontal displacement (fig. 7.16):

$$\delta_q Q = \frac{2aQ}{EA}, \qquad (A = ab) \tag{13}$$

and at angular deflection:

$$\alpha_q Q = \frac{0{\cdot}6Q}{EI}. \tag{14}$$

The coefficients in these formulae are average values only, as they depend on the ratio b/a.

7.2.5 Displacements of a foundation surface. Modulus of elasticity of rock at depth and of stratified rock

The general equation for average displacements of the loaded surface can also be used to calculate vertical displacement u_p of a point located outside the circular loaded area. In this case the equation for uniform load on a circular plate can be expressed in the form:

$$u_p = C_p \frac{P(1 - v^2)}{aE}$$

in which C_p is a coefficient dependent on the ratio ρ/a, where a represents the radius of the loaded area and ρ the distance to point P, outside the loading plate.

The value of C_p is given by (Waldorf *et al.* 1963)

$$C_\rho = \frac{4}{\pi^2(a/\rho)} \left[\int_0^{2\pi} \left(1 - \frac{a^2}{\rho^2} \sin \theta \right)^{1/2} d\theta \right.$$

$$\left. - \left(1 - \frac{a^2}{\rho^2} \right) \int_0^{2\pi} \left(1 - \frac{a^2}{\rho^2} \sin \theta \right)^{-1/2} d\theta \right], \tag{15}$$

in which the two integrals are elliptical and for which mathematical tables can be consulted.

The displacement of surface points located outside the loading area give a more accurate measurement of the real modulus of undisturbed rock below the surface zone shattered by blasting. Points outside the plate area are affected by deformations at greater depth. Points located at depth z below the centre of the loading plate are characterized by the following values.

Displacement:

$$U_z = p \frac{z}{E} \left[(1 + v) \left(1 - \sqrt{\frac{z^2}{a^2 + z^2}} \right) + 2(1 - v^2) \left(\sqrt{\left(\frac{a^2 + z^2}{z^2} \right)} - 1 \right) \right]. \tag{16}$$

Vertical stress:

$$\sigma_z = p \left[\frac{z^3}{(a^2 + z^2)^{3/2}} - 1 \right]. \tag{17}$$

For points located at a distance the exact theory of elasticity has to be used (Waldorf *et al.*, 1963) (fig. 7.17).

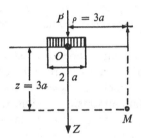

Fig. 7.17

At a depth $z = 3a$ under the plate the displacements and stresses are compared to the value for $z = O$,

under the plate, at $\rho = 3a$,

$u_z = 30\%$ of $U_{z=0}$ $U_{\rho z} = 93\%$ of $U_{\rho z=0}$,

$\sigma_z = 13\%$ of $\sigma_{z=0}$

This table confirms that deflections under the plate up to 70% are due to the elasticity of the upper layer of rock for $z = 3a$ whereas at a distance $\rho = 3a$ from the centre of the plate these same upper layers cause only 7% of the subsidence. From tables published by Waldorf *et al.* (1963) the modulus of elasticity at depth may be as much as 15% to 30% higher than on the surface (the values were measured inside galleries).

Scale effect and its importance when the foundations are not homogeneous. If the foundations are not homogeneous (for example if a surface zone of low modulus occurs under a loading plate, or if there is a weak joint filling at shallow depth) Boussinesq's formula is no longer valid. Taking the average pressure on the plate area A as $p = P/A$ then the average vertical displacement under the plate is

$$\bar{u} = \frac{mp(1 - v^2)\sqrt{A}}{E},\tag{18}$$

or a for a circular plate:

$$\bar{u} = \frac{\overline{m}pa(1 - v^2)\sqrt{\pi}}{E}.$$

Therefore, when a rock mass with modulus E is loaded with a uniform load, p_1, over an area A_1 during a load test and a deflection, u_1, is measured, the deflection, u_2, of the prototype foundation, where an area A_2 is loaded with a p_2, is given by the 'scale effect'

$$\frac{u_2}{u_1} = \frac{p_2}{p_1} \sqrt{\left(\frac{A_2}{A_1}\right)}\tag{19}$$

(equations (18) and (19) being written for homogeneous foundations). This is valid for deflections. Stresses and strains will be proportional to loads: $\sigma_2/\sigma_1 = \epsilon_2/\epsilon_1 = p_2/p_1$.

In the paper under discussion, a pressure test was made with a circular plate (radius $a = 0.225$ m) on a weak rock layer (1·0 m deep) overlying harder rock, the modulus of which was $E_2 = 163\ 700$ kg/cm². The thickness of the weak layer under the dam abutment was 5 m. Deflections were measured under the test plate, and it was assumed that they were partly due to deformation of the weak layer and partly to deformation of the more rigid rock. Deformation in the weak layer was estimated using equation (17). The calculation was carried out in detail, going through all the values U; for both model and prototype. It was shown that the effective (or combined) modulus E_{eff} is given by the simple relation

$$\frac{1}{E_{eff}} = \frac{f_1}{E_1} + \frac{f_2}{E_2},\tag{20}$$

in which E_1 and E_2 are moduli of surface layer and rock mass respectively, and f_1 and f_2 represent fractional components of surface displacement due to surface layer and the sound rock mass beneath.

7.2.6 The effective modulus E_{eff} of fissured rock *in situ*

The presence of joints considerably reduces the *in situ* modulus and Poisson ratio. The joints constitute surfaces in partial contact. Elastic compression and shearing displacements of the contact areas have to be added to the deformations within blocks. Waldorf *et al.* suggest the following model: with joints dividing the mass of rock into roughly cubical blocks. There are scattered contact areas over the joint faces, elsewhere the joints are free. The area of each contact surface is considered small in comparison with the dimensions of the block.

Initially, the area of contact is nh^2 based on n contact areas, each area h^2. The side of the block is d, the average stress on the block being σ, the load on the cubical block is σd^2; causing a compression. The closing of the faces under the contact area will be

$$\frac{2\bar{c}\sigma d^2 (1 - v^2)}{nhE},$$

and the total compression δ:

$$\delta = \frac{\sigma d}{E} + 2\bar{c}\,\frac{\sigma d^2(1 - v^2)}{nhE}$$

and the effective modulus:

$$E_{\text{eff}} = E\,\frac{1}{1 + 2\bar{c}(1 - v^2)d/nh}.$$

Waldorf takes $\bar{c}(1 - v^2)$ as approximately 0·9 and

$$E_{\text{eff}} = E\,\frac{1}{1 + 1\cdot8(d/nh)}.$$

In a particular case investigated, he found:

$$n = 10 \text{ and } h/d = 0\cdot18,\ E_{\text{eff}} = E/2.$$

The pressure on contact areas is $\sigma(d^2/h^2)$ or approximately 30σ. Assuming $\sigma = 70$ kg/cm² (1000 lb/in²) and a joint spacing $d = 0\cdot5$ m, in a rock with $E = 300\,000$ kg/cm² (5×10^6 lb/in²) a joint closure of 0·1 mm would occur. This would be sufficient to reduce the effective modulus to $E_{\text{eff}} = 175\,000$ kg/cm². Conversely if E and E_{eff} have been measured and the joint spacing is known to be d, then nh can be calculated and the joint survey could be checked, as suggested by the Austrian method. Increased pressure causes an increase in contact areas. This increase can be expressed as

$$\delta = C_2 \left(\frac{P}{E}\right)^{2/3},$$

in which δ is the closing together of the two bodies, C_2 is a constant. The basis developed here for estimating the ratio E_{eff}/E has previously been used to introduce a different v_{eff} coefficient from that of the Poisson ratio v.

7.3 Fissured rock masses

7.3.1. Stability conditions for fissured rock masses

Two main types of rupture of rock material have been described previously in chapter 4: brittle fracture, and visco-plastic rupture by shear, which can be represented on the Mohr circle. In addition, the case of fissured rock samples has been discussed, and rupture conditions of fissured and of intact rock samples compared on fig. 4.10.

Similar types of rupture of fissured rock masses were mentioned in section 6.5.5 and the relevant equations developed in section 7.1. A more systematic analysis, based on the direction of the fissures and joints, is possible (Bray, 1967; Naef, 1969).

The whole rock mass is supposed to be stressed and σ_1 and σ_2 are the two principal stresses at point O. The stability condition along a possible surface of rupture, at an angle β to the first principal stress, is expressed by the Coulomb's stability condition:

$$\tau = c + \sigma \tan \phi.$$

This equation can be introduced in the basic stress equations:

$$\sigma = \frac{\sigma_1 + \sigma_2}{2} + \frac{\sigma_2 - \sigma_1}{2} \cos 2\beta,$$

$$\tau = \frac{\sigma_2 - \sigma_1}{2} \sin 2\beta.$$

Assuming that $c = 0$ for conditions near rupture by shear, a simple calculation yields:

$$R = \frac{\sigma_1 + \sigma_2}{\sigma_1 - \sigma_2} = \frac{\sin (2\beta + \phi)}{\sin \phi}.$$

A diagram $R = R(\beta)$ can be traced (fig. 7.18a), the interpretation of which is easy on Mohr circles.

(1) Point A on fig. 7.18a. For $\beta_A = 0$, $R = 1$, the compression stress σ_1 is parallel to the shear plane ($\beta_A = 0$). Rupture occurs by brittle fracture when σ_2 is $= 0$ or negative (σ_2 is then a a tensile stress).

(2) Point B on fig. 7.18a. For $\beta_B = 45° - \phi/2$, rupture occurs for the conditions described in classical soil mechanics. The Coulomb straight line is tangent to the Mohr circle (fig. 7.18b) and replaces the intrinsic curve. On fig. 7.18a, the point B corresponds to such conditions.

(3) The point C on fig. 7.18b. C corresponds to $\beta_C = 90° - \phi = 2\beta_B$. For this point $R = 1$, and again $\sigma_2 = 0$. It corresponds to the case where rock is sliding along the joint with no lateral restraining stress.

The point B is the only one for which rupture occurs for conditions similar to those described in classical soil mechanics. As can be seen on fig. 7.18a, from point A to point C, the angle β corresponding to the inclination of the plane of rupture varies from $0°$ to $90° - \phi$ and the inclination therefore varies from $90°$ to ϕ, which is steeper than the angle of friction ϕ. This proves the greater stability of stressed fissured rock masses.

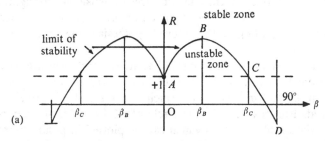

(a)

Fig. 7.18a Stability conditions for fissured rock masses. $R = (\sigma_1 + \sigma_2)/(\sigma_1 - \sigma_2)$;
$$\beta_A = 0°; \beta_B = 45° - \phi/2; \beta_C = 90° - \phi = 2\beta_B.$$

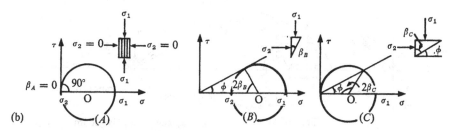

(b)

Fig. 7.18b Stability conditions for fissured rock masses:
$$R = \frac{\sigma_1 + \sigma_2}{\sigma_1 - \sigma_2} = \frac{\sin (2\beta + \phi)}{\sin \phi}.$$

(A) $\beta_A = 0$, $R = 1$, $\sigma_2 = 0$. (B) $\beta_B = 45° - \phi/2$, $R > 1$. (C) $\beta_C = 90° - \phi$.

These remarks can be extended to a system of several families of jointing planes. The resistance of the mass is then reduced. When there is a greater number of jointing planes and/or when the distance between planes is small, the rock mass can be assimilated to a loose soil.

7.3.2 Strength analysis by Pacher and Müller

A basis for analysing the strength of jointed rock, developed by Pacher and Müller (see Müller, 1963a), has several very practical applications in civil and mining engineering. An excellent account of this has been given by John (1962).

In this approach the resistance quotient (R_q) is defined as 'the ratio between the resistance R' of a jointed rock mass along any section against tensile, shear or frictional failure, and the stressings along the same section by tensile and shear stresses' ($R_q = R'/S$). Resistance R' depends on the geomechanical properties of the anisotropic rock mass (as described in sections 3.1 and 3.2), and the load pattern applied. Stressings depend mainly on the latter. In a particular instance the resistance quotient R_q is determined for any section, thus resulting in a two-dimensional or three-dimensional configuration of R_q. When this ratio becomes less than unity the yield limit is reached or rupture occurs, depending on the failure hypothesis being applied, along the respective section.

Figure 7.19 analyses tensile failure. (The stress conditions are reduced to two dimensions for clarity.) The circle of Mohr is traced for the point under consideration, σ_3 and σ_1, being the principal stresses. The intrinsic curve C is also traced. For any angle β the circle of Mohr yields the compression stress σ_C or the tensile stress σ_T which can be plotted in polar co-ordinates. The same intrinsic curve C yields for any angle β on the Mohr circle the corresponding tensile strength S_T. The resistance to tensile failure is $R' = (\sigma_C + S_T)$ and the resistance quotient is $R_q = R'/S_T = (\sigma_C + S_T)/S_T$.

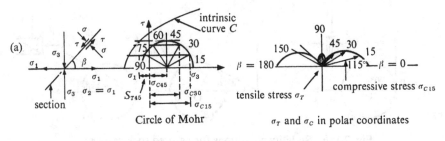

Circle of Mohr σ_T and σ_C in polar coordinates

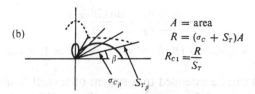

$A = $ area
$R = (\sigma_C + S_T)A$
$R_{C1} = \dfrac{R}{S_T}$

Fig. 7.19 Concept of the 'resistance quotient for tensile failure of a rock mass': (*a*) stress conditions; (*b*) determination of resistance quotients (after John, 1962).

In fig. 7.19 the conditions for non-jointed isotropic material are represented. When the material is jointed, the extent of the joint must be determined. Assuming the joint area to be $A_1 + A_2 + \ldots$ a factor

$$\kappa_2 = \frac{A_1 + A_2 + \ldots}{A}$$

can be determined, where $A = $ total area. The section through rock is now $A(1 - \kappa_2)$ and the tensile strength normal to the joint is $S_T(1 - \kappa_2)$. The tensile strength has to be calculated for all angles β and obviously this depends on the inclination of the joints, their density and extension. Polar co-ordinates are best used to represent σ_C, σ_T, S_T and R_q values.

A similar method is used to calculate the 'resistance quotient' to shear failure (fig. 7.20). Shear and compression stress are read on a Mohr circle

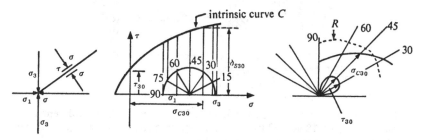

Fig. 7.20 Concept of the 'resistance quotient' for shear failure of a rock mass (after John, 1962).

and plotted on polar co-ordinates. The shear strength S_s of the rock material is given by the intrinsic curve C and the resistance quotient $R_q = S_s/\tau$ calculated and plotted. Figure 7.20 refers to unjointed rock. In jointed rock, the joint area must be calculated for all angles β and the shear strength estimated accordingly.

7.4 The clastic theory of rock masses

7.4.1 Laboratory tests on clastic and discontinuous models

According to Trollope (1968) 'a clastic mass comprises an assembly of units, each unit having a finite physical shape, e.g. spherical, cubical, ellipsoid. Depending on the shape of the boundary, the units will tend to pack in groups wherein some systematic arrangement will dominate and, in general, the mass is made up of varying arrangements of systematically packed zones.' This clastic model is in direct opposition to the accepted mathematical model of the 'continuum'.

Krsmanović & Milic (1963) simulated a rock foundation on a model $100 \times 100 \times 17$ cm made up of horizontal layers of parallelepipedic blocks $4 \cdot 0 \times 4 \cdot 0$ cm wide and 16 cm deep. The load, transmitted to the blocks through a slab 15×16 cm, was varied from 3 to 30 kg/cm², the modulus of elasticity of the 'rock' from $E = 40\,000$ to 23 500 and 15 670 kg/cm². The angle of internal friction of intact 'rock' was about 50° to 56°, the angle of friction in the joints about 30° to 36°. Cohesion of intact hard rock was low, i.e. $0 \cdot 12$ kg/cm² and compression strength 65 kg/cm². Stresses and deformations (settlements) were measured for typical points of the model. Stress distributions (figs. 7.21 and fig. 7.22) were shown to depend largely

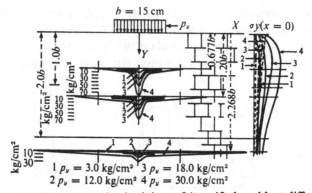

Fig. 7.21 Normal stresses σ_y in the joints of 'stratified rock' at different depths for loads $p_y = 3, 12, 18$ and 30 kg/cm². Concentrated central force applied through an elastic slab (after Krsmanović & Milic, 1963).

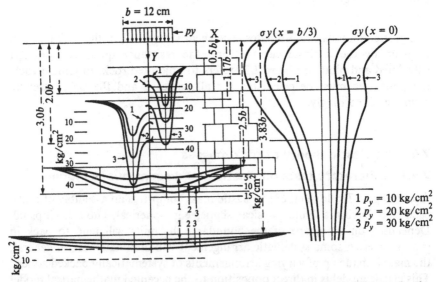

Fig. 7.22 Normal stresses σ_y in the joints of 'stratified rock' at different depths for loads $p_y = 10, 20$ and 30 kg/cm². Load applied through a plate of great rigidity. Free surface of model loaded uniformly with $p_y^{\prime} = 3$ kg/cm for consolidation before beginning the tests (after Krsmanović & Milic, 1963).

on: (1) the stiffness of the slab transmitting the load to the 'rock' model, (2) the amount of vertical prestressing σ_0 given to the jointed rock model. It was found that the nature of the rupture in a discontinuum composed of jointed rock was much more complex than that in a grained semispace. Rupture of the model occurs in very different circumstances.

The French laboratory of the École Polytechnique (Maury, 1970a) started experimental research on a similar problem of stress transmission through discontinuous media. A first series of tests were based on the transmission of a load P through a plate to an elastic body resting on a stiff base of steel. The second series concerned a similar load transmission through several elastic bodies piled on top of each other. The tests were carried out using a new photo-elastic material characterized by an elastic modulus $E = 29\,500\ \text{kg/cm}^2$ and a Poisson ratio $\nu = 0.43$. This material shows an excellent linear response to stresses. During the tests no lateral stress was applied to the elastic body, which could expand freely in a lateral direction. The purpose of the tests was to measure shear stresses for different angles of friction either between two layers of plastic material or between these plastics and the steel basis.

An angle of friction ϕ of only $1.5°$ was obtained when the smooth surfaces of the plastics were protected with thin layers of 'Teflon (R)' and 'Lanoline'. Loading and unloading the plastic body up to twenty times did not change this angle at all. Uncoated plastics gave an angle $\phi = 24°$ which was not altered when loading the model up to twenty times. An angle $\phi = 33°$ was obtained when using very fine sand between the polished surfaces of the plastics. Using the same sand on the polished surface between plastics and steel resulted in an angle of friction $\phi = 37°$. The small steel plate transmitting the load had a width $a = 30\ \text{mm}$ and a thickness $e = 10\ \text{mm}$. It was very rigid compared to the plastic body of height h.

Other tests, carried out with an elastic plate of the same material as the tested body ($e = 20\ \text{mm}$ thick), were for varying values $h/a = 7$, or 3 or 1.5 (fig. 7.23a). The shear stress τ_B on the base was compared to the stress

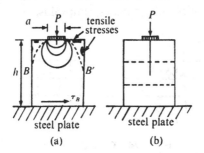

Fig. 7.23

τ_{Bo} in the continuous half-space, at the same level. Table 7.5 shows the ratios τ_B/τ_{Bo} obtained for different conditions. The stress distribution in the test model is more disturbed for small friction factors $\phi = 1.5°$, than for $\phi = 24°$ or $\phi = 33°$.

When the test piece is cut by several planes a similar effect is achieved. The lower the angle of friction ϕ, the greater the disturbance in the stress pattern (fig. 7.23b). With a low ϕ value, the compression stresses are more

Table 7.5

		$h/a = 3$	$h/a = 7$
$\phi = 1.5°$	$\tau_B/\tau_{Bo} =$	2·0	1·9
$\phi = 24°$	$\tau_B/\tau_{Bo} =$	1·3	1·1
$\phi = 33°$	$\tau_B/\tau_{Bo} =$	1·0	1·0

concentrated. Tensile stresses also develop which did not exist in the continuous half-space. The use of discontinuous models to represent fissured rock masses has been further developed when testing dam models. Several laboratories represent rock abutments on the models by using a plastics material where the joints and fissures are represented to scale (Fumagalli, 1967). By varying the friction factor along these joints, it can be seen that weakening of the rock mass by jointing depends on the cohesion of the rock mass. More information on this technique will be given in the chapters concerning dam abutments.

These preliminary results show the danger and complexity of tensile stresses which may develop although the theory of the continuous half-space does not indicate such a possibility.

7.4.2 The clastic theory

Figure 7.24*a*, *b*, *c* and *d* are 'clastic models'. Assuming that there is no friction between the elements, spheres, ellipsoids, cubes, at the points of

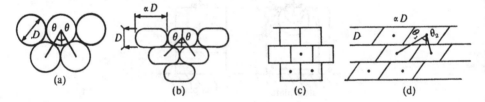

Fig. 7.24   Clastic models.

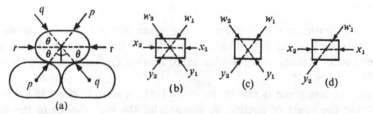

Fig. 7.25   No-arching condition (after Trollope, 1968).

contact, it can be shown that for static equilibrium the contact forces p, q, r in fig. 7.25a need to go through the element's centre of gravity.

Trollope (1957–68) introduces the degree of arching as a major variable in any problem. Figure 7.25b represents the general case where all the forces are greater than zero. In fig. 7.25c the no-arching condition is represented as $X_1 = X_2 = 0$; in fig. 7.25d the full-arching condition is $W_2 = Y_1 = 0$.

In addition the α constant is the type of element and θ the distribution angle. For particles of width D (fig. 7.24b), distribution angle θ and weight w, the average density for γ is (depth of the model $B = 1$)

$$\gamma = \frac{2w \tan \theta}{D^2}.$$

The distance between successive layers is $(d/2) \cot \theta$. It may be readily checked that for squares (fig. 7.24c) $\tan \theta = \frac{1}{2}$ and $\gamma = w/D^2$ (depth of the model $B = 1$). Expressions can be developed for stresses caused by the self-weight of the elements; others for the transmission of external forces from one layer to the next, assuming different values for $k = q/p$ (fig. 7.25a) or $k = Y_1/Y_2$ (fig. 7.25b), k being the arching factor. Trollope tried to analyse Krsmanović experiments using the clastic theory that he had developed for a load P.

Case of no arching ($k = 1$): In this case the forces are transmitted along 'distribution lines'. There are no horizontal contact forces and no relative movements of the blocks. The magnitude of the forces transmitted along the distribution lines will be $P/2 \cos \theta$. In the seventh layer the pressure distribution is as indicated in fig. 7.26.

In the case of 'full arching' ($k = 0$) forces $P/\cos \theta$ and $P/2 \cos \theta$ are transmitted along the distribution lines and the vertical stress distribution is different from the previous case (fig. 7.27). Neither distribution corresponds exactly to the tests by Krsmanović, which most probably was for

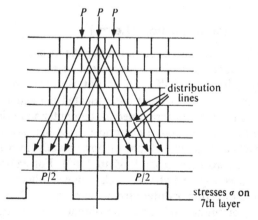

Fig. 7.26 No-arching condition (after Trollope, 1968).

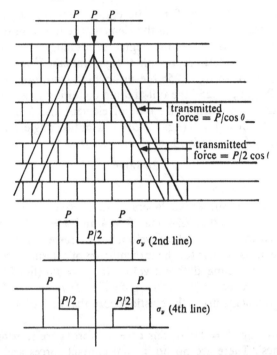

Fig. 7.27   Full arching condition (after Trollope, 1968).

partial arching ($0 < k < 1$). The tests used by Krsmanović, Habib, Fumagalli and others all include friction.

The clastic theory has been developed to analyse various mining problems, including the pressure which develops on lined galleries, pressure on excavations, rock subsidence and the stability of slopes.

7.5   The finite element method (f.e.m.)

Neither the classical mechanics of continua nor the mechanics of discontinua can be adapted to some situations encountered in rock mechanics. In particular, problems where the half-space is not isotropic, or with complex boundary conditions, require a more sophisticated mathematical approach, and a method which is easily programmed for computer analysis. The recently developed 'finite element' method of numerical stress analysis can be extended to deal with particular forms of anisotropy, mainly in rock mechanics (Zienkiewicz and co-workers, from 1965 to 1968). For example, problems can be solved concerning orthotropic rock tunnels, where the modulus of elasticity E_1 in the direction x differs from the modulus E_2 in a direction at right angles to x, or tunnels in curved non-homogeneous strata (fig. 7.28). Dam foundations on nonhomogeneous rock masses, cut by faults or weak

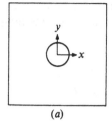
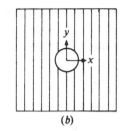
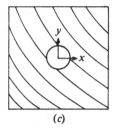

(a)   (b)   (c)

Fig. 7.28 Configuration and loading problems analysed with the finite element method: (a) isotropic: $E_1 = E_2$, $\nu_1 = \nu_2$. (b) orthotropic: $E_1 = 1$, $E_2 = E_3 \neq E_1$. (c) curved, non-homogeneous strata (after Zienkiewicz, Mayer & Cheung, 1966).

strata can be investigated; stress calculations where rock tensile strength is nil can be compared to that when it is not nil.

A case of anisotropy, generally referred to as transverse isotropy, is where the material is isotropic in the yz plane but non-isotropic with respect to directions normal to this plane (fig. 7.29). In a completely general three-dimensional, anisotropic, elastic situation the six stress components and the

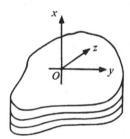

Fig. 7.29 A stratified (transversely isotropic) material.

six strain components can be related by a six by six matrix of coefficients. Using cartesian co-ordinates:

$$
\begin{Bmatrix} \epsilon_x \\ \epsilon_y \\ \epsilon_z \\ \gamma_{xy} \\ \gamma_{yz} \\ \gamma_{zx} \end{Bmatrix} =
\begin{bmatrix}
a_{11} & a_{12} & a_{13} & a_{14} & a_{15} & a_{16} \\
a_{21} & a_{22} & a_{23} & - & - & - \\
- & - & - & - & - & - \\
- & - & - & - & - & - \\
- & - & - & - & - & - \\
- & - & - & - & - & -
\end{bmatrix}
\begin{Bmatrix} \sigma_x \\ \sigma_y \\ \sigma_z \\ \tau_{xy} \\ \tau_{yz} \\ \tau_{zx} \end{Bmatrix}
\tag{1}
$$

This matrix is symmetrical and therefore 21 elastic constants are sufficient to explain the behaviour of any material. In an isotropic material, the number of constants is reduced to two: E and ν.

Referring to fig. 7.29 which shows the orientation of axes assumed in a stratified material, it can be shown that the independent constants remaining in the strain–stress relationship are (Zienkiewicz, 1965).

$$
\begin{Bmatrix} \epsilon_x \\ \epsilon_y \\ \varepsilon_z \\ \gamma_{xy} \\ \gamma_{yz} \\ \gamma_{zx} \end{Bmatrix} = \begin{bmatrix} a_{11} & a_{12} & a_{12} & 0 & 0 & 0 \\ & a_{22} & a_{23} & 0 & 0 & 0 \\ & & a_{22} & 0 & 0 & 0 \\ & & & a_{44} & 0 & 0 \\ & & & & 2(a_{22} - a_{23}) & 0 \\ & & & & & a_{44} \end{bmatrix} \begin{Bmatrix} \sigma_x \\ \sigma_y \\ \sigma_z \\ \tau_{xy} \\ \tau_{yz} \\ \tau_{zx} \end{Bmatrix} \tag{2}
$$

Alternatively, using more conventional definitions of elastic constants, by analogy with the isotropic case, we can write

$$
\left.\begin{aligned}
\epsilon_x &= \frac{1}{E_1}\sigma_x - \frac{\nu_1}{E_1}\sigma_y - \frac{\nu_1}{E_1}\sigma_z \\[2mm]
\epsilon_y &= \frac{1}{E_1}\sigma_x + \frac{\nu_1}{E_2}\sigma_y - \frac{\nu_2}{E_2}\sigma_z \\[2mm]
\epsilon_z &= -\frac{\nu_1}{E_1}\sigma_x - \frac{\nu_2}{E_2}\sigma_y + \frac{1}{E_2}\sigma_z \\[2mm]
\gamma_{xy} &= \frac{1}{G_1}\tau_{xy} \\[2mm]
\gamma_{yz} &= \frac{1}{G_2}\tau_{yz} = 2\frac{(1+\nu^2)}{E_2}\tau_{yz} \\[2mm]
\gamma_{zx} &= \frac{1}{G_1}\tau_{zx}
\end{aligned}\right\} \tag{3}
$$

When considering only plane strain problems (see Zienkiewicz, Cheung & Stagg, 1966):

$$
\epsilon_z = \gamma_{yz} = \gamma_{zx} = 0 \quad \text{and} \quad \frac{E_2}{E_1} = n, \tag{4}
$$

the σ_z becomes

$$
\sigma_z = \frac{\nu_1 E_2}{E_1}\sigma_x + \nu_2\sigma_y,
$$

resulting in the following relationships in the plane xy:

$$\left.\begin{aligned}
\epsilon_z &= \frac{1}{E_1}\left(1 - \frac{\nu_1^2 E_2}{E_1}\right)\sigma_z - \frac{\nu_1}{E_1}(1 + \nu_2)\sigma_y \\[2mm]
\epsilon_y &= -\frac{\nu_1}{E_1}(1 + \nu_2)\sigma_z + \frac{1}{E_2}(1 - \nu_2^2)\sigma_y \\[2mm]
\gamma_{zy} &= \frac{1}{G_1}\tau_{zy}.
\end{aligned}\right\} \quad (5)$$

It is more convenient to express the stresses in terms of strains, therefore:

$$\left.\begin{aligned}
\sigma_z &= \frac{E_1}{(1 + \nu_2)(1 - \nu_2 - 2n\nu_1^2)}\{(1 - \nu_2^2)\epsilon_z + n\nu_1(1 + \nu_2)\epsilon_y\} \\[2mm]
\sigma_y &= \frac{E_2}{(1 + \nu_2)(1 - \nu_2 - 2n\nu_1^2)}\{\nu_1(1 + \nu_2)\epsilon_z + (1 - n\nu_1^2)\epsilon_y\} \\[2mm]
\tau_{zy} &= G_1\gamma_{zy}.
\end{aligned}\right\} \quad (6)$$

Equation (6) can be conveniently written in matrix form

$$\{\sigma\} = \begin{Bmatrix} \sigma_z \\ \sigma_y \\ \tau_{zy} \end{Bmatrix} = [D]\begin{Bmatrix} \epsilon_z \\ \epsilon_y \\ \gamma_{zy} \end{Bmatrix} = [D]\{\epsilon\}. \quad (7)$$

(In the case of linear visco-elasticity the matrix (D) is one of differential or integral operators.)

To analyse the stress in a two-dimensional body it is first sub-divided into small triangular (or rectangular) elements (fig. 7.30) which are assumed to

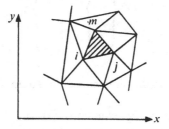

Fig. 7.30 Subdivision of a two-dimensional body into finite elements.

be connected to each other only at nodal points corresponding in this instance with the apices of the triangles. If at each node, such as i, the displacements in directions of the x and y co-ordinates are listed as

$$\{\delta_i\} = \begin{Bmatrix} u_i \\ v_i \end{Bmatrix}, \quad (8)$$

and if the displacements of an element i, j, m, are defined as

$$\{\delta^e\} = \begin{Bmatrix} \delta_i \\ \delta_j \\ \delta_m \end{Bmatrix}, \tag{9}$$

then it is possible to associate theoretical displacement forces which act at the nodes as

$$\{R^e\} = [K^e]\{\delta^e\}. \tag{10}$$

It is possible to set up a series of simultaneous equations at each node to ensure equilibrium. Then, for example, at a node i we have

$$\Sigma \{R_i^e\} = \Sigma \{F_i^e\}, \tag{11}$$

in which $\{R_i^e\}$ is the internal force contributed by an element, $\{F_i^e\}$ is the external force contributed by an element and summation concerns all the leements meeting at a particular node. The external load $\{F_i^e\}$ may be due either to concentrated loads acting at such a point or may be caused by distributed load acting through the element. As all the internal forces are dependent on the nodal displacements the equilibirum conditions result in a system of simultaneous equations:

$$[S] \begin{Bmatrix} \delta_1 \\ \delta_2 \\ \cdot \\ \cdot \\ \cdot \end{Bmatrix} = \begin{Bmatrix} F_1 \\ F_2 \\ \cdot \\ \cdot \\ \cdot \end{Bmatrix}, \tag{12}$$

from which the displacements and hence the stresses (equation (7)) can be determined.

Displacements are given by equations (8) and (9). The displacements within the element are assumed to vary linearly as

$$\{\delta\} = \begin{Bmatrix} a_1 + a_2 x + a_3 y \\ a_4 + a_5 x + a_6 y \end{Bmatrix} = \begin{bmatrix} 1 & x & y & 0 & 0 & 0 \\ 0 & 0 & 0 & 1 & x & y \end{bmatrix} \begin{Bmatrix} a_1 \\ \cdot \\ \cdot \\ \cdot \\ a_6 \end{Bmatrix}. \tag{13}$$

The six constants a can be determined by equating the six nodal displacement components with x and y co-ordinates taking on appropriate values as

$$\{\delta^e\} = \begin{Bmatrix} u_i \\ v_i \\ u_j \\ v_j \\ u_m \\ v_m \end{Bmatrix} = \begin{bmatrix} 1 & x_i & y_i & 0 & 0 & 0 \\ 0 & 0 & 0 & 1 & x_i & y_i \\ 1 & x_j & y_j & 0 & 0 & 0 \\ 0 & 0 & 0 & 1 & x_j & y_j \\ 1 & x_m & y_m & 0 & 0 & 0 \\ 0 & 0 & 0 & 1 & x_m & y_m \end{bmatrix} \begin{Bmatrix} a_1 \\ - \\ - \\ - \\ - \\ a_6 \end{Bmatrix} = [A]\{a\}. \quad (14)$$

By equation (13):

$$\{\delta\} = \begin{bmatrix} 1 & x & y & 0 & 0 & 0 \\ 0 & 0 & 0 & 1 & x & y \end{bmatrix} [A]^{-1}\{\delta^e\}.$$

The strains are defined as

$$\{\epsilon\} = \begin{Bmatrix} \epsilon_x \\ \epsilon_y \\ \gamma_{xy} \end{Bmatrix} = \begin{Bmatrix} \dfrac{\partial u}{\partial x} \\ \dfrac{\partial v}{\partial y} \\ \dfrac{\partial u}{\partial y} + \dfrac{\partial v}{\partial x} \end{Bmatrix} = \begin{bmatrix} 0 & 1 & 0 & 0 & 0 & 0 \\ 0 & 0 & 0 & 0 & 0 & 1 \\ 0 & 0 & 1 & 0 & 1 & 0 \end{bmatrix} [A]^{-1}\{\delta^e\} = [B]\{\delta^e\}.$$

The stresses are given by

$$\{\sigma\} = [D]\{\epsilon\}. \quad (7)$$

The finite element method is being extensively used for the analysis of strain–stress patterns around large tunnels and underground excavations. Residual rock stresses, the k-value, rock jointing, rock faults and progressive fissuring of the rock mass can be introduced in the programs. Similarly the stability of rock slopes and many other problems of rock mechanics can be solved with the finite element method.

In sections 10.2, on special problems for mining and tunnel engineers, 10.5.3, on Fenner's equation, 10.10, on the estimate of the required rock support, 10.11 on underground hydro-electric power stations and 16.3 on Waldeck II, the use of the finite element method for solving special problems will be amply discussed.

Cundall (1971; 1974) has developed a computer program that can model behaviour of assemblages of rock blocks and visually display this behaviour on the screen of a cathode-ray oscilloscope. There is no restriction in block shapes and no limits to the magnitude of displacements and rotations that are allowed. The rock geometry is specified by the user, who draws lines on the screen of the cathode-ray oscilloscope which is connected to a mini-computer. The programme allows the blocks to move relative to one another, under the action of gravity and forces specified by the user, as a function of time. Joint properties may be specified.

8 Interstitial water in rock material and rock masses

8.1 General remarks

The importance of testing rock materials for permeability and the current methods being used have already been mentioned in section 4.9. The problem of interstitial water and how it influences rock masses is extremely complex. It involves the chemical reactions of water on rock and rock on water; the physical characteristics of water; its behaviour in porous and in fissured rock; the formation of underground caverns, caves and rivers and also the stability of rock masses. Some of these points will be dealt with more fully in subsequent chapters on rock slopes, tunnels and dams. This chapter is a general summary on the behaviour of water in rock masses.

8.2 Some general equations on the flow of water in fissured rock

The flow of water in a porous solid, like concrete or rock, results from hydraulic gradients S from one point to the next:

$$S = \text{grad}\,(z + p/\gamma) = \text{grad}\,U, \tag{1}$$

z being the level and p the pressure at the point and γ the specific weight of the water. U is the hydraulic potential. Changes in water pressure from one point to another create differentials in surface forces in the pores which are equivalent to body forces. It is normally accepted that the flow of water, and other fluids (oil) through pervious solids follows the Darcy law (Darcy, 1856; Muskat, 1937; C. Jaeger, 1956), which states that the velocity of percolation v is proportional to the gradient S of the hydraulic potential:

$$v = kS = k\,\text{grad}\,U, \tag{2}$$

k being the permeability of the solid to water usually expressed as centimetres per second. This law has been generalized for the case of the flow of homogeneous fluids in porous media as

$$k = \frac{\kappa D^2 \gamma}{\mu} = K\frac{\gamma}{\mu}, \tag{2a}$$

where D is the effective diameter of the openings (pores or joints) of the solid, or a characteristic dimension of its texture; μ the dynamic viscosity;

γ the specific weight of the liquid and $K = \kappa D^2$ the physical coefficient of permeability, κ, being a quantity without dimensions depending on the geometry of the pores.

If an array of fissures of opening, e, with parallel faces separated by a distance, d, is assumed, the following expression is obtained (Serafim & del Campo, 1965; Talobre, 1957) for the filtration through that array

$$v = \frac{e^3 \gamma}{12 d \mu} S \qquad (3)$$

and the flow through a single joint of constant thickness e is expressed as

$$v = \frac{e^2 \gamma}{12 \mu} S. \qquad (3a)$$

(The viscous flow in a circular tube is given by Poiseuille's formula, $v = \gamma R^2 S / 8 \mu$, with $R = D/2 =$ radius of the tube.)

Some authors suggest differentiating between the two types of flow in fissured rock: the primary flow through the rock pores and the secondary flow through the rock fissures, assuming that the flow through the first is much slower than through the second. This may be correct for compact, widely fissured granite, but there are some porous rocks, with void index not higher than $i = 5\%$ to 10% for which the flow through the pores may be quite substantial. The degree of correlation between perviousness k and compressive stress depends on the shape of the voids and minute canals in the rock (sections 4.9 and 4.10). In most cases when permeability of the rock results from open fissures and fractures the factor k depends on the directions along which it is measured. If x, y and z are the three principal directions of the anisotropy (Serafim 1968):

$$\{v\} = [K] \{\text{grad } U\}, \qquad (4)$$

where $[K]$ is a three-by-three matrix defined by nine numerical coefficients. The three components of $\{v\}$ are:

$$v_x = k_1 \, \partial u / \partial x$$
$$v_y = k_2 \, \partial u / \partial y \qquad (5)$$
$$v_z = k_3 \, \partial u / \partial z.$$

When analysing the steady flow in anisotropic rock it can be stated that the weight of a liquid which enters in a unit volume of the porous body in a unit of time is equal to the quantity which flows out of that volume. Therefore:

$$\text{div } (\gamma v) = \frac{\partial}{\partial x} (\gamma v_x) + \frac{\partial}{\partial y} (\gamma v_y) + \frac{\partial}{\partial z} (\gamma v_z) = 0. \qquad (6)$$

Assuming the liquid (or gas) is not compressible we find that the two previous equations yield:

$$\frac{\partial k_1}{\partial x}\frac{\partial u}{\partial x} + \frac{\partial k_2}{\partial y}\frac{\partial u}{\partial y} + \frac{\partial k_3}{\partial z}\frac{\partial u}{\partial z} + k_1\frac{\partial^2 u}{\partial x^2} + k_2\frac{\partial^2 u}{\partial y^2} + k_3\frac{\partial^2 u}{\partial z^2} = 0, \quad (7)$$

when writing

$$\bar{x} = \frac{x}{\sqrt{k_1}}, \qquad \bar{y} = \frac{y}{\sqrt{k_2}}, \qquad \bar{z} = \frac{z}{\sqrt{k_3}}, \quad (8)$$

omitting the first term of the previous equation containing

$$\frac{\partial k_1}{\partial x}, \qquad \frac{\partial k_2}{\partial y} \quad \text{and} \quad \frac{\partial k_3}{\partial z}.$$

we obtain

$$\frac{\partial^2 u}{\partial \bar{x}^2} + \frac{\partial^2 u}{\partial \bar{y}^2} + \frac{\partial^2 u}{\partial \bar{z}^2} = 0, \quad \text{or} \quad (\Delta^2 u = 0). \quad (9)$$

In the case of a homogeneous field of percolation, with $k_1 = k_2 = k_3 = k$, the meaning of the last equation is obvious. The solution for such a Laplace equation is well known. Used in classical soil mechanics, it treats groundwater problems, percolation in loose soils and foundations. The methods of potential flow are used in most cases, or approximations based on the same method (Jaeger, 1949). They could therefore be used in rock mechanics whenever a homogenitic percolation is assumed. After using this method, Pacher & Yokota (1962/63) mention that *in situ* borehole measurements made at the Kurobe IV dam site correspond with those estimated from model tests and the electric analogue (chapter 11, sections 11.3, 4).

If during the flow, some water is retained or originates in a given unit volume around a point, unsteady flow results. In such a case the left member of equations (6) and (7) cannot be equal to zero and (7) can be written:

$$\frac{\partial q}{\partial t} = \left[k_1\frac{\partial^2 p}{\partial x^2} + k_2\frac{\partial^2 p}{\partial y^2} + k_3\frac{\partial^2 p}{\partial z^2} + \frac{\partial p}{\partial x}\frac{\partial k_1}{\partial x} \right.$$
$$\left. + \frac{\partial p}{\partial y}\frac{\partial k_2}{\partial y} + \frac{\partial p}{\partial z}\frac{\partial k_3}{\partial z} + \gamma\frac{\partial k_3}{\partial z} \right] dV$$

where dV is a volume and γ is the specific weight of the liquid, which is assumed to be constant. It can be assumed that the weight of water contained in the volume dV is proportional to the percentage of the volume of voids or porosity n, less the percentage of the volume of air a. Then:

$$q = \gamma(1 - a)n \times dV.$$

If the liquid is compressible, its specific weight γ at the pressure p is equal to

$$\gamma = \gamma_0\beta p + \gamma_0,$$

where γ_0 is the specific weight at the pressure $p_0 =$ atmospheric and β the volumetric compressibility. The theory can be developed further on such lines (Serafim, 1968).

8.3 Effective stresses in rock masses

There is a great deal of evidence that the law of effective stresses used in soil mechanics:

$$\tau = \tau_0 + (\sigma - p) \tan \phi = \tau_0 + \sigma' \tan \phi,$$

where p is pore pressure, can frequently be applied to rock mechanics. Handin & Hager (1957) and others found that depending on the effective confining pressure there is no difference in the ultimate strength of triaxial conditions

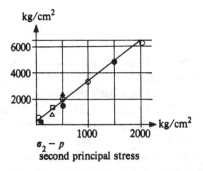

Fig. 8.1 Triaxial tests on Berea sandstone with various internal pressures (p) (values in kg/cm²): $\bigcirc$ $p = 0$; $\bullet$ $p = 500$; $\triangle$ $p = 1000$; $\blacktriangle$ $p = 1500$; $\square$ $p = 1750$; $\blacksquare$ $p = 2000$ (after Handin & Hager, 1957).

for various rocks (e.g. siltstones, sandstones, limestones, shales) with or without pore pressure (fig. 8.1). For example Handin & Hager found that for Berea sandstone

$$\tau = 153 \cdot 8 + (\sigma - p) \tan 29° 15'.$$

It was shown that the Griffith theory of rupture agrees with experimental data when the pore pressures are deduced from the principal stresses.

It seems that in brittle porous rocks, pore pressures have no influence (Serafim, 1968). This means that their strength is mainly due to the strength of the bonds between solid material or grains, and not to the shear strength. Pore pressures produce two separate effects: they compress the solid matter, thus reducing the volume of the solid matrix; and create body forces proportional to the variations of the pressure dp. Serieys (1966) has shown that the elastic modulus of Jurassic limestone varies with porosity, thus confirming some remarks by Serafim and Hamrol (discussed in section 4.8).

8.4 Flow of water in rock masses with large fractures

The analysis of water percolation developed in the preceding chapters implicitly assumes the rock to be crossed by a narrow net of fissures. A different approach is suggested when large rock fractures forming a wide net are present.

8.4.1 The equations of flow in fractures

Wittke (1966; and Wittke & Louis, 1969) suggests the following.
(1) For laminar type of flow in rock fissures (Poiseuille type of flow):

$$q_1 = \frac{g}{12\nu} (2a)^3 S.$$

(2) For turbulent flow in smooth fissures (Blasius type of flow), the most likely to occur:

$$q_1 = \left\{ \frac{g}{0 \cdot 079} \left(\frac{2}{\nu}\right)^{1/4} (2a)^3 \right\}^{4/7} S^{4/7}.$$

(3) For fissures filled with sandy material:

$$q_1 = 2a \times k \times S.$$

In these equations, q_1 is the water discharge in the fissure; g the acceleration due to gravity; ν the kinematic viscosity $= 1 \cdot 2 \times 10^{-6} \text{ m}^2/\text{s}$ for a temperature $t = 13 \text{ °C}$; $2a$ the width of the fissure in metric units and S the gradient of the energy line (or in this case, the pressure line) of the flow. Additionally, the Darcy formula may have to be considered:

$$S = \lambda \frac{1}{R_h} \frac{v^2}{2g},$$

where R_h is the hydraulic radius $R_h = 4a$, λ being here the friction factor of Darcy or Nikuradse. (For more information on these four equations, consult a textbook on fluid mechanics (C. Jaeger, 1956).)

8.4.2 The computer solution

Wittke applies these equations to a (two- or three-) dimensional system of rock fissures. Figure 8.2a represents a two-dimensional system of fissures or fractures; K_1 is the main system, K_2 the secondary system. At any point of junction N of the fissures the algebraic sum of all discharges arriving or departing from the point is zero; $\Sigma_N q = 0$. Similarly, the sum of all pressure losses occurring in fissures along a closed circuit, like the circuit R in fig. 8.2a, must be zero.

The boundary conditions to be introduced are the same as those in an equivalent soil mechanics problem (C. Jaeger, 1956). When the problem is

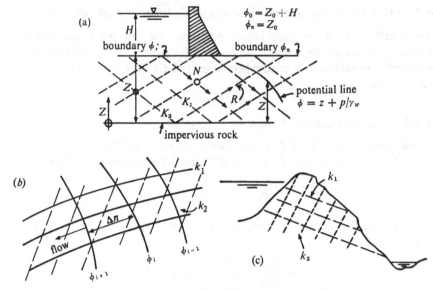

Fig. 8.2

correctly written up, there are as many equations as unknowns. Equations are solved on a computer. Lines (or surfaces) can be traced which are the equivalent of potential lines (or surfaces) in the theory of water percolation through homogeneous porous media.

8.4.3 Transmission of forces from water to rock masses

'Potential lines' ϕ_{i-1}, ϕ_i, ϕ_{i+1} are traced in fig. 8.2b. The forces acting on the volume of rock between two such lines are: the uplift force:

$$U = \gamma_w \quad \text{(vol)},$$

where γ_w is the specific weight of the water, and the pressure force P in the direction of the flow:

$$P = \gamma_w \frac{\Delta_\phi}{\Delta_n} \quad \text{(vol)},$$

with $\Delta_\phi = \phi_i - \phi_{i+1}$ and Δ_n distances between potential lines.

8.4.4 Examples

Wittke checked his method on models. He measured the pressures and traced the 'potential' lines which were compared to the results of calculations. There were only small differences between calculations in formulae for potential flow (Poiseuille) and turbulent flow (Blasius).

A typical problem which can be solved with this method is represented in fig. 8.2c. A fissured rock shoulder is retaining water in an artificial storage basin. Pressures on the rock mass and water losses can be calculated. Drainage of the rock shoulder was necessary so that a reasonable stability of the rock mass could be achieved.

Other problems in the stability of rock slopes or the pressure on the rock surrounding empty tunnels will be dealt with in chapters 9 and 10.

8.5 Physical and physico-chemical alteration of rock by water

As is shown in section 2.1 some rocks react badly when soaked with water. A French team (Bellier, 1964) found that a moderately clay-like sandstone expanded by about 0.5% when soaked with water and that its crushing strength dropped by about 50% (fig. 8.3). The swelling pressure of this rock measured on a special odometer, increased to about $60 \, kg/cm^2$ on the saturated face but to only $30 \, kg/cm^2$ on the perpendicular face before the

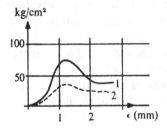

Fig. 8.3   Triaxial compression of very clayey sandstone. (1) Dry rock; (2) soaked rock.

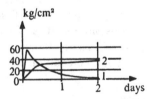

Fig. 8.4   Swelling pressure as measured in a special odometer during saturation on (1) the saturated face, and (2) on the perpendicular face (after Bellier *et al.*, 1964).

rock disintegrated (fig. 8.4). Some other rocks (e.g. English Channel chalk), which have a crushing strength of over $100 \, kg/cm^2$ when saturated, disintegrate when drying. A substantial drop of the Young's modulus, E, has been observed on some rock laboratory samples and on the same rock mass *in situ*.

The French team investigated alterations to acid rocks formed at very high temperatures (siliceous). Percolation water at normal temperatures

Table 8.1

	Si	Al	Mg	Na	K
Chemical composition in %	67	13	2·7	3·4	3
Elements entrained by water in mg/l	2270	35	1	120	57

Characteristics of the percolation water:
Before the test: pH 6, Resistivity ohm/cm 473 000
After the test: pH 10·3, Resistivity ohm/cm 1 960

caused alkaline reactions and rock deterioration. Table 8.1 concerns tests with pure water at 80 °C over a period of 24 hours.

Chemical reactions which cause alteration to rocks are extremely complicated. Their bulk effect on the sample can be observed by checking any changes in the discharge and observing any chemical changes of percolation water. Decrease of the percolation discharge may be due to swelling of the altered material inside the microfissures and microfractures.

In a summary of their research the French team decided that three rock groups are suitable for dam foundations:

(a) Massive gneiss and granite not disturbed by tectonic movements remain practically unaltered. A few macrofractures can be easily detected.

(b) Hard sedimentary rocks (quartzites, hard schists, massive limestone, etc.): general stratification can easily be followed.

(c) Metamorphic rocks, having had many changes in pressures and tectonic shocks, often have a complex pattern of fissures. Many schists and gneiss rocks belong to this variety. It is often very difficult to give a proper diagnosis of these rocks and a thorough examination of possible alteration must be carried out. Percolation tests are of great importance.

8.6 Aging of rock masses

The physical and physico-chemical alteration of rock masses due to interstitial water in pores, fissures and fractures is one aspect of the more general phenomenon known as 'aging of rock'.

(1) *Climatic conditions.*   Temperature variations contribute to rock aging. The superficial deterioration of massive granite which slowly decomposes to sand is a typical example of this process. It is described in all textbooks on geology, and has caused many difficulties to dam designers (Spain, Portugal, etc.). In the Alps, glacial erosion has often cleared the weathered granite away.

(2) *Strains and stresses. Relaxation from residual stresses.* Relaxation from residual stresses near the rock surface sometimes causes rock fissures and fractures parallel to the rock surface (Kieslinger, 1958). This creates difficulties in dam construction (Vajont dam rock abutments). Geologists have described large slabs of rock which become detached and slide along the rock surface. Slow rock slides may cause very high strains and stresses deep inside the rock mass (Jaeger, 1968). The Vajont slide is an example of this phenomenon. Seismic measurements taken showed decreasing wave velocities over time which proved that the rock was deteriorating through age. Artificial strains and stresses from structures built on or inside the rock (dams, tunnels and cavities) may also cause fissures, fractures and aging.

Sometimes straining and/or fissuration slow down with time. Dam foundation displacements can come to a standstill as shown for years of systematic surveys (Rossens dam displacements, measured by Gicot). But in some cases the straining progresses to rupture (Malpasset dam, chapter 13). Similarly, fissuration about cavities comes to a natural standstill (Rabcewicz and Talobre, chapter 10); but mines have been completely closed by rockfalls or squeezed by overstrained rocks. Aging is an irreversible deterioration of physical, chemical and mechanical properties.

Part Three
Rock mechanics and engineering

During the past twenty years engineers and geologists, specializing in rock mechanics and its problems, came to the conclusion that *in situ* tests would yield progressively more answers to questions on rock stability and strength. More recently, they realized the importance of precise and detailed laboratory research. Difficulty in recognizing the real cause of the Malpasset dam collapse has partly brought about this change in attitude.

Despite the expense, exhaustive *in situ* tests are vital where there are specific engineering problems. To emphasize the link between civil engineering and rock mechanics, Part Three is devoted to problems of: rock slopes, slope stability; tunnel design, stability of cavities; dam design and foundations.

---

9 Rock slopes and rock slides

Any mention of rock slopes and rock slides should include the remarkable paper published in 1932 by A. Heim. He systematically describes and analyses, as a geologist, all slides known to him and classifies them into over twenty types. One of his more important findings concerns the distinction between slowly progressing slides, where conditions of stability are only slightly disturbed and the very rapidly accelerating slides where concentrated energy and momentum cause rock masses to reach exceedingly high velocities.

His analysis of the geological cause for rock imbalance which starts slides, like weakening and aging of rock masses, water percolation, springs and the action of man, brings him no nearer to the true mechanics of a slide. This is not a geological problem but one of rock statics and dynamics.

Stability of slopes, superficial rock falls and deep-seated lines of rupture will be dealt with in this chapter. The complex dynamics of accelerated discontinuous rock slides will be illustrated in chapter 14, statics of rock faces will be analysed more closely in chapter 15.

Terzaghi (1962*b*) has published an interesting paper called 'Stability of steep slopes . . .' after many years of research. Essays by Stini, Müller and others have produced additional information, but the necessity for a more reasoned analysis of the problems of statics and dynamics became apparent after the Vajont rock slide. Terzaghi suggests that rock slopes should be classified by the type of rock, the analysis of the mechanisms of rupture, the action of water in the pores, fissures and cracks. It is also essential to try to classify falls and slides into type, depth, volume involved and dynamics (see Hoek, 1972).

9.1 Terzaghi's theory; the 'critical slope'; stability of rock faces

Slopes in open pits, open mines, in quarries, along open cuts for railways and roads must be stable. The first problem to be analysed concerns the 'critical slope' for superficial stability; according to Terzaghi no rock falls will occur along slopes at an angle of less then 35° (a statement which geologists may not accept).

In the early stages of erosion most jointed rocks possess considerable effective cohesion c_i of rock masses:

$$c_i = cA_s/A,$$

(A_s = total area of solid rock area, A = total area, c = cohesion of intact rock material) and, as a consequence, they can form vertical or quasi-vertical slopes. The seat of most cohesion of the jointed rock is located in the 'gaps' which interrupt the continuity of the joints. As the height of the side walls of the erosion valley increases the shearing stresses in the rock adjacent to the walls increase correspondingly. When splitting of the gaps occurs cohesion, c_i, decreases. Local stress concentration causes new surfaces of failures to be superimposed on the local system of joints. Depending on the joint pattern, the sides of the valley may vary between 30° and 90°.

Sheeting is a form of joint parallel to the valley, occurring mainly in granite.

Typical types of delayed slope failure are the relatively superficial *rock falls*, and the more deep-seated *rock slides*. Falls may be connected with the weakening effect of frost, whereas slides involve rock masses located below the limit of frost action. The shearing resistance, τ_r, at a given point, P, of a potential sliding surface in a porous and saturated material is given by the well-established empirical law:

$$\tau_r = c_i + (\sigma - p) \tan \phi.$$

According to Terzaghi, all intact and jointed rock masses with effective cohesion have the mechanical properties of brittle materials. Failure of brittle material slopes starts at the point where the shearing stress, τ, becomes equal to τ_r. Stresses on the surrounding rock masses increase and *progressive failure* occurs by brittle shear fracture. If the rock has a random pattern of jointing the shear resistance equation is valid for any section in any direction. The rock behaves in an analogous manner to an unjointed stiff clay. It has been established that the steepest stable rock slopes are S-shaped, similar to the profile through root and tongue of a clay slide. The critical slope angle ϕ'_c decreases with increasing height of the slope, but ϕ'_c remains larger than ϕ. In regularly jointed rock the value of ϕ depends on the type and degree of interlock between the blocks on either side of the sliding surface. The effective cohesion, c_i, of rock masses is very much smaller than the cohesion, c, of rock material. Because of progressive failure c_i tends towards zero and it is safe to assume that $c_i = 0$.

9.1.1 Slopes in unstratified jointed rock

Rocks in this category (granite, marble) are divided by continuous random joints into irregularly-fitting blocks locally interconnected. This macro-structure of the rock mass is a large-scale model of the microstructure of intact crystalline rock. Such rocks were tested by von Karman (1911) and by Ros & Eichinger (1930). More recent tests by Borowicka (1962) revealed that the angle of friction of crystalline rocks varies from $\phi = 40°$ at 100 kg/cm² pressure, to $\phi = 25°$ at 1000 kg/cm² pressure. According to Terzaghi, the critical angle ϕ'_c for slopes with underlying hard massive rock masses with a random joint pattern is about 70°, provided seepage is not acting upon the walls of the joints.

9.1.2 Slopes in stratified sedimentary rock

Stratified sedimentary rocks have layers varying in thickness between a few inches and many feet. These are separated from each other by thin films of material different from that of the rest of the rock. The bedding planes are almost invariably surfaces of minimum shear-resistance and are likely to be continuous. (In Terzaghi's paper they are referred to as bedding joints.) The cross-joints are generally nearly perpendicular to the bedding joints, and they are commonly staggered at these joints. The cohesive bond along the walls of the cross-joints is equal to zero. The intersections between the cross-joints and the bedding planes may be more or less parallel in one or more directions. Less frequently they may be nearly randomly orientated.

Because of the almost universal presence of bedding- and cross-joints stratified sedimentary rock with no effective cohesion ($c_i = 0$) has the mechanical properties of a body of dry masonry composed of layers of more or less prismatic blocks which fit each other. The cohesion across the joints between all the blocks of each layer is zero. The stability of a slope will depend primarily on the orientation of the bedding planes with reference to the slope. This relationship is illustrated by figs. 9.1 and 9.2. Cross-joints are assumed to be staggered and perpendicular to the bedding joints. The angle, ϕ_f of friction along the walls of all the joints is assumed to be 30°. If the bedding planes are horizontal, no slide can occur, and the critical slope is vertical: $\phi_c = 90°$.

9.1.3 Bedding planes dipping into the mountain

In fig. 9.1 the bedding planes dip into the mountain at an angle α. The line *A–A* cuts the rock mass at an angle 90° – α to the horizontal. If 90° – $\alpha < \phi_f$ no failure could occur along planes *A–A* (ϕ_f = angle at friction). If the cross-joints are parallel to *A–A*, but staggered, the position of the critical slope depends on the average value of the ratio C/D between the average length of the offset *C* between cross-joints and the average spacing *D* between bedding joints (fig. 9.1). For any value of α smaller than $90 - \phi_f$ the critical

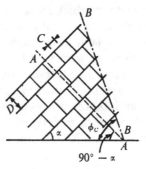

Fig. 9.1 Diagram illustrating the inclination, ϕ_c of the critical slope, *B–B*, in stratified rock (after Terzaghi, 1962*b*).

slope angle is equal to that of the line *B–B* in the figure. At any given value of α, the critical slope angle, ϕ_c, increases with increasing values of the ratio C/D and at a given value of C/D it decreases with decreasing values 90° – α until 90° – $\alpha = \phi_f = 30°$.

At this point the critical slope angle abruptly increases to 90° because the slope angle of the cross-joints becomes smaller than the angle of friction $\phi_f = 30°$ along the joints. However, as 90° – α further decreases and α approaches 90° the danger of a failure by buckling of the layers located between bedding joints increases. Cohesion along the bedding joints increases the critical slope angle for any value of α smaller than 90° – ϕ_f. If the stratum is steeper, cohesion practically eliminates the possibility of a failure of the exposed stratum by buckling.

9.1.4 Bedding planes dipping towards the valley (fig. 9.2)

If the bedding planes dip towards the valley at an angle smaller than the angle of friction, $\phi_f = 30°$, the critical slope is 90°. For values of α greater than 30° the critical slope angle is equal to α. If the slippage along bedding joints is resisted by effective cohesion c_i in addition to friction, the steepest stable slope is no longer plane. Up to a certain height H it will be vertical as shown by fig. 9.2 and above it the slope will raise at an angle α.

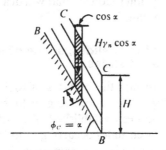

Fig. 9.2 Greatest height, H, of a rock cliff. The rock cohesion, c_i, is not negligible (after Terzaghi, 1962b).

Let H be the height of the vertical part of the slope. The force which tends to produce a slip along a bedding joint through the foot of the slope is (γ_r = unit weight of the rock):

$$\gamma_r H \cos \alpha \sin \alpha,$$

per unit of area of the bedding joint and the force which resist the slip is

$$c_i + \gamma_r H \cos^2 \alpha \tan \phi_f.$$

Hence the vertical slope as shown in fig. 9.2 will not be stable unless

$$H \leqslant \frac{c_i}{\gamma_r \cos \alpha (\sin \alpha - \cos \alpha \tan \phi_f)}.$$

An increase in height of the vertical slope would be immediately followed by a slide along the bedding plane B–B through the foot of the slope.

9.2 Rock slides; deep-seated lines of rupture

The previous section dealt with the stability of rock faces and the critical slope along which rock falls may occur. Rock slides may be of any size and are caused by ruptures located deep in the rock mass. The statics and dynamics of a very large slide may be completely different from those of a smaller one and sometimes they are most difficult to analyse.

There are many varieties of rock slides, both man-made and from natural causes. Cuttings for roads or railways is one cause. It was the cause at the construction site of the Lötschberg railway in the centre of the Alps. Here the south ramp to the main terminal was too sharply inclined and rock slides were a constant hazard. They also occur as a result of excavations for dam foundations (Bort dam in France) and open-cast mines (Hoek, 1970, 1972, 1973; Londe *et al.*, 1969). But by far the most common man-made cause is the varying water levels in artificial storage reservoirs.

The result of water and ice percolating through rock will be dealt with in the next section, and the section following that will deal with water percolating from a hydraulic pressure tunnel.

9.2.1 The formation of rock slides, shape of sliding surfaces and the progress of slides

The rupture of a rock mass and the progress of rock slides is very different to, and more complex than, slides occurring in loose soils, for which relatively simple theories have been established. As early as 1882, Heim had stressed the difference between slides occurring in loose soils, rock slides and mixed slides, a classification which he maintained in his main paper, *Bergsturz und Menschenleben* (1932), in which he discussed various types of large and very large slides, as they occurred in the Alps up to 1932. He analysed conditions in the area where the ruptures first occurred and followed the progress of the slides along the mountain slopes until the movement ceased. Figure 14.1, from Müller on Vajont, illustrates perfectly a type of rock slide with downturning movement of the rock tops which could not occur in loose material.

According to Mencl (1967), confirming Terzaghi's remarks on the stability of rock faces, the pattern of the main sets of joints determines the rupture of rock masses and the shape of most rock slides. Mencl shows how the movement of the whole mass of rocks sometimes depends on the inclination of a near-vertical set of joints which may be 'positive' (inclined towards the valley) or 'negative' (inclined away from the valley).

Figure 9.3 shows the different stages of a rock slide starting with tensile fractures on the top of the cliff and progressive macrofractures within the rock masses. According to this theory, final rupture would occur at the base of the slope.

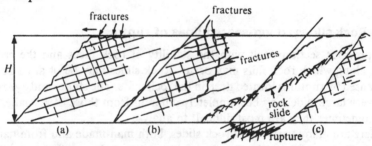

Fig. 9.3 Progressive deep fracture of the rock mass causing a deep rock slide. (*a*) Tensile stresses cause fissures at the top of the cliff; (*b*) progressive rock fractures in the deeper layers; (*c*) rupture at the basis of the slope causing rock slide (after Müller, 1963*a*).

Some recent tests on photoelastic models show that tensile stresses develop at the base of the slope – at depth inside the rock. If these findings are substantiated then it may be that tensile stresses activate the rupture. Two-dimensional models of a problematic rock bank at Kurobe IV dam (Japan) were constructed of highly elastic material with a gelatinous base. First the isotropic case was investigated: then a model featuring a system of joints was tested. The results provided valuable data within the complex of stress distribution (Müller & John, 1963).

Stability analysis of rock masses in doubtful equilibrium along an inclined plane follows the simple rules for statics. In fig. 9.4 the weight W of the rock

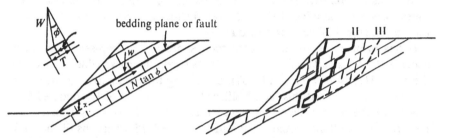

Fig. 9.4 Rock slide along an inclined bedding plane or fault ($\phi < \alpha$).

Fig. 9.5 Rock fall, possible lines of rupture: I or II.

mass above a possible plane sliding surface (bedding plane or rock fault) is decomposed in a triangle of forces. When the weight component, $T = W \sin \alpha$, is larger than the friction force, $N \tan \phi$, there is danger of a slide unless there is high rock cohesion along the sliding surface which compensates for the differences between positive and negative forces. However, the stress analysis discussed in section 7.3.2 usually applies. In fig. 9.5 rupture is supposed to occur within the rock mass, probably following the bedding planes and surfaces of fractures (lines I or II). Rupture would also happen along slip line III, fracturing the rock and cutting through the bedding planes (Vajont).

In a high mountainous area bisected by deep valleys, the rock underlying the valley slopes may be the seat of shearing stresses. If these stresses are already close to the shearing resistance of the rock, any increase in the valley depth, caused perhaps by accelerated erosion, would result in a general rock slide.

In fig. 9.5, the curved line III–III is supposed to develop progressively in the mass of fissured rock, but cases have also occurred where a curved smooth rock stratum, lying deep below the rock surface, offered an easy start to a rock slide. The Vajont rock slide followed such geological stratifications in the upper, steeply inclined, parts of the slope, but cut through other stratifications on the lower part.

Heim has described several major Alpine rock slides which could be classified under either the case illustrated in fig. 9.4 or in fig. 9.5.

Bernaix (1975) suggests that instability of a rock mass could be due to progressive shearing of a rock zone well below the surface. In a high mountainous area bisected by deep valleys, the rock underlying the valley slopes may be the seat of shearing stresses. If these stresses are already close to the shearing strength of the rock, any increase in the valley depth, caused perhaps by accelerated erosion, would result in a general rock slide.

Many problems concerning stability of rock slopes (open-cast mines) are three-dimensional problems. Such problems were systematically analysed by Londe *et al.* (1969), John (1970) and Hoek (1972).

After rupture of the rock masses, the slide progresses downwards. Heim made a distinction between slowly progressing slides and rapidly accelerating slides. Modern examples of slow and rapid slides are the Pontesi (slow) and the Vajont (rapid) slides (see section 9.6).

It has been observed that major rock slides along deep-seated surfaces occur with very small relative movements between the rocks which constitute the mass. Harrison & Falcon (1937), describing the Saidmarreh landslide in south-west Iran, noted this fact. Müller (1961) on p. 203 of his detailed first report on the Vajont slide, compared this rock slide to snow avalanches. He writes of the 'thixotropic behaviour of masses'. Photographs of the rock cliffs taken before and after the slide, when the rocks had shifted 300 to 400 m across the gorge and lifted 140 m upon the opposite side, show hardly any change in the stratifications.

Moving rock masses absorb energy. Mencl (1966b) analysing the Vajont rock slide, used models to simulate slides moving along sharp bends of 'chair-shaped' sliding surfaces. Some authors estimate that considerable energy losses occur locally at such bends and suggest that the equivalent friction loss at Vajont could have been as high as $\tan \phi' = 0.05$ ($\phi' \cong 3°$).

Model tests (Fumagalli & Camponuovo, 1975) and *in situ* measurements have contributed to our knowledge of statics and dynamics of steep rock slopes. Different types of ruptures were detected, analysed and even model-tested; it was found that the time factor is vital (see chapter 14 on Vajont). Extrapolation of precise measurements of displacements permits the accurate prediction of the final rupture and of the start of the slide (Jaeger, 1969a; Kennedy, 1970). The method has been used systematically for checks on the stability of the steep high slopes of open-cast mines.

9.2.2 Deep-seated sliding surfaces

It is now evident that some of the geomorphological features of the Alps, formerly ascribed to ice action, could be surface manifestations of deep-seated rock slides. At several dam sites in exceptionally narrow sections of deep valleys it was found that the rock forming one of the slopes had advanced towards the valley in a more or less horizontal direction, whereby the rock involved in the movement remained relatively intact (Ampferer, 1939; Stini,

1942). At one of these sites ground moraine was encountered in a boring at a depth of about 300 ft below the surface of what appeared to be rock *in situ* (Stini, 1952*b*). It is known that several very large slides with deep lines of rupture occurred in Switzerland in historical times.

Heim (1932) and Müller (1964) have produced lists of all the known major rock slides which have occurred throughout history. The most tragic was the Vajont rock slide (Italy) which happened during the night of 9 October 1963. A very high dam had been built across the narrow canyon of the Vajont River, with the dam crest at a level of 722·50 m. The fissured dolomite rock had to be reinforced on both sides of the gorge to withstand the thrust of the dam abutments. In 1960, rock masses on the left bank started to move very slowly. At the same time the water level rose in the larger reservoir. The slide reached as high as 1250 m on the slopes of Mount Toc. In 1963 it was observed that there was some connection between the different water levels in the reservoir and the rock movements. Geologists and rock experts predicted a progressive rock slide which after gradually filling the reservoir would come to a natural standstill. But tragically, they were wrong. The rock slide was unexpectedly violent and it reached extremely high velocities. The resulting water wave was so tremendous that it spilled over the dam crest and flooded the little town of Longarone (province of Venice) causing 2400 deaths. (See chapter 14 for a more detailed account of this disaster.)

It is worth while quoting some of Terzaghi's remarks:

Practically nothing is known concerning the mechanism of these deep-seated large-scale rock slides. It is not known whether the slides took place rapidly or slowly, and it is doubtful whether they are preceded by important deep deformation of the rocks located within the shear zone. However, it is known that the rock located above the surface of sliding has been damaged at least to a moderate extent. Existing joints have opened and new ones have been formed. Hence the compressibility and secondary permeability of the rocks has increased. Furthermore, in the immediate proximity of the surface of sliding, the rock is completely broken or crushed. Hence a site for a high concrete dam should not be considered suitable unless there is positive evidence that the underlying rock has never been subject to displacement by a deep-seated rock slide.

Terzaghi wrote this about a year before the Vajont–Longarone tragedy.

9.3   Effect of interstitial water on slope stability

9.3.1   Cleft water pressure

Terzaghi, in his theory of slopes, uses the term 'secondary permeability' for the permeability which results from the pressure of open and continuous cracks and fissures in the rock. It depends on the width and spacing of these passages. Primary permeability is found in the voids of the intact rock located between the fissures and is so minimal compared with secondary permeability that a tunnel driven through intact rock below the water-table appears to be dry.

If secondary permeability of the rock were uniform, the water-table would assume a more or less parabolic shape. But rock masses just behind the slope are frequently more fissured than those at a distance, making water percolation easier.

If water were immobile in the fissures, the water-table would be horizontal, the uplift forces on the rock would be vertical, and lateral forces balanced. When there is a flow of water a lateral pressure is exerted on the rock mass. This is called 'cleft' water pressure and it is proportionate to the difference in water level on both sides of the rock mass. It can be calculated by using the methods for calculating water pressure in soils. Additionally, the water percolating through fissures filled with clay-like material will reduce the friction forces along these fissures. Combined effects of this type may weaken the foot of some slopes and cause a rock slide.

Bjerum (Norway) submitted a detailed survey of local conditions to the Austrian Society of Rock Mechanics in Salzburg (1963), which proved a definite connection between climate and rock slides. The area under observation has severe winters with very heavy snowfalls. There is also a lot of rain during autumn. Most of the major slides take place in April. At that time the melting snow feeds large quantities of water into the rock joints, which are still plugged with ice. The resulting pressure build-up causes a rock fall or slide. There is a second peak-period in October–November.

The report also gives details of the rock slide which occurred at the south end of Loen lake (south-west coast of Norway) on 13 September, 1936. The slope is located on granite gneiss with poorly developed foliation and rises at an average angle of 50° to a height of about 3600 ft above lake level. The middle portion of the slope, at a height of about 1600 ft, is nearly vertical. The planes of foliation dip at an angle of about 65° towards the lake. The rock behind the vertical face is weakened, probably by sheeting joints situated parallel to the face, but the rock between the joints is practically impervious. During a heavy rainstorm the rock between the vertical face and one of the vertical joints, about 1·3 million yd³ in volume, dropped from the cliff and fell into the lake, causing a 230-ft-high wave which destroyed a village and killed many people.

Although Terzaghi believes that the effects of 'primary permeability' are negligible, there are cases where they are quite significant. Some rocks may absorb pore water quite easily. Phyllite for example, when soaked in water over a period of three days, may have its modulus of elasticity reduced by 50 or 70%. This can only be explained if water penetrates in all the pores of the rock. Similarly, microfissured rocks (gneiss) may absorb water, even if it is at a slow rate. Some samples have disintegrated completely within a year of being put in water.

9.3.2 Statics of immersed or partly immersed rock

It is accepted in soil mechanics that the pore water reduces the shear strength of the soaked soil. When u is the pore water pressure, Coulomb's law becomes:

$$\tau = c + (\sigma - u) \tan \phi.$$

This law is also valid for rock as seen in chapter 8. In the case when rock cohesion is negligible:

$$\tau = (\sigma - u) \tan \phi.$$

This very general remark can be used when discussing the stability of rock slopes immersed or partially immersed in water.

In fig. 9.6 the volume of rock, V_1, is immersed in water, V_2 is above water. The stability of the rock mass, $\gamma_r(V_1 + V_2)$ is considered where γ_r is the

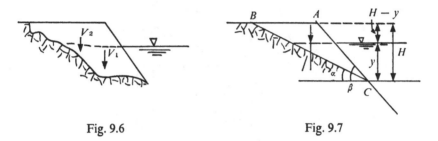

Fig. 9.6   Fig. 9.7

specific weight of the solid rock materials, voids excluded, and γ the specific weight of the water. Assuming n to be the volume of the pores then the vertical uplift force, U, on the rock mass is

$$U = \gamma V_1(1 - n),$$

and the weight is

$$W = \gamma_r(1 - n)(V_1 + V_2),$$

and the vertical resultant is

$$W - U = (\gamma_r - \gamma)(1 - n)V_1 + \gamma_r(1 - n)V_2.$$

There is no horizontal component from the water pressure. Figure 9.7 refers to conditions along the slopes of a lake or reservoir. There is a possibility for the rock mass BAC to slide along the plane sliding surface BC. The weight of the mass of rock BAC is

$$W = \tfrac{1}{2}\gamma_r(1 - n)H^2(\cot \alpha - \cot \beta) = \tfrac{1}{2}\gamma_z(1 - n)H^2 \frac{\sin(\beta - \alpha)}{\sin \beta \sin \alpha}.$$

The uplift is

$$U = \tfrac{1}{2}\gamma(1 - n)y^2 \frac{\sin(\beta - \alpha)}{\sin \alpha \sin \beta},$$

for $y < H$ where y is a variable water depth, and

$$U = \tfrac{1}{2}\gamma(1 - n)H^2 \frac{\sin(\beta - \alpha)}{\sin\alpha \sin\beta}, \quad \text{for} \quad y \geqslant H.$$

The force acting normal to the plane surface CB is $(W - U)\cos\alpha$ and the tangential force is $(W - U)\sin\alpha$.

The condition for rock stability is $\alpha < \phi$ for all cases from $y = 0$ to $y = H$ and $y > H$ and does not depend on the uplift U.

Conditions of stability along a circular surface are very similar. O is the centre of the arc of circle CB with radius r and $W - U$ the resultant of the weights and uplift forces. On a small element of circle Δs the normal and

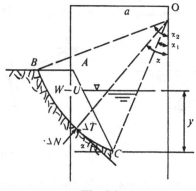

Fig. 9.8

tangential forces are ΔN and ΔT. The three equations for equilibrium of masses are (fig. 9.8) (Lotti & Pandolfi, 1966a):

$$\Sigma \cos\alpha\Delta N + \Sigma \sin\alpha\Delta T = W - U,$$

$$\Sigma \sin\alpha\Delta N - \Sigma \cos\alpha\Delta T = 0,$$

$$r \Sigma \Delta T = (W - U)a.$$

The condition for stability is:

$$\frac{\Delta T}{\Delta N} \leqslant \tan\phi$$

at any point of the circular surface. This solution is identical to the approach used in soil mechanics. There are others which are also acceptable and these can be worked out using the general theorems of the mechanics of the centre of gravity of rock masses.

Hydrostatic forces acting on the rock masses never have any horizontal component. When there is a flow of interstitial water, hydrodynamic forces

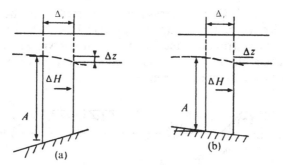

Fig. 9.9   Horizontal component of forces, ΔH, transferred to rock mass.

have a component of forces ΔH in the direction of the flow, (fig. 9.9) and
the force ΔH transferred to the rock is equal and opposed to the restraining
force ΔR on the flowing water, which is then:

$$\Delta H = -\Delta R = \Delta zA = \frac{\Delta z}{\Delta x} A \, \Delta x = SA \, \Delta x,$$

and

$$\sum_1^n \Delta H = \sum_1^n SA \, \Delta x,$$

on the whole length of the water-table to be considered, where S is the slope
of the water-table and $A = $ cross-section area. The slope of the sliding surface
does not matter and the force $\Delta H = -\Delta R = 0$ when $\Delta z = 0$ (hydrostatic
conditions, no flow).

Tests and measurements made *in situ* have proved that in most cases
seepage of water in fissured rock masses can be simulated by the flow through
porous media (sand and gravel), and a seepage factor can be calculated from
in situ measurements.

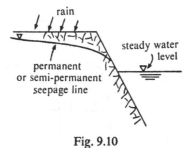

Fig. 9.10

Theories developed for ground water flow can be used for fissured rock
masses. For example, fig. 9.10 shows the ground water-table caused by water
percolating in the rock masses. The horizontal component of the forces

$\Sigma\Delta H$ is usually small. Greater forces may be exerted when the level of the water in storage reservoirs varies rapidly by $-\Delta H_1$ or $+\Delta H_2$ as shown in figs. 9.11a, b. This theory from soil mechanics is well known. Figure 9.11c is a method for calculating the forces transmitted by a flow of water to the rock using the approach developed in section 8.2.

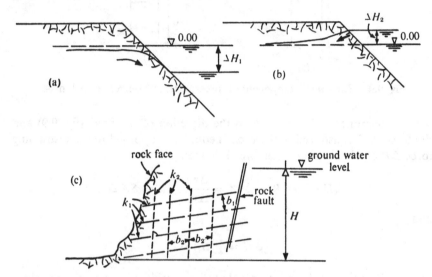

Fig. 9.11 k_1 and k_2 = perviousness of primary and secondary joint; b_1 and b_2 = distances between joint systems.

9.4   External forces which load a slope

Typical examples are the foundation, on an inclined slope, of the pillar of a bridge (Naef, 1969) or of a dam abutment (Krsmanović, 1967). Conditions as shown in fig. 9.12 are to be analysed. The Boussinesq theory on stress

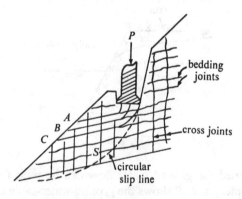

Fig. 9.12   Force, P, loading a slope. $A, B, C \ldots S$ = possible lines of rupture.

distribution under a concentrated load P perpendicular to a horizontal plane yields, for a Poisson ratio $\nu = 1/m = 0.5$, a simple radial stress distribution (fig. 9.13):

$$\sigma_r = \frac{3P}{2\pi r^2} \cos \delta.$$

The two other stresses σ_2 and σ_3 are zero.

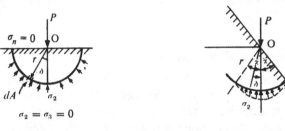

Fig. 9.13   Fig. 9.14

A simplified approach to the problem of a force P loading an inclined slope surface necessitates the assumption in this case of similar radial distribution of stresses (fig. 9.14):

$$\sigma_r = cP(1/r^2) \cos \frac{\pi}{2\alpha} \delta,$$

with the condition:

$$\int \sigma_r \cos \delta \, dA = P,$$

which yields the value of the constant c (α = angle of the slope with the vertical).

For a concentrated load P table 9.1 has been calculated for c (Naef, 1969).

Table 9.1				Table 9.2		
α	$\pi/2\alpha$	c		α	$\pi/2\alpha$	c^*
90°	1	0·48		90°	1	0·64
60°	1·5	0·80		60°	1·5	0·83
45°	2	1·28		45°	2	1·06
30°	3	2·65		30°	3	1·34

Similarly, for a line load p a formula:

$$\sigma_1 = \sigma_r = c^* p(1/r) \cos (\pi/2\alpha)\delta$$

has been developed, where c^* is given by table 9.2.

The stresses due to the concentrated or the line load are added to the stresses developing in the rock mass under its own weight. The safety conditions along different lines of possible rupture, like lines *A, B, C, . . . S* in fig. 9.12, are analysed.

In soil mechanics (loose soils) it is accepted that full safety should be reached at any point. In rock mechanics an average safety for the whole mass is usually acceptable, with a convenient factor of safety. More detailed calculations are possible, using the more recent computing methods (Zienckiewicz, 1968).

9.5   The dynamics of rock slides

The preceding two sections refer to the static conditions which prevail at the moment $t = 0 - \epsilon$ just before the rock slide starts. Heavy rainfall, the thawing of interstitial ice in rock fissures, the variation of the water level in a lake or reservoir may cause a disturbance in the stability of the rock masses. Newton's law expresses that along a plane surface (fig. 9.15):

$$M \frac{d^2 x}{dt^2} = W \cos \alpha - R,$$

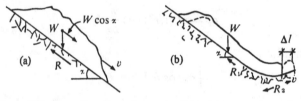

Fig. 9.15   Progressive acceleration of a rock slide: (*a*) on a plane sliding surface; (*b*) on a partly curved surface.

where *M* is the mass of rock starting to slide, with an acceleration d^2x/dt^2 caused by the positive force $W \cos \alpha$ but retarded by friction and resistance force *R*. The acceleration is positive as soon as $W \cos \alpha > R$. The velocity dx/dt which is finally reached after a time $T = \int dt$ is rarely very high, as the difference of forces $W \cos \phi - R$ at the time $t + \epsilon$ is forcibly small. When the sliding surface is curved at its bottom (chair-like surface) (fig. 9.15*b*) the positive weight component $W \cos \alpha$ on the incline decreases as the slide progresses, whereas the resistance forces *R* on the more or less horizontal end of the 'chair' remain the same. The slide comes to an early standstill. It is probable that with increasing velocity dx/dt the friction factor of rock on rock decreases but this is probably not substantial, unless the velocities are high – possibly several metres per second. Such a description of a 'classical slide' with slow start and reasonable acceleration, does not fit the conditions and characteristics of the Vajont rock slide. The sliding surface was typically 'chair-shaped'. *A posteriori* calculations were carried out by several people, assuming typical shapes for the most probable sliding surface. They found that, at the start of the slide, the tan ϕ and ϕ values must have been between tan $\phi = 0.316$ to 0.525 and $\phi = 17.5°$ to $27.7°$. (Usual assumed values for rock-on-rock friction $= 35°$ to $75°$.) Assuming the friction factor remained more or less constant during the slide the tremendous acceleration can hardly be explained. Another calculation based on the height of the waves which developed on the surface of the storage reservoir, concludes that the 'average value' of the friction angle must have been as low as $10°$. Assuming this value to be correct at the time $t = 0 + \epsilon$ it is not possible to explain why the rock masses were still stable at the time $t = 0 - \epsilon$. The only conclusion is that the rock slide of Vajont was not a 'continuous slide' of the classical type but a 'discontinuous slide'.

Discontinuous types of flow are known in fluid mechanics. A typical example is when a frozen river thaws and the sudden break up of ice generates discontinuous waves of ice and water downstream at dangerously high speeds (Jaeger, 1968).

From the velocities of rock displacements measured in boreholes at different points of the slow sliding rock masses before the final rupture at Vajont, it can be assumed that rupture of the rock along the upper reaches of the steep slopes occurred some time before the catastrophe. The ruptured upper rock masses were bearing down on a 'tooth' of solid rock at the bottom of the slope which was being progressively crushed until it suddenly failed by brittle fracture. The tremendous potential energy accumulated in the rock masses was suddenly released and the final rupture occurred like an explosion (Jaeger, 1969).

9.6 Classification of rock falls and rock slides

9.6.1 Rock falls

Rock falls concern rock masses falling free or nearly free from the rock face.

(a) The most common type occurs in rocky mountains when rock masses of any size, located high up steep mountain slopes, are loosened and fall down to the valley.

(b) Large rock falls have occurred along the steep cliffs bordering Norwegian fjords. In the autumn, cleft water fills vertical rock fissures; this freezes in winter and the ice causes the fissures to widen and break. In the spring, thawing of the ice causes large rock slabs to fall in the fjord water, resulting in waves high enough to have drowned small villages.

9.6.2 Rock slides

Referring to the comments in section 9.2, rock slides could be classified according to the shape of the area of rupture which can be:

(*a*) A plane bedding surface.

(*b*) A curved bedding surface.

(*c*) A dented line of rupture.

(*d*) The rupture can be a slow progressive shearing of rocks or a brittle fracture occurring suddenly over a large area, triggering the slide.

(*e*) A further possible classification of rock slides can be based on their size. There are minor slides (100 m^3 to a few thousand m^3) and major slides. The 1959 Pontesei slide (3 million m^3) and the first Vajont slide (700 000 m^3) are large. The final slide at Vajont in 1963 (250 million m^3) and several in prehistory, described by Heim (1932), like that of Glärnisch (Switzerland, 800 million m^3) and Flims (Switzerland, 1200 million m^3), are very large among major slides.

(*f*) The angle of the slope along which the slide occurs is another possible classification. Many occur on slopes of 10° to 12° up to 20° which is far smaller than the so-called angle of friction of rock on rock. This indicates that conditions along the sliding surface are very different from the simplified assumptions of rock-on-rock friction.

(*g*) Finally, slides could be classified as uniform continuous or non-uniform discontinuous (Jaeger, 1968*a*, *b*). Most geologists aptly describe them as 'continuous rock creep'. It is immaterial how fast the rock movement is during the last phase, provided acceleration remains constant or rises steadily. The total time usually taken is about several minutes.

The Pontesei rock slide, described by Walters (1962) which occurred in an area not far away from the Vajont gorge, involved about 3 million m^3 of loose material; rocks, gravel and sand, clay and rocks and was most probably 'continuous', lasting several minutes. So was the first slide just upstream from the Vajont dam. Because of this geologists and engineers were convinced that the larger slide they expected to occur further upstream in the Vajont gorge would also be continuous. On the contrary it was discontinuous, reaching an extremely high velocity after a rock (brittle?) fracture occurred.

It is worthwhile mentioning again the very detailed analysis by Heim of all the rock slides, some of them very spectacular, which occurred in the Swiss Alps before 1932.

9.6.3 Special problems

Many contributory reasons for rock falls and slides have already been mentioned, but in most cases the real cause is an imbalance of static forces, strains and stresses which cause local ruptures and progressively worsening cohesion of fractured rock masses. Rock sheeting and rock stress relaxation causing fissures parallel to the valley are special cases of strain conditions. Excavation, e.g. dam or bridge foundations, galleries and tunnels or rock cuts for roads and railways, may also create imbalance and therefore slides.

Slow rock creep on the surface can be linked to strained rock conditions deep in the mountain. In the case of the Malgovert tunnel (see section 2.2) the 14-km tunnel was excavated downstream of the high-arch dam of Tignes (France), and very high pressures developed about the gallery immediately after excavation which crushed timber and steel reinforcements. On one section, about 4 km long, tunnelling was slow work and sometimes very difficult. From the outside the mountain looked like a typical slow-creeping rock mass. Geologists estimated that this was only superficial and they did not foresee the great difficulties which were encountered deep within the mountain. But they should have realized from the surface signs that there was a possibility of rock strain along the tunnel trace. (Similar conditions have been observed at many places including the Andes.) In these cases it is absolutely essential to determine the actual depth of the sliding masses before excavation. The conditions created in pressure tunnels will be examined in the next chapter.

The Kandergrund tunnel was also mentioned in section 2.2. In this case rupture was caused by standing pressure waves (water hammer). It was probably the eleventh harmonic of the 4-km-long tunnel which progressed, but rupture of the concrete lining occurred at only one point where the rock was of mediocre quality. Water seeping from the fissured pressure tunnel caused a severe landslide and two people were killed.

9.7 Supervision of potential rock slides; stabilization of slides

There is no known method for stopping a major rock slide, but minor rock slides or rock falls can often be stabilized.

9.7.1 Supervision of potential rock slides

Preceding chapters have emphasized the need for scientific investigation and survey of large potential rock slides, to define their characteristics and estimate their actual danger. Potential rock slides can sometimes be detected by the surface appearance of the slope: some signs are trees growing at an angle, isolated rocks beginning to roll or slide, and the downstream edge of the slope becoming unstable

To survey a creeping rock mass requires precise measurements of the position of lines of fixed points on the surface and checking their displacement in horizontal and vertical directions. These measurements must be recorded versus time and then systematically analysed. Some slow rock slides in the Alps have been traced back seventy years with the help of early topographic maps of the area which have been compared to new triangulations. Others have been observed by direct measurements for over fifteen years.

The permissible errors should not be greater than 10% of the expected displacement to be measured. For example a displacement of 5 cm should be measured to an accuracy of about ± 5 mm (Kobold, 1968). The position of a point on the map can be determined by angular measurements within ± 5 mm over a distance of 1 km if the basis of the triangulation is known and the fix points of the basis of the triangles are not moving. The measurement of a length is more difficult. Electronic methods allow an accuracy of ± 11 mm over 1 km, the accuracy of a laser is far greater. Surface measurements must be supplemented with those from inside the mass of moving rocks (boreholes and galleries). The depth of the sliding surface should be estimated and the volume of the moving masses calculated on the basis of the measurements.

An interesting example of measurements made inside moving rock masses was at the semicircular railway tunnel of Klosters (Alps), excavated in slightly unstable rocks. Accurate geodesic measurements were made along the rails inside the tunnel and at its two portals in 1952, 1956 and 1966. The portals moved 15 to 20 cm in fourteen years. Inside tunnel displacements were lower, about 5 to 10 cm, maximum 12 cm, and 18 cm near the portals.

Any acceleration of the displacements must be interpreted as a possible warning of ruptures inside the mass. Differences in the velocity profiles may indicate discontinuous creep of rock masses and may possibly point to the formation of a discontinuous rock slide (Jaeger, 1968a, b; Müller, 1964).

During the Fifteenth Symposium on Rock Mechanics at Salzburg (September 1964) displacements varying from 0·25 cm to 50 cm a day were described by several experts.

Seismic tests may detect rock fissuration or fracturation at different depths in the creeping masses. At Vajont a sharp drop in wave velocities was observed about two years before the rock burst. Other experts measure the noise level, which increases before a rock fall occurs and decreases as stabilization progresses. Measurements of the water-table and its variations and correlating this with rock displacements are an easy method to check what changes are occurring deeper inside the rock.

9.7.2 Stabilization of potential rock slides

(a) *Minor rock slides.* Potential minor rock slides can be stabilized with a concrete retaining wall or buttress, with anchor bolts or stressed cables anchored in sound rock or with a combination of both. But proper drainage of the rock is always a first requirement, and sometimes cleaning and concreting fractures may be useful. The design of any reinforcement must be based on a detailed study of the rock masses; their fissures and fractures and potential sliding surfaces. The final choice of the retaining system is then a matter of statics of forces and weights.

Proposals for concreting a rock fracture are shown in fig. 9.16 – there are many possible alternative methods. The statics of a retaining wall are shown in fig. 9.17. The weight W of the rock mass has a component $W \sin \alpha$ in the direction of the potential slip line. The friction force, $F = W \cos \alpha \times \tan \phi$,

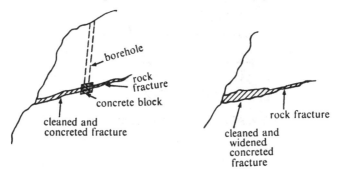

Fig. 9.16 Cleaning and concreting of rock fractures (after Müller, 1963*a*).

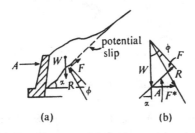

Fig. 9.17 Retaining wall (*a*) and polygon of forces (*b*). *W*, weight of rock mass; *A*, reaction from the wall; *R*, total reaction force on slip line; *F*, friction component due to *W* only; ϕ, angle of friction.

would be unable to withstand this component without a reaction force *A* from the retaining wall, which has a component $F^* = A \cos \alpha$ in the direction of the slip. The reaction *A* from the wall becomes an active force only when there is an incipient sliding movement of the rock mass. Post-tensioned cables are immediately active. *P* is assumed to be the force in the cable after being jacked up (fig. 9.18). Possible combinations of retaining walls and

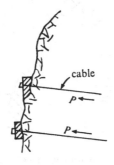

Fig. 9.18 Steep slope reinforced with anchored cables.

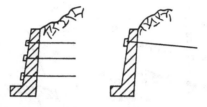

Fig. 9.19   Anchored retaining walls.

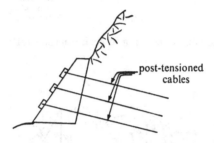

Fig. 9.20   Anchored buttress.

buttresses with anchored cables are shown in figs. 9.19 and 9.20. An interesting reinforced structure with cable anchorage is shown in fig. 9.21.

Proper drainage of the rock mass is always vital.

Force Z applied to an anchored cable. The cable is prestressed by applying a force F which causes the steel or reinforced concrete plate to compress the

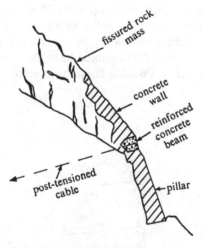

Fig. 9.21   Reinforced concrete structure with cable anchorage for protection of the Passan–Wegscheld road (after Müller, 1963a).

rock surface (fig. 9.22), q_0 is the unit load, s the side of the plate, $F = q_0 s^2$ the force, L the cable length and A its cross-section, the surface displacement being $\delta = aF = q_0 s/E_r$. A force Z is applied to the anchorage; whereby

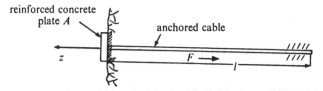

Fig. 9.22   Tensile force, z, on an anchored cable.

a force Z_2 is transmitted to the cable, causing a cable lengthening $\Delta_2 = bZ_2 = Z_2 L/EA$. A force $Z - Z_2 = Z_1$ causes a relief in the rock compression by the plate; the surface displacement decreases by $\Delta_1 = aZ_1 = \Delta qs/E_r$. We must have

$$Z = Z_1 + Z_2,$$

$$\Delta_1 = \Delta_2 = aZ_1 = bZ_2; \qquad Z_1 = \frac{b}{a}Z_2 \text{ and } Z_2 = kZ.$$

The force in the cable is now

$$F + Z_2 = F + \frac{a}{b}Z_1 = F + \frac{a}{b}(Z - Z_2).$$

When Z_1 reaches a value R the plate gets loose from the rock and the whole force Z is transmitted to the cable. A short calculation yields $k = 1/(1 + E_r sL/EA)$ and $R = F/(1 - k)$.

(b) *Stabilization of large rock masses.* Chapter 15 is entirely devoted to the analysis of the stabilization of the Baji–Krachen rock spur and the 300 m high rock abutment at the Tachien dam.

10 Galleries, tunnels, mines and underground excavations

10.1 Introduction

Tunnelling engineering techniques are very closely related to rock mechanics and mining. It is not possible to summarize here the vast amount of information now available on this subject, therefore only specific aspects will be discussed.

First to be mentioned among tunnels is the Thames tunnel built between 1826 and 1843 by Brunel. The excavation of the 12 849-m-long Mont Cenis tunnel between France and Italy, in 1870–1, was one of the major undertakings at the end of the last century. It was rapidly followed by others in America (Hoosac tunnel, 1873, Canadian tunnels on the Canadian Pacific Line, 1875) and in England with the Severn tunnel in 1873 and the Blackwall tunnel in 1891. The first very large Alpine tunnel, the 14·98-km-long St. Gotthard tunnel was opened in 1882, followed shortly by the Arlberg tunnel (10 240 m) in 1884. The longest are the two Simplon tunnels both of 19 803 m, built in 1906 and 1922. The twentieth-century underground railway systems and hydro-power developments gave fresh impetus to tunnelling techniques. More recently, large road and power tunnels have been built through mountain ridges. The Gotthard Road tunnel (Switzerland), now under construction, will be 16·32 km long and the Orange–Fish power and irrigation tunnel in South Africa, 82·45 km.

Considerable progress has been made during the past twenty years in drilling techniques, the quick removal of soil and in lining methods. Roof bolting has been introduced on a large scale.

In this chapter, only a limited number of theoretical and practical problems in relation to rock mechanics will be examined. Some additional information can be found in sections 2.2 on engineering geology, 5·4 and 5·5 on residual stresses and 6.2 on *in situ* tests in galleries.

10.2 Additional information on stresses around cavities

10.2.1 Special problems for mining and tunnel engineers

A classical problem of mining engineering is how to estimate the stability of a self-supporting underground structure in which the stresses are carried on the walls, pillars or other unexcavated parts of the openings, rather than on linings or steel supports. Whereas hydro-power and railway engineers tend towards a more or less circular shape for tunnels and galleries, mining engineers frequently have to deal with rectangular openings, with rounded corners, or multiple parallel excavations.

Assuming the square opening to be in a vertical field of stresses p, the sides of the square being vertical and horizontal, the maximum tangential stress σ_t, will depend on the ratio r/B, where r = radius of the rounded corner

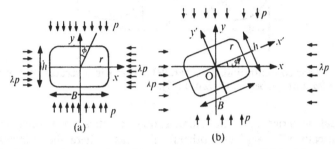

Fig. 10.1   Rectangular opening with rounded corners in a field of forces p and λp. (*a*) Field of forces parallel to sides of rectangle; (*b*) field of forces inclined to sides of rectangle.

and B = width of the square (fig. 10.1). The maximum stress σ_t occurs for an angle ϕ^* (ϕ angle with the vertical). Then σ_t maximum is given by table 10.1

Table 10.1   *Values of σ_t for $h = B$ (square) and a vertical stress field = p*

$r/B =$	$\frac{1}{2}$	$\frac{1}{4}$	$\frac{1}{8}$	$\frac{1}{36}$
$\phi = 0$	3	1·8	1·78	1·68
$\phi = \phi^*$	—	2·78	3·35	4·6
$\phi = 90°$	−1	−0·8	−0·8	−0·8

The case $r/B = \frac{1}{2}$ obviously corresponds to the circle $r/D = \frac{1}{2}$.

For rectangular openings the maximum relative stress concentration σ_t/p for $r/B = 0\cdot10$ rises to about 5 for $h/B = \frac{1}{4}$, where h = height of rectangle, width B.

Assuming two parallel cylindrical openings (fig. 10.2) such that the distance B between the holes is equal to the diameter D of the circles (the distance between the centres of the circles is therefore $2D$). It can be shown that the maximum circumferential stress σ_t rises from $\sigma_t = 3p$ (one opening) to $\sigma_t = 3\cdot4p$. The vertical stress in the middle of the distance B is then about $1\cdot7p$ (p = vertical uniform stress field in the rock). Such a result could be expected to be due to the rapid decrease of the circumferential stresses σ_t with the distance from the circle (see section 5.4).

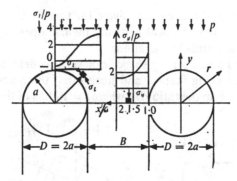

Fig. 10.2 Stress concentration. Two parallel galleries in a vertical field of forces p ($B = D = 2a$) (after Obert & Duvall, 1967).

Elliptical galleries and three-dimensional cavities have been analysed mathematically (Terzaghi and others). The case where the principal axes of the cavity are inclined towards the field of stresses has been solved equally (fig. 10.1).

The use of photoelastic methods yields rapid results which can be compared with field measurements.

Direct calculations of underground cavern systems (like machine-hall and transformer-hall excavations, downstream surge tanks) using the finite element method yield results differing from the simple analytical approach (Benson, Kierans & Sigvaldson, 1970). This divergence of the results can be explained by the special shape of the excavations. For Ruacana on the Cunene River (South West Africa), the upper part of the vertical downstream surge tank had to be reshaped completely to reduce tensile stresses obtained when calculating the case $k = 0.40$. Power-house excavations or downstream surge tanks may have a height:width ratio > 1; in the case of so-called 'rock pillars' separating caverns calculations disclosed tensile stresses in a direction perpendicular to the high walls (for $k < 1$). For the analysis of such systems of caverns, the use of the finite element method analysis is essential.

10.2.2 Subsidence and caving

The first manifestation of subsidence may be convergence of the walls and roof of the gallery or a succession of local failures in the rock surrounding the openings. This phase of the process is termed *sub-surface subsidence*, as opposed to *surface subsidence*, which causes a depression in the overlying surface (fig. 10.3a). Subsurface subsidence is largely an uncontrolled process. *Caving* is a form of subsurface subsidence which is at least partly controlled by the mining method.

When excavation occurs in relatively thin-bedded deposits with overlying weak sedimentary rocks, surface displacements occur. Any point at the

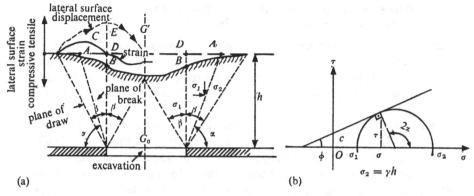

Fig. 10.3 (*a*) Idealized representation of trough subsidence (after Rellensman 1957). (*b*) Circle of Mohr for subsidence conditions.

surface has a vertical downward displacement and a horizontal displacement versus the centre G_0 of the excavation. The vertical surface displacement is a maximum at G' over the G_0 of the excavation. The horizontal surface strain is tensile at points outside the limits of the excavation and compressive within the limits (fig. 10.3*a*). D is a point of zero strain and there is an inflection point B at the surface of the soil. α is the angle of break and point A corresponds to the maximum of tensile strain at the surface. β is the angle of draw.

Table 10.2

	Angle of friction ϕ (measured)	Angle of break (calculated)
Clay	15–20°	52·5–55°
Sand	35–45	62·5–67·5
Moderate shale	37	63·5
Hard shale	45	67·5
Sandstone	50–70	70–80
Coal	45	67·5

In fig. 10.3*b* the angle of break α determines the line along which rupture occurs. τ is the shear stress along this line and $\tau = c + \sigma \tan \phi$. Tracing the circle of Mohr for such conditions yields:

$$2\alpha = 90° + \phi.$$

Seldenrath (1951) quoted by Obert & Duvall (1967) published table 10.2.

10.3   Stresses around tunnels and galleries caused by hydrostatic pressure inside the conduit

Stresses around cavities caused by *in situ* or residual stresses in the rock have already been dealt with (sections 5.4 and 10.2). The stresses to be analysed in this chapter are those induced in the rock masses by the hydrostatic (or hydrodynamic) pressure p of the water or fluid (gas) filling the tunnel gallery or cavity.

Analysis of these stresses usually starts with the theory of stresses in thick pipes, which is then extended to the case of the unlined tunnel and finally to the case of the lined tunnel (Jaeger, 1933, 1949b). The theory of stress–strain measurements in circular tunnels under a radial load (Seeber, 1961) applied along the whole circumference of a tunnel (Austrian technique on rock testing) can easily be related to the equations obtained for the pressurized tunnel. Similarly, these theories can be used to establish safe overburden conditions above a tunnel under a pressure p (Jaeger, 1961b). Finally, the basic equations to be developed in this chapter are used in the theory of water hammer waves which develop in pressure tunnels and shafts.

10.3.1   Theory of thick elastic pipes

The internal and external radius of the pipe is respectively b and c (fig. 10.4).

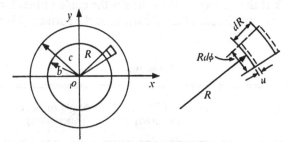

Fig. 10.4   Deformation of a thick pipe.

R is the radius of a small element $R\,\mathrm{d}\phi\,\mathrm{d}R$ inside the pipe, with $b < R < c$. When the pipe is filled with a liquid or gas under pressure, the radius R increases by u and the circumference increases from $2\pi R$ to $2\pi(R + u)$. The specific length increase in a circumferential direction is

$$\delta_t = \frac{2\pi(R + u) - 2\pi R}{2\pi R} = \frac{u}{R}.$$

In a radial direction R increases by u and becomes $R + u$; similarly, $\mathrm{d}R$ increases by $\mathrm{d}u$ and becomes $\mathrm{d}R + \mathrm{d}u = \mathrm{d}R[1 + (\mathrm{d}u/\mathrm{d}R)]$. The specific radial increase in length is

$$\delta_r = \frac{\mathrm{d}R + \mathrm{d}u - \mathrm{d}R}{\mathrm{d}R} = \frac{\mathrm{d}u}{\mathrm{d}R}.$$

On the other hand δ_t and δ_r depend on σ_t and σ_r as follows ($m = 1/\nu$):

$$\delta_t = \frac{u}{R} = \frac{1}{E}\left(\sigma_t - \frac{1}{m}\sigma_r\right),$$

$$\delta_r = \frac{du}{dR} = \frac{1}{E}\left(\sigma_r - \frac{1}{m}\sigma_t\right);$$

or

$$\sigma_t = \frac{mE}{m^2 - 1}\left(m\frac{u}{R} + \frac{du}{dr}\right), \tag{1}$$

$$\sigma_r = \frac{mE}{m^2 - 1}\left(m\frac{du}{dR} + \frac{u}{R}\right), \tag{2}$$

E being the modulus of elasticity. It can be shown that u is a solution of the equation:

$$R^2\frac{d^2u}{dR^2} + R\frac{du}{dR} - u = 0, \tag{3}$$

the integral of which is

$$u = BR + \frac{C}{R}. \tag{4}$$

A rapid checking yields:

$$\frac{du}{dr} = B - \frac{C}{R^2} \quad \text{and} \quad \frac{d^2u}{dR^2} = \frac{2C}{R^3}, \tag{5}$$

which, when introduced in equation (3) give $0 = 0$. As u is known the stresses σ_t and σ_r are now:

$$\sigma_r = \frac{mE}{m - 1}B - \frac{mEC}{(m + 1)R^2} = B' - \frac{C'}{R^2}, \tag{6}$$

$$\sigma_t = \frac{mE}{m - 1}B + \frac{mEC}{(m + 1)R^2} = B' + \frac{C'}{R^2}. \tag{7}$$

The constants B and C depend on the boundary condition for $R = b$ and $R = c$.

A thick pipe is loaded by internal static pressue p_i and external pressure p_e (fig. 10.5).

For $R = b$:

$$\sigma_r = B' - \frac{C'}{b^2} = p_i.$$

For $R = c$:

$$\sigma_r = B' - \frac{C'}{c^2} = p_e.$$

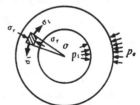

Fig. 10.5 Thick pressure pipe.

This yields:

$$C' = \frac{(p_e - p_i)}{c^2 - b^2} c^2 b^2;$$

$$B' = \frac{p_e c^2 - p_i b^2}{c^2 - b^2}.$$

Therefore, for $R = b$:

$$\sigma_r = \frac{p_e c^2 - p_i b^2}{c^2 - b^2} - \frac{(p_e - p_i)}{c^2 - b^2} c^2 = p_i,$$

$$\sigma_t = \frac{p_e c^2 - p_i b^2}{c^2 - b^2} + \frac{(p_e - p_i)}{c^2 - b^2} c^2 = -p_i \frac{b^2 + c^2}{c^2 - b^2} + \frac{2c^2 p_e}{c^2 - b^2}.$$

For $R = c$:

$$\sigma_r = \frac{p_e c^2 - p_i b^2}{c^2 - b^2} - \frac{(p_e - p_i)}{c^2 - b^2} b^2 = p_e$$

$$\sigma_t = \frac{p_e c^2 - p_i b^2}{c^2 - b^2} + \frac{(p_e - p_i)}{c^2 - b^2} b^2 = -p_i \frac{2b^2}{c^2 - b^2} + p_e \frac{c^2 + b^2}{c^2 - b^2}.$$

10.3.2 Case of a pressure tunnel in sound rock (fig. 10.6)

The boundary conditions on a wall of a tunnel in sound rock are:

$$\text{for } R = b, \qquad p_i = p.$$

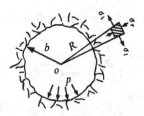

Fig. 10.6 Pressure tunnel in sound rock.

The other boundary conditions are:

$$\text{for } R = \infty, \qquad \sigma_r = p_e = 0.$$

This second condition yields for $R = \infty$:

$$\sigma_r = B' = 0, \qquad B = 0 \quad \text{and} \quad \sigma_t = 0$$

and for any value $0 < R < \infty$:

$$\sigma_r = -\frac{mEC}{(m+1)R^2} = -\sigma_t.$$

For $R = b$:

$$\sigma_r = -\frac{mEC}{(m+1)b^2} = p \quad \text{and} \quad C = -\frac{pb^2(m+1)}{Em},$$

$$\sigma_t = +\frac{mEC}{(m+1)b^2} = -p.$$

At any point inside the rock:

$$\sigma_r = -\sigma_t = -\frac{mEC}{(m+1)R^2} = p\frac{b^2}{R^2}.$$

The stresses σ_r and σ_t decrease rapidly inside the rock. At a distance $R = 2b$, they are only 25% of what they are on the tunnel wall.

10.3.3   Concrete-lined pressure tunnel

(1) *Unfissured sound rock* (fig. 10.7). The internal radius of the tunnel lining is b, its external radius is c. If p is the hydrostatic pressure inside the

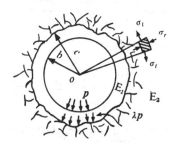

Fig. 10.7   Concrete-lined pressure tunnel in sound rock.

tunnel a certain load, $p_c = \lambda p$, is transmitted from the concrete lining to the rock. Furthermore, it is assumed that there is no gap between concrete and rock. For $c < R < \infty$, the stress σ_r is given by:

$$\sigma_r = \frac{m_2 E_2}{m_2 - 1} B_2 - \frac{m_2 E_2}{m_2 + 1}\frac{C_2}{R^2},$$

the subscript '2' for m_2, E_2, B_2, C_2 referring to rock. The condition $\sigma_r = 0$ for $R = \infty$ yields $B_2 = 0$ and $\sigma_r = -\sigma_t$ for the whole mass of rock. For $R = c$ on the rock side;

$$\sigma_r = -\frac{m_2 E_2}{m_2 + 1}\frac{C_2}{c^2} = p_c = \lambda p,$$

and

$$C_2 = -\frac{p_c c^2(m_2 + 1)}{E_2 m_2}, \quad \text{with } B_2 = 0.$$

The radial displacement u for $R = c$ in the rock is

$$u_{R=c} = B_2 R + \frac{C_2}{R} = \frac{p_c c}{E_2} \frac{m_2 + 1}{m_2}.$$

In the concrete when $b < R < c$, we introduce subscript '1' to represent the concrete.

For $R = b$:

$$\sigma_{R=b} = \frac{m_1 E_1}{m_1 - 1} B_1 - \frac{m_1 E_1}{m_1 + 1} \frac{C_1}{b^2} = p.$$

For $R = c$:

$$\sigma_{R=c} = \frac{m_1 E_1}{m_1 - 1} B_1 - \frac{m_1 E_1}{m_1 + 1} \frac{C_1}{c^2} = \lambda p,$$

and

$$C_1 = - \frac{m_1 + 1}{m_1 E_1} \frac{c^2 b^2}{c^2 - b^2} (1 - \lambda) p,$$

$$B_1 = - \frac{m_1 - 1}{m_1 E_1} \frac{(b^2 - \lambda c^2)}{c^2 - b^2} p.$$

Equating the elastic displacements
for $R = c$:

$$u_{R=c} = B_1 c + \frac{C_1}{c} = B_2 c + \frac{C_2}{c}$$

$$\frac{m_1 - 1}{m_1 E_1} \frac{(b^2 - \lambda c^2)}{c^2 - b^2} cp + \frac{m_1 + 1}{m_1 E_1} \frac{cb^2}{c^2 - b^2} (1 - \lambda) p = \frac{m_2 + 1}{m_2} \frac{c}{E_2} \lambda p.$$

Out of this equation we get:

$$\lambda = \frac{p_c}{p} = \frac{\dfrac{2b^2}{E_1(c^2 - b^2)}}{\dfrac{m_2 + 1}{m_2 E_2} + \dfrac{(m_1 - 1)c^2 + (m_1 + 1)b^2}{m_1 E_1(c^2 - b^2)}}.$$

The stresses in the concrete lining are:
for $R = b$:

$$\sigma_{rb} = p,$$

$$\sigma_{tb} = - \frac{c^2 + b^2 - 2\lambda c^2}{c^2 - b^2} p;$$

for $R = c$:

$$\sigma_{rc} = \lambda p,$$

$$\sigma_{tc} = - \frac{2b^2 - \lambda(c^2 + b^2)}{c^2 - b^2} p.$$

(2) *The concrete lining is fissured.* If the concrete lining the tunnel walls were uniformly fissured in a radial direction, a pressure:

$$p_c = \frac{b}{c} p,$$

would be directly transmitted to the rock and the stresses on the rock surface would be

$$\sigma_r = -\sigma_t = \frac{b}{c} p.$$

(With water penetrating the radial cracks the pressure transmitted could be as high as $p_c = p$.)

(3) *The fissures penetrate in the rock to depth d* (fig. 10.8). Along the rock surface the radial pressure is

$$p_c = (b/c)p \quad \text{and} \quad \sigma_t = 0; \qquad \sigma_{rc} = p_c = (b/c)p.$$

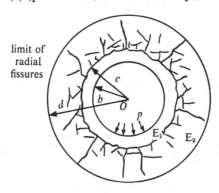

limit of
radial
fissures

Fig. 10.8   Lined tunnel in radially fissured rock ($R \leqslant d$).

At any depth $R < d$ inside the fissured rock mass $\sigma_t = 0$; $\sigma_r = (b/R)p$. At the limit of the sound rock the pressure is $p_d = (b/d)p$. Inside sound rock ($d < R < \infty$):

$$\sigma_r = -\sigma_t = p \frac{b}{d} \frac{d^2}{R^2} = \frac{bd}{R^2} p.$$

10.3.4   Steel-lined pressure tunnels and shafts

(1) *In sound rock* (fig. 10.9).   The hydrostatic pressure inside the tunnel is p. A pressure $p_b < p$ is transmitted from the steel shell to the concrete and a pressure $p_c < p_b$ from the concrete to the rock.

The elastic deformation u_b of the steel lining is

$$u_b = \frac{(p - p_b)}{E} \times \frac{b^2}{e} = p(1 - \lambda_1) \frac{b^2}{Ee}$$

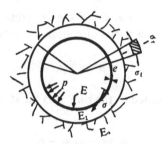

Fig. 10.9   Steel-lined tunnel.

where E = modulus of elasticity of the steel plate, e = thickness of the steel plates and $p_b = \lambda_1 p$. The deformation u_b of the steel shell must be equal to the elastic deformation of the internal face of the concrete lining and the deformation of the external face of the concrete must equal the yielding of the rock surface.

Detailed calculations (Jaeger, 1933) show that:

$$\lambda_1 = \frac{p_b}{p}$$

$$= \frac{b^2/Ee}{(b^2/Ee) + [b/m_1 E_1(c^2 - b^2)][(m_1 - 1)(b^2 - \lambda_2^* c^2) \atop + (m_1 + 1)(1 - \lambda_2^*)c^2]}$$

and

$$\lambda_2^* = \frac{p_c}{p_b} = \frac{2b^2/E_1(c^2 - b^2)}{(m_2 + 1)/m_2 E_2 + [(m_1 - 1)c^2 + (m_1 + 1)b^2]/m_1 E_1(c^2 - b^2)}.$$

(2) *Rock fissured radially* (fig. 10.8).   The deformation u_b of the steel plate is now:

$$u_b = \frac{p - p_b}{E}\frac{b^2}{e} = p(1 - \lambda_3)\frac{b^2}{Ee} \quad \text{with } p_b = \lambda_3 p$$

and of the concrete lining and rock (d = length of radial rock fissures):

$$u_b = -\frac{(p_b + p_d)}{2}\frac{(b - d)}{E_1} + p_d\frac{d(m_2 + 1)}{m_2 E_2}.$$

Equating the two values of u_b yields:

$$\lambda_3 = \frac{b^2/Ee}{(b^2/Ee) + [(d^2 - b^2)/2dE_1] + [(m_2 + 1)b/m_2 E_2]}.$$

10.3.5   Boundary conditions

Implicitly the theory developed in 10.3.3 and 10.3.4 assumes perfect adherence of concrete to rock, or steel lining to concrete. Equating rock radial deformations to concrete radial deformations is then possible. In most practical cases, rock has already deformed elastically or plastically before the concrete support is able to bear some radial load. Shrinkage of the concrete has also to be considered. Some designers cause special boundary conditions to occur at the contact of concrete and rock when grouting the contact area, with the purpose of creating a radial pressure on the concrete lining to balance stresses due to inside hydraulic pressure in the gallery.

Lombardi (1974) claims that the correct assessment of boundary conditions is more important than sophisticated calculating methods using the finite element method techniques. His approach will be dealt with in section 10.10.

10.3.6 Steel liner buckling

Natural cleft water pressure or pressure from water infiltrating from the upper reaches of a shaft (surge tank area) can cause heavy outside pressure on the steel liner, pushing it inwards. When the pressure shaft is full of pressurized water it can be assumed that the inside water pressure balances the outside cleft water pressure. When the steel-lined shaft is emptied, inward buckling of the liner may occur. Similar conditions may occur during grouting of the rock behind the lining.

These problems have been considered by Jaeger (1955a, and discussion of the paper). Amstutz (1950, 1953) has dealt extensively with the theory of the stability of steel liner against buckling.

10.4 Minimum overburden above a pressure tunnel

10.4.1 General remarks: basic design assumptions

The problems in determining minimum overburden required over pressure tunnels are of great importance to hydro-power engineering practice. The amount of elastic pressure waves (water hammer waves) allowed to penetrate in the tunnel and the design of the surge tank depend on the rock conditions about the tunnel (Jaeger, 1948a, 1955a, b).

Assuming a pressure tunnel with diameter $2R$ located at a depth H under the horizontal rock surface (fig. 10.10) the hydrodynamic pressure in the tunnel is $p/\gamma = \lambda H$, where H and p/γ are in feet or in metres and γ = specific density of the water. The highest value of λ for safety against uplift has to be determined. A former 'rule of thumb' method assumed for the highest acceptable pressure p the condition: $p/\gamma \leqslant \frac{1}{2}H$ or $\lambda \leqslant \frac{1}{2}$, but some pressure tunnels have been safely designed for $\lambda = 1$ or even 2. This pressure rule is based on the very crude assumption that the weight of a slice of rock with a width $B = 2R$, a height H and a length $= l$, should be equal to or larger than, the vertical uplift on the same area $2R \times l$, an additional factor of safety equal to 5 being included (rock density, $\gamma_R = 2 \cdot 5$).

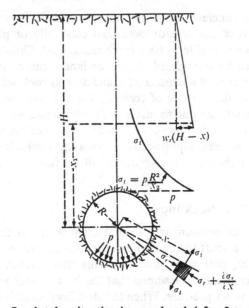

Fig. 10.10   Overburden situation in sound rock (after Jaeger, 1961*b*).

Other criteria sometimes adopted by tunnel designers are listed as follows.

(1) *Old rule.* $p/\gamma = \frac{1}{2}H$, very high safety factor against lifting of the rock.

(2) *Some American unlined pressure tunnels* (Haas tunnel; Nantahala tunnel): $p/\gamma = H$, it is implicitly assumed that water seeping from the unlined pressure tunnel will not reach the rock surface through rock fissures.

(3) *Sydney water supply tunnel.* $p/\gamma = 2\cdot4H$, the depth of cover is such that the weight of a column of rock over the tunnel should be equal to the maximum water pressure. The same rule was adopted by Sir William Halcrow when he designed the Glen Moriston tunnel (Livishie hydroelectric development), believed to be the highest head for a non-steel-lined tunnel in Europe; and by the designers of the Ashford Common Water Supply tunnel, through London Clay.

(4) *Terzaghi* suggests that the depth of cover should be half the water-head $p/\gamma = 2H$, a condition very similar to the previous rule.

(5) *Spray hydroelectric scheme* (Calgary, Canada). Especially notable is the high-pressure tunnel in this scheme. The rock overburden is $H = 216$ ft and the water pressure $p/\gamma = 1220$ ft, about $p/\gamma = 5H$.

10.4.2 Some common failures of pressure tunnels

Precise information on pressure tunnel failure is limited, but a few case histories have been published.

(1) *The Ritom hydro-power pressure tunnel* of the Swiss Federal Railway, built in 1920, started leaking shortly after being commissioned. The geological strata were inclined towards the valley to which the power tunnel was parallel. The tunnel designers, aware of the dangers of a possible rock slide caused by leakage, emptied and repaired the tunnel.

The tunnel was horseshoe-shaped and thin fissures were mainly concentrated at its rounded corners. Pressure tests (section 6.2) carried out in the Amsteg power tunnel, located on the north of the Alps, proved the rock to be elastic. It was assumed that because of its horseshoe shape the Ritom tunnel was unfavourable as a pressure tunnel excavated in indifferent rock and that the concrete lining, not reinforced, was too stiff for a tunnel adjacent to a rock slope. The tunnel was repaired without major damage being caused to the rock strata.

(2) *A rupture of the Kandergrund tunnel* (fig. 10.11) caused a rock slide. A forest was destroyed, a farmhouse crushed and two people killed. Three

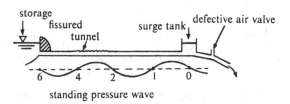

Fig. 10.11   Rupture of the Kandergrund pressure tunnel caused by water hammer waves (after Jaeger, 1948*a*).

2·5-cm-thick fissures about 30 m long, along the tunnel crown and the springs of the tunnel soffit, had caused water to leak. Leakage was probably far more substantial than in the previous case (no figures were available). The cause of the disaster was traced back to a water hammer pressure wave which penetrated the tunnel and lifted the lined soffit in a region where rock was definitely weak (Jaeger, 1948*a*; section 2.2).

Several similar rock slides caused by leaking tunnels occurred in the Andes, but detailed information is not available.

(3) *The diversion gallery of the El Frayle dam* (Peru) (fig. 10.12) was ruptured when the rock abutment gave way. Seepage of water through the rock was mentioned as the cause of the damage (section 13.2).

All these cases are characterized by the proximity of the pressure gallery to a rock slope.

(4) *Sydney Water Supply tunnel* (New South Wales). The damage caused to this was quite different. This concrete-lined tunnel was designed so that the maximum water pressure of about 500 ft (150 m) would not exceed the

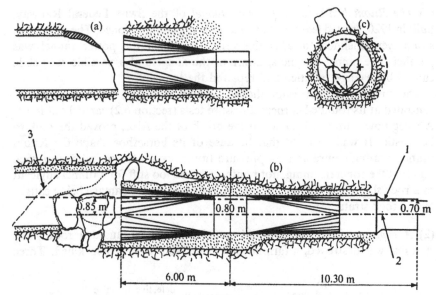

Fig. 10.12 The Frayle discharge gallery. Shearing of the diversion gallery. (*a*) Longitudinal section; (*b*) plan view; (*c*) cross section. (1) Axis of gallery before rupture; (2) axis of gallery after rupture; (3) fracture (after Mary, 1968).

weight of the overlay, in sandstone rock. Grouting was carried out in two stages to 300 lb/in² (24 kg/cm²). On completion the tunnel was divided into watertight sections and each section tested by progressively increasing the water pressure. In one section the lining ruptured when the pressure reached approximately the design value; in another failure occurred at about a quarter of the designed pressure. The ruptures took the form of a horizontal crack about half way up the tunnel walls. On account of the high permeability of the sandstone, water losses were high. It is probable that failure was due to the relatively high compressibility of the sandstone; cement grouting rarely stiffens sandstone rocks with a very low modulus of elasticity.

(5) *The Bryant Park tunnel* (New York) failed, it seems, because there was a connection from it to a horizontal fissured zone near the surface and the ground above this was lifted.

10.4.3 Types of rocks and galleries; drainage

Concrete dams are built on competent or acceptable rock whenever possible. On indifferent rock, rock-fill dams would be the preferred alternative. Tunnels may have to cross different types of rock, some of which are of poorer quality. Difficult problems can seldom be avoided along the whole trace of a long power tunnel.

A tunnel excavated in competent rock, whether lined or unlined, offers few hazards. A simple theory, which will be developed in this chapter, solves the problem of the minimum depth of overburden for a tunnel excavated below a horizontal plain or near an inclined slope. Some of the very large tunnels in the Province of Quebec are of this type. They are concrete-lined, mainly in order to reduce friction losses.

In fissured, well-drained rock, conditions are similar. The stress calculation takes into account the reductions in strength caused by rock fissures, which are assumed to be in a radial direction – the most unfavourable direction. A concrete lining is required which can be considered to be watertight provided its permeability is low compared with that of the rock masses. Considerable progress has been made in obtaining watertight linings with the Austrian techniques of shotcrete: a very flexible thin concrete layer, often reinforced with wire mesh, with or without normal concrete lining for additional strength. Soft rock with a low modulus of elasticity, E_r, may cause the concrete lining to be fissured when the tunnel is filled with water under pressure. Use of shotcrete behind the lining, reinforced or not, may be the cure. Rock squeezing inwardly when under excavation has been described for the Malgovert tunnel (section 2.2). These are dangerous rock conditions requiring special analysis.

Where there is no natural drainage of the rock masses, then full account has to be taken of water pressure in the rock fissures or dangerous uplift conditions may occur. Alternatively, the rock masses may be artificially drained. This was the case in the Gondo pressure shaft in Switzerland, and the Kemano pressure shaft in Canada. The Gondo, with a steel-lined pressure shaft (fig. 10.13) crosses a geological fault, maximum surge level reaching

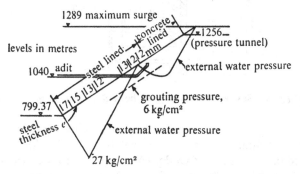

Fig. 10.13   Gondo pressure shaft: diagram of external water pressure; drainage through adit.

1289 m. An adit at level 1040 m was used for draining the fault. Several boreholes were drilled in the direction of the fault to assist the drainage of the rock and to relieve substantially the cleft water pressure on the outside of the steel lining. The gross head on the Kemano steel-lined pressure shaft is

790 m (fig. 10.14). The steel lining is backed by 'Prepakt' concrete, reputed to be impervious. In order to reduce the outside water pressure on the very high shaft, an intermediary adit was designed to drain the rock masses.

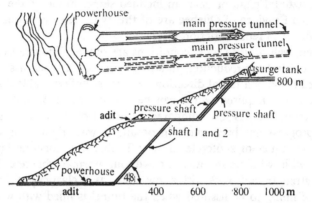

Fig. 10.14   The Kemano pressure shaft. The adit is used for decreasing the pressure of the interstitial water.

Drainage of rock near vertical unlined shafts leading to an underground power-house is also recommended in order to avoid build up of hydrostatic pressures on the power-house wall (fig. 10.15). The design of drainage

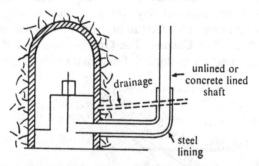

Fig. 10.15   Suggested drainage of the rock near a pressure shaft.

systems requires great care, as proved by the following case history. The most spectacular failure recorded in recent times in the Alps, was that of the steel-lined pressure shaft of Gerlos (Austria). The rock consists of schists, mostly of doubtful quality, and the overburden is meagre. A drainage gallery was running in the rock parallel to the 1·70-m-diameter conduit. Ruptures occurred at four points on the lower part of the conduit, in October 1945 and September 1948, where the static pressure was about 413·5 m (highest point of rupture) and 628·0 m (lowest point). Although the steel lining was designed to withstand the full hydrostatic pressure without reaching yield point it was, in all four cases, cracked along lines parallel to the drainage gallery. A considerable quantity of rock was washed away leaving large excavations in the rock masses (fig. 10.16). The generating sets were flooded to the point of immersion. It is believed that this rupture was caused

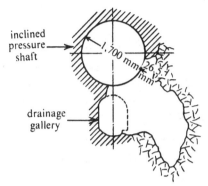

Fig. 10.16   Gerlos: damaged pressure shaft. Rupture occurred along the drainage gallery.

by uneven subsidence of the concrete packing along the drainage gallery, which allowed bending stresses to develop in the steel lining.

Power tunnels or discharge galleries passing through the rock abutments of high concrete dams create special problems when they are under full hydrostatic pressure, in spite of shallow rock cover. Such problems have to be analysed with special care in relation to the general stability of the dam abutment. Grouting and drainage systems must be adapted to local conditions (see section 13.2).

In recent years experts have considered the possibility of storing gas at very high pressures in underground excavations, old mines or galleries. Gas-tightness can be achieved with plastic materials. Extremely high pressures could be transferred to rock which may be overstressed and crushed to some depth round the cavity.

10.4.4   Pressure galleries located under horizontal rock surface

(1) *Case of competent, unfissured rock* (fig. 10.10) (Jaeger, 1961b).   At a depth $(H - x)$ under the surface, the rock is assumed to be loaded by its own weight $\gamma_R(H - x)$. A gallery would then be located in a vertical field of stresses $\gamma_R(H - x) = \sigma_v = p^*$ where p^* is the so-called residual rock stress. In a horizontal direction there is a second field of stresses:

$$\sigma_h = k\gamma_R(H - x), \qquad \text{with } k \leqslant 1 \text{ (Heim's hypothesis)}.$$

On the other hand, at a distance x from the centre of the tunnel, filled with water at a pressure p, the circumferential tensile stress σ_t is

$$\sigma_t = p(R/x)^2 = -\sigma_r,$$

where σ_r = radial compression.

As the rock is not fissured, an arbitrary condition for the balance of stresses at a depth $(H - x_1)$ can be assumed, so that

$$-\sigma_t \leqslant \sigma_h \quad \text{or} \quad p(R/x_1)^2 < k\gamma_R(H - x_1). \tag{1}$$

Writing that $x_1 = H/n$ and $p = \lambda\gamma H$ then:

$$\lambda \leqslant (H/R)^2 k(\gamma_R/\gamma)(n - 1)/n^3. \tag{2}$$

This analysis is valid provided the rock is unfissured. The rock's permissible tensile strength σ_3 at the depth $(H - x_1)$ is neglected.

Assume, for example, that $n = 3$, $x_1 = H/3$, $\gamma_R/\gamma = 2\cdot5$ and $k = 0\cdot7$; then $\lambda = 0\cdot13(H/R)^2$, and for

$$H/R = 5, \qquad \lambda \leqslant 3\cdot2$$
$$H/R = 10, \qquad \lambda \leqslant 13\cdot0$$
$$H/R = 100, \qquad \lambda \leqslant 1300.$$

It is obvious that λ must depend on H/R. The condition that at a depth $H - x_1$ no lifting of the rock should occur, the safety factor being equal to 1, yields:

$$p = \lambda\gamma H \leqslant \gamma_R H \quad \text{or} \quad \lambda \leqslant (\gamma_R/\gamma), \tag{3}$$

a condition usually more severe than condition (1). Equivalence of the two conditions would occur for $H/R = 4\cdot35$, at a very shallow depth.

(2) *Case of radially fissured rock* (fig. 10.17) (Jaeger, 1961*b*). It is permissible to assume that in rock with radial fissures starting from the tunnel,

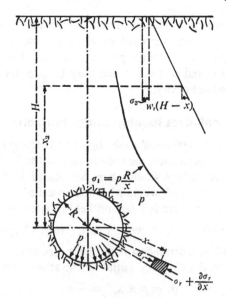

Fig. 10.17 Overburden situation in radially fissured rock (after Jaeger, 1961*b*).

the stresses decrease inversely with x/x_2. The condition now to be considered is that at a height $x_2 = H/n$ above the tunnel centre line tensile circumferential stresses should not cause new fissures to expand (theory of Griffith and Hoek), or

$$p(R/x_2) = \lambda_2 n\gamma R \leqslant k\gamma R(H - x_2) - \sigma_3,$$

where λ_2 corresponds to case (2),

or

$$\lambda_2 \leqslant k \left(\frac{\gamma_R}{\gamma}\right)\left(\frac{H}{R}\right)(n - 1)/n^2 - \sigma_3/n\gamma R, \qquad (4)$$

where σ_3 is the permissible rock stress; it is negative when representing the tensile strength of the rock. Neglecting σ_3 the ratio λ/λ_2 becomes $(\lambda/\lambda_2) = (H/nR)$. When a tunnel is steel lined, the pressure to be considered is that transmitted from lining to rock.

The two cases analysed here assume that rock is prestressed in the horizontal and in the vertical direction ('residual stresses') and drained.

10.4.5 Pressure tunnels adjacent to rock slope (fig. 10.18)

A similar approach can be used to assess the situation created for a pressure tunnel situated near a rock slope at angle α. The stress conditions at a point P, at a distance x from the centre O of the tunnel are to be analysed, with

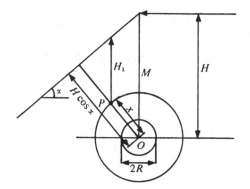

Fig. 10.18 Pressure tunnel adjacent to rock slope.

$x = H \cos \alpha/n$, where n is a freely chosen parameter. If $\sigma_H = k\sigma_z$ is the horizontal component of the field of residual stresses at the point O, at the point P the stress will be (σ_z = vertical stress):

$$\sigma_{H_1} = k_1\sigma_z, \quad \text{with} \quad \sigma_z = \gamma_R H_1 = \gamma_R H \frac{n - 1}{n}.$$

It is to be expected that the horizontal stress component coefficient k_1 decreases progressively from O towards P as $k = 0$ at the surface of the rock slope (stress relieving). From fig. 10.18:

$$H \cos \alpha = H_1 \cos \alpha + x.$$

Furthermore the residual stress, σ_α, at angle α is

$$\sigma_\alpha = \frac{\sigma_z + \sigma_{H_1}}{2} - \frac{\sigma_z - \sigma_{H_1}}{2} \cos 2\alpha,$$

so that at point P

$$\sigma_\alpha = \frac{\gamma_R}{2} \left\{ H \frac{(n-1)}{n} (1 + k_1) \right\} - \frac{\gamma_R}{2} \frac{H(n-1)}{n} (1 - k_1) \cos 2\alpha.$$

On the other hand, the circumferential stress, σ_t, in P caused by the hydrostatic pressure, $p = \lambda_3 \gamma H$, in the tunnel is

$$\sigma_t = p \frac{R^2}{x^2} = p \frac{R^2 n^2}{H^2 \cos^2 \alpha} = \lambda_3 \gamma \frac{HR^2 n^2}{H^2 \cos^2 \alpha}.$$

It is assumed that in order to obtain safe stress conditions at the point P we must have

$$\sigma_\alpha \geqslant |\sigma_t|,$$

where the circumferential stress σ_t is a tensile stress, and σ_α a compression stress. The tensile strength of the rock is neglected.

This ultimately leads to a condition for the safe value of λ_3:

$$\lambda_3 \leqslant \frac{\gamma_R}{\gamma} \frac{(n-1)}{2n^3} \left(\frac{H}{R}\right)^2 [(1 + k_1) - (1 - k_1) \cos 2\alpha] \cos^2 \alpha.$$

A similar condition can be developed assuming the rock to be radially fissured. In this case

$$\sigma_t = p \frac{R}{x} = \lambda_4 \gamma H \frac{Rn}{H \cos \alpha} = \lambda_4 \gamma \frac{Rn}{\cos \alpha}.$$

We assume that a condition $\sigma_t < \sigma_\alpha - \sigma_3$, where $\sigma_3 (= \sigma_{\text{tens}})$ is the tensile strength of the rock (negative value), is a safe condition whereby rock fissure will not progress deep into the rock mass. This yields:

$$\lambda_4 \gamma \frac{Rn}{\cos \alpha} < \frac{\gamma_R}{2} H \frac{(n-1)}{n} \{(1 + k_1) - (1 - k_1) \cos 2\alpha\} - \sigma_3,$$

or

$$\lambda_4 < \frac{\gamma_R}{\gamma} \frac{H}{R} \frac{(n-1)}{2n^2} \{(1 + k_1) - (1 - k_1) \cos 2\alpha\} \cos \alpha - \frac{\sigma_3 \cos \alpha}{\gamma Rn}.$$

The conditions at point M, vertically above the tunnel centre must also be checked, as they may be less favourable than the stresses at P, mainly as a result of low k values.

10.4.6 Other cases

(1) *When the horizontal surface is covered with loose alluvium* (fig. 10.19); here the depth of alluvium is assumed to be h' and the total height of overburden is $H = h' + h''$. At the boundary of rock and alluvium the vertical

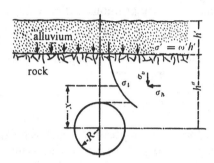

Fig. 10.19 .The horizontal rock surface is covered with loose alluvium (after Jaeger, 1961*b*).

stress is $\sigma' = \gamma'h'$. ($\gamma' =$ specific weight of the alluvium.) At a distance x from the centre of the tunnel the vertical stress is $\sigma_v = \gamma'h' + \gamma''(h'' - x)$ and the horizontal stress is $\sigma_h = k\sigma_v$. At the boundary of rock and alluvium the σ_t value for $x = h''$ should be less than the permissible tensile strength in the rock mass.

(2) *When the rock suffers plastic deformations* (fig. 10.20). Some sandstones have a very low modulus of elasticity, some claylike rocks are plastic rather

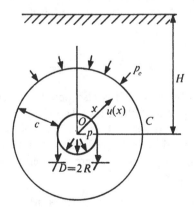

Fig. 10.20 Model for analysis of a gallery in plastic rock (after Naylor).

than elastic. Overstressed rock may have the characteristics of a plastic mass rather than a brittle one. An alternative approach to elastic rock deformations under strain and stress may therefore be desirable.

Deep galleries or excavations can be used for storing gases under very high pressures, but this might strain the rock beyond its limit of elasticity causing rupture. Several theories have been developed to deal with such problems.

Naylor considers a gallery excavated in plastic rock and calculates the radial equilibrium of a cylinder of rock at radius x (fig. 10.20):

$$(\sigma_r + d\sigma_r)(x + dx)\, d\theta - x\sigma_r\, d\theta = (\sigma_t + \tfrac{1}{2}d\sigma_t)\, dx\, d\theta,$$

c is the thickness of a plastic cylinder, p_c being the radial load on it.

Neglecting second order terms gives

$$\sigma_r - \sigma_t = -x\,\frac{d\sigma_r}{dx}. \tag{1}$$

When the hydrostatic load p in the pressure tunnel is great in relation to the radial pressure p_c sufficient to put the whole ring c into a state of plastic shear, then the stress at every point within the ring is represented by a Mohr circle in contact with a failure envelope. This can be represented by a straight line through the origin at an inclination ϕ to the abscissa (law of Coulomb when rock cohesion is nil). From the geometry of the Mohr circle:

$$\sigma_t = \sigma_r \tan^2 (45° - \phi/2). \tag{2}$$

Substituting (2) in (1) and writing $[1 - \tan^2 (45° - \phi/2)] = a$ gives

$$a\,\frac{dx}{x} = -\frac{d\sigma_r}{\sigma_r}.$$

Integrating between the limits:

$$x = \tfrac{1}{2}D, \quad \sigma_r = p \quad \text{and} \quad x = \frac{D}{2} + c, \quad \sigma_r = p_c,$$

gives

$$p = \left(1 + \frac{2c}{D}\right) ap_c.$$

According to an essay by Naylor it is also possible to examine the effect of water pressure in the pores of the rock. Assume the pore pressure at any point x of the cylinder at radius x be $u(x)$. The soil mechanics concept of effective stress can be used in rock mechanics as shown by several authors. Therefore (fig. 10.20), $\sigma = \sigma' + u$, where $\sigma' =$ effective stress. The basic equation (1), now becomes

$$\sigma'_r - \sigma'_t = -x\,\frac{d}{dx}(\sigma'_r + u).$$

Introducing the law of Coulomb:

$$\frac{d\sigma'_r}{dx} + \frac{a}{x}\,\sigma'_r = -\frac{du}{dx}.$$

Multiplying through by the integrating factor x^a gives

$$\sigma'_r x^a = \int x^a \left(-\frac{du}{dx}\right) dr + \text{constant.}$$

In order to integrate it is assumed, for example, that the function $u(x)$ is given by $u = A \log_e x + B$ and $du/dx = A/x$. A can be determined by introducing arbitrary boundary conditions, for example $u = p$ for $x = D/2 = R$ and $u = 0$ for $x = c + D/2$ which gives $A = p/[\log_e (1 + 2c/D)]$.

Substituting for du/dx in the differential equation and solving for the same boundary conditions gives

$$p = \frac{\kappa}{\kappa - 1} \log \kappa p_c,$$

when

$$\kappa = \left(1 + \frac{2c}{D}\right)^{1 - \tan^2(45 - \phi/2)}.$$

Another approach which is probably acceptable when rocks are being crushed under very high pressures, introduces the theory of plasticity of cylindrical tubes (Jaeger, 1961b). It is assumed that there is a plastic inner region extending from a = inner radius of the tunnel to d ($a < r < d$) and an elastic region from d to b ($d < r < b$). Before the strain in the cylinder develops, $a = a_0$ and $b = b_0$ (initial values).

Within the elastic tube (fig. 10.21) the following equations are valid.

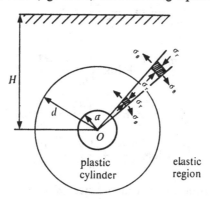

Fig. 10.21   Plastic cylinder theory (after Jaeger, 1961b).

The elastic equations in cylindrical co-ordinates (r, θ, z) are

$$E\epsilon_r = E\frac{\delta u}{\delta r} = E\left(A - \frac{B}{r^2}\right) = \sigma_r - \nu(\sigma_\theta + \sigma_z),$$

$$E\epsilon_\theta = E\frac{u}{r} = E\left(A + \frac{B}{r^2}\right) = \sigma_\theta - \nu(\sigma_z + \sigma_r),$$

$$E\epsilon_z = \sigma_z - \nu(\sigma_r + \sigma_\theta).$$

The boundary conditions $\sigma_r = p$ for $r = a_0$ and $\sigma_r = 0$ for $r = b_0$ give

$$A = \nu \epsilon_n - \frac{(1 + \nu)(1 - 2\nu)}{E(b_0^2/a_0^2 - 1)}p,$$

$$B = -\frac{(1 - \nu)b_0^2}{E(b_0^2/a_0^2 - 1)}p.$$

The final expressions for the stresses are then:

$$\sigma_v = p(b_0^2/r^2 - 1)/(b_0^2/a_0^2 - 1);$$

$$\sigma_\theta = -p(b_0^2/r^2 + 1)/(b_0^2/a_0^2 - 1);$$

$$\sigma_z = -E\epsilon_z - 2\nu p/(b_0^2/a_0^2 - 1).$$

In the present case, only that part of the rock mass situated beyond $r > d$ is elastic, for which the stress equations are

$$\sigma_r = C(b_0^2/r^2 - 1); \qquad \sigma_\theta = -C(b_0^2/r^2 + 1);$$

$$\sigma_z = -E\epsilon_z - 2\nu C = \text{constant}.$$

In the plastic region Tresca's criterion (1864) is adopted, which introduces a yield stress: $Y = \sigma_r - \sigma_\theta$, everywhere in the plastic region. Hill (1950) demonstrates that in the plastic region the equation of equilibrium leads to

$$\frac{\partial \sigma_z}{\partial r} = \frac{\sigma_r - \sigma_\theta}{r} = Y/r,$$

and

$$\frac{\sigma_r}{Y} = \frac{1}{2} + \log_e \left(\frac{d}{r}\right) - \frac{d^2}{2b_0^2},$$

$$\frac{\sigma_\theta}{Y} = -\frac{1}{2} + \log_e \left(\frac{d}{r}\right) - \frac{d^2}{2b_0^2},$$

$$\frac{p}{Y} = \log_e \left(\frac{d}{a}\right) + \frac{1}{2}\left(1 - \frac{d^2}{b_0^2}\right).$$

For the particular problem considered here $d = H/n$, $k = 0.7$ and $p/\gamma = \lambda_p H$, which yields the corresponding λ_p ratios. Numerical calculations have been made for comparing λ_1, λ_2 and λ_p, assuming $x = H/n = H/3$ and $k = 0.7$. The result of this investigation is shown in table 10.3. This table shows the importance of rock characteristics and strength in determining a 'safe overburden' for protecting a pressure tunnel. 'Plastic rock' may be substantially less safe than fissured rock. There are several other methods of dealing with this problem which are described in works by Mintchev (1966), Reyes & Deere (1966), Obert & Duval (1967) and Kastner (1962). The finite element method is the one most likely to be used for further research.

Table 10.3

H/R	λ_1	λ_2	λ_p
5	3·2	1·95	1·88
10	13	3·9	3·4
100	1300	39	8·2

10.4.7 Unstable tunnel designs: danger of rupture of rock masses

The previous calculations apply to rock masses which are reasonably dry or drained and where a horizontal component of the residual field of stresses (rock prestressing) substantially contributes to the stability of the rock. The problem there was to find the stress conditions and so avoid a progressive rupture of the rock by tensile stresses. The following paragraphs concern more general aspects of unstable tunnel designs.

Figure 10.22 represents conditions caused by pressure grouting around a tunnel. A tunnel with a diameter $2R$ is grouted, the borehole length being l

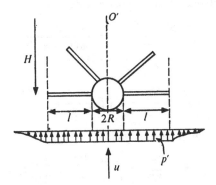

Fig. 10.22   Uplift force U caused by grouting fissured rock. $L =$ length of borehole (tunnel length to be considered = space between borehole lines) (after Jaeger, 1961*b*).

on each side. Grout penetrates to a certain depth in the rock, and $D' > 2(R + l)$ is the estimated width of the area on which uniform grout pressure p' is supposed to be applied. If the rock is fissured, there is a tendency for the rock to be lifted. According to fig. 10.23 a pressure p' uniformly distributed on a width $2D'$ causes at a point O' at a distance $z = D' \cot \alpha$ above O, a vertical stress

$$\sigma_1 = \frac{p'}{\pi}(2\alpha + \sin 2\alpha),$$

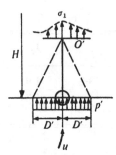

Fig. 10.23   Uplift force $U = 2p'D'$ acting on boreholes or on rock fissures. Stresses σ_1 at point O'.

or

$$\sigma_1 = \frac{p'}{\pi} 2\alpha \left(1 + 1 - \frac{(2\alpha)^2}{3} + \ldots\right) \cong \frac{4\alpha p'}{\pi}\left(1 - \tfrac{2}{3}\alpha^2 \ldots\right).$$

In order to avoid uplift near point O we must have

$$\sigma_1 < \gamma_R(H - z),$$

where H = depth of overburden. Therefore:

$$\frac{p'}{\gamma_R(H - z)} = \lambda^* < \frac{\pi}{4\alpha(1 - \tfrac{2}{3}\alpha^2)}.$$

On the other hand,

$$\tan \alpha = \frac{D'}{2z} = \alpha \left(1 + \frac{\alpha^2}{3} + \ldots\right).$$

Combining the last two equations yields:

$$\frac{p'}{\gamma_R(H - z)} = \lambda^* < \frac{2\pi z}{4D'} \times \frac{1 + \alpha^2/3}{1 - 2\alpha^2/3} = \frac{\pi z}{2D'}(1 + \alpha^2 \ldots),$$

$$\frac{p'}{\gamma_R(H - z)} = \lambda^* < \frac{\pi}{2}\left(\frac{z}{D'} + \frac{D'}{4z}\right),$$

which yields an approximate estimate of the distance z at which the rock overburden neutralizes the uplift stress, σ_1, by its own weight. A similar calculation should be carried out for the horizontal stress so that

$$\sigma_n < k\gamma_2 H,$$

and the occurrence of vertical fissures is prevented.

When the rock stratification in the rock mass is inclined, a similar calculation has to be carried out in any potentially dangerous direction, at an angle β (fig. 10.24).

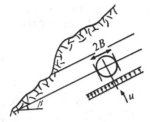

Fig. 10.24 Uplift caused by interstitial water in rock fissures.

There is no known case of rock rupture caused by grouting a deep tunnel (there are several cases recorded which occurred during dam foundation grouting). This proves that grout pressures are being correctly chosen by tunnel designers and that the technique of grouting by stages at different pressures is efficient.

The methods used here for analysing the required overburden for rock grouting can be used for calculating the stability of rock slopes under the pressure of water percolating from pressure tunnels. Figure 10.25 assumes

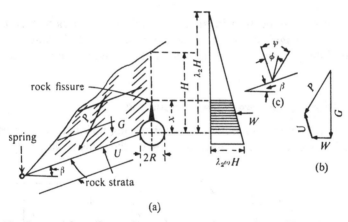

Fig. 10.25 Seepage from fissured pressure tunnel causes instability of rock masses along rock strata (after Jaeger, 1961*b*).

that the rock is stratified, the strata dipping towards the valley with an inclination β. The diagram shows the resultant P of the weight G, the hydrostatic thrust W, and the uplift $U = m_r \lambda_r \gamma H/2$ (where $m_r \leqslant 1$ is a coefficient taking into account the fact that only part of the seam in the stratified rock is under uplift). Depending on the angle ψ between the resultant P and the normal to the inclined strata (angle β), the shaded mass of rock will be stable if $\psi < \phi =$ angle of friction, and unstable if $\psi > \phi$ as shown in fig. 10.25.

In plastic or near-plastic material or in deeply fractured rock (as often found in the Andes), graphical stability analysis could be carried out on principles similar to those used in soil mechanics (fig. 10.26). Both figs. 10.25 and 10.26 assume that the cohesion of the rock is negligible ($c = 0$).

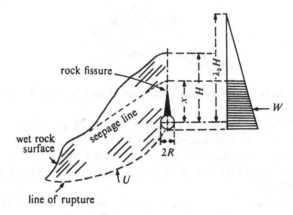

Fig. 10.26 Seepage from fissured pressure tunnel causes instability of rock slope (after Jaeger, 1961*b*).

This is unlikely to be true in all cases, and rock cohesion may provide additional safety against sliding, which can be estimated. Seeping water may lubricate sliding planes and substantially reduce the friction factor tan ϕ.

10.4.8 Cleft water pressure on galleries

The method developed in section 8.2.1 for tracing flow nets in fissured rock can also be used for estimating the pressure on the walls of an empty gallery,

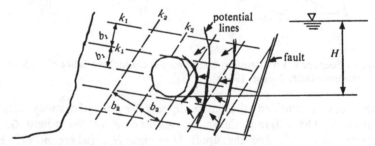

Fig. 10.27 Cleft water converging on an unlined empty gallery. k_1 and k_2 = permeability factors of the primary and secondary joints.

fig. 10.27. The primary and secondary joints are characterized by their permeabilities k_1 and k_2. Potential flow lines are traced and the pressures on rock calculated.

10.4.9   Storage of oil in large underground caverns

Rock in Sweden tends to be good, even very good, so that oil can be stored in large caverns. Massive 20 m × 35 m high sections, running hundreds of metres, are cut to give up to 3 Mm³ of storage capacity. The caverns are not lined; linings crack and leak. Swedish experts estimate that caverns are cheaper than tanks for a required capacity in excess of 20 000 m³ (excavation cost £7 per m³, 1975 prices). The break-even point is about 0·5 Mm³ in France in poorer rock (excavation cost about £24 per m³).

There must be a consistent and small ground-water flow, to prevent the leakage of the stored product (oil). If the water flow drops too low, it has to be force-fed through extra feeder tunnels (2 to 3 m diameter) cut in the rock above the caverns. Force feeding may be necessary anyway if the caverns are operated as separate units: an empty cavern will draw down the water-table dangerously. When properly managed caverns are said to be maintenance-free and leak-, fire- and explosion-proof.

It has also been suggested that atomic power plants be located underground. One small unit has been built in a cavern in Switzerland. Water seepage was no problem and there was no environmental pollution.

10.5   Overstrained rock about galleries

Experience has shown that shortly after excavating a gallery in pressurized rock, the surrounding rock shows signs of being strained or even overstrained. Rock bursts occur and deformations are observed. Interpretation of these rock deformations depend on the basic approach to the general state of residual stresses in rock masses.

10.5.1   Talobre's interpretation

Talobre assumes that the rock is in a hydrostatic state of stress (Heim's hypothesis) and that the horizontal residual stress, σ_h, is equal to the vertical stress, σ_v, which is equal to p^*, i.e. $\sigma_h = k\sigma_v$, with $k = 1$. According to the theory developed in section 5.4 for circular galleries the circumferential stress σ_t at the periphery of the rock excavation is then $\sigma_t = 2p^*$, at whatever angle ϕ (fig. 10.28a). This is correct so long as $\sigma_t = 2p^*$ is smaller than the elastic limit σ_{el} of the rock.

Whenever the circumferential stress at a distance r from the centre of the gallery reaches this limit of elasticity, the rock is plastically deformed or even crushed (fig. 10.28b), which relieves the stresses as indicated on the figures. At the same time there is a small inward displacement of the rock. Crushing of the rock will occur if $\sigma_{max} - \sigma_{min} > \sigma_{cr}$ where σ_{cr} is a critical value obtained from the Mohr circle. It is likely that the rock crushing leads progressively to stabilization of the deformations of the rock and the tunnel.

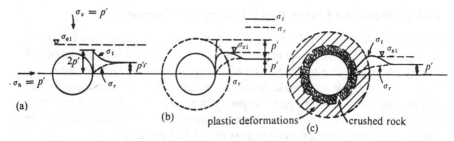

Fig. 10.28 Stresses about cavities in isotropic rock, uniform residual stresses ($\sigma_v = \sigma_r = p'$): (*a*) $2p' < \sigma_{e1}$; (*b*) $2p' > \sigma_{e1}$; (*c*) crushed rock, σ_{e1} = elastic limit (after Talobre, 1957).

It should be understood that these three sketches give only qualitative information, as they cannot take into account the lack of homogeneity of natural rock.

10.5.2 Rabcewicz's interpretation (1957, 1964, 1965)

Rabcewicz assumes that the gallery is excavated in rock strained by a vertical field of forces $\sigma_v = p^*$ (with $k = 0$). A new equilibrium is progressively obtained.

Stress rearrangement generally occurs in three stages (fig. 10.29), provided the rock in the neighbourhood of the cavity has not been disturbed by earlier tunnelling. At first, wedge-shaped bodies on either side of the cavity shear off along the Mohr surfaces and move towards the cavity, the direction of the movement being perpendicular to the main pressure direction (fig. 10.29*a*). The increased span thus produced causes the roof and floor to start

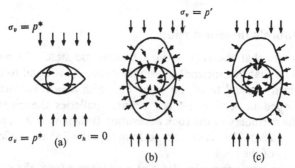

Fig. 10.29 Stresses and strains about a cavity in rock. Initial residual stresses uniaxial ($\sigma_v = p^*$, $\sigma_h = 0$). Progressive redistribution of stresses and rock displacements towards the cavity (after Rabcewicz, 1964, 1965).

converging (fig. 10.29*b*) and lateral pressure develops in a horizontal direction. In the next stage (fig. 10.29*c*) movement is increased and the rock buckles under continuous lateral pressure and may protrude onto the cavity.

Squeezing pressures – the last stage – though common in mining, are seldom encountered in civil engineering.

This description is consistent with diagrams developed for the case of a vertical potential field of parallel forces. It is also consistent with experience and observations published in much earlier textbooks on this subject.

The theories developed by Schmidt, Fenner and Terzaghi could also be used to analyse the more general case when $\sigma_h = k\sigma_v$, with $k < 1$. It leads to conclusions similar to those of Rabcewicz and indicates that dangerous compression stresses develop along the horizontal tunnel diameter. Such stresses have been observed in many tunnels, in particular in the large Mont Blanc road tunnel.

The theories developed here assume a reasonable degree of homogeneity and isotropy of the rock masses. The image of the rock deformations as sketched by Talobre and Rabcewicz are fundamentally altered when the rock is fissured, fractured or stratified. Jaecklin (1965b) said that high tensile stresses develop near the tunnel soffit in horizontally stratified rock to a far greater degree than indicated by the theory of stresses about cavities in isotropic rock. Similarly Talobre (1957) and Rabcewicz (1964) emphasize the great differences between isotropic and fractured rock masses concerning the behaviour of rock about tunnel excavations.

10.5.3 Fenner's equation and comments

Talobre and Rabcewicz in estimating the stress distribution about a gallery excavated in overstrained rock use Fenner's equation (1938):

$$p_i = -c \cot \phi + [c \cot \phi + p_0(1 - \sin \phi)] \left(\frac{r}{R}\right)^{2 \sin \phi / (1 - \sin \phi)}$$

where

  $r =$ radius of the cavity,
  $R =$ radius of the protective, overstrained zone,
  $p_i = \sigma_r^r$ the required radial 'skin resistance' (fig. 10.30) and p_0 the uniform residual rock stress ($\sigma_v = \sigma_h = p_0 = p^*$).

Talobre assumes that usually $p_i = 0$ and the equation gives R, when c and ϕ are known. Talobre gives the following example for a 6-m-diameter tunnel at 1500 m depth:

$$p_0 = 400 \text{ kg/cm}^2 = 4000 \text{ t/m}^2, \qquad p_i = 0,$$

$$\sin \phi = 0.50 \quad \text{and} \quad c \cot \phi = 50 \text{ kg/cm}^2 = 500 \text{ t/m}^2.$$

Fenner's equation becomes

$$-500 + [500 + 4000 \times \tfrac{1}{2}] \left(\frac{r}{R}\right)^2 = 0,$$

or

$$\frac{r}{R} = \sqrt{\tfrac{1}{5}}.$$

With $r = 3{\cdot}0$ m, $R = 3\sqrt{5} = 3 \times 2{\cdot}23 = 6{\cdot}70$ m $= 3 + 3{\cdot}70$ m.

The width of the crushed protective zone would be 3·70 m. Rabcewicz assumes that p_i can be positive, corresponding to the 'skin resistance' of the shotcrete layer or to the radial stress due to rock bolting.

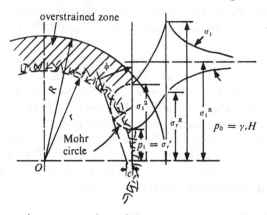

Fig. 10.30 Schematic representation of Fenner's equation (after Kastner, 1962).

Fenner's equation has been criticized, because, in some cases, it fails to yield acceptable values. In section 10.10, on the estimate of the required rock support for tunnels, solutions differing from Fenner's equation will be developed, based on Kastner's and Lombardi's suggestions (Jaeger, 1975, 1976).

Fenner's equation and the alternatives proposed by others are important to many basic engineering problems to be discussed in detail in other chapters. Figures 10.28 and 10.30 could possibly explain the large local deformations of rock vaults measured *in situ*. The measured total settlement of the soffit of a large excavation in the Swiss Alps has been found locally to total about 20 cm since the beginning of the excavations, well in excess of what had been expected (Buro, 1970). Such a large displacement of a cavern roof in rock with no preferred joint direction could be explained by assuming locally a very low modulus of elasticity, E, of the rock at the limit of the excavation (see Kujundzic, Jovanovic & Radosavljevic, 1970). Very recently Kujundzic (1970) has tested the rock about a similar gallery with seismic waves and shown that, at some distance from the gallery, the rock is compressed and the dynamic modulus of elasticity consequently higher than in the virgin rock mass, whereas some decompression occurs near the gallery, where the modulus of elasticity drops to lower values. The same compression and decompression effects have been checked by a team of French experts about a cavern excavated in chalk, but because the cavern had a flat roof and so was not circular the pattern of stresses and the compressed and decompressed zones varied with the progress of excavations.

Hayashi and his colleagues (Hayashi & Hibino, 1970; Suzuki & Ishijima, 1970) in Japan have established correlations between the stresses in rock material and the E modulus. The modulus is shown to rise with compression and to drop in relaxed rock, whereas the Poisson ration ν follows the opposite trend. These opposite trends of E and ν have already been established by Jaeger (1966a). Hayashi introduces into a computer program the variables E and ν for progressively deteriorating rock conditions, whereas Müller & Baudendistal (1970) have developed a program for finite element analysis where displacements are assumed to occur along weak joints characterized by low values of tan ϕ. With both methods, important rock strains and large settlements of the cavern roof are obtained for mediocre rock conditions, which are nearer to values really observed in difficult underground excavations.

10.6 Stress and strain measurements in galleries

It is important to follow the development of strains about excavated cavities exactly. (Details on the Austrian loading tests (Seeber, 1964) have been given in section 6.2 and the borehole extensometers and deflectometers are amply described in section 5.5.) Figure 10.31 shows the arrangement of borehole extensometers in an American tunnel.

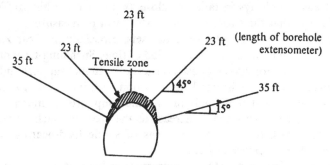

Fig. 10.31 Roof and wall strain measurements obtained in moderately supported locations of the Straight Creek tunnel (USA) (after Hartmann, 1966).

Successful measurements of the load actually transmitted from the rock to the support systems have been achieved recently. 'Terrametrics' (Hartmann, 1966) in the USA use load cells installed locally between rock and support. Such a method has been used for measuring the thrust from a dangerously inclined stratified rock mass on a steel support. Recent investigations in American tunnels and precise measurements on support systems indicate that many of them are over-dimensioned.

The most common method of construction support used in American tunnel practice is the structural steel arch. Measurement of the compressive loads which occur on such supports provides a direct comparison between actual loads and support design strength. Load cells with a range equal to the yield strength of the support should be used. Total deformation leading to failure in a brittle rock mass is often less than 0·15 in. Hartmann recommends a reliable measuring sensitivity of 0·005 in (0·12 mm). Figure 10.32 shows

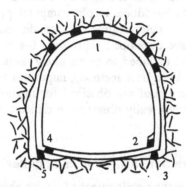

Fig. 10.32 Load cell (1–5) installation for measuring horizontal and vertical loads on a horseshoe set, and uplift on an invert strut (after Hartmann, 1966).

the arrangement of load cells recommended for different types of rock support. Cells are always installed as close as possible (within 20 ft) to the blasting face so that the entire set load history can be measured.

Load cells can also be installed under sets, especially in shear zones, to measure structural loads over a long period of time. Such long-term measurements are desirable for revising design criteria for tunnel support and lining. Support load data also indicates load variation with time and face position. It has also been noted that loads on installed supports as much as 200 ft from a shear zone show a sharp increase as the face approaches shear zone areas. Repair work disturbances in areas of stable load-bearing supports also cause a sharp increase in load.

More information on modern measurement techniques is given in section 10.11.5.

10.6.1 Loads transmitted from rock to concrete or steel lining

There is not much information on loads transmitted from rock to concrete linings and few figures are available on conditions after relaxation has occurred in a fractured shear zone. An interesting publication by Habib, Bernède & Carpentier (1965) describes measurements taken from pressure cells behind the lining of the Monaco railway tunnel. Initial pressures were low and uniform, but after six months they had increased and were unevenly concentrated (fig. 10.33).

In zones requiring moderate support, figures obtained from loads on provisional steel sets could be used for the design of concrete linings. Goffi &

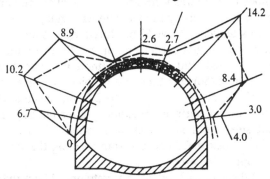

Fig. 10.33 Monaco railway tunnel: rock pressures measured behind the tunnel lining with pressure cells. - - - - Initial contact pressure (1 May 1962), – – – pressure 6½ months later, before grouting (14 Nov. 1962), ——— 4 months after the tunnel entirely excavated (31 Aug. 1964) (after Habib, Bernède & Carpentier, 1965).

Oberti (1964) proposed a method for direct measurement of strains in a concrete lining (fig. 10.34). But the final decision usually rests with the amount of possible fissures and leakage caused by hydrostatic pressure inside

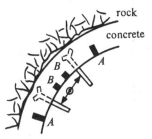

Fig. 10.34 Method for measuring load transmitted from rock to concrete lining. (*A*) Total stress-relief grooves; (*B*) partial stress-relief grooves (after Goffi & Oberti, 1964).

a pressure gallery, rather than rock loads. The same can be said about steel linings designed for inside water pressure or against buckling from interstitial water pressure.

10.7 New tunnelling techniques in relation to rock mechanics

Conventional tunnelling methods are dealt with in many specialized textbooks and it is not proposed to discuss these here (Morgan, 1961; Kastner, 1962; Pequignot, 1963).

10.7.1   The new Austrian tunnelling method (NATM); examples

In papers introducing a method he called the 'new Austrian tunnelling method', L. von Rabcewicz (1957, 1963, 1964) remarked that by the old tunnelling methods two months were needed to excavate and line a 9-m-long section of a double-track railway tunnel. Timber logging could not generally be removed and its tendency to yield produced violent loosening pressures, which frequently caused roof settlement up to 40 cm or more before the masonry could be closed. But the new method using shotcrete and rock bolts (Austrian patent 1956) cuts time and problems considerably. Shotcrete is a mortar containing aggregates of up to 2·5 cm, pneumatically and violently applied to the rock surface immediately after blasting. Rock bolting, with or without steel mesh and steel bars, often reinforces the shotcrete layers. This reduces rock deformations, transforming the surrounding rock mass into a self-supporting arch.

According to Rabcewicz, a layer of shotcrete only 15 cm thick, applied to a tunnel of 10 m diameter, can safely carry a load of 45 tonnes/m² corresponding to a burden of 23 m of rock, more than has ever been observed with roof falls. If a steel support structure incorporating 'number 20' type wide flanged arches at 1-m centres were used under these conditions, it would fail at about 65% of the load carried by the shotcrete lining.

Shotcrete becomes strong very quickly (fig. 10.35), which is very important when a rapid high-bearing capacity is needed. Its immediate flexural–tensile

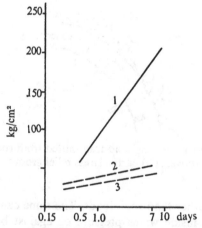

Fig. 10.35   Shotcrete strength: (1) crushing strength; (2) shear strength; (3) tensile strength.

strength amounts to 50% and 30% of the compressive strength after 2½ days. A recently introduced hardening–accelerating admixture based on silicification gives even better results. In one of the tunnels at the Kaunertal hydro-electric scheme a 2-cm jet of water was satisfactorily plugged with shotcrete alone, without even installing a relief pipe. Bernold recommends the adjunction to the shotcrete of steel needles, 2·5 cm long, 0·4 mm in diameter, in a proportion of 2 to 3% of the solid concrete weight.

Rabcewicz writes that a shotcrete layer applied immediately after opening up a new rock face turns a rock of doubtful stability into a stable one. The close interaction between shotcrete and rock leaves the neighbouring rocks almost unaltered, enabling them to participate effectively in the arch action. It also absorbs the tangential stresses which build up to a peak close to the surface of a cavity. The zone of arch action can be increased at will by rock bolting.

Russian experts tested shotcrete for the large Inguri tunnel (Masur, 1970). Drilling through the shotcrete and the rock, they checked that the shotcrete really forms a solid bond with the rock. They tried grouting behind the shotcrete layer with pressures up to 15 or 20 kg/cm², which was entirely successful. Fissures and joints were well cemented by the injections.

The first successful application of a shotcrete lining was carried out at Lodano–Mosagno tunnel (Maggia hydroelectric power system) in Switzerland in 1955. Rabcewicz and others (see Suttcliffe, 1969; Abel, 1970; Cecil, 1970) acquired considerable experience in the new methods. Some Austrian engineers were slow in adopting them, but the technique rapidly spread from Austria to Switzerland, Scandinavia, France, Australia, South Africa, Canada and the United States.

Rabcewicz himself commented on the probable sequence of events which take place during the loosening of rock masses around a tunnel (see section 10.5.2). A better description of the behaviour of rock masses around a tunnel or gallery was given many years before Rabcewicz's proposals in a brilliant paper by Robert Maillart (1922), commenting on some remarks by the great geologist Albert Heim (1878) and some subsequent comments by Charles Andrea (1926) (one of the builders of the Lötschberg tunnel). Further theoretical research by Talobre (1957), Kastner (1962), Lombardi (1970, 1972), Egger (1973) and Duffaut & Piraud (1975) has contributed much to the understanding of the efficiency of combining rock bolting and shotcreting.

In their papers, Maillart (1922, 1923) and Andrea (1926, 1961) carefully examine the experience gained during excavation of the Simplon tunnels, under well over 2000 m overburden. They explain how the rock mass, and sometimes the horse-shoe shaped masonry supporting it, deforms, the rock being progressively decompressed as decompressed rings are being formed above the tunnel. Sometimes rock spalls at the rock surface, mainly when the tunnel radius is too large, but finally settles, a relatively thin lining (0·40 m for the Simplon tunnel) being sufficient as rock support, in spite of the fact that the crushing strength of the concrete is less than the crushing strength of the rock. This analysis which is more realistic than Rabcewicz's approach could have led to an early discovery of the NATM. It did not.

A very lucid analysis of the problems of excavations and rock supports and of the historical development of the theories and methods has been given by Lombardi in his address to the Lucerne Symposium (1972).

Shotcrete, especially when combined with rock bolting, has proved excellent as a temporary support for all qualities of rock. In very bad cases steel arches are used for reinforcement of the weaker tunnel sections. Rabcewicz mentions a tunnel of 8 m² cross-sectional area for a hydroelectric scheme in the Austrian Alps. Originally it had been driven without shotcrete, using only steel arches and steel lagging. When the tunnel reached a zone of kaolinized gneiss under an overburden of 250 m with heavy water inflow of 35 l/s, the pressure became so heavy that the arches were deformed and their footing forced into the ground. The heavy inflow of water could only be relieved slightly, as the discharge pipes became clogged shortly after placement. Excavation had to be stopped. After re-driving the roof in the deformed portion, new steel arches had to be placed at 60-cm centres on heavy wooden sills and another arch interposed between each set. As soon as a set was placed the surface was immediately shotcreted to a complete ring. This difficult situation, aggravated by unsuccessful attempts at driving, was brought under control using these methods.

The Kaunertal tunnels and the pressure shafts are another example worth studying. The 5–15-cm thick shotcrete is now part of the complete thickened final lining. At some places the shotcrete lining failed because the 'Perfo-anchors' and bottom bracing were not applied in good time (delays up to one full year). The Serra Ripoli super highway twin tunnel (Italy) is another example where conventional tunnelling methods failed, but the shotcrete method reinforced by sectional steel was successful.

Experience by others than Rabcewicz can be mentioned. The Mont Blanc tunnel (France) (Panet, 1969) is a typical example of rock bolts being used mainly to avoid rock bursts. Under a rock overburden of 2000 m and with a k-value of only 0·40, the Mont Blanc granite was overstressed. Bolts 1·50 to 3·50 m long, designed for a maximum load of 14 tonnes and a minimum spacing of 0·50 m were able to stabilize roof and walls. Rock bolting can be used either for pinning locally unstable rock masses or as a systematic rock support. Talobre (1957) and Rabcewicz & Rescher (1968) have developed methods of designing rock bolts for the purpose of stabilizing rock vaults. The natural tendency for a 'sustaining arch' to be formed in overstressed rock masses, as shown by Hayashi's finite element method analysis (1970) explains the success of this practical method. Rock bolts, when stressed, apply an average pressure p^* on the surface of the rock and consolidate the natural arch formed in the mass. Rock bolts designed to cause pressures $p^* = 0·70 \text{ kg/cm}^2$ to $1·5 \text{ kg/cm}^2$ are usually satisfactory. Higher values are often required for p^* for large excavations like underground power houses, where anchored cables are being used. Veytaux underground power station (Rescher, 1968) and Lago Delio (Mantovani & Bertacchi, 1970) are good early examples of the NATM; Waldeck II (chapter 16) is more recent.

The NATM has also been widely used in the construction of underground railways, where difficult local tunnelling problems had to be convincingly resolved. The construction of instrumented test sections provided the most reliable design data for major tunnel schemes. This enabled the consulting engineers to control the tunnel work rigorously. Almost negligible settlements and consequently no damage to buildings is an essential condition of such tunnelling. Furthermore this method makes it possible to compare the calculation routine of the project stage with the reality of the final phase. In Frankfurt, a 150-m-long test section was installed. In Munich, the test section was 60 m long. It showed that temporary ground stability could be achieved in difficult formations by groundwater lowering, combined with the instant installation of ground anchors and the application of a protection skin of shotcrete of 50 mm thickness, which was soon afterwards strengthened by a further layer of pneumatically applied concrete of 200 to 250 mm thickness.

Information on the use of the NATM in highly stressed rock formations (Tauern mountain chain in Austria) was given at the 24th International Symposium of the Austrian Society for Geomechanics, Salzburg, 1975 (summarized in *International Construction*, 1975, pp. 29–41). Experience has shown that under extreme stress conditions wire mesh, steel arches and shotcrete layers provide only protection against rock falls, water seepage and atmospheric influences, etc., while rock anchors attain the most dominant structural function by including the rock itself into the support system up to anchor length at least. The implication is that for every type of rock there is an optimum number of anchors of a certain length to suit the particular tunnel dimensions. This anchor system can safely compensate for the disturbed equilibrium of forces which was caused by the tunnel excavation. Progressive rock movements up to 400 or 500 mm have been tolerated. In order to allow for such deformations, gaps in the shotcrete skin are left between the side walls and the tunnel roof. There the steel mesh can deform and the steel arches can be adjusted. Before the rock comes to rest, it is convenient for a second shotcrete lining to be installed which can accept and resist stresses in a similar manner to the first.

Similar news concerning the use of the NATM comes from the large Gotthard road tunnel (Lombardi, 1974) now under construction in Switzerland, from the Snowy Mountains (Australia), from the 82-km long Orange–Fish tunnel (South Africa) and from Ruacana power station (South West Africa).

Very recently the construction of a section of the Munich Underground, in the middle of the town, in 'soft ground' (Golser *et al.*, 1977), using the NATM techniques, has allowed engineers to develop some of their advanced theories (Rabcewicz *et al.*, 1972). The very large Munich Underground tunnel was excavated with top heading and a bench near the head. In order to get to work on a nearly circular shotcrete-lined section as quickly as possible a temporary shotcreted invert was built on top of this bench, full excavation proceeding later.

The radial deformations of the excavations were measured systematically during the three successive phases of the lining:

(a) a first layer of shotcrete, including the supporting steel arches,
(b) a system of anchoring,
(c) the second layer of shotcrete.

The importance given to systematic measurement of radial deformations is also characteristic of Lombardi's methods which will be discussed in the next chapters.

The new NATM techniques touch the whole field of modern tunnelling. It has been used in the most diverse situations, but there is no agreed theory explaining how the thin shotcrete layer works so efficiently. Some authors give credit to its bending strength, others believe its efficiency is due to its shearing strength combined with the consolidation of the rock vault or rock walls, an opinion which seems to be corroborated by tests in the Inguri tunnel and others by Bernold. The effect of short rock bolts keeping rock blocks solidly together forming an arch, is easier to understand: the Romans knew how to build masonry arches without even using mortar. (See section 7.4 on the clastic theory of rock masses and section 10.7.2 on the Bernold sheet system and its failure by shearing.)

10.7.2 The Bernold sheet system (Wohlbier, 1969; Bernold 1970; Jaeger, 1975/76)

Bernold has developed a technique of thin concrete lining behind perforated steel sheets, which is applied immediately to the rock surface after blasting, before rock deformations due to internal rock ruptures and plastic deformations develop. With this method rock strains are maintained within limits and a radial rock pressure is transmitted at an early stage to the concrete lining. This technique is important when watertightness is imperative for hydraulic tunnels.

The technique developed by Bernold which has been used for all types of tunnels or galleries, some of them very large, as in the Gotthard road tunnel now under construction, is based on similar but not identical principles to the NATM. The basic idea is to build a thin concrete lining immediately behind the heading, the vertical face of the heading stiffens the rock vault nearby during early stages of concreting.

Figures 10.36 and 10.37 illustrate the technique favoured by Bernold. A ribbed reinforcement sheet (fig. 10.36) provided with slots or weep holes, rests provisionally on steel arches, supporting the steel sheet during concreting of the lining (fig. 10.37). The concrete lining is 15 to 25 cm thick. A rule of thumb given by Bernold suggests a ratio $d/R = 1/15$ or even $1/20$

(d = thickness of the concrete lining, R = radius of the concrete vault). During concreting, about 3 per cent of the concrete is lost through the weep holes. After hardening of the lining, grouting the contact zone with the rock is essential to fill any void. Then the arches are removed, but the steel sheet remains a permanent part of the lining.

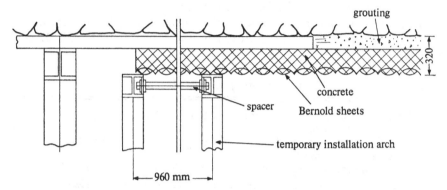

Fig. 10.36 Longitudinal section of the Bernold concrete lining system (dimensions in mm) (after Bernold).

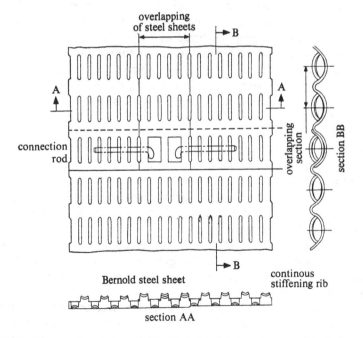

Fig. 10.37 Plan and sections of the Bernold sheet system used to help prevent deformations (after Bernold).

Bernold categorized rocks in the following way:
(*a*) slightly ficable (shearing strength larger than tangential tensions);
(*b*) ficable;
(*c*) very ficable (shearing strength smaller than tangential tensions);
(*d*) rock under stress,
The temporary stability of the rock roof is estimated at about 24–48 h for (*a*), 8–18 h for (*b*), 4–12 h for (*c*) and 0–2 h for (*d*).

Table 10.4 summarizes some technical data concerning the reinforcement sheet, of standard size 1200 × 1080 mm (= 1·296 m²) and with 12 cm overlap.

Table. 10.4   *Reinforcement sheet data*

Sheet thickness	1 mm	2 mm	3 mm	5 mm
Total weight (including connection rod) per m²	11·0 kg	21·0 kg	31·0 kg	52·0 kg
Steel cross-section of one reinforcement rib	0·57 cm²	1·09 cm²	1·62 cm²	2·7 cm²

The thin concrete lining has been tested in Japan at the Public Works Research Institute, Ministry of Construction. The test arch had a radius of 4·128 m and a thickness of 25 cm with 2 mm steel sheet (1st test series). Other tests were with thickness 15 cm, with 2 mm steel sheet (2nd test) and 25 cm without sheet. Rupture occurred by shear near the soffit of the arch. Similar ruptures have been observed *in situ* on tunnel linings.

Theoretical research confirmed that a thin lining can be ruptured by high tangential compression stresses, σ_t, in the arch, by buckling, bending or by shear. The most likely cause of failure is shear, and Kurt W. Weirich from the Bernold Office has published diagrams which explain how a lining has to be developed against possible rupture by shear (fig. 10.38).

Rupture by excessive tangential stress σ_t is not likely to occur. It will be remembered that the finite element method analysis of rock stressing is supposed to yield the σ_t values around the excavation, which, according to Bernold is not the most important. Buckling and rupture by bending would occur by lack of bondage of the lining with the rock, which justifies the grouting of the concrete-rock contact zone.

Gunite is used to cover the rough steel sheet and smooth the tunnel surface. In some cases the Bernold linings have been protected with a second concrete ring, concreted at a later stage. An impervious sheet of bitumen was laid between the two linings.

The Bernold system has been used mainly for large or very large tunnels but also for hydro-power galleries. It is very important to emphasize that a Bernold concrete lining put into place a few hours after rock excavation is really put under compression by rock deforming radially. When hydraulic

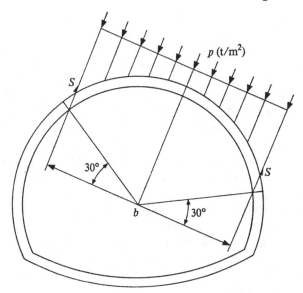

Fig. 10.38 Possible method of failure of a lining by the rupture of the lining due to shearing.

pressure is applied to a pressure tunnel, these compression stresses can balance some of the tensile stresses induced in the lining.

10.7.3 Tunnel-boring machines

Some 25 years ago (Oahe Dam, 1953) tunnel-boring machines were developed mainly for softer rock formations up to a rock strength of about 1000 kg/cm² to 1250 kg/cm². More recently machines capable of boring much stronger rocks, including igneous rocks, with a strength of up to 3500 kg/cm² have been developed. Machines ranging in capacity from miniborers (e.g. the the pilot gallery of the Sonnenberg tunnels has a diameter of only 3·5 m) to those capable of boring tunnels over 11·0 m in diameter are available on the market. These larger machines are obviously supplemented with sophisticated equipment for mucking and for tunnel roof support near the heading. Correct alignment of the boring is obtained with a laser-type directional control.

There are problems in determining the physical rock characteristics capable of influencing the progress of excavations: rock uniaxial crushing strength, tensile strength, *E* modulus, and resistance to sawing are the most important. These are determined experimentally, as are also the penetration of the borehead and the wear of different types of cutters and boring heads (Descoeudres & Rechsteiner, 1973; Rutschmann, 1974). The type of cutters, the thrust on the boring head, the type and size of the whole machine are determined on the basis of such tests, including extensive field tests.

There is a variety of rock-cutter bits. Many designs use roller bits or disc cutters mounted across the full diameter of the machine. For softer rock types, it is possible however to use the drag-pick type of cutter with tungsten carbide inserted tips capable of breaking out sedimentary rocks up to a strength of 1000 to 1250 kg/cm² (Pirrie, 1970). Rapid and easy transport and change of cutters is essential. Roof shields have been designed which provisionally stabilize the rock vault above the tunnelling machine.

For the excavation of the twin, 1300-m-long, Sonnenberg road tunnels (Beusch, 1972, 1974) it was decided to apply the tunnel reaming method for the first time. At first the pilot borehole (3·5 m diameter) was made. This pilot borehole was enlargened by two reaming machines, via a stage of 7·70 m to a final diameter of 10·46 m. This fully mechanical driving of the tunnel was competitive with conventional drilling and tunnelling methods. As in any excavation, it was important to discover if the rock type could stand without support for a considerable time. Rock bolting and shotcreting were used for final rock support.

The problems at Mangla Dam are typical of those encountered in situations where a variety of rock types have to be bored (Binnie *et al.*, 1967). Five 11-m-diameter tunnels, 488 m long, had to be excavated through indifferent clayish bedrock. In general, the material had a fairly low compressive strength, but within some sandstones occasional layers exist which have a strength of 422·5 kg/cm². It was these hard layers which led to the original selection of the disc-type cutters on the machine's head. During the early part of the excavation it became apparent that, except for the hard bands, the majority of the material was much weaker than had been assumed when the disc cutters were selected. Considerable difficulty was experienced with fall-outs into the 30 cm clear space between the leading edge of the cutters and the solid diaphragm of the machine, despite the 1000 h.p. available to turn the cutting head. Fortunately the motors were of the three-speed constant horsepower design with high torque and it was possible to operate the cutter head in reverse. Successful modifications to the boring machine included removing all the disc cutters and installing a complete set of adjustable 'drag bits'. Propulsion rams were added to enable the machine to move ahead without the use of extreme hydraulic pressures, and the sliding 'shoes' were made larger to compensate for the softness of the rock at the bottom of the tunnels.

Most tunnel-boring machines have been designed to produce tunnels with a circular cross-section. 'Mini fullfacers' have been developed to produce non-circular cross-sections. With these machines, rock is cut in a radial direction by means of a number of separate arms, each equipped with tungsten carbide tools.

Manoeuvrability of the machine, small turning radius, continuous control of steering while boring, convenient control of all functions of the machine from the operator's cab and accessibility of all parts are important design features. Most modern borers have laser-type directional control.

10.8 Rock bolting

The principal function of rock bolting is to reinforce and support partially detached, thinly laminated or incompetent rock which otherwise would be subject to failure. Bolts introduce additional stresses and strains in the rock mass, which should improve the general stability.

10.8.1 Types of bolts and cables

Rock-bolts are produced in a wide variety, by numerous manufacturers; some of the bolts are protected by patents.

The slotted type (fig. 10.39).   These have a steel bolt threaded on one end, a swaged or flame-cut slot in the other, with a wedge, a bearing plate and a

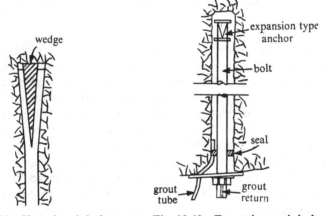

Fig. 10.39   Slotted rock-bolt.   Fig. 10.40   Expansion rock-bolt.

nut and washer. The bolt diameter may be $\frac{20}{32}$ in to 1 in, its length from 2 to 10 ft. A $1\frac{1}{4}$-in borehole is drilled to a depth 2 in less than the bolt length. The wedge is placed in the slot and the bolt driven over it with an impact hammer. The bearing plate, washer and nut are positioned and the bolt tensioned to about 10 000 lb.

Expansion-type bolts (fig. 10.40).   These consist of a head bolt threaded onto an internally tapered one- or two-piece shell. Tightening the bolt causes the shell to expand, providing an effective anchor. Hole diameters range from $1\frac{3}{8}$ to $2\frac{1}{4}$ in, the length of the hole being longer than the bolt. A $\frac{3}{4}$-in bolt in D mild steel has a yield stress of 14 000 lb.

Groutable rock-bolts. The 'Williams' groutable rock-bolt consists of a hollow-cored high-strength alloy steel reinforcing bar, threaded at each end. An expansion shell anchor is installed at the end of the borehole. A plastic grout tube is placed in position and plastics grouting material infiltrated. A 1-in diameter Williams bolt when fully grouted in a $1\frac{5}{8}$-in hole will develop an ultimate strength of up to 40 000 lb and support a working load of 3000 lb. Standard bolt lengths are 8, 10 and 15 ft.

The 'Perfo-bolt' system consists of perforated steel half-sleeves filled with mortar, tied together and injected in the drill hole. A steel bar is then pushed through the sleeve, forcing the mortar through the perforations and completely filling the drill hole. This system provides a non-tensioned installation and can be used in soft rock. Similar systems have been developed in Sweden, Germany and other countries.

Epoxy-grouted rock-bolts have been widely used instead of expansion rock-bolts in recent years. Bolts of this type, 3 m long, have been used for supporting the Kariba North Bank rock vault of the Machine Hall. For supporting the Tachien Dam steep rock face above the right-hand abutment, the contractors Torno-Kumagai and their specialized sub-contractors replaced the steel rods by epoxy grouted cables 3 to 10 m long, called 'tendons'.

Prestressed anchored cables. When great forces are required for rock support (for supporting the vault of a large underground excavation, or its walls, or for fixing steep rock faces) prestressed anchored cables are used. Forces up to 100 or 170 tonnes can be applied with one cable. There are many different types of cables available on the market. Some contractors even prefer to manufacture them on the site, buying only the more special equipment. Cables can be 10 to 30 m long. They consist of a varying number of threaded steel wires, diameter 7 mm to 16 mm. Special high-tensile steel is used. Such cables often have to remain anchored and stressed for very long periods; protection of the wires against rusting and any damage is therefore of the greatest importance. Protection can be provided by vaseline or bitumen coating. A polyethylene or PVC sleeve envelops the steel wire bundle. Flexibility of the cable during transport and ease of penetration in the borehole are important. For Waldeck II underground station, different types of cables proposed by manufacturers were systematically tested on the site, in a special small gallery, for ease of transport and handling. Cables differ in the quality of the steel used, the way they are anchored to the rock, the length of the anchorages, the shear stresses transmitted to the rock and the way the force is transmitted to the rock surface, the design of the cable head and the possibility of measuring at regular intervals the force in the cable with a load cell, adjusting it when required. The cables are usually tested up to 65% of the steel tensile strength and permanently loaded up to 50% of this strength. When rock deformations cause the force in the cable to exceed the accepted permanent load, the tensile force in the cable must be released (see: Veytaux underground power station, Rescher, 1968).

Figure 10.41 shows a VSL prestressed cable used at the Veytaux underground power-house (Switzerland). Cables of 170 tonnes (34, 8-mm wires),

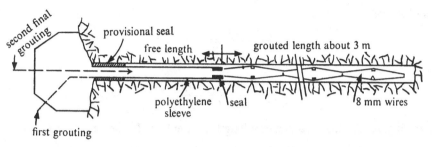

Fig. 10.41   VSL prestressed cable.

and 125 tonnes (24, 8-mm wires) were used, length varying between 11 m and 13 m, the length used for anchoring the cable being about 3 m. A special cement grout was used for anchorage, the strength of which was increased by the shape of the wires. The grout had a ratio of water to cement ($= W/c$) of 0·4, plus a special hardener (2% of the cement weight).

10.8.2   Calculating rock-bolts and anchored cables

(1) *Reinforcement of a horizontally laminated roof*

(a) *Completely suspended rock lamina* (fig. 10.42). A lamina of length L and width B, thickness t and unit weight γ_r is to be bolted. There are n_1 rows of

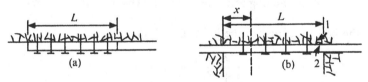

Fig. 10.42   Laminated roof, supported by bolting (after Obert & Duvall, 1967)

bolts, containing n_2 bolts per row. If the lamina is completely suspended by the bolts, the load per bolt is

$$W = \frac{\gamma_r tBL}{(n_1 + 1)(n_2 + 1)}.$$

(b) *The lamina is held at the edges* (fig. 10.42b). It acts as a clamped beam, The lamina 2 is bolted to lamina 1. Let q_1 and q_2 be the loads per unit length, E_1 and E_2 the moduli of elasticity, and I_1 and I_2 the moments of inertia of the two beams. The deflections y_1 and y_2 of the two beams must be identical so that

$$y_1 = y_2 = \frac{(q_1 + \Delta q)x^2(L - x^2)}{24E_1I_1} = \frac{(q_2 - \Delta q)x^2(L - x^2)}{24E_2I_2}$$

and therefore:

$$\frac{(q_1 + \Delta q)}{E_1I_1} = \frac{q_2 - \Delta q}{E_2I_2},$$

or

$$\Delta q = \frac{q_2E_1I_1 - q_1E_2I_2}{E_1I_1 + E_2I_2},$$

which gives the load Δq per unit length on the bolt. In the case that $I_1 \cong \infty$

$$\Delta q = q_2.$$

(2) *Bolting an inclined fissure or fracture plane.* F is supposed to be the force parallel to the rock surface, ϕ the angle of friction along the rock fracture, and α the angle the normal to the joint plane makes with the surface (after Obert & Duvall, 1967). B is the force in the bolt.

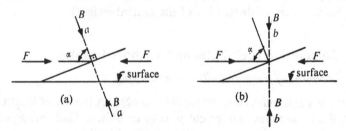

Fig. 10.43   Bolting of a rock fault (after Obert & Duvall, 1967).

When the bolt is normal to the joint (fig. 10.43a) then:

$$\frac{F \sin \alpha}{B + F \cos \alpha} < \tan \phi,$$

or

$$\frac{B}{F} > \sin \alpha \, (\cot \phi - \cot \alpha).$$

If $\alpha < \phi$ no bolt is necessary.

When the bolt is normal to the surface (fig. 10.43b), then

$$\frac{F \sin \alpha - B \cos \alpha}{F \cos \alpha + B \sin \alpha} < \tan \phi.$$

In the two examples, bolting is not effective unless F is small.

(3) *Bolting the roof of a gallery* (after Talobre 1957). Talobre assumes implicitly in his approach to the problem of rock bolting that the gallery is excavated in a rock where horizontal and vertical components of the residual stresses are identical ($\sigma_h = k\sigma_v$; $k = 1$) and there is no preferred direction of stratification. When the tunnel roof is circular, it can be assumed that bolts consolidate a circular rock arch within the rock mass. This arch is compressed radially by the mass of rock and circumferential stresses develop in it. Talobre (1957) gives an example (shown in fig. 10.44).

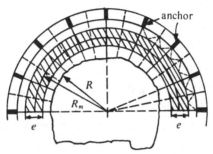

Fig. 10.44   Rock-bolts reinforcing a circular roof (after Talobre, 1957).

Tunnel internal radius	$R = 2\cdot50$ m
Thickness of the supporting arch	$e = 1\cdot00$ m
Radius of the supporting arch	$R_m = 3\cdot75$ m
Load on the arch (assumed)	$p = 3\cdot0$ t/m²
Thrust N in the arch	$N = 3 \times 3\cdot75 = 11\cdot25$ t
Circumferential stress	$\sigma_t = (N/e) = 11\cdot25$ t/m²
Assuming the intrinsic curve (Mohr circles are to be traced) for the rock to require a radial component σ_r minimum for maintaining rock stability	$\sigma_r = 2$ t/m²
Area covered by one bolt	$1\cdot5 \times 1\cdot5 = 2\cdot25$ m²
Required force in one bolt	$T = 2\cdot0 \times 2\cdot25 = 4\cdot5$ t
Tension force used for one bolt	$2 \times 4\cdot5 = 9$ t

(4) *Designing rock-bolts after L. v. Rabcewicz* (fig. 10.45). The rock fault has the direction *a–a* at an angle α to the horizontal and the rock-bolts are

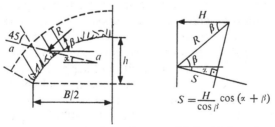

$$S = \frac{H}{\cos \beta} \cos (\alpha + \beta)$$

Fig. 10.45   Rock-bolts reinforcing a roof (after Rabcewicz, 1957).

inclined at an angle of 45° to *a–a*. *R* is the force in the arch formed by the roof, *H* the horizontal component of *R*, β is the angle of *R* to the horizontal and $R = H/\cos \beta$. The angle of friction on the rock fault is ϕ. The thickness of the arch is *e*, *B* its width and *h* its height. The force in the bolts cut by the fault *a–a* is ΣQ_b and *S* the shear force along *a–a*.

From fig. 10.45 we take that

$$S = \frac{H}{\cos \beta} \cos (\alpha + \beta)$$

$$= \frac{H}{\cos \beta} \sin (\alpha + \beta) \tan \phi + \frac{\Sigma Q_b}{\sqrt{2}} (1 + \tan \phi),$$

and

$$\Sigma Q_b = \frac{H}{\cos \beta} [\cos (\alpha + \beta) - \sin (\alpha + \beta) \tan \phi] \frac{\sqrt{2}}{1 + \tan \phi}.$$

Assuming horizontal fissures and $\alpha = 0$, Rabcewicz gives the following example:

$$\alpha = 0, \quad B = 14 \text{ m}, \quad h = 4 \cdot 5 \text{ m}, \quad e = 5 \text{ m},$$

$$\tan \phi = 0 \cdot 70, \quad \tan \beta = 0 \cdot 70, \quad H = 40 \text{ t}$$

$$\Sigma Q_B = H(1 - \tan \beta \tan \phi) \frac{\sqrt{2}}{1 + \tan \phi}$$

$$= 40(1 - 0 \cdot 49) \frac{1 \cdot 4}{1 \cdot 7} = 17 \text{ t}.$$

Neglecting the friction $\tan \phi$ along the fissure ($\tan \phi = 0$) yields

$$\Sigma Q_B = 40\sqrt{2} = 57 \text{ t}.$$

(5) *Anchored prestressed cables; Veytaux underground power-station (Switzerland)*, (Rescher, 1968). The underground power-station at Veytaux has been excavated in average to indifferent quality limestones and marl of the Dogger formation (80% to 90% limestones, 10% to 20% schists). The strata, 0·20 m to 1·50 m thick, are nearly horizontal. They are cut by three different systems of faults and fractures, some of them filled with clay-like material or with mylonites. The rock is characterized in that a gallery wider than 3·50 m would not be stable without support. Many vertical fractures, parallel to the axis of the cavern, and some crushed zones were encountered. The main cavern is 136·50 m long, 30·50 m wide and 26·65 m high.

The excavation was obviously difficult, and support had to be adequate. Several alternatives were considered; reinforcement of the roof and sides with prestressed cables was chosen. The design of the bolting was based on a modulus of elasticity for the stable loads $E = 100\,000$ kg/cm^2, a friction factor $\phi = 31 \cdot 5°$ and a cohesion factor $c = 3$ kg/cm^2. The problem was assumed to be two-dimensional, stresses in the direction of the length axis of the cavity were neglected, and a ratio $k = \sigma_h/\sigma_v = 0.33$ assumed.

Cables (38 wires of 8 mm, 11 to 13 m long) were placed in 115-mm-diameter boreholes, tensioned to 170, then grouted under a 140 tensile force. Special Danish cement was used which hardened after five days, the cable was then finally tensioned. Other cables of 125 tonnes maximum strength (24 wires of 8 mm) in 102-mm-diameter boreholes were also tensioned with a force of 115 tonnes. In all, 652 cables were used. Additionally, 1700 smaller bolts of 3·5 to 4·5 m length, capable of carrying 15 tonnes were used between the larger meshes of the larger cables (the mesh 4·30 × 2·90 m for the roof).

The whole design was analysed as a two-dimensional problem with the mathematical method of finite elements of Zienkiewicz, checked by a series of photoelastic models. The last of these was a fissured model, with the main geological joints observed on site reproduced. The tan ϕ on the model was only 0·19 but stability of the arch was nevertheless obtained by annular compression.

10.8.3 Grouting rock about cavities; case histories

(1) Rock grouting techniques were developed mainly to achieve imperviousness of rock abutment and dam foundations. At the same time they were used to improve the watertightness of hydro-power pressure tunnels. Between 1920 and 1925 grout pressures of 4 kg/cm² were used, and 4 kg/cm² are still used in some cases; for example, the free-flow tailwater tunnel of Santa Massenza, Italy, was grouted at low pressure. Pressures of 7 kg/cm² are usual although high pressures of 40 kg/cm² have also been known in some cases.

Grout closes rock fissures as it hardens. In addition to lowering rock perviousness, it creates at least temporary zones of compression in the rock and increases the rock modulus of elasticity. Therefore, grouting may be used for consolidating the rock about a cavity or for creating a ring of compressed, more compact rock, round the concrete lining of a pressure tunnel. Some of the compression stresses may be transmitted from the rock to the lining, and extra care is necessary when grouting behind a steel lining to avoid the danger of buckling it (Amstutz, 1950/53).

The two following case histories illustrate some of the more interesting aspects of rock grouting around hydro-power pressure tunnels.

(2) *The two pressure tunnels of the Mauvoisin hydro-power scheme (1957).* The 4·7-km-long tunnel of Mauvoisin–Fionnay has an internal diameter, when concrete lined, of 3·20 m. It withstands static pressures varying from 166 m at the inlet to 194 m near the surge tank. Dynamic pressures may rise to 204-m. The lining consists of a 25-cm concrete ring. The whole tunnel was pressure-tested to locate weak sections, which were then reinforced with an

additional gunite lining 7 cm thick, bringing the internal diameter down to
3·06 m (shotcrete was not in general use at that time).

Because of the high hydrodynamic pressure, the tunnel was designed well
inside the mountain, and the intermediary adit is 710 m long. It divides the
tunnel into two geologically very different sections. When testing the upstream section with a hydrostatic pressure of 16 kg/cm^2, average water
losses were 246 l/s. Testing the downstream section at 17 kg/cm^2 caused
losses of only 0·3 l/s – a thousand times smaller. The upstream section was
grouted and reinforced with gunite, the lower section was not. (Gunite is a
mortar projected from an airgun.) Grouting was in three stages: 2·5-m-deep
drill holes were grouted at 6 kg/cm^2, then 1·5-m-deep holes were grouted at
40 kg/cm^2 and finally 4-m-deep holes were grouted at 20 kg/cm^2.

From the geological point of view the upstream tunnel section can be
divided in three sub-sections:

Bündner-schists   158 m long
Quartzite zone   260 m long – very fissured
Casanna-schists   2650 m long – few fissures.

The Bündner-schists (Jura formation) consist of limestones, marbles and
dark phyllites. The Casanna-schists contain sericite, chlorite and gneisses.
The absorption of cement was as shown in table 10.5. (A 20-m-long tunnel

Table 10.5

Pressure	6 kg/cm^2	40 kg/cm^2	20 kg/cm^2
Bündner-schists	148 kg/m	110 kg/m	88 kg/m
Quartzite	400 kg/m	680 kg/m	800 kg/m
Casanna-schists	201 kg/m	52 kg/m	

section in quartzite absorbed 60 000 kg of cement!) About 2100 m of tunnel
in the upstream section were gunited, the steel reinforcements consisted
either of 16 rings of 18 mm diameter (heavy reinforcements), 16 rings of
14 mm diameter, or 12 rings of 14 mm diameter (light reinforcements). After
three years the tunnel was inspected and the gunited sections were found
to be practically without fissures.

The 14·72-km-long Fionnay–Riddes tunnel had to cross difficult rocks.

At one place in the carbon formation the tunnel heading had to be closed
with a concrete wall and the rock consolidated by grouting with cement and
chemicals under pressure of 100 kg/cm^2. In some sections steel supports had
to be used as the interstitial water pressure was very high. Grouting was
carried out at 16 kg/cm^2 and 40 kg/cm^2, cement absorption varied between
90 kg/m and 7400 kg/m. In the carbon zone it was about 1000 to 2000 kg/m.
About 100 small valves (consisting of small rubber spheres) were built in to
allow cleft water to penetrate into the tunnel. 110 l/s of water penetrated

in the empty tunnel, flow decreasing to 70 l/s, when the tunnel was under full pressure (80 to 90 m water pressure).

Altogether 3700 m of the tunnel had to be protected with gunite. A 100-m-long tunnel length had to be abandoned, and the tunnel axis is displaced by 50 m.

(3) *The Blenio tunnels (Switzerland)*. The total length of tunnels and galleries of the Blenio power system is about 77 km, out of which 2 km are in the Trias formation. The plan was such that it avoided much of the Trias areas which were expected to be of indifferent or even poor quality. But the upper aduction gallery, 2·20 × 2·60 m in size, had to cross Trias on a length of about 500 m. This was judged acceptable because the overburden was only 100 to 150 m and drainage was possible in a lateral direction. Very wet powdered dolomite was crossed with a 1·00 × 1·50-m gallery, which was then widened to the full 2·2 × 2·6-m section. It took thirty working days to cross the 3 m of carboniferous rock. The following 50 m in dry dolomite were no problem, but a second zone of very wet, unstable powdery dolomite had to be crossed over several hundred metres. A horizontal borehole was drilled in the tunnel axis in order to get more precise information about the rock quality and also to drain it. Two additional holes were drilled from the rock surface to locate the Trias zones.

It was suggested that the freezing method should be used to cross the poor area but it was rejected because the water discharge of 30 to 35 l/s was estimated to be too great. The horizontal borehole showed that 60% of the powdery dolomite had grain sizes between 0·06 and 0·2 mm. A mixture of 1 litre potassium silicate in 1 litre of water with a coagulant consisting of 50 g of sodium aluminate in 1 litre of water in the ratio 1·5 to 1·0 was found to give good results, a gel being formed in 17 minutes. The grouting reduced the permeability factor from 2×10^{-3} cm/s to 2×10^{-6} cm/s and increased the crushing strength of the samples to 1·1 kg/cm². Additional grouting was carried out with a mixture of cement, dolomite powder and a coagulant 'Sika'. The 46-mm-diameter boreholes were 12 m long. Altogether 67 tonnes of cement, 34 tonnes of dolomite powder, 4300 litres of silicate and 200 litres of Sika were used for grouting. The grout holes were arranged like an umbrella around the gallery.

Excavation of the dangerous rock was carried out with the help of a 5·90-m-long, 3·04-m-diameter, circular shield and precast segments. A force of 1400 tonnes (14 × 100 tonnes) was needed to force the shield forward.

The costing figures are interesting (Swiss francs):

Excavation in good rock	784 Fr/m
Excavation in the trias	6700
Excavation with the shield	5715

This case history can be compared to that of the Malgovert tunnel in section 2.2.

(4) *Pressure grouting the Rama tunnel.* In an effort to create tangential compression stresses inside a circular concrete lining and increase its resistance to fissuring by internal hydraulic pressure, Yugoslav designers of the Rama hydropower tunnel have used high-pressure grouting of the rock mass around the tunnel.

The Rama pressure tunnel (Kujundzic, 1970) is 5 m in diameter and the internal water pressure reaches 10 kg/cm². Extensive preliminary exploratory works were carried out; engineering geological and geophysical investigations, rock modulus of elasticity and its distribution around the tunnel at different depths, rock deformations and permeability characteristics were investigated. Detailed investigations were undertaken on an experimental section 29·5 m long, and 5 m diameter.

The effects of the grout pressure in the lining were measured with 24 electro-acoustic and 8 mechanical extensometers. The measured compression stresses varied between 3 and 26 kg/cm² as can be seen from fig. 10.46. This

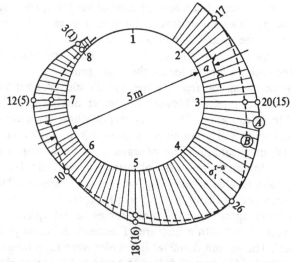

Fig. 10.46   Normal pressure stresses in a section of the concrete lining of the Rama tunnel. *A* indicates the maximum boundary stress at a grout pressure of 20 kg/cm² and *B* the final boundary stress (after Kujundzic, 1970).

is less than expected, possibly because the grout could not penetrate into the rock and between lining and rock, because of the limited number of fissures. After a while the stresses relaxed as can also be seen in the figure. Radial deflections of the rock were measured (between 7·4 and 10·4 × 10⁻³ cm).

The grouting reduced the leakage of water from 130 l/s to 62 l/s or from 4 l/s/1000 m² to 1·9 l/s/1000 m² thus proving the efficiency of the method.

Many other interesting examples could be mentioned. The purpose of the grouting of the Mauvoisin pressure tunnels was to obtain watertight pressure conduits in water-bearing rock formations. For the 13-km-long Inguri tunnel

(USSR), with a diameter of 9·5 m, the internal water pressure was designed to be $p = 16·5$ to 17 kg/cm² (Masur, 1970). Grouting to consolidate the rock was carried out in three stages: stage 1 to a depth of 1·5 m with grout pressure 10 kg/cm²; stage 2 to depth 4·0 m, grout pressure 20–25 kg/cm²; and stage 3 to a depth of 2·5 m with a grout pressure of 30 kg/cm². Similar consolidation techniques have been successfully used for the 9·5 km long pressure tunnel of Chute des Passes, in Canada ($D = 10·5$ m, $p = 18$ kg/cm²), the Bersimis I tunnel ($L = 12$ km, $D = 9·5$ m and $p = 13$ kg/cm²) and many other tunnels. A tunnel lining with precast concrete bricks provided with grooves permitting easy grouting of the space between bricks and rock was developed by Kieser (1960).

10.9 Classification of jointed rock masses for tunnelling. Estimate of required rock support based on rock characteristics

Bieniawski's *Engineering Classification* classifies rock masses irrespective of the engineering job to be done. Most other classifications of jointed rock masses are centred on tunnel design and construction, the main problem being the estimate of the required rock support, a problem of ever-growing importance. In 1872 the contractor Louis Favre started the excavation of the St Gotthard railway tunnel: the tunnel lining represented 25 % of the tunnelling costs. For the St Gotthard Road tunnel, presently being excavated under the same mountain pass, the cost of the lining represents 45 % of the costs (Lombardi, 1972).

Two different lines of approach have been developed for correlating the characteristics of rock masses and the required tunnel support. A first attempt could be described as the correlation of geomechanical parameters (unconfined compressive strength, defect systems) with types of rock support. Statistical analysis of case histories is important. A second approach could be called the engineering approach. It is based on the calculated probable deformations of the rock masses – depending on rock characteristics, shape of the tunnel and location of the relevant tunnel section relative to the tunnel heading at the time the rock support becomes efficient – and the deformation of the loaded and strained rock support.

(*a*) An early approach by Lauffer (1958) correlates rock types, the active unsupported span of rock and the time that this span takes to fail (stand-up time). An active unsupported span is the width of the tunnel or the distance from support to the face, if this is less than the width of the tunnel. Figure 10.47 reproduces, with some modifications by Bieniawski, Lauffer's diagram. Lauffer mentioned different factors influencing rock mass stability during tunnelling. They are schematically reproduced in fig. 10.48 (after Lauffer, 1958 and Müller, 1963). Lauffer's interesting suggestion on conditions for unsupported rock stability is too restrictive. The necessity for considering rock supports is obvious.

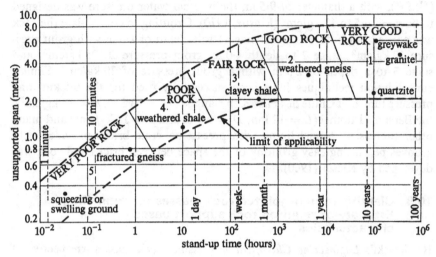

Fig. 10.47 Rock mass classification for tunnelling (modified after Lauffer).

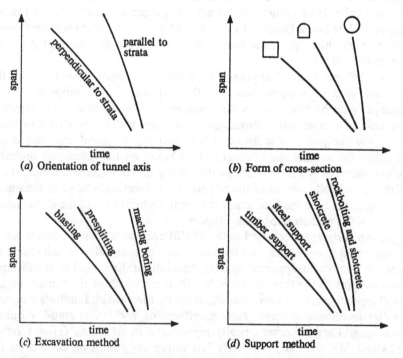

Fig. 10.48 Factors influencing rock mass stability during tunnelling (schematically after Lauffer and Müller).

(*b*) Before Lauffer, Terzaghi (1946) was interested in an inquiry concerning rock supports in railway tunnels supported by steel ribs with wooden blocking. Some experts believe the figures given by Terzaghi to be over-conservative in the better quality of rock, but the figures appear quite relevant to present-day practice when excavating medium-size tunnels in very difficult rock conditions, and are in fact quite widely used. In table 10.6 the support pressures have been tabulated for nine classes of rock mass defined by Terzaghi. It is assumed that the table refers to tunnels with height $H = B$, the tunnel width, $B = 5$ m and 10 m. The last column of the table refers to the RQD (Deere) and the mass quality Q, defined by Barton (1974), to be discussed in a further paragraph.

(*c*) Rock support in tunnels was also one of the major concerns of Bieniawski, when developing his Geomechanics Classification (see section 6.7). Table 10.7, reproduced from his 1973 paper, summarizes his recommendations concerning tunnel supports.

(*d*) A Norwegian team, headed by Barton (1974) was not convinced by the efforts of Terzaghi, Bieniawski and others, who published their findings at about the same time. Barton *et al.* produced an alternative rock classification and an Excavation Support Ratio (ESR) based on a very great number of case histories of tunnels, in particular an extensive, detailed survey by Cecil (1970) and another by Cording (1972). This information is analysed in a long document which, because of the great many details it gives, is very difficult to summarize.

Barton *et al.* chose six parameters to describe the rock mass quality, Q:

$$Q = \frac{\text{RQD}}{J_n} \cdot \frac{J_r}{J_a} \cdot \frac{J_w}{\text{SRF}}$$

Where:

RQD $=$ rock quality designation (Deere, 1963);

$J_n =$ joint set number;

$J_r =$ joint roughness number;

$J_a =$ joint alteration number;

$J_w =$ joint water reduction factor;

SRF $=$ stress reduction factor.

The authors' suggested values for these six parameters are given in table 10.8. The stress reduction factor, SRF, is an important parameter, when calculating the rock mass quality, Q. It takes into account special features of rock weaknesses which have a severe weakening effect on the whole rock mass. Geologists and geophysicists alike know the dangers of some isolated weak rock seams and of contact zones between seams. The SRF does account for them (Cecil, 1970).

The quotient (RQD/J_n) represents the overall structure of the rock mass.

Table 10.6 Estimates of roof support pressures (after Terzaghi and Barton)

Designation	Rock load (m)	Support pressure in kg/cm²		RQD	Q
		$B = H = 5$ m	$B = H = 10$ m	(after Barton)	
(1) Hard, intact	—	0	0	100	>1200
(2) Hard, stratified	0 to 0.5 B	0 to 0.6	0 to 1.3	100	20–10
(3) Massive, moderately jointed	0 to 0.25 B	0 to 0.3	0 to 0.6	100	50–25
(4) Moderately blocky and seamy	0.25 to 0.35 $(B + H)$	0.3 to 0.9	0.6 to 1.8	80	6–2
(5) Very blocky and seamy	$(0.35$ to $1.10)(B + H)$	0.9 to 2.9	1.8 to 2.9	50	1–0.4
(6) Crushed	$1.10 (B + H)$	2.9	5.7	20	0.08–0.04
(7) Squeezing rock	$(1.10$ to $2.10)(B + H)$	2.9 to 5.5	5.7 to 10.9	20	0.03–0.01
(8) Squeezing rock, great depth	$(2.10$ to $4.50)(B + H)$	5.5 to 11.7	10.9 to 23.4	0	0.004–0.001
(9) Swelling rock	up to 80 m	up to 20.0	up to 20.0	0	0.003–0.001

The two last columns are Barton's estimates, based on his formula for RQD (Deere, 1963) and Q (Barton, 1974)

Table 10.7 *Geomechanics classification: guidelines for selection of primary tunnel support; tunnel sizes: 5–12 m; construction by drilling and blasting* (after Bieniawski, 1973)

Rock mass class	Average stand-up time at unsupported span	Rock-bolts*		Shotcrete			Steel sets	
		Spacing	Additional support	Crown	Sides	Additional support	Type	Spacing
1	10 years 5 m	Generally not required						
2	6 months 4 m	1·5–2·0 m	Occasional wire mesh in crown	50 mm	Nil	Nil	Uneconomic	
3	1 week 3 m	1·0–1·5 m	Wire mesh, plus 30 mm shotcrete in crown as required	100 mm	50 mm	Occasional wire mesh and rock-bolts, if necessary	Light sets	1·5–2·0 m
4	5 hours 1·5 m	0·5–1·0 m	Wire mesh, plus 30–50 mm shotcrete in crown and sides	150 mm	100 mm	Wire mesh and 3 m rock-bolts at 1·5 m spacing	Medium sets plus 50 mm shotcrete	0·7–1·5 m
5	10 min 0·5 m	Not recommended		200 mm	150 mm	Wire mesh, rock-bolts and light steel sets. Seal face. Close invert.	Heavy sets with lagging, immediately 80 mm shotcrete	0·7 m

* Bolt diameter 25 mm, length ⅓ tunnel width. Resin bonded fully.

Table 10.8   (*abbreviated, after Barton* et al.)

---

(1) Parameter RQD; as in table 6.4 (after Bieniawski).

(2) Joint set number, J_n
    (*a*) Massive, few joints   0·5– 1·0
    (*b*) One-joint set   2
    (*c*) Two-joint sets   4
    (*d*) Three-joint sets   9
    (*e*) Three-joint sets, plus random   12
    (*f*) Four or more sets   15
    (*g*) Crushed rock   20

(3) Joint roughness number, J_r
    (*a*) Discontinuous joints   4
    (*b*) Rough, irregular, undulating   3
    (*c*) Smooth, undulating   2
    (*d*) Smooth, planar   1

(4) Joint alteration number, J_a ϕ_2
    (*a*) Tightly healed   0·75
    (*b*) Unaltered joint walls   1·0   25°–35°
    (*c*) Slightly altered   2·0   25°–30°
    (*d*) Silty coatings   3·0   20°–25°
    (*e*) Soft clay   4·0   8°–10°
    (*f*) Over-consolidated clay   6·0–8·0 12°–16°
    (*g*) Swelling clay   8·0–12   6°–12°
          (tan ϕ_2 = roughness coefficient)

(5) Joint water reduction factor, J_w
    (*a*) Minor inflow   1·0
    (*b*) Medium inflow   0·66
    (*c*) Large inflow   0·5
    (*d*) Exceptionally large inflow   0·2–0·05

(6) Stress reduction factor, SRF
    (*a*) Multiple weakness zones   10·0
    (*b*) Single weakness zone containing
         clay, depth less than 50 m   5·0
    (*c*) Same, depth more than 50 m   2·5
    (*d*) 'Sugar cube' rock   5·0
    (*e*) Heavy swelling rock pressure 10·0–15·0

When boreholes are not available, the RQD can be estimated using the following empirical formula:

$$RQD = 115 - 3·3J_v$$

(J_v = total number of joints per m³ of rock mass)

with RQD = 100 for $J_v < 4·5$.

    Barton's formula does not agree with Deere's definition of the RQD, which may cause confusion for $J_v < 4·5$.

The quotient (J_r/J_a) represents the roughness and degree of alteration of the joint walls or filling materials. The authors write: 'Quite by chance it was found that the function $\tan^{-1} (J_r/J_a)$ is a fair approximation to the actual shear strength that one might expect of the various combinations of wall, roughness and alteration products.'

Table 10.9 is an extract from a table calculated by Barton *et al.*

Table 10.9 *Values of* $\tan^{-1} (J_r/J_a)$

J_r	$J_a = 0.75$	4	12
4	79°	45°	18°
2	69°	27°	9.5°
1	53°	14°	4.7°

Pressure support P. Barton suggests the following formula for the pressure support, P (in kg/cm²) for the roof:

$$P_{\text{roof}} = \frac{2 J_n^{1/2} \, Q^{-1/3}}{3 J_r} \quad \text{(in kg/cm}^2)$$

or approximately

$$P_{\text{roof}} = \left(\frac{2 \cdot 0}{J_r}\right) Q^{-1/3} \quad \text{(in kg/cm}^2).$$

Barton's classification includes in its tables parameters representing the roughness of the joints and the strength of the joint filling, which are absent from Bieniawski's tables. All these figures are based on an extensive analysis of case histories, but such a system of correlations remains empirical.

10.10 Estimate of required rock support based on rock deformations

10.10.1 General remarks

Tunnelling engineers in charge of large excavations may feel the geomechanicist's approach discussed in section 10.9 to be not entirely convincing. Some of them (Kastner, 1962; Lombardi, 1971, 1974; Egger, 1973; Jaeger, 1973, 1975) advocate a completely different approach to rock-support estimates based on rock mass deformations, either measured or predicted on the basis of measured rock parameters.

In section 10.3, the transmission of the inside water pressure of a hydro-power pressure tunnel to the tunnel lining and to the rock mass has been analysed. It is easy to reverse the problem and to analyse the transmission of rock loads to the tunnel lining, or more generally to the tunnel support.

Dealing first with the simplest problems, equations of elasticity can be used to analyse deformations around a circular gallery cut through homogeneous elastic rock masses. The rock load will be transmitted to the circular lining or to the rock support. Other situations have to be considered too: is the rock behaving elastically or is it fissured and deforming plastically? How deep do fissures, caused by over-stressing of the rock around the tunnel, penetrate

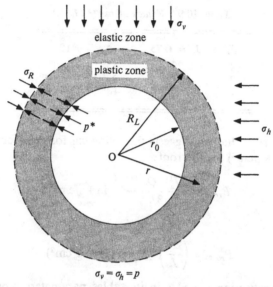

Fig. 10.49   Tunnel in plastic and elastic rock masses.

inside the rock mass (value of R_L in fig. 10.49)? How can a balance of deformations be established between rock and rock support? When does the rock support really react to the progressive rock deformations? Is there a delaying action to be considered, depending on time and on location of the considered tunnel section, distant from the tunnel face?

None of these questions, important to engineers, can be answered by the previous discussion on engineering classification rock masses.

10.10.2 The required rock parameters

Any analysis of an underground structure requires the knowledge of the following parameters:

(1) The uniaxial crushing strength σ_c of intact rock.
(2) The modulus of elasticity E of the rock mass in a direction parallel to the stratification and in a direction perpendicular to it.
(3) The angle of friction, ϕ, along a rock fissure.
(4) The cohesion c of the rock along a fissure.

(5) Number of joint sets, spacing of joints.
(6) Detailed description of the joints.

When mechanical rock cutting with a tunnel borer is considered, the following additional information is important:

(1) The Brazilian tensile strength test.
(2) The tangent modulus of elasticity for $\sigma_c/2$.
(3) Some cutting tests on rock material.

10.10.3 Action and reaction between rock mass and rock support. Case of a circular tunnel excavated in a homogenous rock mass

(*a*) It is assumed that the natural residual stress in the rock mass is isotropic, therefore $\sigma_v = \sigma_h = p$. If it were possible, when excavation and blasting occur, to replace immediately the failing residual stress, p, by an equivalent rock support pressure, p^*, nothing would happen to the rock mass: radial deformations, δ, would be impossible and δ would be zero.

In fact, invariably rock begins to deform in a radial direction, and displacements occur. For a certain time, the deformations probably remain elastic. Internal ruptures inside the rock mass will eventually cause the deformations to become 'plastic'. A curve of rock deformations, δ, versus the support pressure p^* can be traced, as on fig. 10.50*a* and *b*. When this curve

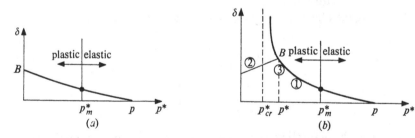

Fig. 10.50 Plastic and elastic rock masses surrounding the tunnel bore.

cuts the vertical axis, $p^* = 0$, rock mass stability can be achieved without rock support. But stability is also obtained for any intermediate point, with $p^* > 0$. In fig. 10.50*b* the curve does not cut the axis $p^* = 0$. Stability is not possible without a rock support causing a radial pressure $p^* > p_{cr}$. It is assumed that this reaction of the rock support can be represented by a curve (2), which cuts curve (1) at a point B, representing the accepted rock deformation occurring for a support pressure p^*. Depending on the choice for a rock support, there may be many possible solutions like B. Figure 10.50 represents the general situation for rock stressing when $p^* > 0$ and $R_L > r_0$.

(*b*) *Elastic rock deformations. Stable solution.* Assuming elastic rock deformations and stable rock masses, the equations developed in section 5.4 can be used for the stress–strain developing about a cavity, caused by residual rock pressure (residual stresses). These equations can be combined to the equations developed in sections 10.3.1 to 10.3.3.

10.10.4 Plastic deformation of the fissured rock mass

(*a*) *No rock support* (after Kastner, 1962 and Jaeger, 1975/76).   In most cases the overstressed rock around the tunnel or cavity gets fissured when excavation proceeds. Rock deforms plastically. In plastically deforming rock masses the following general equations, in polar coordinates, can be used (see Duvall & Obert, 1967, p. 75):

$$\left.\begin{aligned}
\sigma_r &= \frac{1}{r^2}\frac{\partial^2 F}{\partial \psi^2} + \frac{1}{r}\frac{\partial F}{\partial r} \\[2mm]
\sigma_t &= \partial^2 F/\partial r^2 \\[2mm]
\sigma &= \frac{\partial}{\partial r}\left(\frac{1}{r}\frac{\partial F}{\partial \psi}\right)
\end{aligned}\right\} \tag{1}$$

where r and ψ are polar coordinates and F an Airy function, still to be determined. σ_r, σ_t and τ are the stresses in radial and circumferential direction, τ being a shear stress.

Assuming circular symmetry ($\sigma_v = \sigma_h = p$), $\partial F/\partial \psi = 0$ and:

$$\left.\begin{aligned}
\sigma_r &= \frac{1}{r}\frac{dF}{dr} \\[2mm]
\sigma_t &= d^2 F/dr^2 \\[2mm]
\tau &= 0.
\end{aligned}\right\} \tag{2}$$

In plastically deforming rock, the basic equation of Coulomb (Hook) applies to the whole fissured rock mass and

$$\tau = c + \sigma \tan \phi \tag{3}$$

is a condition for all the values (σ, τ).

The Coulomb straight line (3) is traced on fig. 10.51. All Mohr circles are tangent to it and

$$\sin \phi = \frac{\sigma_t - \sigma_r}{\sigma_t + \sigma_r + 2\bar{\sigma}} \tag{4}$$

with $$\bar{\sigma} = c \cot \phi.$$

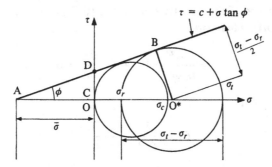

Fig. 10.51 Circles of Mohr.

For $\sigma_r = 0$, $\sigma_t = \sigma_c$ (uniaxial compression strength of intact rock). Therefore

$$\sin \phi = \frac{\sigma_c}{\sigma_c + 2\bar{\sigma}};$$

$$\bar{\sigma} = \frac{\sigma_c}{2} \cdot \frac{1 - \sin \phi}{\sin \phi}. \tag{5}$$

Introducing equation (5) into (4) yields

$$\sigma_t - \frac{1 + \sin \phi}{1 - \sin \phi} \sigma_r - \sigma_c = 0. \tag{6}$$

Writing

$$\zeta = \frac{1 + \sin \phi}{1 - \sin \phi} \tag{7}$$

equation (6) becomes

$$\sigma_t - \zeta \sigma_r - \sigma_c = 0. \tag{8}$$

In order to get the Airy function F, we introduce equations (2) into (8) and obtain in succession

$$\frac{\mathrm{d}^2 F}{\mathrm{d}r^2} - \zeta \frac{1}{r} \frac{\mathrm{d}F}{\mathrm{d}r} - \sigma_c = 0 \tag{9}$$

$$\left.\begin{array}{l} F = C_1 \dfrac{r^{\zeta+1}}{\zeta + 1} - \dfrac{\sigma_c}{\zeta - 1} \dfrac{r^2}{2} + C_2 \\[2mm] \dfrac{\mathrm{d}F}{\mathrm{d}r} = C_1 r^{\zeta} - \dfrac{\sigma_c}{\zeta - 1} r \\[2mm] \dfrac{\mathrm{d}^2 F}{\mathrm{d}r^2} = C_1 \zeta r^{\zeta-1} - \dfrac{\sigma_c}{\zeta - 1} \end{array}\right\} \tag{10}$$

At the rock surface of the circular tunnel (fig. 10.50) $r = r_0$ and $\sigma_r = 0$ (no rock support). Therefore:

$$\sigma_r = \frac{1}{r_0}\frac{dF}{dr} = 0 \quad \text{and} \quad \frac{dF}{dr} = 0.$$

Equations (10) yield

$$C_1 = \frac{1}{r_0^{\zeta-1}} \cdot \frac{\sigma_c}{\zeta - 1}$$

$$C_2 = 0$$

and for $r > r_0$

$$\frac{dF}{dr} = \frac{1}{r_0^{\zeta-1}} \cdot \frac{\sigma_c}{\zeta-1} \cdot r^{\zeta} - \frac{\sigma_c}{\zeta-1} r$$

$$\frac{d^2F}{dr^2} = \frac{1}{r_0^{\zeta-1}} \cdot \frac{\sigma_c}{\zeta-1} \cdot \zeta r^{\zeta-1} - \frac{\sigma_c}{\zeta-1}$$

and from (2) within the plastic zone ($r < R_L$):

$$\left. \begin{aligned} \sigma_{rp} &= \frac{\sigma_c}{\zeta-1}\left[\left(\frac{r}{r_0}\right)^{\zeta-1} - 1\right] \\ \sigma_{tp} &= \frac{\sigma_c}{\zeta-1}\left[\left(\frac{r}{r_0}\right)^{\zeta-1}\zeta - 1\right] \\ \tau_p &= 0 \end{aligned} \right\} \tag{11}$$

where the subscript p means 'plastic zone'.

It is important to determine how deep the fissures penetrate into the rock mass. Beyond the limit of the plastic and elastic zones, for $r > R_L$, the elastic rock stresses caused by the residual stresses $\sigma_v = \sigma_h = p$ are (see section 5.4):

For $r \geqslant R_L$:

$$\left. \begin{aligned} \sigma_{re_1} &= p\left(1 - \frac{R_L^2}{r^2}\right) \\ \sigma_{te_1} &= p\left(1 + \frac{R_L^2}{r^2}\right) \\ \tau_{e_1} &= 0 \end{aligned} \right\} \tag{12}$$

(the subscript e referring to the 'elastic zone').

Additionally at the limit $r = R_L$ a radial pressure, σ_r, is transmitted from the plastic zone to the elastic zone. This radial stress σ_r causes inside the elastic zone the stresses

$$
\left.\begin{aligned}
\sigma_{re_2} &= \sigma_R \frac{R_2^L}{r^2} \\
\sigma_{te_2} &= -\sigma_R \frac{R_L^2}{r^2} \\
\tau_{e_2} &= 0.
\end{aligned}\right\} \tag{13}
$$

The total stresses inside the elastic zone are therefore:

$$
\left.\begin{aligned}
\sigma_{re} &= \sigma_{re_1} + \sigma_{re_2} = p\left(1 - \frac{R_L^2}{r^2}\right) + \sigma_R \frac{R_L^2}{r^2} \\
\sigma_{te} &= \sigma_{te_1} + \sigma_{te_2} = p\left(1 + \frac{R_L^2}{r^2}\right) - \sigma_R \frac{R_L^2}{r^2} \\
\tau &= \tau_{e_1} + \tau_{e_2} = 0.
\end{aligned}\right\} \tag{14}
$$

For $r = R_L$, we equate 'elastic' and 'plastic' stresses.

$$
\sigma_{RL} = \sigma_{rp} = \sigma_{re} \quad \text{and} \quad \sigma_{tp} = \sigma_{te}
$$

and finally

$$
\left.\begin{aligned}
\frac{\sigma_0}{\zeta - 1}\left[\left(\frac{R_L}{r_0}\right)^{\zeta-1} - 1\right] &= \sigma_{RL} \\
\frac{\sigma_c}{\zeta - 1}\left[\left(\frac{R_L}{r_0}\right)^{\zeta-1} \zeta - 1\right] &= 2p - \sigma_{RL}
\end{aligned}\right\} \tag{15}
$$

σ_{RL} can be eliminated by addition of the two equations (15):

$$
\left(\frac{R_L}{r_0}\right)^{\zeta-1} (\zeta + 1) = \frac{2p(\zeta - 1)}{\sigma_c} + 2 \tag{16}
$$

and

$$
R_L = r_0 \left[\frac{2}{\zeta + 1} \cdot \frac{p(\zeta - 1) + \sigma_c}{\sigma_c}\right]^{1/\zeta-1}. \tag{16a}
$$

Equations (16) and (16a) obviously refer to the particular case where $\sigma_r = \sigma_h = p$, $k = 1$ and $p^* = 0$ (no rock support).

Some interesting examples on how to use equation (16a) are given in Kastner's and Rabcewicz's publications. From Kastner (1962), (see Jaeger, 1975/76) we take the following examples:

(i) The rock is supposed to be characterized by

$$
\sigma_c = 20 \text{ kg/cm}^2, \quad \phi = 30°, \quad \sin \phi = 0.5, \quad \gamma = 2.0 \text{ t/m}^3.
$$

The case $p = \sigma_v = 120 \text{ kg/cm}^2$ considered by Kastner, corresponding to an overburden of 600 m, yields $R_L = 2.55 r_0$.

The case $R_L = r_0$, with $\sigma_c = 2p$, would correspond to an overburden of only 10 kg/cm² or 50 m rock for $\gamma = 2$ t/m³.

(*ii*) This second case concerns a very high overburden with $H = 1200$ m, and $p = 312$ kg/cm² for a rock specific weight $\gamma = 2\cdot6$ t/m³. The tunnel is assumed to have a radius $r_0 = 2\cdot0$ m and the rock is very good, characterized by:

$$\sigma_c = 500 \text{ kg/cm}^2, \ c = 60 \text{ kg/cm}^2, \ \gamma = 63°,$$

The R_L value would be $R_L = 2\cdot026$ m and

$$R_L - r_0 = 2\cdot026 - 2\cdot00 = 0\cdot026 \text{ m only.}$$

The circumferential stress $\sigma_t = 2p = 624$ kg/cm²: there is a possibility for very thin overstressed rock layers around the tunnel to spall. The case mentioned by Kastner is similar to the Mont Blanc Road tunnel under 2000 m head, where $k = 0{,}4$ was assumed. Spalling occurred and was stopped with short rock-bolts.

(*iii*) A case analysed by Rabcewicz concerns a rock characterized by:

$$\sigma_c = 120 \text{ kg/cm}^2, \quad c = 30 \text{ kg/cm}^2, \quad \phi = 36\cdot9° \quad \text{and} \quad p = 312 \text{ kg/cm}^2$$

which yields for $r_0 = 2\cdot00$ m, $R_L = 3\cdot04$ m and $R_L - r_0 = 1\cdot04$ m.

The length of the required supporting rock-bolts should be larger than $R_L - r_0$. The required bolt pressure required for safe rock support will be estimated in the following paragraphs.

(*b*) *Rock support, p_m, required to avoid visco-plastic rock deformations.*

(*i*) Before analysing the general problem of supported plastic rock masses, it is convenient to investigate which rock support $p_{\max}^* = p_m^*$ is required to avoid any rock fissures and rock plasticity. Such conditions are as represented in fig. 10.52.

Let us first consider a pressure p_m^* acting on a plane surface. Assuming $\sigma_v = p$ to be the vertical pressure in the mass, and np the pressure parallel to the rock surface, fig. 10.52 shows that:

$$\sigma_r = p_m^*, \qquad \sigma_t = np \tag{17}$$

$$\sin \phi = \frac{\frac{1}{2}(np - p_m^*)}{\bar{\sigma} + \frac{1}{2}(np + p_m^*)}$$

and

$$p_m^* = np \frac{1 - \sin \phi}{1 + \sin \phi} - \sigma_c \frac{1 - \sin \phi}{1 + \sin \phi}$$

$$p_m^* = (np - \sigma_c) \frac{1 - \sin \phi}{1 + \sin \phi} = \frac{np - \sigma_c}{\zeta} \tag{18}$$

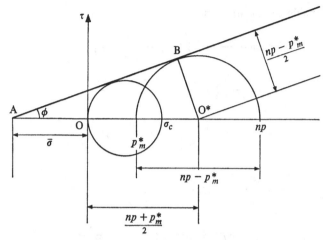

Fig. 10.52   Circle of Mohr for $\sigma_t = np$ and $\sigma_r = p_m^*$.

If we relate p to the rock crushing strength, σ_c,

$$p = s\sigma_c, \qquad s = p/\sigma_c$$

$$p_m^* = \frac{ns - 1}{\zeta}\,\sigma_c$$

(ii) When the rock surface is circular, with radius r_0, the pressure $\sigma_r = p_m^*$ on the rock surface causes inside the rock mass tangential rock stress $\sigma_t = -p_m^*$. The total tangential rock stress is therefore $\sigma_t = np - p_m^*$ and

$$p_m^* = \frac{(np - p_m^*) - \sigma_c}{\zeta} \tag{19}$$

or

$$p_m^* = \frac{np - \sigma_c}{\zeta + 1} = \frac{ns - 1}{\zeta + 1}\,\sigma_c \tag{20}$$

In the case of a circular tunnel excavated in homogeneous rock mass where $\sigma_v = \sigma_h = p$, $n = 2$ and $p_m^* = (2s - 1)\sigma_c/(\zeta + 1)$.

(c) *Plastic deformation of a fissured rock mass: rock support required, general case (circular symmetry)* (fig. 10.50). The groups of equations (1) and (2) are still valid. For

$$r = r_0, \qquad \sigma_r = p^* < p_m^*$$

and

$$\frac{dF}{dr} = r_0\sigma_r = r_0p^* = C_1r_0^\zeta - \frac{\sigma_c}{\zeta - 1}\,r_0$$

$$C_1 = \frac{1}{r_0^{\zeta-1}}\left(p^* + \frac{\sigma_c}{\zeta - 1}\right).$$

This gives inside the plastic zone:

$$\sigma_r = \frac{1}{r}\frac{dF}{dr} = \frac{1}{r}\left(C_1 r^\zeta - \frac{\sigma_c}{\zeta - 1}r\right)$$

$$= \frac{r^{\zeta-1}}{r_0^{\zeta-1}}\left(p^* + \frac{\sigma_c}{\zeta - 1}\right) - \frac{\sigma_c}{\zeta - 1}$$

$$\sigma_r = \frac{\sigma_c}{\zeta - 1}\left[\left(\frac{r}{r_0}\right)^{\zeta-1} - 1\right] + p^*\left(\frac{r}{r_0}\right)^{\zeta-1} \tag{21}$$

Similarly:

$$\sigma_t = \frac{d^2 F}{dt^2} = C_1\zeta r^{\zeta-1} - \frac{\sigma_c}{\zeta - 1}$$

$$= \frac{1}{r_0^{\zeta-1}}\left(p^* + \frac{\sigma_c}{\zeta - 1}\right)\zeta r^{\zeta-1} - \frac{\sigma_c}{\zeta - 1}$$

$$\sigma_t = \frac{\sigma_c}{\zeta - 1}\left[\left(\frac{r}{r_0}\right)^{\zeta-1}\zeta - 1\right] + \left(\frac{r}{r_0}\right)^{\zeta-1}\zeta p^*. \tag{22}$$

At the limit of the overstressed zone, for $r = R_L$, we get

$$\sigma_{R_r} = \frac{\sigma_c}{\zeta - 1}\left[\left(\frac{R_L}{r_0}\right)^{\zeta-1} - 1\right] + p^*\left(\frac{R_L}{r_0}\right)^{\zeta-1}$$

$$\sigma_{R_t} = 2p - \sigma_{R_r} = \frac{\sigma_c}{\zeta - 1}\left[\left(\frac{R_L}{r_0}\right)^{\zeta-1}\zeta - 1\right] + \left(\frac{R_L}{r_0}\right)^{\zeta-1}\zeta p^*.$$

Adding these two equations and eliminating the unknown σ_{R_r} stress:

$$2p = \frac{\sigma_c}{\zeta - 1}\left[\left(\frac{R_L}{r_0}\right)^{\zeta-1}(\zeta + 1) - 2\right] + p^*\left(\frac{R_L}{r_0}\right)^{\zeta-1}(\zeta + 1)$$

or

$$\left(\frac{R_L}{r_0}\right)^{\zeta-1}(\zeta + 1) + \frac{\zeta^2 - 1}{\sigma_c}p^*\left(\frac{R_L}{r_0}\right)^{\zeta-1} = 2 + \frac{(\zeta - 1)}{\sigma_c}\cdot 2p \tag{23}$$

an equation relating R_L/r_0 to p^*. This equation can be transformed and finally yields:

$$\left(\frac{R_L}{r_0}\right)^{\zeta-1} = \frac{2}{(\zeta - 1)}\cdot\frac{(\sigma_c + (\zeta - 1)p)}{(\sigma_c + (\zeta - 1)p^*)} \tag{24}$$

which is identical to an equation published – without demonstration – in 1962 by Kashner in his great treatise *Statik des Tunnel- und Stollenbaues* (see equation (7), p. 175 of Kastner).

For entirely elastic conditions (case *b* (*ii*)),

$$R_L = r_0, \quad p^* = p^*_{max} = p^*_m$$

yielding

$$p_m^* = \frac{2p - \sigma_c}{\zeta + 1} = \frac{2s - 1}{\zeta + 1}\,\sigma_c$$

as found before.

Equation (23) is supposed to supersede the well known equation of Fenner (1938)–Talobre (1957) which in some cases gives improbable results.

Remark. When the rock support can be adjusted at will (for example rock prestressing anchors) the use of equation (23) solves the problem. In most cases (for example concrete lining) the reaction p^* from the support depends on relative radial deformations of the rock surface and of the support, a problem which can be solved by iteration.

10.10.5 Some results obtained by more detailed calculations

(*a*) *Using the finite element or similar methods.* The theories and methods developed in paragraphs (*a*), (*b*) and (*c*) of section 10.10.4 assume homogeneous rock masses and $\sigma_h = \sigma_v = p$, conditions for residual stresses. Such simplified theories permit, nevertheless, very interesting approximations for the solutions of many important problems and for rock support estimates. They show that p^* depends on p, and on the angle of friction, ϕ, implicitly included in the parameter ζ. The p_m^* value is also shown to depend on the ratio $s = p/\sigma_c$, the physical interpretation of which is obvious for many tunnelling problems.

There are many other cases where this simplified approach cannot cope with the local geological or geomechanical situation, for example when the ratio $k = \sigma_h/\sigma_v$ is very different from unity, or when a weak rock seam crosses the cavity at an unfavourable angle. In such cases, the finite element method is the tool to be used.

Papers by Zienkiewicz *et al.* (1964/65), Hayashi (1970), Müller *et al.* (1970), Daemen (1975) etc., discuss typical examples and analyse important case histories. Lombardi (1970), sometimes critical of the finite element method, has published a series of graphs obtained with a method equating radial deformations and stresses at the boundary of the elastic and the plastic rock zones and at the contact of rock mass and rock support. They confirm that R_L depends on ϕ and on p, as indicated by the theoretical analysis, but also on $k = \sigma_h/\sigma_v$ and on the direction of the joints and faults (see section 16.3 on Waldeck II). A few typical diagrams of Lombardi (1974/75) are reproduced.

Figures 10.53 and 10.54 show how R_L depends on p^* and ϕ. Figure 10.55, from the same author, refers to an underground power-station of 30 m width, its arch being secured by means of stressed rock anchors. The forces in these anchors are not constant over the rock surface nor are they always radial. They are adapted to local conditions. The failure zone is mainly at

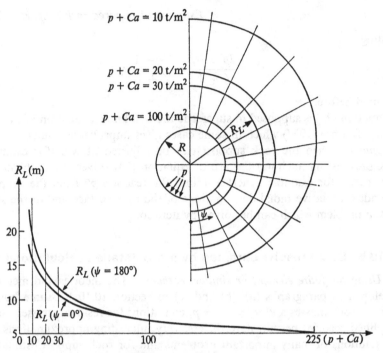

Fig. 10.53 Extension of the failure zone as a function of the reaction pressure; (after Lombardi, 1970), Ca = rock cohesion.

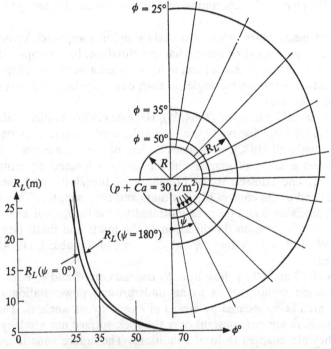

Fig. 10.54 Extension of the failure zone as a function of the friction angle (after Lombardi, 1970).

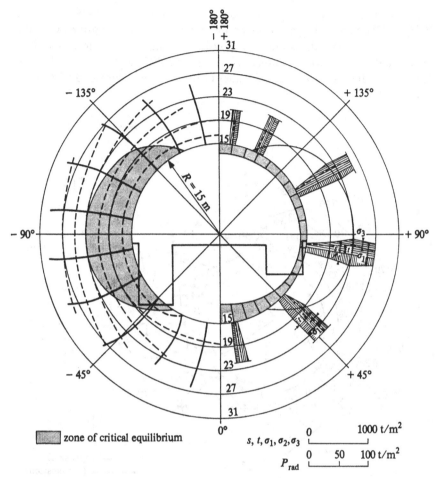

Fig. 10.55 Subterranean power station with roof secured by rock anchors. Anchor forces, failure zone and distribution of stress (after Lombardi, 1970).

the bottom of the excavation. The distribution of the stresses about the cavity are illustrated on the figure, so are the stress trajectories and the sliding lines.

(*b*) *Advanced research on the progressive expansion of the visco-plastic zone inside the rock mass surrounding a cavity being excavated.* Most rock deformations are not instantaneous. A first elastic rock deformation occurs very rapidly after blasting. Further deformations follow, progressing with the excavation. Visco-plastic deformations develop within the rock mass, which, in large excavations, may take several months to achieve equilibrium. In some cases, engineers are faced with a measurable creep of the rock mass.

A first attempt at solving these difficult problems consists in developing mathematical models for plastic rock deformations with time (Hayashi, 1968/70, Daemen, 1970), and solving these most complex equations with the finite element method. Some new physical parameters are introduced in such equations for which numerical values are difficult to guess. Hayashi decided to adjust such numerical values until his results coincided with some

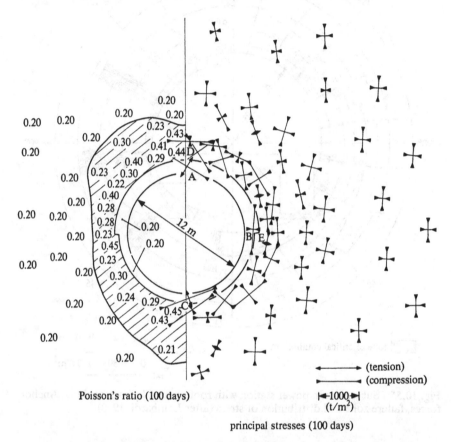

Poisson's ratio (100 days)

→———→ (tension)
►———◄ (compression)

|◄–1000–►|
(t/m²)

principal stresses (100 days)

Fig. 10.56 Stresses (right) and Poisson's ratio (left) in a tunnel lining and in visco-plastic rock mass (after Hagashi, 1968/70 and Jaeger, 1976).

measured curves of deformations. Figure 10.56 illustrates a relatively simple problem of progressive deformation of a 12-m diameter tunnel reinforced with a 1·20-m-thick concrete lining, depending on progress of excavations, for the case $k = 0·4$, after a period of 100 days. The formation of a compressed rock cylinder around the cavity can be seen on the right of the figure; on the left varying Poisson ratios can be noticed, confirming the remarks made in section 5.3.2. Comparing these stress patterns with the classical analytical solution, ignoring lining, fundamental differences appear. The classical solution shows, for $k = 0.40$, the circumferential stresses to be:

on the horizontal diameter $\sigma_\theta = 2.0p$
on the vertical diameter $\sigma_\theta = 0.2p$
(p = vertical component of residual stress.)

(c) *The engineering approach.*   Several engineers in charge of large tunnel designs advocate an 'engineering approach' to the problem of rock supports, to which Lombardi (1970/74) is an important contributor. They do not accept, without restrictions, either the geophysical classifications, or the results mathematicians obtain with the finite element method. Lombardi (1972) states that a correct description of all boundary conditions at the contact of rock and support and within the rock mass, at the edge of the visco-plastic mass, is more important than the method chosen for solving analytically or numerically the basic equations of elastic or visco-plastic deformations. Figure 10.57 shows rock deformations near the tunnel heading: a

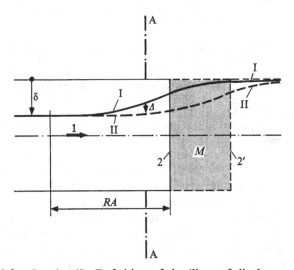

Fig. 10.57 (after Lombardi). Definition of the 'lines of displacement'. Influence of blasting the mass M on the position of the 'lines of displacement'. $1 =$ tunnel axis; $2-2' =$ heading before and after blasting; A–A Control section; $\delta =$ radial displacement at the excavation edge; I–I line of displacement before blasting; II–II line of displacement after blasting; $\varDelta =$ additional displacement; $RA =$ radius of influence of the heading 2.

three-dimensional problem, which, at some distance before or after the heading, is assumed to become two-dimensional. On fig. 10.57, the lines I–I and II–II represent the radial deformations δ of the rock surface before and immediately after the blasting of the mass M. The heading itself should be represented as a disc stressed and strained radially.

Figure 10.58 illustrates the evolution of the deformations with time and the progress of excavations, supposed to be continuous. Part of the rock deformation is said to be 'elastic' and occurs rapidly after blasting, the rest of the so called 'visco-plastic' deformation is due to the progressive loosening of the fissured rock mass. As shown in fig. 10.58, the deformation b of the rock support is only a fraction of the total deformation δ, and d is that part of rock settlement which the rock support prevents.

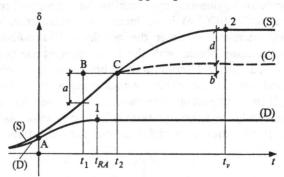

Fig. 10.58   (after Lombardi). Radial displacements at tunnel edge in a given control section as a function of time. t = time since the heading passed the control section; δ = radial displacement; D–D displacement versus time neglecting viscosity (continuous advance of the heading assumed); S–S displacement versus time including rock viscosity; t_{RA} = time elapsed until the distance from the heading to control section RA on Fig. 10.57; t_1 = installing support structure; t_2 = support begins to work after closing of any gap; b = deformation of supports; d = absorbed rock displacement.

The basic idea of the approach of Lombardi and his team has been illustrated in Fig. 10.50a and b. Characteristic lines of rock deformations $\delta = (p^*)$ are traced. They concern the elastic and the visco-plastic rock deformations. When the characteristic line cuts the ordinate $p^* = 0$, there is no necessity for rock support; the tunnel is self supporting. Otherwise a rock support developing a radial pressure $p^* > p_{cr}$ is required, p_m^* corresponding to the radial pressure required for keeping all rock deformations elastic.

Figure 10.59 shows rock characteristics (1) for an excavated gallery (tunnel bore), (2) for an elasto-plastic rock disc cut at the heading, (3) is the characteristic line relative to the rock core and (4) to the artificial rock support. δ' represents the sum of all rock deformations occurring before the rock support takes over some load and becomes active. This line (4) cuts (1) at point A, causing a radial pressure $p^* > p_{cr}$ on the rock surface. Without this support pressure p^*, the tunnel would collapse. Line (3) cuts line (2) at B. If not, the heading would collapse. There are therefore four possible stability or instability cases, depending on whether the tunnel roof or tunnel heading are stable or not.

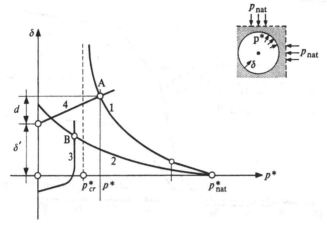

Fig. 10.59 (after Lombardi). Characteristic lines of deformations versus radial load for: (1) excavated gallery (tunnel bore); (2) rock disc at the heading; (3) rock core; (4) artificial rock support. δ' = sum of all rock deformations not absorbed by support; d = deformation of rock support corresponding to load p^*.

Any estimate of rock support based on rock deformations, like the method described in this paragraph, requires extensive *in situ* measurements of rock deformations. Figure 10.60 shows the arrangement adopted for measuring displacements of the rock inside the Gotthard Road Tunnel, now under construction in the Swiss Alps. Figure 10.60*a* shows the displacements of the points e_1 and e_2 of the horizontal extensometer. As can be seen on the diagram, displacements begin even before the excavation has reached the relevant measuring cross-section. Diagram *b* of the same figure refers to the displacements D of the inclined deflectometer. On this diagram too, the deformations D are recorded against the distance from the measuring section. In Fig. 10.60*c* the same deformations D are recorded versus time. This is the type of information required for tracing characteristic lines. The information obtained during the construction of the Gotthard Road Tunnel is being used for the design of the projected, far longer, St Gotthard Railway Basis tunnel. Figure 10.61 represents some of the many characteristic curves used for that project on the basis of *in situ* measurements.

Lombardi and his team have developed in detail a method which allows the correct prediction of expected rock displacements depending on the size and shape of the cavity and the physical characteristic of the rock mass. A rock stress diagram similar to the one shown in fig. 10.55 is obtained by trial and error, assuming boundary conditions for the rock support reaction on the rock surface, and for possible boundary conditions at the limit of the visco-plastic and the elastic zones inside the rock mass. Such assumptions must usually be corrected several times until an acceptable answer is obtained by iteration. The method used for tracing the stress–strain diagrams inside

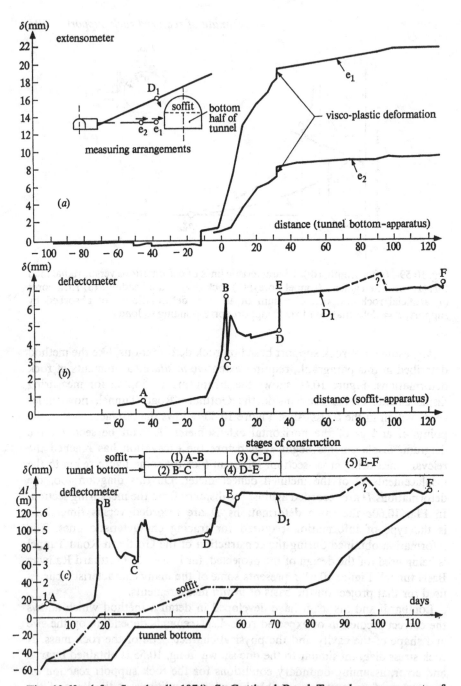

Fig. 10.60 (after Lombardi, 1974). St Gotthard Road Tunnel, measurements of rock displacements in granite. (a) Displacement of points e_1 and e_2 of extensometer versus excavation progress; (b) deflectometer displacements versus excavation-progress; (c) same versus time. Progress of excavations: A–B soffit excavation to relevant measuring cross-section; B–C excavation of bottom part of relevant section; C–D following up with excavation. Δl = position of bottom and soffit relative to measuring equipment.

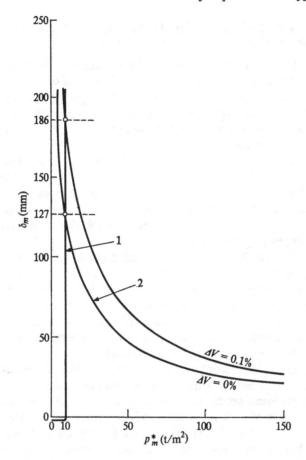

Fig. 10.61 (after Rutschmann, 1974). Rock characteristic lines (deformations δ versus radial load pressures p^*) for a tunnel diameter 11 m and an overburden $H = 1000$ m (Gotthard Railway Basis Tunnel project). Rock parameters: elastic zone: $\phi = 40°$, $c = 1.5$ kg/cm², $E = 4 \times 10^5$ kg/cm²; ruptured zone: $\phi = 35°$, $c = 0.5$ kg/cm², $E = 2 \times 10^5$ kg/cm². $\Delta V =$ increase in volume of loose rock mass. 1 = characteristic line for half rock core (heading); 2 of the tunnel wall.

the visco-plastic and elastic rock mass can be either the finite element method, or any other less sophisticated mathematical method the designer has evolved for that purpose.

A similar approach has been developed by Egger in a Ph.D thesis of Karlsruhe University (1973) and by Duffaut & Piraud (1975). The 'engineering approach' as the method based on rock deformations could be called, is attracting the interest of tunnel designers and is worth developing more systematically.

10.10.6   The St Gotthard tunnels. An engineering approach to the design of a very long railway tunnel

The old, 14·98 m long, St Gotthard railway tunnel connecting Germany to Italy through the Swiss Alps via Basle and Olten was opened in 1882. The contractor, Louis Favre, was one of the very great pioneers of his time. The tunnel enters under the high Gotthard mountain range on the North Side at Göschenen and reaches Airolo on the South Side of the Alps, at a level of about 1154 m above sea level.

At the moment a very large road tunnel is being built, 16·321 km long, also from Göschenen to Airolo. This tunnel will be 11·0 m wide and 8·7 m high. A service gallery, 2·5 m × 3·10 m, runs parallel to the main tunnel. Excavation work started in July 1970; the teams working on the service gallery from the south end and from the north end met in March 1976. Work on the main tunnel was proceeding normally.

The Swiss Federal Railways are considering the construction of a second railway tunnel, called the St Gotthard Basis Tunnel, at a lower level, about 500 m above sea level. Its length should be over 40 km, the diameter of the tunnel being 11 m, with a 4 m diameter service and aeration tunnel running in parallel to the main tunnel. Overburden thickness will reach 2500 m. Consultants for the design are Electro-Watt Engineering Services and Lombardi, who were also responsible for the Road Tunnel.

Extensive geological, geophysical and rock mechanical research is presently being carried out to prepare designs and cost estimates for the Basis Tunnel. The problems are the more difficult, as on such a length, and at such a depth, prediction of rock characteristics is difficult. Furthermore, at various points along the tunnel, larger excavations for service and overtaking of trains, will have to be constructed, and each of these will not only create stability problems of their own, but also may have some influence on the stability of the tunnel. The following work is being done:

(*i*) An overall geological study of the area (only partially known from previous excavation of the two other tunnels), based on borings, 1700 m deep, has been carried out. Two possible geological interpretations of the findings, one optimistic, the other pessimistic, have been worked out.

(*ii*) From the knowledge already acquired from similar rock types in that area, probable jointing and fissuring of the rock masses have been assumed for the different rock zones.

(*iii*) A series of geomechanical tests have been carried out to determine the rock modulus of elasticity, the uniaxial crushing strength of sound rock samples, rock cohesion along fissures or joints and the angle of friction along the joints. Rock samples were taken from the surface of the mountain, inside exploration galleries, adits and deep boreholes.

It was assumed that the rock parameters at the level of the deep tunnel would be about the same as those of the collected samples. Several typical assumptions were made concerning the residual stresses and the prestressing of the rock at great depth.

Rock stability estimates were on the lines of the methods developed in the preceding paragraphs. Figure 10.60 shows the type of information obtained during the construction of the Road Tunnel, and fig. 10.61 shows one of the

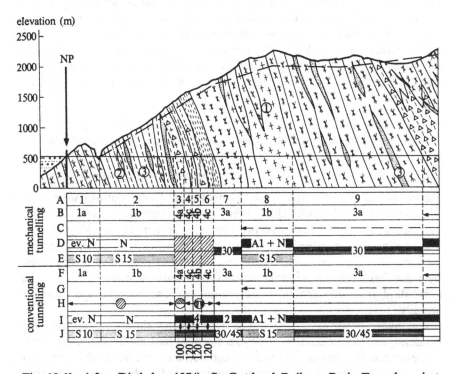

Fig. 10.62 (after Diethelm, 1974). St Gotthard Railway Basis Tunnel project. Excavation method and type and thickness of the lining as a function of rock quality and overburden.

A = tunnel section; B = case of rock stability: cases 1 to 4, degree of stability a, b or c; C = possible spalling of rock; D = provisional rock support; E = permanent rock support; F = case of stability; G = possible spalling of rock; H = type of excavation, full-face or progressive excavation; I = provisional rock support; J = permanent rock support. Type and thickness of rock support S 10 = shotcrete 10 cm; T 20 = precast concrete segments 20 cm; B 30 = concrete lining 30 cm.

many characteristic rock mass curves which were traced on the basis of the rock parameters obtained *in situ*.

Figure 10.62 summarizes the findings for one such exercise. Four cases had to be considered:

case 1: Tunnel bore and tunnel heading stable.
case 2: Tunnel bore stable, tunnel heading not stable.

case 3: Tunnel heading stable, tunnel bore not stable, requiring a rock support.

case 4: Tunnel heading and bore not stable, both requiring support.

According to quality, the rocks were classified in 6 types and the proportion of each type estimated for each tunnel section. As can be seen in fig. 10.62, projects were developed for mechanical excavation and for conventional excavation methods. The rock support was chosen on the basis of the expected rock characteristic lines for all sections along the tunnel. Projects were developed for optimistic and for less optimistic estimates of rock conditions and priced. Figure 10.62 gives other interesting information on the project.

10.11   Rock mechanics for underground hydroelectric power stations

10.11.1   General information on underground power stations

Before World War II, only a few hydroelectric power stations were built underground (Snoqualmie, USA, 1898; Brechbergmuehle, Germany, 1907; Projus, Sweden, 1907; Brommat in France, see Jaeger, 1955a). Shortly after World War II, a number of power stations were designed as underground stations. This was not for safety against military aggression, but, as proved by a detailed inquiry in several countries, for economic reasons. Since those early days the techniques of underground stations have been developing steadily.

Any hydro-power station should be designed to yield the highest financial dividend, and power economics play an essential part in underground works. For example, in some cases the choice between a conventional above-ground design and an underground power station depended on the geology of the site: poor rock at the surface was not suitable for anchoring a conventional pressure pipeline; inside the mountain, rock was competent and suitable for a large excavation. It was found that an underground power station was the better and cheaper solution. Problems concerning the pipelines in the open as against pressure shafts excavated in rock are often vital for final decision. Hydraulic problems may force decisions. High vertical axis pump-turbines, required for pumped storage, must be located well below the downstream reservoir level: in such cases the underground design is often the more acceptable (Waldeck II, chapter 16).

The location of the power house depends on the efficiency of the hydraulic system, the stability of the hydraulics, of the surges, the oscillations in the surge shafts and the pressure waves (Jaeger, 1955a, 1956, 1964c, 1976). Figure 10.63 shows the three main types of underground power stations. It is obvious, from these sketches and those in fig. 10.64, that difficult hydraulic problems have been encountered in the design of such systems.

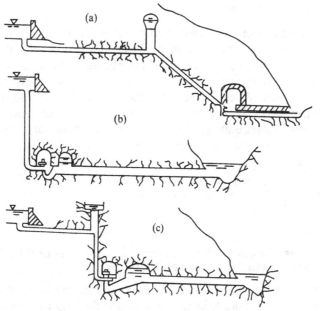

Fig. 10.63 Types of underground power station: (a) head-race arrangement (b) tail-race arrangement; (c) intermediate solution.

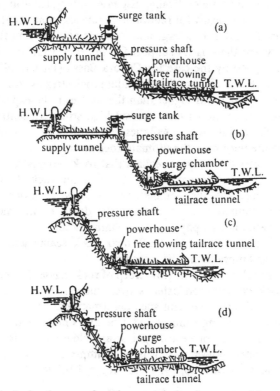

Fig. 10.64 Typical schemes of underground development: (a) upstream surge tank, free-flow tail-race; (b) two surge tanks system; (c) no surge tank; (d) down-5a, stream surge tank. (Jaeger 1949b, 1951956.)

The main element of an underground design is the machine hall: turbines and generators can be either a horizontal axis design or a vertical axis. For pumped storage stations, the horizontal arrangement (Veytaux) is rather an exception. The arrangement of the transformers is always a problem to be closely investigated. Whenever possible, the transformers should be located outside (when the cables can be kept short), but it is often unavoidable to have the transformers underground. The problem of the surge tanks, their location and size is a matter for lengthy hydraulic calculations.

Underground designs allow a high concentration of power. Figures 10.66 to 10.70 refer to typical powerful Canadian designs, most of them in igneous rocks. The more recent Churchill Falls is even far larger than the stations shown on the figures.

10.11.2   Classification of rock types for underground hydro-power stations (Jaeger, 1975/76)

Basically any decision on the design of an underground power station depends on the strain–stress pattern to be expected around the large cavities in relation to the intrinsic curve of the rock mass, *in situ* (see fig. 4.10).

(*a*) In case the most dangerous Mohr circles do not touch the intrinsic curve, the design could, theoretically, dispense with systematic rock support. (Cables could be used locally for reinforcing some less competent rock zones.) This is the case for several underground stations excavated in the Scandinavian granite, where there is no support for the rock vault nor for the rock walls. It is also possibly the case for Cabora Bassa (Jaeger, 1975/76) where rock is excellent. But at Cabora Bassa residual rock stresses stored in the rock mass are extremely high, far higher than the stresses to be expected from the rock overburden. The model tests showed high stress concentrations along the cavern walls, mainly where galleries cut the hall at a right angle, and where stresses are very near the ultimate rock strength.

(*b*) In many cases the designer realizes that rock is overstressed. That this is the case may be due to general weakness of the rock. In this case the approach by Fenner–Talobre, or better the method developed by Kastner will relate the required rock support to the depth of the damaged rock around the cavern and the physical rock characteristics.

Other possible cases concern local weak rock seams and the possible rupture by shear failure along these seams.

(*c*) Creeping rock conditions: slow creep of rock masses has been observed when slow rock deformation extends over many months. Special designs have to be worked out to contain such deformations. The concrete foundations of the heavy rotating runners and generators must be erected with the utmost precision and be protected against any slow rock deformation. The Saussaz underground power station is an example of a large excavation in creeping rock masses (Bozetto, 1974 and Jaeger, 1975/76).

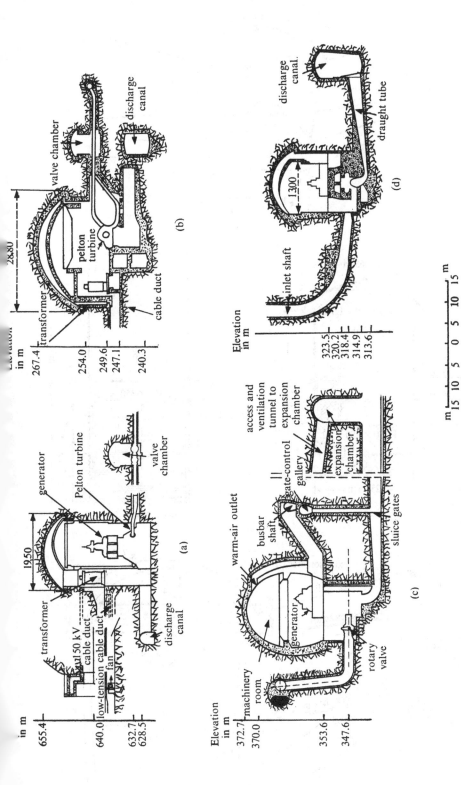

Fig. 10.65 Types of underground power-station: (a) Innertkirchen, Switzerland (gross static head 672 m); (b) Santa Massenza, Italy (gross static head 590/460 m); (c) Santa Giustina, Italy (gross static head 182/95 m); (d) Isere–Arc, France (gross static head 152 m).

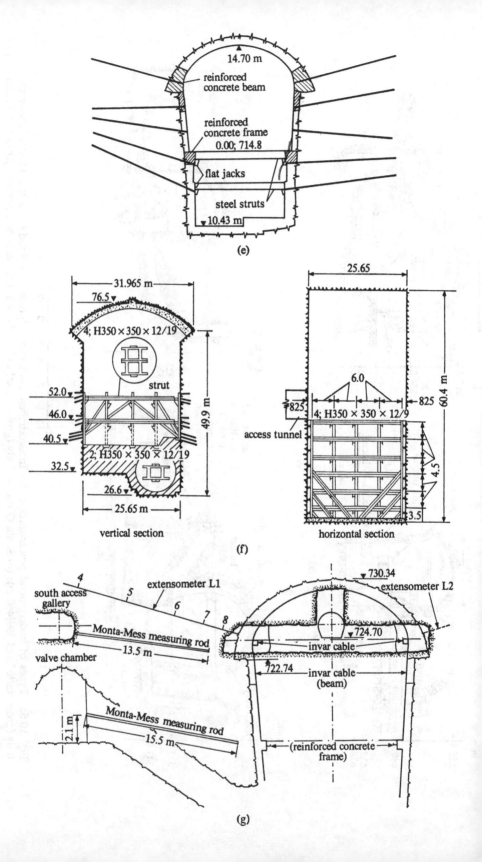

(e)

31.965 m
76.5 ▼
4; H350 × 350 × 12/19
strut
52.0 ▼
46.0 ▼
40.5 ▼
32.5 ▼
2; H350 × 350 × 12/19
26.6 ▼
25.65 m
49.9 m

vertical section

25.65
6.0
825   825
4; H350 × 350 × 12/9
access tunnel
60.4 m
4.5
3.5

horizontal section

(f)

4
south access gallery
5
extensometer L1
6
7
8
730.34
extensometer L2
Monta-Mess measuring rod
13.5 m
724.70
invar cable
valve chamber
722.74
invar cable (beam)
Monta-Mess measuring rod
2.1 m
15.5 m
(reinforced concrete frame)

(g)

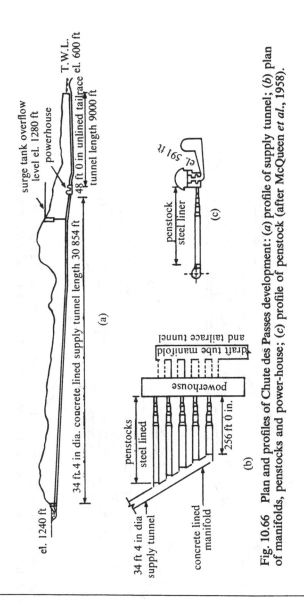

Fig. 10.66 Plan and profiles of Chute des Passes development: (*a*) profile of supply tunnel; (*b*) plan of manifolds, penstocks and power-house; (*c*) profile of penstock (after McQueen *et al.*, 1958).

See opposite

Fig. 10.65 (*e*) cross-section of the machine hall at Saussaz power station showing the steel supporting struts (after Bozetto, 1974 and Jaeger, 1976); (*f*) reinforcement to help prevent deformation at Kisenyama underground power station, showing steel reinforcing frame (after Yoshida & Yashimura, 1970 and Jaeger, 1976); (*g*) the method of measuring rock displacement using Monta–Mess measuring rods at the Saussaz power station (Bozetto, 1974; Jaeger, 1976).

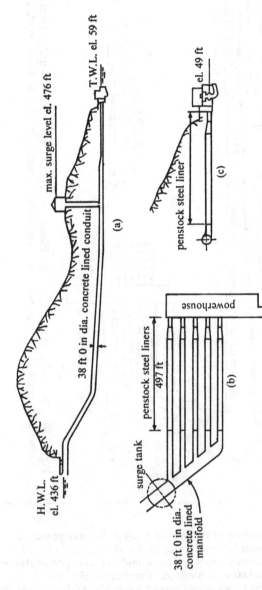

Fig. 10.67 Plan and profiles of Bersimis no. II development. (*a*) profile of supply tunnel and penstock; (*b*) plan of manifold, penstocks and power-house; (*c*) profile of penstock (after McQueen *et al.*, 1958).

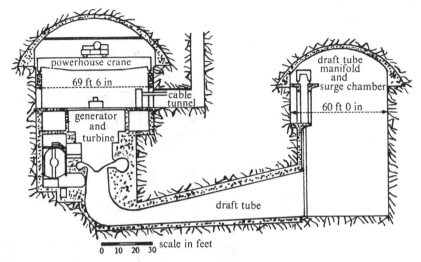

Fig. 10.68 The Chute des Passes power-house and draft tube manifold (after McQueen *et al.*, 1958).

Hayashi has investigated the case of progressive rock deformations penetrating inside the rock mass around a cavern. The deformations are considered as being a function of the time (Hayashi, 1970).

In all these cases (*a*) to (*c*) engineering geology plays a major part in the choice of the site and in many detailed decisions. There is no sound engineering without sound geological advice. The geologist should warn about rock masses being part of an unstable area. Faults have to be carefully investigated for possible unstability. The old rules, often advocated in the past, concerning the position of the cavern long axis relative to the most dangerous family of joints, stratifications and faults, still hold.

(*d*) Despite thorough investigation errors may still occur. The cavern for the underground power-house of Verbano, of the Maggia system (Switzerland) had the correct orientation. During excavations a rock slide occurred at one end and the cavern had to be shifted along its axis and relocated in a better rock mass.

A large geological fault crossed the rock downstream of the surge tank of the Santa Giustina power-station (Italy). The pressure tunnel and surge tank are located in fissured hard dolomite. It would have been possible to locate the power-house in the dolomite and cross the fault with the tail-race tunnel. In the final design the power-house is located downstream of the fault in plastic marl. Because of this decision the pressure shaft crosses the fault. Instead of a conventional steel-lined shaft, the designers decided to build a self-supporting pressure pipeline, capable of withstanding the full hydrodynamic pressures, located in a steeply inclined gallery and wide enough for inspecting the pipeline (fig. 10.65*c*).

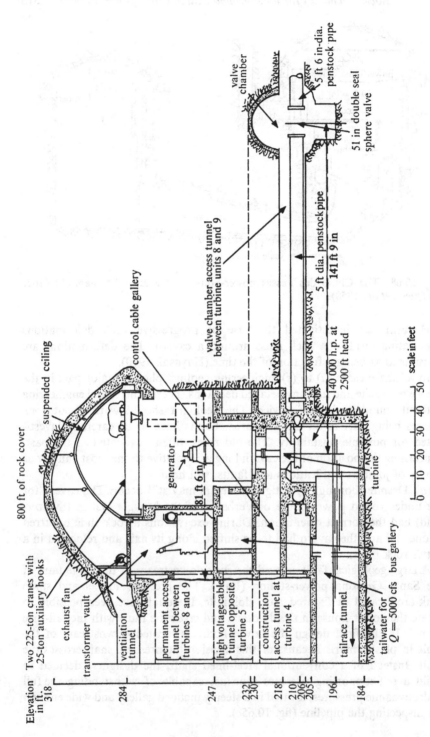

Fig. 10.69 Underground power-station at Kemano, Canada (gross static head 790 m).

Elevation
in ft.
318

800 ft of rock cover

Two 225-ton cranes with
25-ton auxiliary hooks

suspended ceiling

control cable gallery

valve chamber access tunnel
between turbine units 8 and 9

valve
chamber

5 ft 6 in-dia.
penstock pipe

51 in double seal
sphere valve

5 ft dia. penstock pipe
141 ft 9 in

140 000 h.p. at
2500 ft head

generator

81 ft 6 in

turbine

transformer vault

exhaust fan

ventilation tunnel

permanent access
tunnel between
turbines 8 and 9

high voltage cable
tunnel opposite
turbine 5

construction
access tunnel at
turbine 4

tailrace tunnel

tailwater for
Q = 5000 cfs bus gallery

scale in feet

0 10 20 30 40 50

284

247

232
230

218

210
206
205

196

184

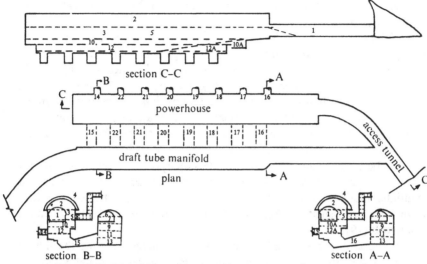

Fig. 10.70   Bersimis no. 1 power-house, excavation sequence.

Because of the pressure which may develop in the plastic marl, the power-house excavation is oval-shaped, heavily lined, with a thick horizontal beam which reinforces the structure at generator floor level (fig. 10.65c). Strong horizontal steel beams were also adopted for stabilization of the vertical walls of the Saussaz machine hall.

10.11.3   Stress–strain analysis of rock masses surrounding large excavations

(*a*) When rock behaves elastically, with constant value of the *E*-modulus, an analytical solution is possible when the cavern is assumed to be circular or oval-shaped. Some problems can be solved in three dimensions (Terzaghi & Richart, 1952). When rock is overstressed, an analytical solution could be developed only for a circular profile (Kastner, 1962; Jaeger, 1975/76).

(*b*) The finite element method (f.e.m.) gives mathematical solutions in case of overstressed rock around a cavity of any shape, even when the rock is jointed or crossed with faults.

(*i*) The results depend on the shape of the excavation. Dangerous stress concentrations occur at any sharp corner (see details in chapter 16 on Kariba North and Waldeck II). Negative stresses are often detected along high, flat, vertical walls and on the flat bottom of the excavations. An ovalized excavation is far safer than a conventional machine hall design with vertical walls. Similarly a low rock vault causes dangerous stress patterns; the rock vault should be well arched. When several parallel caverns are excavated, they react on each other. Depending on the value of the horizontal residual stresses, negative stress areas may develop between caverns. A detailed stress analysis should always include the surge tank area; reshaping of the cavern in order to reduce stresses at the top of the surge shaft has sometimes been necessary (Ruacana).

(*ii*) The ratio of the residual stresses, $k = \delta_h / \partial_v$, is one of the most important parameters of stress analysis, the stress pattern depends widely on this parameter. *In situ* measurements have shown that k varies from about 0·3 to possibly 1·5 or more; there is often a great dispersal of measured values; values of k larger than 1 were found in Scandinavian granite, but also in some parts of Africa and Canada. Designers often hesitate to base the final design on such high values and usually check the stress pattern for lower values as well. Very often $k = 0·4$ is taken as basis for safe designs.

(*iii*) The finite element method of stress analysis becomes most important when the rock mass is jointed, or when it is traversed by faults which substantially modify the pattern of stresses. Examples of such stress analysis have been published by Müller, Malina & Baudendistel (1970) and by Lombardi (1970).

(*iv*) The stress–strain analysis must also be carried out for the different phases of the progressive excavation. When excavating a flat rock vault, negative stresses develop at the crown of the vault; when excavation of the high walls proceeds downwards the negative stress areas move towards the walls. It has also been found that sharp corners, as they develop during excavation cause high stress concentrations which can irreversibly damage the rock mass. When the opening is lined, the lining can be considered as a boundary condition to the rock mass (mainly the rock vault), but very often the concrete lining is done when rock has already widely deformed.

(*c*) *Other methods* (Jaeger, 1975/76). Photo-elastic model tests are often used for checking results obtained with the finite element method or for checking details such as local stress concentrations which may not be obtained correctly on a finite element analysis. The designs of Veytaux, Kafue and Waldeck II were checked with this method.

The finite element method is usually applied assuming a two-dimensional pattern of stresses. The method could be, and has been, extended to three-dimensional models, but the computer effort is considerable. Stresses occurring at Cabora Bassa (Jaeger, 1975/76) have been analysed on a three-dimensional plaster of Paris model. Stress concentrations at the intersection of secondary galleries cutting the main machine hall at right angles were observed. The stresses were exceptionally high because of the very high residual stresses which, at Cabora Bassa, were present in the igneous rock mass.

10.11.4 Design and construction methods

Mining and railway engineers have done most of the pioneering work on the design and construction of underground openings and have provided a basis on which hydro-electric power designers have worked.

(*a*) Most of the problems the designers have to face have been mentioned in the previous paragraphs; summarizing all of this information it can be said that the following must all be closely investigated: geological and geotechnical information including rock strength, the rock intrinsic curve, rock jointing and faults, the residual stress pattern, the absolute value of these stresses and the ratio *k*; the shape of the excavations, the stress–strain pattern for different design alternatives and the different excavation stages; the hydrodynamic conditions, surges and water-hammer waves in the system.

The problem for the designer is to adjust correctly all these data and conditions. A great variety of solutions has been developed. The turbines and generators can be vertical axis type, which is the usual solution. Santa Massenza (Italy) (see Jaeger, 1955) and Veytaux (Switzerland) have horizontal axis arrangements. A classical solution is to locate the transformers in a cavern upstream and parallel to the machine hall, the surge tanks being on the downstream side (Churchill Falls, Ruacana, etc.). Sometimes the transformers are located in an extension of the machine hall, such avoiding two parallel excavations (Waldeck II). Kaech put the transformers inside the machine hall at Inerkirchen, the valves being in a small cavern on the upstream side of the machine hall: for safety reasons in case of a valve burst the flow would discharge direct into the tail-water tunnel (Jaeger, 1955).

Vertical walls have long been adopted as the normal solution in the design of machine halls but in some cases, spalling of the rock has occurred with such a design. An oval machine hall was adopted for Santa Giustina which was excavated in a plastic marl. Recent research using finite element methods of analysis has shown the advantage of oval excavations where there is a possibility of rock spalling (Waldeck II). In any case the rock vault above the cavern should be well arched. Flat vaults are likely to cause trouble.

(*b*) There are great difficulties in analysing the stresses in a concrete arch, if the arch is considered as supporting a rock load. How should the load be estimated? It may depend on the time gap between excavation and concreting. The main difficulty concerns the possible displacements of the arch springs. A flat arch has a tendency to push towards the rock mass, but experience shows that in most cases, during excavation of the high walls, the arch springs move inwardly. The finite element method usually assumes that the concrete vault is just a lining of the rock vault. The fissures which occurred on the Kariba North Bank concrete vault confirm such an assumption to be correct, as will be shown in section 16.2.

The success of the NATM in tunnelling and in supporting difficult galleries has induced designers to use similar methods for supporting rock vaults above the machine halls. Veytaux, Ruacana, Waldeck II are recent examples of this modern technique.

In table 10.10, *a* represents the spacing of the big anchor cables in the length direction of the cavern, *b* the spacing along the circumference of the rock vault, *P* the permanent load (smaller than the test load) on the cables

Table 10.10

	a	b	ab	P	p*	p*_tot
	m	m	m²	tonnes	kg/cm²	kg/cm²
Veytaux (vault)	4·30	2·90	12·5	140	1·123	
Lago Delio (walls)	2·97	3·00	8·9	80	0·897	p* average = 0·38 to 0·45 kg/cm²
Waldeck II (vault)	3·0	4·0	12·0	132	1·1	p*_tot = 1·6 kg/cm²

and p^* the average loading of the rock due to the big anchors. In two cases the total loading p^*_{tot} on the rock caused by anchors and bolts is also given.

At Veytaux 1700 bolts of length 3·5 to 4·5 m, load 15 tonnes per bolt were used in addition to the cables. For Lago Delio, the figures are as follows:

444 cables, $P = 80$ to 100 tonnes, $L = 17$ to 30 m
405 cables, $P = 35$ to 60 tonnes, $L = 16$ to 20 m
309 cables, $P = 19$ to 22 tonnes, $L = 15$ m
6549 bolts, $P = 5$ to 11 tonnes, $L = 3$ to 5 m

For Waldeck, II, for the vault:

890 cables, $P = 132$ tonnes, $L = 28$ to 30 m
 64 cables, $P = 100$ tonnes, $L = 15$ to 20 m
 62 cables, $P = 40$ tonnes
4000 bolts, $P = 12$ tonnes, $L = 4·5$ to 6·5 m

Waldeck II surge tank:

  45 cables, $P = 132$ tonnes, $L = 18$ to 21 m
151 cables, $P = 100$ tonnes, $L = 17$ to 20 m
 22 cables, $P = 35$ tonnes, $L = 12$ to 17 m
1018 bolts, $P = 10$ tonnes, $L = 6$ m

Sometimes cables are used to consolidate the concrete arch springs or the concrete supports for the crane rails. A very good example is the Saussaz power station (Bozetto, 1974). Inward movements of the vertical walls have to be controlled to within reasonable limits. When cables are not sufficient, strong horizontal beams, either of reinforced concrete or steel, sometimes combined with concrete, are used. The case of Santa Giustina has been mentioned. The Saussaz was excavated in creeping fissured sandstone and schists (Bozetto, 1974; Jaeger, 1976). As can be seen from fig. 10.65*e*, deep cables were used for anchoring the arch springings and horizontal reinforced concrete beams immediately below the springings. Major reinforcement was at the generator floor level where a rigid reinforced concrete frame is stiffened by horizontal steel beams. There is a second stiffening frame with steel beams a few metres lower.

Flat jacks are used to transmit the frame deformations to the steel beams. The pressure in these jacks adapts to the forces transmitted to the steel beams. The pressures had to be relaxed twice to avoid buckling the beams. Intermediate rows of deep anchors are used to reduce the horizontal deformation of the rock walls. The 15 -m and 18-m-long prestressed cables were designed to rupture at 160 t and 190 t respectively.

There are other power stations where the rock walls had to be supported with cables, arches, horizontal beams or struts. At Lake Delio, Italy, very high walls had to be supported by permanent prestressed cables and by provisional arches or beams, which were removed after settling of the rock deformations.

At Kisenyama, Japan, dangerous horizontal displacements of the walls were stopped when heavy struts were built in to stabilize them. These struts considerably obstructed the concreting of the turbine foundations (Jaeger, 1976; fig. 10.65*f*).

Excavation methods for large machine halls were initially based on the conventional tunnelling procedures used for railway tunnels. Decisions on the method to be adopted were sometimes left to the contractor. Consulting engineers later adapted the 'Belgian' and 'Austrian' tunnelling procedures to large cavities. A very good example of such a technique is the Kariba South Bank machine hall excavation. In very good rock, the 'quarrying' method was adopted, ignoring the stress and strain distribution and the stored residual stresses around the cavity. The very large Storrnorfors tailrace tunnel (Sweden), excavated, on the whole tunnel width, in exceptionally good Swedish granite, was entirely successful. On the other hand, the large excavation for Kariba II machine hall in gneissic rock excavated nearly full-face encountered exceptional difficulties. The large Kemano machine hall (Canada) was excavated using a complicated system of vertical shafts and a series of horizontal galleries in order to minimize excavation and mucking costs. Comments on Waldeck II (section 16.3) show modern thinking on how to proceed with large excavations in difficult rock.

Decisions on excavation methods should be discussed in detail between geologists, experts on rock mechanics, consultant and contractor.

Drainage of the large caverns is a subject which is often neglected. In most cases the machine hall is below the water-table. In addition, infiltration from the upstream side, sometimes equally from the downstream side, are to be expected. Cleft water pressure on the lining of steel-lined shafts is always a major problem to designers of shafts. All these problems should be analysed as a whole and adequate drainage of a large rock mass area may be the safest – and cheapest – solution.

In some cases the whole surface, rock vault and rock walls, are concrete lined, the lining being essential as rock support. In other cases only the concreting of the vault is essential to rock stability, the walls being just covered with a thin lining. Thin brick walls have also been built, the space between rock and walls being used for rock drainage. In few cases the rock vault too remains bare, sometimes protected by a false thin concrete arch. In some Swedish underground power-stations rock vault and rock walls are entirely bare.

The design and construction of machine halls and transformer halls, which are similar in shape, the latter being slightly smaller, are not the only problems to be solved. Downstream surge tanks are often required, the volume of which may be considerable; their shape and height often cause stress concentrations or negative stresses. Hydraulic problems have to be considered when balancing the advantages and disadvantages of long or short pressure conduits versus shorter or longer tailrace tunnels, free flow or under pressure, and the location of the machine hall along the hydraulic line of conduits. The size of some tailrace tunnels, several kilometres long, may be very considerable (Storrnorfors).

For moderate inside water pressures concrete lining of pressure shafts is often acceptable. Steel lining is the most usual solution. Steel linings should be designed to withstand buckling by outside cleft water pressure (Amstutz, 1950, 1953; Jaeger, 1955a). 'Prepakt' concrete has been successfully used between rock and steel lining for the Kemano pressure shafts. Kaech, at the Maggia power-station used wet, sandy concrete of reduced crushing strength but greater imperviousness to consolidate the steel linings of the pressure shafts and avoid cleft water pressures. Drainage of the rock about the shafts is important too.

Special techniques and equipment had to be developed for the excavation and lining of vertical and inclined shafts. The choice of the angle most favourable for inclined shafts is sometimes chosen to help move the spoil by gravity.

Detailed discussions of the design and construction of three very large underground power stations are to be found in chapter 16.

10.11.5   Modern measurement techniques in underground works (Jaeger, 1971, 1974)

Future success in design and construction of underground works will depend, to a large extent, on improved measurement techniques really adapted to the sequence of operations, excavations, lining and grouting.

The classical jacking tests – on which earlier tunnel designers relied – do not yield very convincing results. The stress pattern (compression stresses) under the jack does not compare with the stress pattern about a pressure tunnel. Several authors favour the 'hydraulic pressure chamber tests' which put the rock under tensile stress in the circumferential direction and yield more convincing values for the E modulus of the rock about the excavation. But this method is costly.

Seismic waves have been used by Kujundzić (Kujundzić, Javanovic &

Radosavljevic, 1970) to check the varying E-value of the rock about a circular tunnel. The result has confirmed the classical Talobre diagram. It is likely that correlations between the dynamic E modulus, measured by the waves, and the static E-value, measured in hydraulic pressure chambers, can be established, as indicated by some remarks of Kujundzić. It is essential to know this E-value for deformations of the tunnel and convergences of the rock to be correctly interpreted.

The k-value can be measured *in situ* either with the 'stress tensor gauge' of the Laboratorio Nacional de Engenharia Civil (Lisbon) (Rocha & Silverio, 1969) or with the simple 'doorstopper' of E. R. Leeman (1969) (South Africa). The stress tensor gauge consists essentially of a plastic cylinder in which electrical strain gauges are embedded. A 7·5-cm-diameter borehole is drilled to the required depth, a coaxial hole with diameter 3·7 cm and a length of about 90 cm is drilled at the bottom of the previous hole, and the stress tensor gauge is cemented to the walls of this hole. With the doorstopper method, a BX borehole is drilled to the required depth and its end is flattened and polished with diamond tools. A strain gauge rosette is glued to the flat end of the borehole and strain readings recorded. Then the borehole is extended with the BX diamond coring crown, thereby overcoring the rosette support and stress-relieving the core. The BX core, with rosette attached, is removed and final strain readings taken. At an early design stage, some *in situ* readings are taken giving preliminary information on the probable k-values. Later measurements from galleries and adits can provide some useful checks on the k-values at selected points of the underground system.

The present attitude of designers is to favour systematic and continuous measurements of rock strains and tunnel deformations. The measured strains should be compared with the local measured E moduli (seismic waves method) and then with the computed values for a series of possible alternatives (assuming the E-value, the jointing and friction factors, etc.).

From recently published case histories, it can be seen that expert designers favour long cable extensometers penetrating deep into the rock and anchored beyond the estimated R_L radius. Such extensometers have been used not only for checking tunnel and cavern roofs, but also in increasing numbers for checking the stability of cavern walls. The measurement of tunnel convergence, commonly used in mines, should be introduced in civil engineering for checking vertical and horizontal displacements of opposite points of tunnel circumferences. The rock jointing and the k-value may cause interesting distortion of the primitive circular tunnel shape.

Ten cable extensometers were used for the Lago Delio cavern (Italy) (Mantovani, 1970 and Fanelli & Riccioni, 1973) and 28 for the difficult Veytaux power cavern (Switzerland) (Rescher, 1968). As previously mentioned, Kujundzic used strain meters in the concrete lining of the Rama tunnel and checked the dynamic E modulus with seismic waves. Extensive measurements have also been carried out in Japanese underground works.

Measurements made inside or about the St Gotthard Road tunnel have been mentioned in section 10.10.6. More information on modern measuring techniques will be given in section 16.3 dealing with Waldeck II underground power-station. Other methods are described in a paper on Saussaz, a station excavated in creeping rock (Bozetto, 1974). Pressure from the vertical rock walls is supported by two horizontal concrete beams, heavily steel reinforced. Rock deformations were measured with an extensive system of extensometers, and checked by direct optical measurements (fig. 10.65g).

There is one vital parameter on which we really lack information: the increase in volume ΔV of loosened 'visco-plastic' rock. Few authors have commented about it and more information is urgently required on this point.

Reports submitted to recent International Congresses show an interesting trend in rock mechanics, which gives a more coherent insight into the behaviour of rock masses. Modern problems have been discussed for their relevance to civil engineering. It is obvious that additional fundamental research, as opposed to the technical approach, on the behaviour of jointed, fissured, fractured rock masses under strain and stress (see Bieniawski, 1969) is required. New, vitally important, fields will have to be investigated in the laboratories, in conjunction with the tests and measurements *in situ*, if real progress is to be achieved.

11 Rock mechanics and dam foundations

11.1 The classical approach to dam foundations

Some remains of very early dams have been found in Egypt, built by the pharaohs for irrigation. Some of them were pervious and could not have stored much water, others might have been washed away by floods. The oldest known arch dam ever to be built was discovered in 1956 by the French engineer Henri Goblot near Kebar, 150 km from Teheran. Some Spanish and French dams still stand (Alamansa, Elche, etc).

Several early dams failed. Others were built on pervious rock like Camarasa and Monte Jaque in Spain, and required expensive repair work to stop the leakages. Between 1910 and 1930 basic rules and techniques were developed for the safe design and construction of dams. They can be summarized as follows:

(1) check the imperviousness of the storage reservoir;
(2) check percolation conditions under the dam foundations and round the abutments;
(3) calculate gravity dams and buttress dams, including uplift and check the shear strength of the rock foundations;
(4) calculate arch dams and check compression and shear stresses along the periphery of the foundations.

In recent times these methods have been improved. The elasticity of the foundations is now included in the estimate of the total deformation of the dam, and more sophisticated methods are used to establish the curvatures of the shell forming the dam (Jaeger, 1958–64). But despite these improvements and the excellent quality of the concrete used on site, and the improved grouting techniques, tragic dam failures still occur (Gruner, 1963; Mary, 1968). Analysis of the conditions of such failures usually points to some hidden weaknesses in the rock.

Details of some of these failures will be given in part four, but special rock problems concerning dam foundations will be dealt with now.

11.2 Shear and horizontal stresses in rock foundations of dams; the tensile stresses

Jimenes-Salas at the Eighth Congress on Large Dams, 1964, Edinburgh, reported on the danger of tensile stresses and tensile fissures in the rock

[325]

foundations at the heel of dams. Italian authors reported to the same Congress on the displacement of dam foundations under full hydrostatic load, which could also be attributed to overstraining of the rock under tensile stress.

11.2.1 Classical approach

Conventional methods are used for estimating the vertical stresses and the horizontal shear stresses on the horizontal concrete foundations of gravity and buttress dams. Similar methods can be used for the foundations at the concrete periphery of arch dams. It is commonly assumed that, when rock masses are capable of withstanding the estimated vertical compression stresses, σ_v, investigation is also required to establish shear stress conditions assuming the law of Coulomb (see section 7.1 formulas (11) and (14)):

$$\tau = c + \sigma \tan \phi < \tau_{\text{ult}}.$$

When σ_1 and σ_2 are the principal stresses in homogeneous rock masses, the law of Coulomb can be written as

$$\frac{\sigma_1 - \sigma_2}{\sigma_1 + \sigma_2 + 2c \cot \phi} = \sin \phi.$$

When shearing along a weaker line of fracture has to be investigated the formula to be used is

$$\sigma_x \cos \beta \sin (\phi - \beta) + \sigma_y \sin \beta \cos (\phi - \beta) + c \cos \phi - u \sin \phi \geqslant 0,$$

where u is the interstitial water pressure, and where β is the angle of the plane of fracture versus the principal stress σ_1 (section 7.1, formula (16)).

This appears to be straightforward, provided σ_1 and σ_2 are known in direction and value, but it is not, because the horizontal stress components σ_h under the dam foundations, inside the rock masses, are not known, but could be measured.

11.2.2 The horizontal stress component σ_h and the shear stress τ under the foundations of a gravity dam

The following simplified approach gives an idea of the real values of σ_h in the rock under the foundations of a gravity dam (fig. 11.1).

If we assume that the vertical stresses form a triangular load distribution from $\sigma_v = 0$ for $x = 0$ to $\sigma_v = p$ for $x = B$ the load at point ζ $(0 < \zeta < B)$ is

$$P_\zeta = p \frac{\zeta}{B}.$$

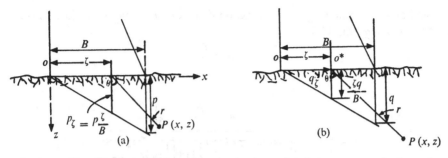

Fig. 11.1 Simplified method for estimating horizontal and shear stresses in the rock foundation under a dam.

At point $P(x, z)$ at a distance r from o^* the stresses caused by the isolated load p_z are:

$$\sigma_z = \frac{2p\zeta}{\pi r} \cos^3 \theta,$$

$$\sigma_x = \frac{2p\zeta}{\pi r} \sin^2 \theta \cos \theta,$$

$$\tau_{xz} = \frac{2p\zeta}{\pi r} \sin \theta \cos^2 \theta.$$

Integration from $\zeta = 0$ to $\zeta = B$ yields

$$\sigma_z = \frac{2pz^3}{\pi B} \int_0^B \frac{\zeta}{[(x - \zeta)^2 + z^2]^2} \, d\zeta$$

$$= \frac{p}{\pi B} \left[x \tan^{-1} \frac{Bz}{x^2 + z^2 - Bx} - \frac{Bz(x - B)}{z^2 + (x - B)^2} \right],$$

$$\sigma_h = \sigma_x = \frac{2pz}{\pi B} \int_0^B \frac{\zeta(\zeta - x)^2}{[(x - \zeta)^2 + z^2]^2} \, d\zeta$$

$$= \frac{p}{\pi B} \left[\frac{Bz(x - B)}{z^2 + (x - B)^2} + z \log \frac{z^2 + (x - B)^2}{z^2 + x^2} \right.$$

$$\left. + x \tan^{-1} \frac{Bz}{z^2 + x^2 - Bx} \right],$$

$$\tau_{xz} = \frac{2pz^2}{\pi B} \int_0^B \frac{\zeta(x - \zeta)}{[(x - \zeta)^2 + z^2]^2} \, d\zeta$$

$$= \frac{pz}{\pi B} \left[\frac{Bz}{z^2 + (x - B)^2} - \tan^{-1} \frac{Bz}{z^2 + x^2 - Bx} \right].$$

Similarly, assuming again a triangular load distribution, an isolated horizontal load q_ζ (fig. 11.16):

$$q_\zeta = q \frac{\zeta}{B}$$

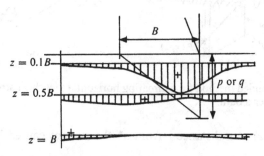

Fig. 11.2 Horizontal stresses, σ_h, caused by a vertical triangular load. The same diagram also valid for shear stresses, τ, caused by horizontal loads (after Del Campo & Piquer, 1962).

causes at point P, stress:

$$\sigma_z^1 = \frac{2q\zeta}{\pi r} \sin\theta \cos^2\theta,$$

$$\sigma_x^1 = \frac{2q\zeta}{\pi r} \sin^3\theta,$$

$$\tau_{xz}^1 = \frac{2q\zeta}{\pi r} \sin^2\theta \cos\theta,$$

and integration from $\zeta = 0$ to $\zeta = B$ yields:

$$\sigma_z^1 = \frac{2qz^2}{\pi B} \int_0^B \frac{\zeta(x - \zeta)}{[(x - \zeta)^2 + z^2]^2}\, d\zeta$$

$$= \frac{qz}{\pi B}\left[\frac{Bz}{(x - B)^2 + z^2} - \tan^{-1}\frac{Bz}{z^2 + x^2 - Bx}\right],$$

$$\sigma_x^1 = \frac{2q}{\pi B} \int_0^B \frac{\zeta(x - \zeta)^3}{[(x - \zeta)^2 + z^2]^2}\, d\zeta$$

$$= \frac{q}{\pi B}\left[3z \tan^{-1}\frac{Bz}{z^2 + x^2 - Bx} - \frac{Bz^2}{z^2 + (x - B)^2}\right.$$

$$\left. - 2B - x \log\frac{z^2 + (x - B)^2}{z^2 + x^2}\right].$$

$$\tau_{xz}^1 = \frac{2qz}{\pi B} \int_0^B \frac{\zeta(x - \zeta)^2}{[(x - \zeta)^2 + z^2]^2}\, d\zeta$$

$$= \frac{q}{\pi B} \left[\frac{Bz(x-B)}{z^2 + (x-B)^2} + z \log \frac{z^2 + (x-B)^2}{z^2 + x^2} \right.$$

$$\left. + x \tan^{-1} \frac{Bz}{z^2 + x^2 - Bx} \right].$$

Figures 11.2 and 11.3 show the horizontal stresses, σ_h, and the shear stresses, τ, caused by a vertical triangular loading, and fig. 11.4 the horizontal shear

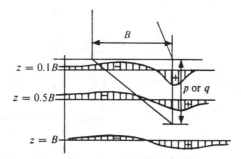

Fig. 11.3   Shear stresses, τ, caused by a triangular vertical load. The same diagram also valid for vertical stresses, σ_v, caused by horizontal loads (after Del Campo & Piquer, 1962).

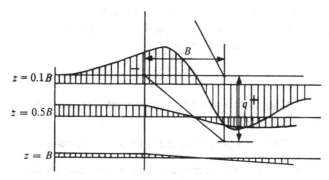

Fig. 11.4   Horizontal stresses, σ_h, caused by a horizontal triangular load.

stresses, σ_h, caused by a horizontal triangular loading. They are also valid for shear stresses, τ, and the vertical stresses, σ_v, caused by horizontal loads. The diagram shows that σ_h and τ are not linear. Figure 11.4 shows clear negative tensile horizontal stresses, σ_h, in the region of the heel of the dam. When vertical and horizontal (hydrostatic) loads act on the dam the corresponding diagrams have to be combined. Horizontal tensile stresses still exist under the dam.

From a rock mechanics point of view these tensile stresses are of major importance as it is sometimes doubtful whether rock masses can stand them.

Knowing σ_v and σ_h the usual formulae allow the calculation of the principal stresses σ_1 and σ_2, for the cases reservoir full and empty.

The basic assumption that vertical and horizontal loadings are triangular is only an approximation. Modern computing methods can be used for direct stress analysis of dam plus rock foundations. For example, Zienkiewicz (1965, 1968) has used the finite element method and assumes that the rock mass stands tensile stresses or that there is a 'no-tension' solution (fig. 11.5).

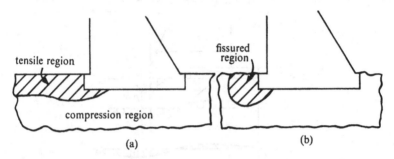

Fig. 11.5   Stress distribution at the heel of a gravity dam: (*a*) initial elastic solution, assuming tensile stresses; (*b*) final 'no-tension' solution, fissured rock (after Zienkiewicz, 1968).

Foundations of gravity dams, buttress dams, arch dams and also anchored dams, have been analysed by this method, and they always show tensile or fissured regions near the heel of the dam.

11.3   Percolating water through dam rock foundations; grouting and drainage

11.3.1   Interstitial water pressure and uplift forces

This is a problem of great complexity and importance to dam designers. Research started after the Bouzey dam in France failed on 27 April 1895. A committee headed by Maurice Lévy was put in charge of technical investigations and his findings indicated, probably for the first time, that interstitial water pressure in the rock and hydrostatic uplift forces under the dam foundations had caused the catastrophe.

The carefully worded recommendations of this committee were later summarized in a practical rule concerning the uplift forces to be assumed under dam foundations.

The uplift pressure at the heel of the dam is equal to γy, where y is the depth of the foundation below water level and γ the specific weight of the water.

In fig. 11.6 it is assumed that on the toe of the dam the water is at level y_0 above the foundation. The uplift force U is equal to the area of the trapezium ($\gamma = 1$)

$$U = \gamma B \frac{(\lambda y + y_0)}{2} \text{ with } \lambda = 1, B = \text{width of dam.}$$

This is equivalent to the so called 'condition of Lévy' the wording of which was slightly different and considered mainly water pressures in fissures in the dam masonry.

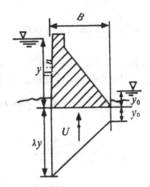

Fig. 11.6   Condition of Lévy for uplift pressures under a dam ($\lambda = 1$).

The rule of Lévy is now generally accepted. The percentage of the area on which uplift acts raises a most important question. If there is grain to grain contact in the concrete or between concrete and rock, and if the uplift only acts in the concrete or rock pores, this would be equivalent to a corresponding reduction in the λ value ($\lambda < 1$).

The following table indicates how the ratio B/H of a gravity dam width B at the base to dam height H varies with λ, to avoid tension cracks:

$$\lambda = 100\% \text{ uplift,} \qquad B/H = 0.84$$
$$\lambda = 67\% \qquad\qquad\quad B/H = 0.76$$
$$\lambda = 50\% \qquad\qquad\quad B/H = 0.72$$

Several research workers have tried to measure uplift pressure in concrete but their results have been inconclusive. Leliavsky (1958), improving on former techniques, forced water under pressure p into the pores of the concrete (fig. 11.7). This may cause the concrete to rupture for a water pressure p'. The uplift force corresponding to this pressure is $z = fAp'$ where A is the cross-section of the concrete sample being tested and f a factor representing the percentage of the area A on which the uplift is supposed to act. Measured values of z have been erratic in the past. Leliavsky applied a compression

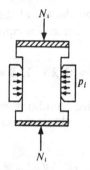

Fig. 11.7   Principle of Leliavsky pore pressure test.

force N to the test sample. Repeating the test for pressures $p_1, p_2, p_3, p_4 \cdots$
of the test water, he obtained the following relations (fig. 11.7).

$$z + N_1 = fAp_1, \qquad z + N_2 = fAp_2,$$

and

$$f = \frac{N_1 - N_2}{A(p_1 - p_2)},$$

$$z + N_3 = fAp_3, \qquad z + N_4 = fAp_4,$$

and

$$f = \frac{N_3 - N_4}{A(p_3 - p_4)}.$$

The average value obtained for f by Leliavsky was 0·91. A technical correction of this factor brought it down to 0·84 which is the value he recommends. Most designers will accept $f = 1·0$ and therefore $\lambda = 1·0$ for undrained
dam foundations. Others will hold to the value $\lambda = 1·0$ even for drained dam
foundations, but many experienced designers will in this case accept $\lambda = 0·80$
to 0·90.

It is advisable to check any design which is based on a value $\lambda < 1$ for
stability or stress increases when the actual value λ is greater than the assumed λ. In most cases the increased stresses in the concrete (or eventually
some tensile stresses in the concrete) are well inside acceptable limits. Rock
stresses must also be checked.

11.3.2   Classical grouting techniques for concrete dam foundations

(1) *The grout curtains.* Glossop (1960, 1961) established that the first
grouting was carried out in 1802 in Dieppe (France) by Charles Bérigny.
The method was a remarkable success and was well known in France long
before the 1832 publication by Bérigny.

In 1933 Lugeon set down rules for grouting of concrete dam rock foundations with a cement grout, still used today on many dam sites. He considered
three different systems.

(*a*) Low-pressure grouting under the whole area of concrete dam foundations, binding concrete to rock and consolidating foundation rock shattered
by excavations (contact grouting).

(*b*) High-pressure deep grout curtain along the heel of the dam.

(*c*) A grout curtain at large (French: *voile au large*) supposed to cut water
percolation round the dam.

An excellent example of Lugeon's technique is his work on the Sautet arch
dam site (France) (fig. 11.8).

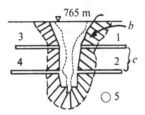

Fig. 11.8   Sautet dam. (*b*) Main grout curtain following the dam heel; (*c*) the
curtain was grouted from adits 1, 2, 3 and 4 and from 5, the diversion tunnel (after
Lugeon, 1933).

Low-pressure grouting under the concrete dam is carried out after the
rock has been loaded by a reasonable thickness of concrete, in order to avoid
lifting of rock masses by the grout pressure. The main grout curtain is
usually done before the dam foundations are concreted. Lugeon suggests
drilling boreholes 5 m deep under the foundation level. The borehole is
sealed at 3 m under this level and the lower 2 m tested under water pressure
and then grouted. The borehole is then redrilled through the hardened grout
and bored down to −10 m, water tested and grouted from −5 m to = −10 m
below foundation level. The drilling, water testing and grouting proceeds
5 m at a time downwards to the required depth. The alternative technique
consists of drilling the boreholes downwards by 15-m steps and grouting them
upwards by 5-m steps, redrilling and then boring another 15-m length.

(2) *The Lugeon unit and the water tests.*   One Lugeon unit corresponds to a
water test where one litre of water per minute is absorbed by the rock by
a one-metre-borehole under a test pressure of 10 kg/cm². According to
Lugeon a rock absorbing less than one Lugeon unit is considered to be
reasonably watertight and no grouting or further grouting is required. One
Lugeon unit is approximately equivalent to $K = 10^{-5}$ cm/s (Darcy's law).

Tests are often carried out on 5-m-long borehole sections, but shorter test lengths are sometimes required for locating isolated fissures. The effective water pressure to be included in Lugeon's formula is, according to fig. 11.9:

$$p = p_{\text{eff}} = p_m + H/10 + \Delta_p,$$

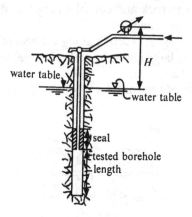

Fig. 11.9   Borehole section for locating isolated fissures.

where p_m is the manometer pressure, H (in metres) the static head above the water-table and Δ_p the pressure losses in pipes and valves.

Assuming a borehole with diameter $d = 2r$ feeding a fissure (e = width of fissure) over an area πR^2, the discharge q through the fissure is then (fig. 11.10):

$$q = \frac{\pi}{6\eta \log_e R/r} pe^3$$

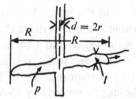

Fig. 11.10   Water discharge through a fissure.

(η = viscosity of the water, $\log_e$ = natural log, q varies little with R/r). For a 5-m test length of the borehole:

    1 Lugeon unit corresponds to $q = 5$ l/min and $e = 0\cdot1$ mm
   10 Lugeon units correspond to $q = 50$ l/min and $e = 0\cdot2$ mm
 100 Lugeon units correspond to $q = 500$ l/min and $e = 0\cdot5$ mm.

When E is the modulus of elasticity of the rock a pressure p causes an increase Δe of the width of the fissure (Sabarly, 1968):

$$\Delta e = \alpha \frac{p}{E},$$

and

$$q = \frac{\pi}{6\eta \log R/r} p \left(e + \frac{\alpha}{E} p \right)^3 \cong A p^4 \quad (A = \text{constant}).$$

The discharge q increases rapidly with the pressure and (for $e \cong 0$) is proportional to p^4. This sharp rise of q is not really due to internal rupture (as sometimes assumed) but to elastic rock compression. The discharge q is inversely proportional to E and larger in deformable rock. (French *claquage*.)

(3) *Grout mixture and pressure; borehole density.* Lugeon recommends the use of normal cement and grout pressues of 35 kg/cm², grout mixture, with a cement to water ratio $C/W = \frac{1}{5}$ to $\frac{1}{10}$. If it is not possible to plug a fissure within twelve hours, he recommends the use of thicker grouts, with $C/W = \frac{1}{2}$ or $\frac{1}{3}$. He suggests grouting should stop when the grout losses are not higher than 25 l/min and meter under a pressure $p = 45$ kg/cm².

Today, special cements are being used, with no grain larger than 45 μm. Lower pressures ($p \cong 10$ kg/cm²) are often efficient with fine cement. American authors recommend that grout pressure should be related to the thickness of the overburden. Their recommendation is 1 lb/in² per foot overburden (1 kg/cm² per 4·2 m). But European experts accept pressures as high as 1 kg/cm² per metre overburden thickness, or about four times as much. For example, for a rock depth of 45 m the American rule would limit the pressures to 11 kg/cm² whereas the European would go as high as 45 kg/cm². The American Task Committee (ASCE) accepts higher pressures than the 1 lb/in² per ft rule. For stratified rock they recommend 22 kg/cm² for 45 m depth and 45 kg/cm² for a depth of 40 m in massive rock.

(This can be discussed on more precise assumptions along the lines developed in section 10.4, figs. 10.22 to 10.30, for uplift forces caused by tunnel grouting.)

In fig. 11.10 pressure p is supposed to be applied over an area πR^2 causing an uplift force $U = p\pi R^2$ on the fissure. How dangerous is this uplift? Some rough figures are available. Tests have been carried out for the Allt na Lairige dam (Jaeger, 1961c), which attempted to rupture the anchorage rock. An uplift force U varying from 1000 tonnes to 4400 tonnes was applied to an area of 4·3 m² (fig. 11.11), using flat jacks. The pressures increased to $p = 23$ kg/cm² and then to $p = 102$ kg/cm² when the flat jacks burst before the rock was severely damaged. Only minor rock movements, possibly caused by internal cracks, had been noticed. The depth at which the flat jacks were anchored was only 18 ft (5·5 m) under the rock surface (Banks, 1955).

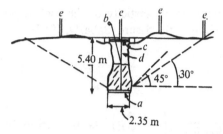

Fig. 11.11 Test anchorage for Allt na Lairige dam: (*a*) six Freyssinet flat jacks each 870 mm in diameter; (*b*) hand pump; (*c*) deflectometer; (*d*) steel tube duct; (*e*) staffs for vertical displacement measurement.

Grouting pressures as high as these would never be applied at such shallow depths and the experiment indirectly proves the resistance capacity of competent rock to grouting pressures; with poor rock, results would probably have been different.

Another approach is to choose the grout pressure on the basis of the potential hydrostatic pressure transmitted to the rock by the storage reservoir when filled, the grout pressure being larger than the hydrostatic interstitial water pressure. Grouting can then be carried out from deep-lying galleries rather than from the surface of the rock foundations. As an example, the grout curtain of the Mauvoisin dam (fig. 11.12) was injected from a deep gallery, at 237 m below dam crest.

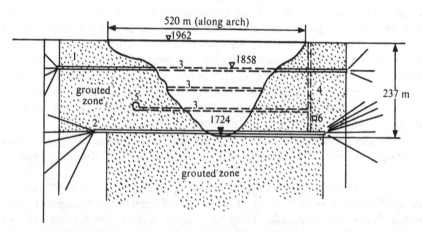

Fig. 11.12 The 237-m-high Mauvoisin arch dam grouting scheme: (1) grouting gallery level 1858; (2) deep grouting gallery, level 1724; (3) inspection galleries; (4) lift; (5) power tunnel to Fionnay; (6) discharge gallery.

The distance between boreholes is chosen on the site according to test results. In competent rock the first boreholes are tentatively drilled 12 m or 6 m apart. A water test log is established along the whole borehole (fig. 11.13).

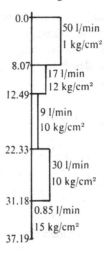

Fig. 11.13   Water loss diagram, Spitallamm dam (after Lugeon 1933).

A similar diagram is traced for the grout absorption along the same borehole. Depending on results, intermediary boreholes are drilled, sometimes the distance between them has to be reduced to 1 m or 0·5 m, or even two rows are drilled. For the Kariba dam, the boreholes through the pervious right rock abutment were arranged in a hexagonal pattern.

A few cases of rock masses being lifted by grout pressure are mentioned in the literature (Zaruba, 1962) with diagrams relating rock deformations to grout pressures and time. Deformations of 0·05 to 0·06 m under pressures of 3 to 6 kg/cm² were measured. Other diagrams by Zaruba mention deformation of 160 mm. He suggests the following empirical values (metric system). For competent rock, nearly vertical main system of fissures and fractures:

$$p = 0.31H + 0.02H^2.$$

For weaker rock masses, horizontal stratifications:

$$p = 0.25H + 0.005H^2$$

(4) *Other types of grouting.* Bitumen, mixtures of cement and silicates, chemicals (silicates), sand, sawdust, ashes, even gravel have been used for grouting. These are specialized techniques extensively dealt with in textbooks (Lugeon, 1933; Cambefort, 1964).

11.3.3 Casagrande's critical opinions; present approach to grouting and drainage problems

(1) *Casagrande's opinions.* In January 1961, A. Casagrande, delivering the First Rankine lecture to the Institution of Civil Engineers, London, systematically questioned grouting procedures and endeavoured to prove that they

are useless for reducing the uplift pressures under dams. He started with an earth dam on loose, pervious alluvium (fig. 11.14a), and related his remarks to concrete dams built on solid competent rock (fig. 11.14b). In this diagram it is assumed that the depth to the impervious soil under the dam is d, and B the width of the dam base.

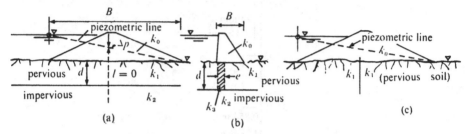

Fig. 11.14   Casagrande's theory on grouting.

Assuming the grout curtain to be a nearly impervious, very thin curtain ($e \cong 0$), then Dachler's well-known theory (1936) shows that it causes only negligible pressure losses. Measurements carried out under earth dams showed that the pressure line is nearly linear, the pressure loss, Δp, caused by the grout curtain being negligible, thus confirming the Dachler–Casagrande theory. The immediate reaction to this theory was to observe that the thickness, e, of a grout curtain should never be negligible. Those under the Serre-Poncon rock-fill dam and the High Aswan dam are quite substantial. In particular, the width e of the grout curtain under a concrete dam with width B is definitely not negligible. As soon as the ratio e/B becomes significant then the pressure losses through the curtain are proportionately greater.

Casagrande's thesis would have been easier to demonstrate if it had been assumed that the soil foundations of the earth dam were pervious at great depth (fig. 11.14c). He published several diagrams similar to fig. 11.15

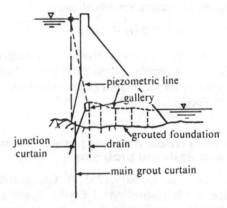

Fig. 11.15   Piezometric line under concrete dam.

showing that the piezometric line, measured under a concrete dam, depends on the position of the drainage system and not on the position of the grout curtain. His conclusions were that expensive grouting is useless and that a much cheaper drainage system is the only efficient method of controlling the piezometric line along the dam foundations.

In fact, the whole theory of uplift pressures and pressure lines had been developed long before Casagrande and it was accepted that they do not depend on the absolute value of the perviousness of the medium, k, but only on the relative perviousness of the grouted to ungrouted medium, k/k_2. But the discharge seeping under a dam is directly proportional to the absolute value of k. It was argued that the purpose of grouting was not really to decrease the uplift pressures (an uplift factor of $\lambda = 0.80$ to $\lambda = 1.0$ being accepted by all dam designers) but to check possible water seepage and losses under or round a dam (Jaeger, 1949b, 1956, 1961b).

It was additionally observed (fig. 11.15) that interstitial water pressure is measured along the foundation line between concrete and rock, which is grouted, so that the k value along the foundation is about the same from heel to toe of the dam. This explains why the pressure line of uplift is linear along the foundations.

(2) *Further research on grouting and draining.* Casagrande's lecture was delivered one year after the rupture of the Malpasset dam. It was a time when many designers were deeply concerned with basic concepts in dam design and they realized that the main problem to be analysed was the bulk stability of the concrete dam plus rock masses modified by the grouting and drainage systems (Jaeger, 1961b, 1964a; Londe & Sabarly, 1966). It was becoming clear that many problems were so intimately linked that none could be discussed without full consideration of the others (fig. 11.16).

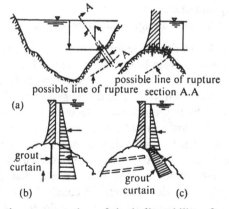

Fig. 11.16 Schematic representation of the bulk stability of a dam, depending on the position of the grout curtain (after Jaeger, 1961b).

Pacher and Müller had been studying the rock abutments of the Kurobe IV arch dam (Japan). (The methods used by Müller and John to determine the density and main directions of the fissures have already been dealt with.) In 1962 (Salzburg), Pacher & Yokota reported on their investigations into seepage problems. Yokota (1963) assumes that water seepage through highly fissured rock is similar to that through a homogeneous porous medium (laminar flow) and that the law of Darcy applies. Comparison of results obtained in nature, by reading piezometric levels in boreholes, with those obtained in the laboratory will show whether this assumption is correct. (All theories developed by Casagrande are implicitly based on the same assumption.) If the law of Darcy is accepted then the use of electrical analogues for rapid research is justified.

Yokota describes how he tested the efficiency of grout curtains using an electric analogue and then made drainage tests on two- and three-dimensional models. For three-dimensional models an agar gel was used. The efficiency of a grout curtain is defined by

$$\eta = (K_1 - K_2)/K_1,$$

where K_1 = the permeability coefficient of the rock mass before grouting and K_2 = the permeability after grouting. The porosity P of the curtain is given by $1 - \eta = K_2/K_1$. A better definition would be to introduce the discharges q_1 and q_2 before and after grouting: $\eta = (q_1 - q_2)/q_1$. The porosity factor then becomes $P = q_2/q_1$. The final purpose of the research was to establish a design for the grout curtain and associated drainage system, for Kurobe IV dam. It was decided that drainage holes with 56 mm diameter and 10 m apart were an acceptable solution.

Tests were carried out on a three-dimensional model:

(*a*) without curtain grouting and without drainage,
(*b*) with curtain grouting only,
(*c*) with grouting and drainage curtain,
(*d*) with so-called direct drainage in addition to (*c*).

The results showed that the reduction of flow due to the grouting curtain was only 10%, while with drain holes added the total flow was increased by 27%; but the level along the downstream rock surface (which is a measure of the pressure) was decreased to 65% of the original flow. Comparisons were made on the prototype by measuring water levels in twenty-four piezometers, when the reservoir was partly full at elevation 1376 m. Yokota summarizes his views as follows: 'Curtain grouting is necessary to plug rather wide openings or cavities in the rock, but it seems very difficult to form a so-called watertight screen.' He found, on the other hand, that drains were very effective in reducing hydrostatic pressure. His conclusions confirm the opinion that grouting curtains and drainage sytems should be used simultaneously (fig. 11.16c) (Jaeger, 1961b, 1964a).

Pacher investigated the general stability of the rock abutments of Kurobe IV dam. Some of his research about the flow lines in grouted, homogeneous rock and the pressure gradients are shown in figs. 11.17 and 11.18. He concluded that the shape of the grout curtain had to be inclined and adapted

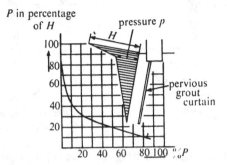

Fig. 11.17 Pressure diagram for a pervious grout curtain, p = pressure as indicated on the diagram (after Pacher, 1963).

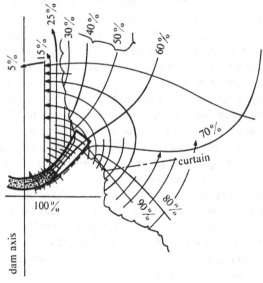

Fig. 11.18 Grout curtain at Kurobe IV dam. Estimated flow and equipotential lines. Symbols: — — — — —, grout curtain; ——, equipotential lines; ——→, flow lines (after Pacher, 1963).

in plan view to local conditions. In fig. 11.18 it can be clearly seen how the curtain is curved in plan view in order to increase the efficient mass of rock forming anchorage for the dam on the downstream side of the curtain.

Research into the same problems was also carried out in France by Londe & Sabarly (1966) and by Sabarly (1968) who arrived at very similar conclusions. To make his point clear, Sabarly published two sketches showing the forces F and F' acting on the rock foundations; one where there is a grout curtain and the other where drainage was used under the dam (fig. 11.19). A simplified calculation shows that $F' = F \sin \alpha / \alpha$. Additionally the direction

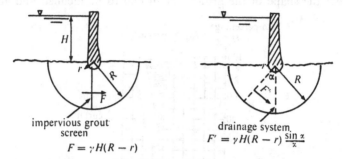

impervious grout screen
$$F = \gamma H(R - r)$$

drainage system.
$$F' = \gamma H(R - r) \frac{\sin \alpha}{\alpha}$$

Fig. 11.19   Simplified sketch comparing the force, F, on a grout screen with the force, F', on a drainage system (after Sabarly, 1968).

of F' is far more favourable than the direction of F. A more detailed calculation could be developed using Pacher's approach. Sabarly also confirms the thesis of Casagrande, by showing that there is a far greater reduction in uplift forces under a dam when drainage is used instead of a grout curtain.

His general conclusions concerning dam foundations are: in the case of an impervious rock mass, a grout curtain is not very efficient. To check uplift pressures he suggests building a drainage system rather than a grout curtain. A theory of drainage system has been developed elsewhere (C. Jaeger, 1956).

In very pervious rock, grouting may be required to reduce the discharge of water seeping through the fractures and fissures. Too high a discharge may saturate the drainage system and make it inoperative or may erode the fissures. Grouting is supposed to check the discharge of percolating water.

A very large grout curtain was required to stop leakage under the Camarasa dam (Spain) as the water losses under pressure could be higher than the average yearly inflow to the reservoir (approx. $11/m^3/s$). A large grout curtain was also required for the Dokan dam (Iraq) (Jones *et al.*, 1958) to stop leakage through very pervious rock. In other cases, stopping the water loss was imperative for economic reasons.

Sabarly recommends inclined grout curtains, similar to those proposed by others (Jaeger, 1961*b*). He mentions the possibility that arch dams might rotate around their foundations. This rotation, in addition to a radial displacement may cause an open fissure under the heel of the dam (fig. 11.20), and a rupture of the impervious grout curtain. (This rupture is not likely to occur on the foundations of double curvature arch dams and of gravity dams where the rotation of the dam base is smaller.)

Sabarly describes the case of an arch dam with a vertical cylindrical upstream face. The vertical upstream grouting curtain followed the heel of

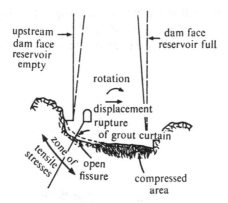

Fig. 11.20   Rotation of the base of an arch dam (after Sabarly, 1968).

the dam. This was reinforced by a short inclined secondary curtain drilled from the gallery (fig. 11.21). The dam functioned well for several years. During a period when the reservoir was empty, piezometric boreholes were drilled from the gallery. When the reservoir was filling up, a few metres

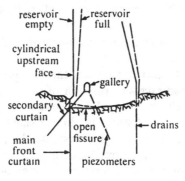

Fig. 11.21   Arch dam with cylindrical upstream face (after Sabarly, 1968).

before reaching its top level the piezometers suddenly began to discharge. Rising water levels made the discharge increase to 1200 litres/min per piezometer, which is quite considerable for small boreholes. Drains were drilled on the toe of the dam (see fig. 11.21) but they discharged very little. It was assumed that rotation of the dam had opened a fissure, probably near the concrete–rock line, and water had penetrated direct into the piezometers, the discharge being

$$q = Ah_1e^3,$$

where e is the width of the fissure, which increases with reservoir levels, and h_1 the available head. Discharge under the dam is increased proportionally to the third power of e and increases with the dam rotation. To stop the water losses and additional uplift pressures, Sabarly suggested that a grout curtain and drainhole should be located as indicated in fig. 11.22. This would stop water losses progressing towards the toe of the dam but would not decrease

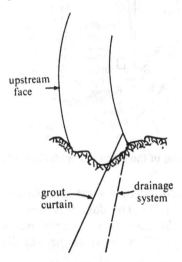

Fig. 11.22   Suggested location of grout curtain.

the uplift pressures, or solve the more important problem of bulk stability of the rock foundations.

It is suggested that the correct choice of the dam profile (double curvature arch dams) would avoid excessive rotation of the foundation and so avoid fissuring of the rock. The Italian 'pulvino' design would be most efficient in such a case. A 'pulvino' is a pressure distribution slab following the periphery of the dam foundations (the dam shell rests on the pulvino). A peripheral joint is provided which is usually grouted after settlement of the rock foundations and before raising the water level in the reservoir (fig. 11.23). But the

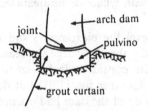

Fig. 11.23   Arch dam with peripheral 'pulvino' (after Semenza).

joint can be left open and provided with an elastic seal. Such a design, strongly advocated in many papers (Jaeger, 1950, 1964*a*, *b*), spreads the compression forces on the rock over a wider field, equalizes them and avoids any major rotation of the foundations, the elastic seal being able to absorb small rotations of the dam shell. A somewhat similar design has been used by Gicot for the Schiffenen dam to compensate for the difference of the elastic modulus of the rock in the river bed and on the natural lateral abutments.

(3) *Rock grouting for consolidation of rock masses.* In considering the problem of drainage-grouting primarily on uplifts, Casagrande has neglected some other very important aspects. The case just mentioned shows the danger of excessive water losses. There are other aspects of the problem requiring close investigation.

Rock grouting can be used for consolidating fissured rock masses. Semenza successfully used it on the rock abutments of the high Vajont arch dam. At the same time the grout pressure increased the rock stiffness and brought the Young's modulus of the rock masses, E_r, to the required high uniform values to be introduced in the elastic analysis of the dam shell.

Several years before Semenza, Coyne (France) used systematic grouting for consolidation of the rock abutments of the Castillon arch dam and the Chaudanne arch dam. More recently, similar techniques were used for the Kariba dam (Rhodesia) and the Dokan arch dam (Iraq). In all these cases rock grouting was vital to the final success of the design.

More information on this will be given in the next chapter.

11.3.4 Geology and rock grouting

Geological problems have not been mentioned much in previous chapters. Rock has been described as 'impervious' or 'very pervious' (Sabarly), and the Lugeon test accepted as the standard criterion. But a far more detailed description of the rock mass is required.

It is assumed that details of the principal families of fissures, faults and fractures are at hand. The penetration of grout in fissures depends not only on the width of the fissures but also on their possible widening. Grout penetrates more easily along fissures normal to the direction of the lowest residual stresses. It is therefore desirable for rock mechanic tests on residua stresses to be carried out in the area of the dam foundation.

Figure 11.24a, b gives a schematic illustration of possible geological situations. In the first case (a) it is assumed that rock provides natural

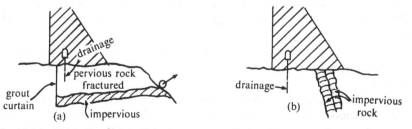

Fig. 11.24   Typical geological situations: (a) favourable to a grout curtain; (b) not favourable unless artificial drainage provided.

drainage under the dam and substantially assists any artificial drainage in reducing uplift pressures under the foundations.

In case (*b*) artificial drainage is absolutely essential if high uplift forces under the dam are to be avoided.

Geology textbooks cover situations like those in fig. 11.25*a*, *b*, *c*. There is an endless list of possible geological situations – from the possibility of water

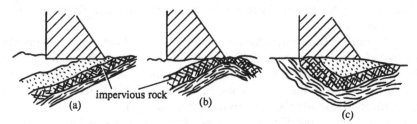

Fig. 11.25   Typical geological situations.

being evacuated to the downstream side of the dam toe, or being retained under the dam, increasing the uplift values higher than $\lambda \geqslant 1$, depending on the relative perviousness of the different rock layers.

This analysis is not sufficient, as the relative natural plasticity of rock masses is of great importance when the general stability of dam abutments is being considered.

11.3.5 General stability of dam abutments

Vital research work has been carried out in recent years on this subject. The concrete abutment of the dam, together with the fissured or fractured rock mass is analysed for general stability, assuming the most unfavourable uplift conditions. This is the only way to decide between a grout curtain or a drainage system. Casagrande's criticism of modern grouting methods has helped dam designers to consider new trends even if many of his sharp conclusions had to be rejected.

The following could be adopted when designing a dam abutment.

(1) Establish the geological pattern of the site, based on the information given by the geologist on stratification, faults and fractures and more detailed information on rock mechanics, fissures, rock elasticity and rock plasticity.

(2) Calculate the 'lines of forces' along which the dam thrust is supposed to be transmitted to the rock masses. Model tests (see next chapter) may be required.

(3) Estimate structural changes which may be caused in the rock masses by the transmission of stresses and strains from the dam to the rock. Tensile stresses may cause zones of rupture where water penetrates easily without loss of pressure. Compressed zones may become impervious, as was clearly

demonstrated by the tests of Bernaix, at the Paris Research Laboratory of the École Polytechnique.

(4) Follow the flow lines of percolating water. In an isotropic zone, the methods developed by Pacher can be used. When the rock becomes anisotropic under strain and stresses, the variations of the perviousness factor k ($k_{tensile}/k_{compression}$) must be estimated and introduced in the calculation. In the case of rupture the hydraulic pressure may be transmitted to its full value deep into the rock masses (fig. 11.16). This is basically the assumption made by Fumagalli (1967) when he tested rock abutments at the ISMES laboratory (see section 11.4.1(4)).

(5) The real uplift conditions on the rock masses forming the rock abutment can be calculated on the basis of the calculations mentioned under (3) and (4). This is a sound basis on which to estimate the bulk stability of dam foundations.

Londe & Sabarly (1966) submitted some general rules for such a calculation to the Eighth Congress on Large Dams, Edinburgh, 1964. The shadow of the collapsed Malpasset dam has formed the background to the intensive research work which led to the formulation of these design rules. Model tests will help the designers and suggest any required improvement to rock conditions by reinforcing and strengthening of rock masses.

11.4 Dam foundation design and construction in recent years: case histories

The broad discussion of design principles developed in sections 11.1 to 11.3 can be compared to the progressive evolution of the design of foundations of large concrete dams. A few case histories will show what has been done in recent years to improve the safety of difficult dams. The technique of using geomechanical models for testing dam abutments has been improved; more attention has been given to drainage systems and, in difficult cases, bold rock reinforcements have been carried out through cable anchorage and pressure grouting.

The following classification of case histories under three different subtitles is only approximate.

11.4.1 Geomechanical models of dam abutments: design and construction

The technique of testing dams, mainly arch dams, on models was first developed in the United States (Bureau of Reclamation) and in Italy and was rapidly adopted by other countries. The first models were plaster, hard light concrete, or rubber, built on rigid hard concrete abutments. Then the technique developed and the rock abutments were represented by a material with a convenient modulus of elasticity. The latest stage is to represent rock stratifications, fissures and faults on the model and test them until rupture occurs. Several typical examples are instructive.

(1) *The Vajont dam.*   The 261·6-m-high Vajont dam (fig. 11.26) is built on highly jointed limestone which can best be described by its modulus of

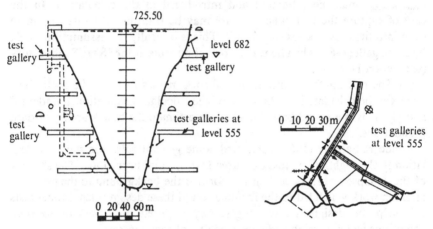

Fig. 11.26   Vajont dam, measurement of rock displacement.

elasticity E which varies from 787 000 to 852 000 kg/cm², measured on laboratory samples, the higher values of E being those for rock at the bottom of the valley. Inside test gallery measurements of E of from 100 000 to 300 000 kg/cm³ were obtained. A model of the dam was built and these varying E values carefully reproduced, so that stresses and strains measured on them could be compared with those obtained on more conventional models of homogeneous abutments (Pancini, 1961a, b).

The main problem at Vajont was not the modulus of elasticity of the rock, which varied between known and reasonable limits, but the joints and fissures in the rock. A model was therefore built in which the abutments were formed by precast parallelepiped blocks of plaster with barium sulphate piled one on top of the other so that the actual contour lines and the three main directions of the joints were correctly reproduced. The joints were filled with different types of glue to represent different types of roughness coefficients. In addition, several rock fractures, some of them parallel to the main direction of the valley were correctly reproduced by sawing the blocks in the appropriate places.

The deflection of the crown was about five times larger on the model with the lowest friction factor than on the model with massive, homogeneous, non-fissured rock abutments. Measurements made on the prototype dam showed the actual deformation to be larger than on the more optimistic model, but still far less than on the worst model. The models were tested to breaking point. One collapsed when, due to the accidental rupture of a watertight rubber sheet protecting the abutment, water seeped through the pile of blocks, simulating the fissured abutments. This may be taken as proof of the very real danger from the seepage of water and full uplift pressure in rock fissures and joints.

In addition to these tests, a considerable number of observations and measurements were made on the site to ascertain the rock properties and rock displacements. Vibrations inside the rock mass were measured during rock blasting and concreting. The wave velocities in rock and dynamic modulus of elasticity were measured before and after grouting. Five test galleries were driven. The modulus E was measured by applying compression forces either with hydraulic jacks or by filling the galleries with water under pressure. Tests were made in galleries to determine the cohesion and friction along weak rock seams. Displacements of the rock abutments were measured both in the direction of the dam thrust and perpendicular to it. Geodetic measurement of any displacement can be supplemented by observation of pendulums hung in vertical shafts in the rock. Displacements of the embankments of the reservoir are also checked from the geodetic net. Cables were used to prestress the rock in the upper part of both abutments, and their length and strain were measured and checked. (The Vajont rock slide will be analysed in chapter 14.)

(2) *Kurobe IV dam, Japan.* Model tests carried out for the Kurobe dam (Oberti, 1960) included those where the correct ratios $E_{concrete}/E_{rock}$ were used on the model and the main rock faults represented. In addition, detailed tests and calculations (with an analogue computer) were carried out to determine the piezometric pressure lines of water percolating through the rock abutments (Yokota, 1962; Pacher, 1963). Particular attention was paid to the grout curtain and to the drainage system, both of which differ in many ways from conventional designs. The grout curtain is displaced in the upstream direction.

An elaborate and costly campaign was launched (John, 1961) to:

(a) record all the rock fissures and their characteristics (direction, dip, thickness, etc.),
(b) test rock under compression and under shear inside galleries,
(c) measure the physical rock properties and its strength near the foundation at the surface of the rock,
(d) measure the modulus of elasticity in the rock.

In fig. 11.27a the fissures on the right abutment of the dam are shown and in fig. 11.27b the arrangements for inclined shear tests on rock *in situ*, isolating a rock block between two parallel galleries. Figure 11.27c shows similar tests carried out at the rock surface. Arrangements for triaxial tests

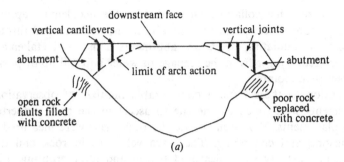

(a)

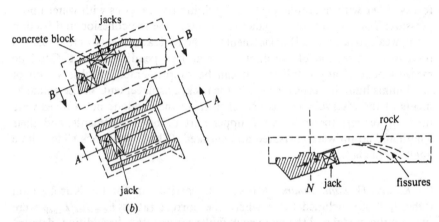

(b)

Fig. 11.27 The Kurobe IV dam: (a) the dam; (b) shear test on a concrete block; (c) shear test at the surface of a rock mass.

on filling material inside a wide rock fracture are similar to those shown in fig. 11.27b.

(3) *Tang-e-Soleyman dam (Northern Iran)*.   The dam has a height of 100 m with a central angle of 73°. The radii of the reference surface are 124 m for the central zone and 211 m for the lateral zones (fig. 11.28). Comparative

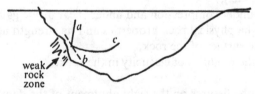

Fig. 11.28 Schematic representation of the failure of the model of the Tang-e-Soleyman dam (Iran): (a) fissure due to bending; (b) shear failure in the concrete, along the dam foundation; (c) horizontal shear fracture (after **Lane & Serafim**, 1962).

tests were carried out on homogeneous and on heterogeneous models. The values of E shown in table 11.1 were measured in kilograms per square centimetre by the Lisbon laboratory (Lane & Serafim, 1962).

Table 11.1

	Sandstone	Mudstone	Limestone	Transitional limestone
	(in 1000 kg/cm²)			
Oven-dried	185	238	798	541
Saturated	115	188	715	535
Jacking tests in a gallery	70 to 570	160 to 500	90 to 420	50 to 60

Tests were first carried out in Lisbon by Rocha and Serafim on two models of a mixture of plaster of Paris (P), diatomite (D) and water (W) in a proportion of $W/P = 2$ and $P/D = 2$. The modulus of elasticity of the foundation rock was assumed uniform and equal to that of the dam. The models were loaded with mercury.

For a second stage of the study two other models were built, the ratios $E_{rock}/E_{concrete}$ being 1/2 for limestone and 1/4 for sandstone and mudstone. The mixtures used were:

$W/P = 2$, $D/P = 0.5$, for the dam.

$W/P = 2.9$, $D/P = 1$, for the limestone.

$W/P = 1.2$, $D/P = 5$, for sandstone and mudstone.

Due to the increased deformability of the foundation, the arch effect became more marked, especially in the upper part of the dam, and the cantilever effect decreased. It is important to note that in such circumstances the maximum compressive stresses in the entire dam decreased, the results showing much smaller bending moments. The compression at the downstream face on the crown of the arches increased, particularly in the upper arches. This was certainly caused by the fact that the upper region of the foundations was more rigid than the region below.

Tests were carried out until rupture of the model dam occurred. Rupture took place either by vertical compression fissures starting in the area of weak foundations when the rock (model) was crushed, and/or by sliding of the foundations in the same area. (This observation caused Serafim to suggest that a similar process might have been involved in the failure of Malpasset.)

It is likely that sudden local weaknesses of the E_{rock} modulus are more dangerous than a uniformly low E_{rock} value. Very thin arch dams may adapt themselves to substantial uniform rock displacements, but less to local overstraining caused by sudden changes in the E_{rock} value along the periphery of the dam foundation. The central angle of the arch in plan is only 73° compared with 110° to 125° which is more usual. The shape was specially studied on models and adapted to the contour lines; the thrust from the arches of the dam is well directed towards the rock mass.

Müller (Thirteenth Congress, Austrian Society for Rock Mechanics, Salzburg) emphasized that he considers the usual central angle value of about 120° often to be too wide, for the arch thrust pushes in a direction unfavourable to the rock abutment. A smaller angle would suit some sites better and increase the stability of the rock abutment. This was done at Tang-e-Soleyman.

(4) *Tests in the ISMES Laboratory, Bergamo (Italy).* Workers at the laboratory have concentrated on further advanced research into arch dam design and the stability of rock abutments. In a few cases they were able to observe the failure of the rock abutment by overloading the model. The friction factor along faults observed *in situ* was as low as $\phi = 25°$ (Cá Selva dam) or even $\phi = 13°$ (Gran Carevo dam). This low ϕ value has been correctly represented on the dam abutment models, together with the stratification and faults of the rock mass as observed *in situ*.

Displacements of the rock on the downstream side of the dam were carefully observed and progressive fissuration of the rock abutments recorded. At first (Cá Selva dam) dislocation and surface cracking may occur in the upper elevations of the arch dam abutment. Surface brittleness is the greater danger because it may cause a general collapse. In such a case, failure is not always preceded by premonitory signs and it may occur abruptly. The rock excavations for dam foundations should always extend sufficiently deep into the abutment rock to avoid such failure (Fumagalli, 1966, 1967, 1968).

The second phase, leading to the ultimate collapse of the model, was determined by the sliding of the lower beds, which occurred only in the final stage of the testing. This plasto-viscous type of failure of the entire abutment rock mass usually presents less unforeseen contingencies. Moreover, its occurrence is presaged by large and slow deformation processes.

After completion of the tests of the Cá Selva dam, the rock buttress was dismantled and analysed layer by layer. It was thus possible to detect the cracking at different elevations. It was confirmed that, at the upper elevations, the dislocation of the blocks was accompanied by a series of surface separation cracks. In a lower zone, a plastic compression process appeared to have occurred near the upstream concrete abutment of the arches, corresponding to a particularly high triaxial state of stresses existing in that area. The influence of the faults was slight. Only two faults moved, which had beddings consistent with the cracks.

It is important to note that the Cá Selva test was carried out by applying the entire water pressure against the vertical grout curtain inside the rock. It is worth while to compare these tests with the comments on the Malpasset dam failure (chapter 13) when the failure was attributed to water pressure on deep rock layers. At Malpasset the rupture of the rock appeared to be by surface brittleness.

Tests on the Place Moulin arch-gravity dam carried out on a 1:70 scale three-dimensional model, showed that the first horizontal cracks to appear on the downstream face of the dam occurred at approximately 4 to 6 times the working load. The general collapse took place at a slightly higher loading (5·1 times the working load) and may be attributed to yielding of the rock along the entire dam foundation (see Malpasset dam case history, chapter 13) (Oberti; Fumagalli, 1967).

In addition some two-dimensional tests were carried out. Plane pumice–cement mortar models were made which corresponded to the most characteristic and heavily loaded arches, in order to investigate the static conditions of the rock in the downstream vicinity of the abutments, where the yieldings brought about the collapse of the three-dimensional model.

Four arches, located at four different levels, from top to bottom of the dams, were tested, first under the 'equivalent hydrostatic' load. On the completion of this test, the loading was increased until the model collapsed: this was done so that the corresponding safety factor could be evaluated. The hydrostatic loading was applied not only to the extrados of the arches, but also to the track of the waterproof grout curtain in the foundation rock. The support given by the rock portion upstream of that curtain was neglected. Strains and stresses in the 'rock' were measured in four directions, and principal stresses were calculated.

11.4.2 Practical examples of rock drainage (arch dams)

The systematic research carried out by Yokota and Pacher for the Kurobe IV dam has been commented on in section 11.3. It is worth while comparing the experience gained here with the actual practice at other sites. According to Casagrande the 'drainage curtain' used in the USA consists of wells about 3 in diameter, 10 ft or possibly 5 ft (3 or 1·50 m) apart. Drainage of the foundations of gravity dams is now an accepted procedure. Little information has been published on the drainage of the rock foundations and lateral abutments of arch dams. It is believed that very often little or nothing is being done in this direction. Usually a drainage gallery built in the concrete mass is provided at the bottom of the dam – and that is all. The following data have been obtained from recent publications on French dams other than gravity dams (Jaeger, 1964*a*).

(1) *Roselend dam (Alps)*. The Roselend dam is characterized by a 150-m-high arch, crowned and supplemented by a buttress section 804 m long. The reservoir top level is at 1557 m and the top of the dam is at 1559 m.

(2) *Roselend buttress section on 'The Spur of Meraillet'*.   Drains were drilled parallel to the plane of the principal watertight screen to a depth of 35 m starting from the right bank extremity as far as buttress 9, and from this element as far as the large arch to a depth of 45 m. The drains were drilled with a diameter of 116 mm (4·5 in), on the basis of three drains to every 20-m-wide element. They are situated 1·85 m downstream of the buttress heads and thus open into the canal of the inspection gallery in the buttress dam.

(3) *Roselend main arch* (150 m high).   On the downstream side of the arch, drains of 116 mm were also drilled with a spacing of 10 m. They penetrate vertically 35 m into the rock. Some 4000 m of drains were drilled before the filling of the reservoir was started. All the drains are equipped with a steel head to protect them and to allow for measuring the resulting flow and the levels of the nappe. Most of the drains on the downstream side of the arch began to show a slight degree of oozing in 1960 when the water level in the reservoir reached a level of 1506 m (about 90 m above foundation level). More recent information is not available.

(4) *Saint Guérin arch dam* (70 m high).   This dam in the Alps is built on homogeneous gneiss. Grout absorption was extremely small and only 40 tonnes of cement were used in the 2200 m of boreholes for the deep screen, and 20 tonnes in the 3000 m of boreholes for contact grouting. The drainage network consists of twenty holes of 116 mm diameter and 20 m depth, which are situated some metres downstream from the structure.

(5) *Lamoux arch dam* (45 m high).   The information obtained on this dam in the Pyrenees shows that ten vertical drains, 76 mm (3 in) in diameter, having a total length of 299 m, were used.

(6) *Neguilhes arch dam on river Orlu in the Pyrenees* (55 m high).   The primary grouting screen consists of thirty boreholes having 470 m total length. They absorbed only 22 tonnes of cement (47 kg/m). The secondary screen (length 405 m) absorbed 29·5 tonnes of cement (73 kg/m). This work was completed by the addition of three drainage holes 25 m deep, drilled several metres downstream from the dam. Water leakage can be gauged.

(7) *Raviège buttress dam* (35 m high).   The drainage gallery in this dam in the Massif Central of France was provided with two networks; a vertical network, approximately parallel to the upstream wall of the concrete dam is intended to keep this wall in good condition. These drains were brought under pressure before filling the reservoir, thus enabling certain defects in the concrete to be noticed and repaired. A second network consisting of a rock drainage system, penetrates 5 m into the rock.

This short enquiry into the drainage of arch dam foundations shows that, in a limited number of cases, great care has been taken to produce a safe design. In others, the number of boreholes used for drainage seems low compared with the data suggested by Yokota and the accepted practice for gravity dams. More disconcerting is the fact that in the majority of cases no information is available on what drainage has been provided, and there is reason to suppose that there may be none.

A point of importance concerns the water pressure at the outlet from the rock along the free rock surface. Before the reservoir is filled, the rock surface may be dry and the rock slopes downstream of the dam are supposed to be stable. When the reservoir is filled, the water pressure builds up behind the rock slopes unless it is relieved by the curtain grouting and/or the drainage. At Kurobe IV, on the model, it reached up to 30% of the static pressure. This build-up of the water pressure may cause a rock slide as in El Frayle (see case history, chapter 12). Positive drainage is recommended.

11.4.3 Reinforcement of rock foundations

Information on the reinforcement of the rock foundations of the Vajont dam and the Kariba dam were given in section 11.4.1. Additional case histories follow.

(1) *The Castillon and the Chaudanne arch dams.* The Castillon dam on the river Verdon creates a reservoir of $149 \times 10^6 \, \text{m}^3$ capacity, about 50 miles (80 km) north-west of Cannes. The height of the dam is 100 m, the length of the crest is 200 m $(L/H = 2)$ and the radius of the face 70 m. The dam is built on fissured limestone of upper Jurassic age; the topographical conditions are very favourable, but the geological conditions much less so. The right abutment is on very broken limestone due to an anticline, the general dip of which is upstream. The fissures are extremely numerous and some exceed 1 m in width. A great amount of exploratory work was done (shafts and galleries, boreholes with water and cement injection trials). In order to consolidate some 200 000 m^3 of rock it was decided to carry out the following work (Jaeger, 1964a; Walters, 1962).

  (a) Inject grout (1300 tonnes) in all fissures where access was possible (filling some 2000 m^3).
  (b) Inject 2650 m^3 of cement grout, inert materials and bentonite. The general grout screen absorbed some additional 5195 tonnes of cement. Some boreholes extended down to no less than 200 m below water level.
  (c) A series of concrete counterforts were built on the right bank, downstream of the dam, each anchored to the rock by five prestressed cables of 1000 tonnes each (fig. 11.29).

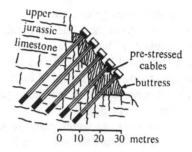

Fig. 11.29   Castillon dam; rock anchorage.

Rock consolidation at the Chaudanne arch dam (56 m high) was carried out on similar lines. Post-tensioned cables penetrate the rock abutment on its whole thickness (fig. 11.30).

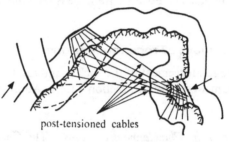

Fig. 11.30   Chaudanne dam.

(2) *The Monteynard dam.*   A large, nearly vertical rock fracture on the right abutment, filled with clayish material, was cleaned and filled with compact concrete. The job was done from vertical shafts and horizontal galleries. This can be classified merely as conventional consolidation work.

(3) *Kariba dam abutment reinforcement.*   In the Kariba gorge, the rock comprising the left abutment of the valley bottom and two-thirds of the right abutment is massive gneiss, whilst the upper part of the right bank (south) is a highly jointed quartzite. The jetting and grouting of the quartzite commenced with holes 10 ft apart. Later, an arrangement of three adjoining hexagons comprising sixteen holes at 5-ft centres was adopted.

In addition, an extensive seam of micaceous material 5 ft thick and at least 100 ft long, some 200 ft into the hillside beyond the dam abutment and across the line of thrust, was discovered. This seam thickened to 30 ft below the surface. It was excavated through shafts and adits and refilled with 16 500 yd³ of concrete. Boreholes were drilled in the grouted hexagonal cells and showed considerable improvement with 80% core recovery, but, nevertheless, signs of soft material still remained in some fissures. Additionally, a series of test galleries and test shafts were driven in the rock. Seismic tests and pressure jacking confirmed the weakness of the quartzite as compared with the gneiss. Further inspection showed that the treatment of the decomposed quartzite by grouting had not produced a fully compact rock. All these results confirmed the engineers' conclusions that the quartzite zone and the weathered gneiss were much too deformable. Additional grouting in order to consolidate the quartzite was prohibitive in cost. The alternative was to construct the gravity abutment block founded on the sound gneiss. The abutment block has been designed as follows (fig. 11.31).

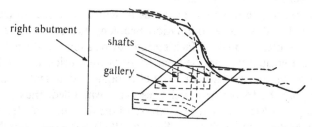

Fig. 11.31 Kariba dam. Reinforcement of right-hand rock abutment.

(a) All the reinforcements have been built beyond construction joint no. 30 and to the right of it. The thrust due to the arch beyond joint 30 was found by trial load analysis to be 94 000 tonnes for a water level of 1560 and 236 000 tonnes for maximum water level of 1605 (dam crest 1606 ft O.D.).

(b) Approximately 30 000 yd³ of additional concrete was placed on the downstream face of the dam and tied to the existing concrete with cables, using 15 000 tonnes of prestress.

(c) Four mass concrete buttresses each 20 ft wide were formed in the quartzite below the abutment and founded on a mass concrete raft resting on the sound gneiss. The buttresses were excavated and concreted from two shafts and a number of adits. A diagonally directed cut-off curtain and retaining wall above it protects the whole area from the risk of leakage from the reservoir.

(d) The treated quartzite between buttresses adds its own weight for safety (Gibb, Coyne& Sogei, 1962).

(4) *Foundation reinforcements of buttress dams.* The trend of ideas concerning the reinforcement of buttress dam foundations gets back to the Beni-Bahdel dam. The original design was by Stucky but its height was subsequently increased. To keep the resultant of all the gravity forces and hydraulic thrust well inside the middle third of the buttress base and to improve the value of $\tan \psi = T/N$, an inclined force pushing in an upward direction was applied at the toe of the dam by means of Freyssinet jacks resting on the foundation rock (Ribes & Butin, 1958).

The same idea was applied to the Erraguene dam on the Oued Djen-Djen in Algeria. One of the reasons for using this technique is the wish to eliminate the always dangerous, permanent and irreversible deformations of the rock. Pressure by the jacks was increased beyond the prescribed value, then released in order to test the elasticity of rock and concrete and to create the best possible conditions of strain distribution in the structure. Depending on the height of the buttresses, the force on the abutment was from 8000 tons to a maximum of 34 000 tons. The force was always adjusted to the water level behind the multiple arch dam.

A similar technique was used at the Sefid Roud dam in Iran (same main contractors). In addition, a foundation problem had to be solved, for the rock was of indifferent quality. Jacking the buttress toe transferred the thrust to rock of better quality, and the weaker zones were bridged. The buttresses are of a more massive design than those of Erraguene dam and it was possible to partly apply the forces before the reservoir was filled, thus giving more time to compensate for the non-reversible and the delayed rock deformations.

The Ben Metir dam in Tunisia (Stucky, 1955) has been described many times as a classical exercise for buttress dam foundations on plastic rock. Two drainage galleries were excavated at some distance under the dam with drains pointing upwards towards the foundations of the buttress. The purpose of this drainage is to accelerate the settlement of the clayey foundations, and according to recent reports settlement has already reached 0·10 m. Galleries have been built in the footing of the buttresses in a direction parallel to the buttresses. In the event of uneven foundation settlement, the base of the buttresses can be reinforced by locating cables in these galleries and post-tensioning them.

Part Four
Case histories

Engineering geology and rock mechanics are applied sciences, the final object of which is to assist in solving practical engineering problems. How they do this is best illustrated by case histories. Some of them have already been discussed in the chapters on engineering geology and rock mechanics; others have been mentioned in the three chapters on rock slopes, galleries, tunnels and excavations and on dam foundations. A few additional cases of special interest have been selected to illustrate the many very different lines of approach vital to design problems; each case stresses a few special points.

Chapter 12 examines cases concerning dam foundations and underground works of American design. Additional information is given about tunnel design and testing. Chapters 13 and 14 describe the two most dramatic failures of large engineering structures in recent times, that of the Malpasset dam and the rock slide at Vajont. Chapters 15 and 16 concern the consolidation of difficult rock slopes and the construction of three large underground power-stations. These five chapters supplement chapters 5 to 11.

---

12   Dam foundations and tunnelling

Examples of particular rock engineering problems have been given in the previous chapters. The following details are taken from authoritative publications. They supplement the information given in parts two and three. It is interesting to note the methods responsible engineers have, in particular cases, employed to test rock materials and rock masses and how they have interpreted the results.

12.1   Rock mechanics for Karadj dam

12.1.1   Description of the project

The Karadj dam is located on the Karadj river, approximately 25 miles from the city of Teheran (Iran). The project will assure adequate and reliable supplies of municipal and industrial water to the capital of Iran. A secondary benefit of the project is hydroelectric peaking capacity.

The drainage area is 764 km²; the average annual run-off at the dam site is estimated at 472×10^6 m³; the maximum recorded flow is 350 m³/s and the spillway design flood is 1450 m³/s. The live storage capacity is 172×10^6 m³. There is a small regulating capacity of 0.6×10^6 m³ on the downstream side of the dam. The additional annual water supply available because of storage is estimated at 115.6×10^6 m³. The main dam is a double-curvature, thin concrete arch. It has a maximum height of 180 m and a crest length of 390 m. The horizontal circular arches have an approximately constant central angle. The radii of the arch centre lines vary from a minimum of 74 m at an elevation of 1612 m to a maximum of 201 m at elevation 1770 m (roadway level). The arch thickness is 32 m at elevation 1612 m, varying to 7.85 m at the roadway. It has been analysed as a symmetrical structure, with base at elevation 1600 m. Below this, and on the right abutment, in the lower portion, massive concrete plugs have been constructed. These plugs are intended to function as foundation rock, and are not analysed as a portion of the arch structure. The volume of concrete is 350 000 m³. The double-curvature design substantially saves on concrete compared to other arch-dam shapes and is approximately 60% less than the amount needed for a gravity dam type for the same site. The power-station will ultimately have an installed capacity of 120 000 kW. The average annual electrical energy delivered will be 149×10^6 kWh.

The dam was first analysed by the crown cantilever method, Further analyses were made with three and then five cantilevers. The maximum computed stresses are 150 lb/in² for tension and 1000 lb/in² for compression. A large programme of model studies was planned to include tests that would incorporate the effect of the differential in the modulus of elasticity of the foundation rock relative to that of the concrete in the arch dam. However, the field tests performed in galleries in the foundation revealed that the modulus of elasticity was quite close to that of the concrete. Measurements ranged generally from 2×10^6 to 2.5×10^6 lb/in² while in the main dam it was 3×10^6 lb/in². The material used to construct both the model of the dam and its foundation was homogeneous plaster. The model studies also answered two economically important design questions: firstly, thrust blocks would be necessary on either abutment; secondly, the opening in the arch dam for the spillway crest piers and gates on the right abutment did not materially affect the stresses in the dam nor the abutment loads. Therefore a thrust beam would not be required across the top of this opening.

A small regulating dam is located 4000 ft downstream of the power plant. It is an arch gravity concrete dam with a centrally located gated spillway (dam crest at level 1600 m; maximum high water elevation 1610 m).

12.1.2 Geology

Waldorf and contributors describe the geology of Karadj dam as follows:

Karadj dam is founded on a diorite sill approximately 360 m thick which was intruded along the bedding planes of the enclosing tuffaceous rocks: the sill strikes nearly parallel with the chord of the arch dam and dips downstream at an angle of approximately 40°. Hydrothermal alteration, which has resulted in kaolinization of the felspars, is found throughout the entire mass of the sill and is particularly evident along fractures spaced approximately 5 metres apart, where zones to a metre wide have been strongly kaolinized and sericitized.

 There are three well-marked joint sets. The most prominent of these strikes roughly parallel with the strike of the sill and dips downstream at an angle of approximately 45°. Spacing of the joints ranges from 0·5 metres to 3 metres. These joints contain calcite and zeolite fillings ranging from a film to 5 cm width.

 Orientation of the three major joint sets is, in general, not adverse to the stability of the abutments under the arch dam loadings and there is no evidence of planes of weakness oriented in a direction that would permit sliding of any part of the abutments.

The characteristic figures for the diorite area are as follows: sound diorite $E = 5\,000\,000$ lb/in² (356 000 kg/cm²) from laboratory tests on sound rock samples; altered diorite $E = 1\,620\,000$ lb/in² (114 000 kg/cm²); compressive strength $\sigma = 4700$ lb/in² (~ 320 kg/cm²).

12.1.3 The test programme

The test programme was established by Waldorf *et al.* (1963) as follows: a circular loading plate 760 mm in diameter, loaded by two hydraulic jacks was used. Total force exerted by the jacks was 400 tonnes. This gave an average surface loading of 1140 lb/in² (~ 80 kg/cm²). The rock surface or loading plane on which the loading plate rested had minimum lateral dimensions of 5 to 5·5 plate diameters. The purpose was to assure that there would be no appreciable effects of the side walls of the test chambers on the rock deflections. Where practicable, the plate loads would be approximately in the direction of the thrust of the dam. The test chambers had to be as near as possible to the plane of the abutment. Three chambers were excavated in each abutment at the elevations 1620, 1680 and 1730 m.

Description of test scheme.   The thrust of the hydraulic jack ram was transmitted by a spherical head to the loading plate which rested against the flat test surface. The gauging system consisted of a number of dials fastened by adjustable arms and cross members to two pipe supports which spanned the full test surface and with a distance between each of five plate diameters. Gauges were positioned to measure movements of the loading plate at four points and at a number of points on the rock surface along the axis passing through the centre of the plate. Measurements were made during both the loading and unloading phases of the cycle. Rock displacements outside the plate would be approximately 10% of the plate displacement, requiring a minimum accuracy of plate displacement measurements of 0·02 mm. The accuracy of the gauges used was actually greater. Close control of the temperature was required.

There were four tests on the normally jointed diorite rock which constituted the bulk of the foundation. Two were on altered rock which appeared in zones and lenses, one of which had close fracturing (chambers 2*A* and 6). A single test was on a wedge of rock formed by a prominent joint intersecting the test face at a flat angle. The massive igneous rock (four tests) was intersected by a system of joints which divided it into blocks generally under a half metre in linear dimensions. Joints were usually tight but were occasionally filled with zeolites and calcite or both. The total thickness of such materials formed only a minute part of the rock. From the evidence of drill cores and drifts the maximum thickness of a zone or lens of altered rock appeared to be under four metres. Accordingly, in the analysis of the effect on foundation modulus a thickness of four metres was assumed for such a zone.

Altogether thirty-seven tests of various kinds were made on seven sites. There were two temperature tests, four plate deflection tests and thirty-one rock modulus tests. A minor correction was made to the plate displacement measurements to account for the compression of the mortar pad on which the plate rested. The mortar pad was normally 50 mm thick and was made of one to one cement–sand mortar. Assuming the modulus of the mortar to be 250 000 kg/cm² ($3 \cdot 5 \times 10^6$ lb/in²) the full load compression would be 0·016 mm for the 760 mm plate and 0·044 mm for the 450 mm plate.

12.1.4 Main results of the tests and measurements: discussion

The corrected effective modulus values, as determined from plate deflections are set out in table 12.1.

Table 12.1 *Effective modulus from plate displacement* (in kg/cm²)

Chamber	Uncorrected modulus	Corrected modulus	Elevation (m)
1	158 500	156 500	1620
2	178 000	174 500	1680
3	127 800	127 800	1730
4	135 500	132 300	1620
5	131 000	127 800	1680
Average	146 300	143 800	

A separate analysis was made for the tests in chambers 2*A* and 6 where rock was altered, and these results are summarized in table 12.2.

Table 12.2 *Effect on modulus of major discontinuities in rock mass*

Types of discontinuity	Chamber 2A (altered, fractured rock)	Chamber 6 (partly open major joint)
Plate diameter	450 mm	760 mm
Measured displacement	1·070 mm	2·142 mm
Corrected displacement	1·096 mm	2·275 mm
Apparent modulus	76 500 kg/cm²	21 800 kg/cm²
Scale: protoype/test plate	51	30
Modulus of altered zone	67 400 kg/cm²	
Modulus of unaltered rock	163 700 kg/cm²	163 700 kg/cm²
E_{eff} for prototype with discontinuity	136 300	106 000 kg/cm²

The authors of the Karadj dam paper estimate that the probable range of the rock modulus is from 143 000 kg/cm² to 183 600 kg/cm². The upper value is representative of the rock mass below the surface zone from excavation operations. These values are computed from points outside the plate. The lower value (143 000 kg/cm²) is computed from plate displacements, and is considered representative of the modulus in the surface zone of disturbed rock (disturbance from excavation operations). The average of the two figures gives a mean value $E = 163\ 700$ kg/cm² ($2·33 \times 10^6$ lb/in²).

Laboratory tests of twenty-three core sections showed moduli ranging from $0·43 \times 10^6$ lb/in² to $11·7 \times 10^6$ lb/in². The crushing strength was:

(*a*) for three highly altered cores   4760 lb/in² (336 kg/cm²)
(*b*) three moderately altered cores   10 860 lb/in² (765 kg/cm²)
(*c*) average for all tests   15 500 lb/in² (1090 kg/cm²)
(*d*) mean value of *E* for core tests $E = 5·05 \times 10^6$ lb/in² or 356 000 kg/cm²

No correlation was found between laboratory tests and the modulus measured in the field. It was found that jointing has more influence on field modulus than variations in the rock itself. But this may not be valid where there is a more extreme variation in rock quality.

Effect of minor jointing and sporadic rock alteration. The mean value of $E = 163\ 700$ kg/cm² quoted before and included in the second table was representative of normal sound rock with joints along which there may be a moderate alteration. But no appreciable effect of this was found. The effect of fairly close minor jointing was included in all the tests (rather uniform spacing of the joints throughout the mass), and this was why the field modulus was low in comparison with laboratory findings.

Effect of major joints and zones of altered and fractured rock.   Waldorf and his colleagues developed a most interesting theory of the 'scale effect' and of the effect of non-homogeneous rock layers. A theory of the effective modulus E_{eff} was similarly developed by these authors (section 7.2). The following figures show how a weak seam modifies the effective modulus E_{eff}. In chamber 2*A*, there was a one-metre-thick uniform zone of altered fractured rock ($E = 67\,400$ kg/cm^2) above the sound rock. Taking 'scale effect' into account, a 4-m-thick superficial zone of altered material under the prototype would yield an $E_{eff} = 137\,300$ kg/cm^2. On a similar basis it was calculated that for $E_1 = 67\,400$ kg/cm^2 altered layer and $E = 163\,700$ kg/cm^2 sound rock mass, for a depth d of sound rock to a 4-m-thick altered zone, the E_{eff} would be (prototype):

$$d = 4 \text{ m}, \qquad E_{eff} = 137\,300 \text{ kg/cm}^2$$
$$d = 10 \text{ m}, \qquad E_{eff} = 144\,000 \text{ kg/cm}^2$$
$$d = 20 \text{ m}, \qquad E_{eff} = 152\,500 \text{ kg/cm}^2$$
$$d = 60 \text{ m}, \qquad E_{eff} = 163\,000 \text{ kg/cm}^2.$$

In chamber 6, a thin, very soft seam with only partial contact was located at a depth of 0·5 m under the test plate. The jacking test showed a (corrected) displacement of $u = 2\cdot275$ mm under a load $Q = 363\,000$ kg. The plate radius was $a = 380$ mm. The formula yields:

$$E = \frac{m_n Q(1 - v^2)}{au \sqrt{\pi}} = \frac{0\cdot96(363\,000)(0\cdot96)}{2\cdot275(380)(1\cdot772)} = 218 \text{ kg/mm}^2 = 21\,800 \text{ kg/cm}^2,$$

where $\sqrt{\pi} = 1\cdot772$ and $v = 0\cdot2$. The average E value for unaltered rock being $E = 163\,700$ kg/cm^2 the deformation of the rock itself should have been

$$u = \frac{0\cdot96(363\,000)(0\cdot96)}{(380)(1\cdot772)(1\cdot637)} = 0\cdot303 \text{ mm}.$$

The compression of the joint can be estimated to be

$$u_1 = 2\cdot275 - 0\cdot303 = 1\cdot972 \text{ mm}.$$

Applying similar reasoning to the prototype (scale 30:1) it was found that

$$E_{eff} = 106\,200 \text{ kg/cm}^2 \text{ against } E = 163\,700 \text{ kg/cm}^2$$

for sound rock.

Assuming the same joint to be located at a depth d, the E_{eff} would have been

$$d = 0\cdot5 \text{ m}, \qquad E_{eff} = 106\,200 \text{ kg/cm}^2$$
$$d = 10 \text{ m}, \qquad E_{eff} = 125\,000 \text{ kg/cm}^2$$
$$d = 20 \text{ m}, \qquad E_{eff} = 142\,500 \text{ kg/cm}^2$$
$$d = 60 \text{ m}, \qquad E_{eff} = 161\,000 \text{ kg/cm}^2$$
$$d = \infty, \qquad E_{eff} = 163\,700 \text{ kg/cm}^2.$$

It is of interest to note that these tests prove that a smaller loading plate and too small a load (like the 50-tonne jack used by Talobre) would be unable to detect a weak rock seam even if it was just below the plate. Figures obtained from such tests may be misleading, because even the geological strata immediately below the loading plate must be known if correct interpretation of the results is to be made.

12.2   Analysis of rock properties at the Dez project (Iran)

12.2.1   Description of the project and geology

The Dez dam (now called the Reza Pahlavi dam) is a double-curvature thin-arch dam, 666 ft high (204 m), built on the Dez river in Khuzestan Province of Southern Iran (Dodds, 1966). The project includes two spillway tunnels, one 47 ft (14·35 m) in diameter and one 42·5 ft (13·0 m) in diameter, and two 34-ft (10·40 m) diameter power tunnels which feed four 13·5-ft (4·13 m) prestressed concrete penstocks. The project is an American design.

All these structures are built on or in a cobble conglomerate (middle to late Pliocene). The conglomerate is a fluviatile deposit of boulders, cobbles, and gravel with some intercalated sandstone lenses and is more than 2000 ft (600 m) thick. There was a high percentage of initial voids which are partially filled by cementing material. The conglomerate is composed of subangular dolomite to well-rounded gravels of limestone, and chert with minor amounts of tuff and mudstone. A matrix of sand, silt and clay partially fills the voids. Calcite ($CaCO_3$) is the principal cementing agent in the rock, but its deposition has not been complete, and open-work pockets are left within the rock mass. According to the American experts in charge of the project, the young age, variable composition and cementation of the rock and the exacting foundation demands made a precise understanding of the physical properties of the rock a necessity.

12.2.2   Testing programme

The exploration and testing programme consisted of the following.

(a) Drilling 11 210 lineal ft of NX boreholes 3 in diameter and 1432 lineal ft of 6-in-diameter boreholes.
(b) Excavating twelve test adits with a total length of 4000 ft.
(c) Testing two 20-ft-long hydrostatic pressure chambers. Measuring *in situ* residual stresses (47 flat jacks). *In situ* bearing-plate tests using 100-ton hydraulic jacks.
(d) Testing 881 6-in-diameter rock cores in the project laboratory for: unconfined and confined crushing strength; sustained load reaction, tensile and shear strength; rock creep.

(1) *The hydrostatic pressure chambers*. These were of the type developed by Oberti and described in chapter 6. Four types of tests were made:

(a) Deformation tests over periods of three and four days at pressures up to 200 lb/in².

(b) Creep tests over a period of sixteen days.

(c) Deformation tests over periods of three and four days at pressures up to 265 lb/in².

(d) Some tests after grouting the surrounding rock.

It was found that the average modulus of elasticity varied between 1 138 000 and 1 561 000 lb/in² (80 000 and 110 000 kg/cm²) with a minimum of 568 000 lb/in² (40 000 kg/cm²). The average secant modulus (total deformation) varied between 710 000 and 994 000 lb/in² (50 000 and 70 000 kg/cm²), depending mainly on the direction of the rock strata.

(2) *Plate loading tests*. These were carried out with a rig similar to the one used by Talobre (1957), but capable of 100 tons. Because these tests are relatively inexpensive and easy to set up, a total of 257 were made and the results analysed by Talobre on the lines developed in his earlier publications. The minimum E values were from $1 \cdot 0 \times 10^6$ to $1 \cdot 8 \times 10^6$ lb/in² (70 000 to 126 000 kg/cm²), the maximum E values varying from $1 \cdot 5 \times 10^6$ to $3 \cdot 8 \times 10^6$ lb/in² (105 000 to 269 000 kg/cm²). The minimum values mentioned here are the secant modulus of the first loading cycle, the maximum values the elastic modulus of the fourth loading cycle.

(3) *Residual stress measurements: flat jack tests*. The method used was based on the work of Tincelin (1952) in the French iron mines using flat jacks located in slots. After deciding on the test location and the inclination of the slot, four studs were cemented on both sides of the future slot along a line at right angles to it 9 to 10 in apart. Twenty-four hours later the initial readings of the distance which separated the studs was made to establish the equilibrium base. A slot approximately 3 in wide by 30 in deep was then cut off the rock and the jack cemented in place. A very dry mix (50% cement, 50% sand) was used. The jacks were usually allowed to set for seven days before testing. During this whole period the measuring studs were covered with paper and mortar, to protect them against bumping.

The residual stresses measured varied between 334 and 2222 lb/in² (23·5 and 156 kg/cm²).

(4) *Laboratory tests*. To develop Mohr's envelopes for the rock so that the '*in situ* measurements' could be correctly correlated to in-place strength, a thorough laboratory testing programme was required. 881 rock cores were tested. To supply specimens thirty-six 6-in diameter drill holes with a total length of 1436 ft were drilled approximately equally spaced down both abutments.

Unconfined and confined crushing tests were carried out on 6-in diameter 12-in-high dry and wet cores, the confinement pressures, p, were 350 and 750 lb/in^2 (24·5 and 53 kg/cm^2). It was found that the confinement pressure, p, improves the crushing strength according to the following relationship:

$$\sigma_{cr} = 2621 + 2 \cdot 68p, \quad \text{(in lb/in}^2 \text{ values).}$$

Tensile strength was measured by the Brazilian method, and the shear strength computed from Mohr envelopes.

12.2.3 Foundation analysis

The results of all the foundation tests for the Dez project were evaluated by Talobre.

The rock is free from major structural defects (faults or joints); lenses and pods of below average strength rock are local and can be adequately treated by: (*a*) removal and back filling with concrete, (*b*) special grouting, (*c*) additional reinforcing of the abutment.

Talobre estimated that sliding failure in the dam abutments could only occur if new failure surfaces developed from crushing of the rock. The ultimate bearing capacity of the Dez abutments depended on the crushing strength of the rock masses.

Twenty-eight determinations of internal stress present in the rock at the abutment surface were made using flat jacks. The average measured was 560 lb/in^2, more than 50% of the measurements were over 470 lb/in.^{2}

Talobre used the very conservative figure of 300 lb/in^2 for the residual compression present in the rock at the abutment surface. Correcting this value for the direction of dam thrust gives actual minimum confining loads of 339 and 450 lb/in^2 for the downstream and the upstream edges of the dam shoulder. The laboratory formula now becomes (average values): for the downstream edge of the saddle block $\sigma_{cr} = 2621 + (339 \times 2 \cdot 68) = 3530$ lb/in^2; for the upstream edge of the saddle block $\sigma_{cr} = 2621 + (450 \times 2 \cdot 68) = 3810$ lb/in^2. (These are average values; local values were usually higher.) Comparing this against local stresses, it was found that safety factors against crushing of the rock were between 4·9 and 11·8.

It was established that the Dez river water is completely saturated with $CaCO_3$ and is therefore not able to dissolve an additional amount of calcite. The effect of the pore pressure under the dam is to lower the effective confining pressure, causing a loss of rock crushing strength, and also to lower the effective load on the rock abutment (uplift). This was taken into account when calculating local safety factors at different elevations.

12.3 Foundation investigations for Morrow Point dam and power-plant and Oroville dam and power-plant

Research for the Karadj dam centred on a more precise interpretation of loading tests on heterogeneous stratified rock and on problems of scaling up results to prototype conditions. For the Dez project, effort was put into estimating the rock-crushing strength under triaxial stress conditions. The main object of the tests conducted by Müller for Kurobe IV dam (see chapter 11) was to determine the shear strength of the fissured rock masses. Three cases – three different sets of problems.

Foundation investigations for the Morrow Point and Oroville sites were for a combined dam site and underground power-plant. The approach here is different from that in previous case histories.

12.3.1 Description of the project and geology at the Morrow Point site

The Gunnison River Development (part of the U.S. Bureau of Reclamation Colorado River Project) is primarily a water-storage and hydroelectric power development (Dominy & Bellport, 1965). Two dams have been constructed: the Blue Mesa rock-fill dam and the Morrow Point dam. A third dam, the Crystal dam, is planned.

Morrow Point is the Bureau's first concrete, double-curvature thin-arch

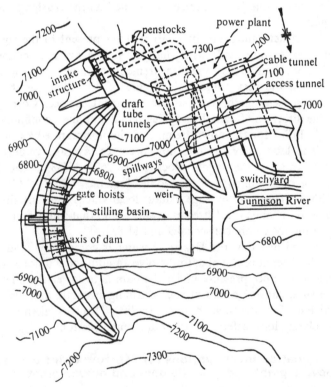

Fig. 12.1 General plan Morrow Point dam and power-plant (after Brown & Morgan, 1965).

dam. It is 465 ft (141 m) high, 52 ft (15·80 m) thick at the base and 12 ft (3·66 m) thick at the crest. It has a crest length of 720 ft (220 m) and a volume of 360 000 yd³ of concrete. The spillway consists of four orifice-type openings in the top central part of the dam, providing a free-fall discharge of over 350 ft to the concrete stilling basin at the toe of the dam. Maximum capacity of the spillway is about 41 000 cusecs (fig. 12.1).

The power-plant chamber is excavated in the canyon wall in the left abutment at about 400 ft below ground surface. It is 205·5 ft long and 57 ft wide with a height ranging from 65 to 134 ft. There are two 60 000 kW generators driven by two 83 000 h.p. turbines. The power penstocks consist of 13½-ft-diameter steel liners in 18-ft-diameter tunnels (fig. 12.2). The reservoir

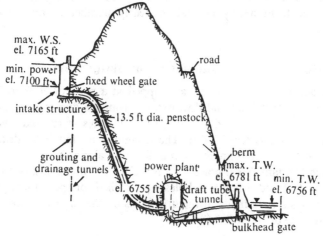

Fig. 12.2   Morrow Point underground power-plant (after Wolf, Brown & Morgan, 1965).

behind Morrow Point dam has a capacity of 117 000 acre ft at a maximum water level; the active storage being 42 120 acre ft.

Morrow Point dam and power-plant and the reservoir basin are located entirely within pre-Cambrian metamorphic rocks. The dam site is in a narrow section of the Black Canyon gorge, which is only 200 ft wide at river level and about 550 ft at crest elevation. The abutments consist of nearly vertical to overhanging cliffs separated by narrow shelves. There is very little overburden. Alluvial material in the valley floor consists of rounded sand, gravel, cobbles and boulders mixed with large angular talus blocks. The rock at the dam site consists of alternating lenticular and irregular beds of micaceous quartzite, quartz–mica schists, mica schists and biotite schists, all of which have been intruded by granitic veinlets along both foliation and jointing. A considerable portion of the lower left abutment consists of granite pegmatite. The quality of the rock types varies considerably, the hardest being micaceous quartzite and granite pegmatite with variations of hardness and strength down to the weaker biotite schists. Of the several rock types, only the biotite schists display moderate weathering.

The direction of the thrust from the dam is across the upstream dip of bedding, which precludes slippage along schist beds in the abutments. The rock contains stress relief jointing which generally parallels the canyon walls.

The underground power-plant is located in uniformly high-quality metamorphic rock intruded by irregular small bodies of granite pegmatite. The bedding strikes nearly normal to the power-plant alignment and dips upstream at angles of 15° to 60°, average about 35°. This is favourable for bridging across arches and helps prevent separation along contacts. Two shear zones cross through the power-plant area. Widely spaced irregular joints are present in the rock, but they are not continuous for any great distance. Strikes range from nearly normal to nearly parallel to the power-plant alignment.

12.3.2 Geological and geophysical investigations

The area occupied by the dam was mapped at a scale of 20 ft to 1 in; the area extending 1000 ft up and downstream from the dam was mapped at a scale of 50 ft to 1 in. 108 diamond core drill holes were made with total footage of approximately 8000 ft. The technique of 'fan hole' drilling was used to investigate the foundation in a direction along the principal lines of thrust from the dam. Most holes were NX size (3-in diameter), but some BX (2⅜-in diameter) and a few 6-in holes were drilled for special test purposes. Percolation tests were made in the drill holes by the rubber packer method, using a pressure of 100 lb/in² (fig. 12.3).

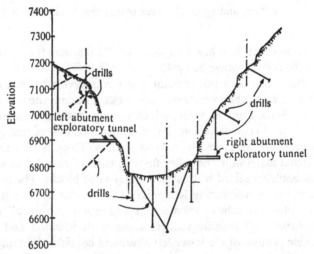

Fig. 12.3 Morrow Point dam site showing angle drill pattern and exploratory tunnels (after Wallace & Olsen, 1965).

Five exploration tunnels, three of which were excavated during preliminary investigations and two during construction, were completed. Four dam-site tunnels, two in each abutment, extend about 100 ft into the rock. One tunnel extends along the entire length of the underground power-house near the top of the proposed machine hall excavation (fig. 12.4). The abutment

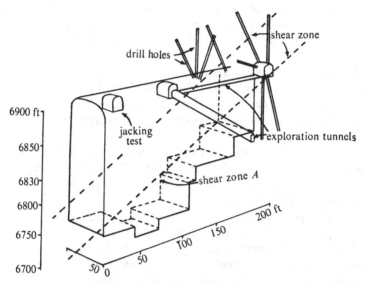

Fig. 12.4 Morrow Point underground power-plant. Exploration tunnels and drill holes in the area of the underground power-plant (after Wolf, Brown & Morgan, 1965).

tunnels provided data on the attitude and severity of the surface and stress relief jointing, and were also used as stations for seismic shots and sites for jacking tests. The lower tunnel on the right side was located along a soft biotite schist bed to investigate what was considered to be the weakest rock at the site.

Drill holes were extended from the power-house tunnel in various directions and television examination was conducted in several of them. This explored geological features not seen from the drill core. Numerous seismic field measurements of the elastic modulus of *in situ* rock were carried out at Morrow Point. For estimating the E modulus the following simplified formula was used:

$$E = (12v)^2 \times 9{\cdot}3 \times 10^{-5} \times d.$$

In this formula E is expressed in pounds per square inch, v is the wave velocity in feet per second and d is the specific gravity in grammes per cubic centimetre. This value E can be corrected for the effect of Poisson's ratio.

The measured wave velocity, v, was 12,800 ft/sec and the Poisson ratio, ν, was 0·2. For quartzite, E was found to be:

$$E = 5·3 \times 10^6 \quad \text{(with } d = 2·70 \text{ g/cm}^3\text{)}.$$

According to Dominy & Bellport, the general test procedure with seismic waves can be applied in several different ways.

(*a*) Using a number of vibration detectors (to a maximum of 12 and 4 explosion points) measurements can be made at different locations along a line to ascertain the depth and wave velocity values in layers at and below the surface.

(*b*) Measurements of wave speed can be made, using one vibration detector at different depths in a drill hole with successive explosions at the mouth of the hole. This procedure can be reversed, i.e. having the detector at the mouth of the drill hole and detonating explosive charges at different depths within it.

(*c*) Measurements can also be made between drill holes by using a single drill hole detector and exploding charges at different depths. In some cases a charge in one hole and a detector at the same depth in another may be used to study velocity distribution.

(*d*) It is also possible to measure wave velocity from the bottom of a drill hole by exploding a charge, the effect of which is registered on a detector located at a distant point in an exploration tunnel. This procedure was employed at the Morrow Point site to ascertain the elastic modulus of the rock in which the penstock tunnels would be excavated.

(*e*) Wave velocity can be measured between an explosion point at the end of an abutment tunnel and a detector on the surface (fig. 12.5).

(*f*) To check the depth of blast damage the Bureau of Reclamation measured wave speed and layer depths at some twenty points on the walls of the power-plant access tunnel which had been driven using the 'smooth wall' blasting technique. In 1962 they also checked the validity of the seismic wave velocity measurements over a section of an abutment. In fig. 12.5, showing the right dam abutment, a seismic line F extended from a shot point, 95 ft from the portal of an exploration tunnel, upward through a section of gneiss and schists for a distance of 176 ft to a vibration detector on the surface. The measured seismic velocity along line F was 10 000 ft/s and with a computed density of 2·75 from tests of samples, the seismic elasticity of the rock was determined to be $E = 3·3 \times 10^6$ lb/in^2. A drill hole was later put down along the seismic line F. The elastic moduli of rock cores were measured and found to be $E = 0·86 \times 10^6$ lb/in^2 for biotite schists, $E = 1·2 \times 10^6$ lb/in^2 for mica schists and $E = 4·14 \times 10^6$ lb/in^2 for micaceous quartzite or quartz–mica schists. The apportioned average laboratory value from static tests of cores was calculated to be $E = 3·1 \times 10^6$ lb/in^2; which compared favourably with the seismic value $E = 3·3 \times 10^6$ lb/in^2.

The whole foundations of the dam and underground power-plant site

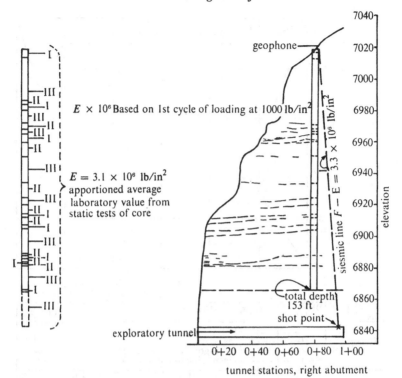

Fig. 12.5 Morrow Point dam. Correlation between seismic measurements of rock elasticity and static laboratory tests (after U.S. Bureau of Reclamation). Rock grouping: (I) biotite schist, average 0·86; (II) mica schist, average 1·20; (III) micaceous quartzite or quartz-mica schist, average 4·14 × 10^6.

were explored with boreholes and exploratory galleries and covered with a dense net of seismic lines. On the basis of the many seismic tests supplemented by jacking tests a most detailed geological mapping of the whole area was possible.

12.3.3 Laboratory and field tests

Petrographic analysis of rock specimens classified rocks into five types (fig. 12.6). They were given the numbers I, II, III, IV and V, rock number III being the more frequent. Type I is biotite, the weakest rock type. Type II is a mica schist with well-developed but irregular foliation and some jointing. Type III is a micaceous quartzite with somewhat less developed foliation but a more prominent system of macro-jointing.

Shear and sliding friction tests. Originally it was planned to conduct all the tests *in situ* in the underground power-plant exploratory tunnel near the dam, but the contractor's work schedule delayed access to the tunnel. Therefore,

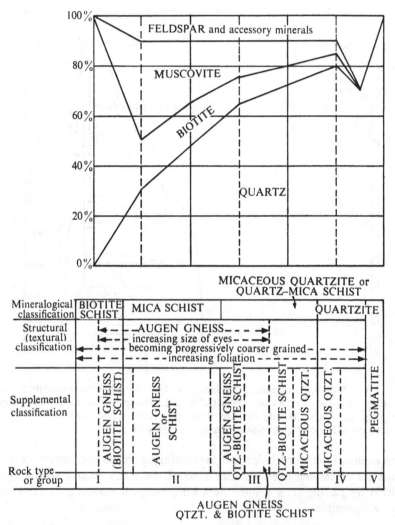

Fig. 12.6 Morrow Point dam area rock classification relationship (after U.S. Bureau of Reclamation).

only two shear tests and ten sliding friction tests were performed in the field, while five shear tests and forty-nine sliding friction tests were conducted in the Denver laboratory of the Bureau of Reclamation, utilizing the large specimens which retained the *in situ* joints and cracks. Included in the laboratory tests were one shear and six sliding friction tests conducted on a concrete block cast on type II rock. The equipment and procedure simulated that used in the field tests. Results from direct shear tests were compared with those obtained from triaxial laboratory tests on NX cores.

Rock specimens were prepared for the laboratory from large blocks trimmed into cubes of about 30 in by line drilling with a diamond core drill. These cubes were encased in concrete about 36 in in size. Next a 15 × 15 × 8-in-high projection was cut with a 36-in diamond saw. Rocks were orientated so that foliation of the shear specimen was parallel to the shearing load. A steel frame was placed around the upper block. A 5 000 000-lb-capacity machine was used for the tests. Two hand-operated hydraulic pumps each capable of producing a pressure of 10 000 lb/in² were used to activate the normal and the inclined jacks.

As explained in section 6.3.1, Serafim, in Portugal, employed two failure criteria, one based on maximum shear stress and the other on inversion of the vertical displacement (see fig. 6.16). In the Portuguese tests, the shear stress at inversion was near, but less than, the maximum shear stress reached; in the tests for Morrow Point dam, the inversion was not so pronounced, being one-half, or less, of the maximum shear stress. This may have been due to the difference in rock types and the fact that the loads in the Bureau tests were considerably larger. The Portuguese granite was weathered, maximum shear ranging from 48 to 159 lb/in², while the rocks tested by the Bureau reached 917 to 1511 lb/in².

Immediately after a block was sheared, a series of sliding friction tests were conducted using normal loads of 800, 600, 400 and nearly 0 lb/in² corresponding to normal loads of 180 000, 135 000, 90 000 and 0 lb. (The actual area of contact was less than the theoretical 16 × 16 = 256 in².) Tests were carried out *in situ* and in the laboratory, in the forward direction and in reverse, on dry and on moist surfaces. The angles of friction varied from $\phi = 27°$ to over $\phi = 40°$ (tan $\phi = 0·51$ to $0·84$).

Laboratory tests on foundation rock cores. The extensive programme just described was supplemented by tests on rock cores. Approximately 250 specimens of foundation rock cores selected from 10 dam site drill holes were tested to determine the main physical properties of the foundation: absorption porosity, specific gravity, modulus of elasticity, E, Poisson's ratio, v, and the tensile, compressive and triaxial strength. Most of the relevant tests were conducted along classical lines. Mohr circles and Mohr envelopes were traced for the typical rock types I to V.

Interesting experiments were carried out to determine the modulus of elasticity on 6-in cores with axial hole. A steel chamber was constructed so that hydraulic oil pressure could be applied to the outside cylindrical surface of a 6-in-diameter sample of rock containing an EX hole. The outside of the core was wrapped in plastic to prevent oil penetrating into the rock samples. The borehole gauge was inserted into the EX hole; deformations versus hydraulic pressures were observed. The modulus of the core was computed by use of the equation:

$$E = \frac{2D^2d}{(D^2 - d^2)} \frac{p}{\delta}$$

where D = outside diameter of the core, d = inside diameter of the core, p = external pressure, δ = change in inside diameter.

Field-jacking tests of foundation rock. Load was applied vertically, then horizontally, to the rock by two jacks, 3 ft apart, each having a capacity of 200 tons. Loads of 200, 400 and 600 lb/in^2 were applied to the rock for a period of approximately one week, maintained by automatic control. Three gauges were installed to measure vertical deformations at each point caused by the various test loads. These consisted of a rock deformation gauge which incorporated a special 'Carlson joint meter', under each of the two jacking loads to take measurements through approximately 15 ft of foundation rock (fig. 6.6). In addition one tunnel diameter gauge was placed midway between the jacking loads.

Residual rock strains. These were measured with the borehole gauge method and with Wittmore strain gauges. The equations required to reduce the strains to stresses in plane stress conditions for a 45° rosette of gauges (Merrill & Peterson, 1961) measured in the borehole are:

$$S + T = \frac{E(U_1 + U_3)}{2d},$$

$$S - T = \frac{E[(U_1 - U_2)^2 + (U_2 - U_3)^2]^{1/2}}{2d\sqrt{2}},$$

and

$$\tan 2\theta_1 = \frac{2U_2 - U_1 - U_3}{U_1 - U_3},$$

where S and T = the principal perpendicular inherent stresses, U_1, U_2 and U_3 = the measured deformations across the diameters 45° apart, E = the modulus of elasticity of the 'homogenous rock', d = the diameter of the EX hole and θ = the angle from S to U_1 measured counter clockwise.

12.3.4   Rock tests at the Oroville dam and power-house

A detailed description of the Oroville dam and underground power-house rock tests was given in a paper by Stoppini & Kruse (1964) from which the following information has been taken.

An extensive exploration programme was conducted at the power-house site as a preliminary to the construction of the underground works. An exploration adit and cross drifts within the left abutment of the dam were used. Similar to the Karadj tests, the Oroville plate-bearing tests used a circular plate with four dial gauges placed at equal intervals around its edge. And, as with the Karadj tests, the deflection pattern of the loaded plate failed to conform to any standard formula found in textbooks: it revealed a tendency for the plate to tip.

Three jacks, each of 100 tons capacity, were placed in line along a reinforced WF beam which in turn delivered the thrust along a short length of 6-in bar stock to a 1¼-in bearing plate 6⅞ in in diameter. It was intended to obtain high loading intensities. The adit was of minimum size (5 ft by 7 ft), unsupported, and driven on a slight gradient into the power-house site. Similarly, from the main adit several drifts and cross cuts were driven. The plate-bearing tests were performed in these and they provided access for extensive core drilling throughout the power-house site. They were also used to measure *in situ* stresses with flat jacks and boreholes. Flat jacks were also used to determine the rock modulus.

Rock relaxation tests were also conducted in the first of the two diversion tunnels (circular tunnels with a finished diameter of 35 ft), approximately 200 ft north-west of the power-house area. The rock mass here is homologous to that in the power-house area and the depth of overburden is virtually the same. Rock bolt extensometers were installed in a radial pattern in five rings of five bolts each, at intervals of approximately 110 to 190 ft. They were anchored at the base and grouted securely at the collar of the borehole. Measurements were to an accuracy of 0·001 in.

Since most of the convergence occurs within the first one or two heading advances following installation, it was important to install these extensometers as near to the heading as drilling would permit. The diversion tunnel was excavated with top heading full width, at which time the extensometers were installed. When the lower bench was removed, further convergence occurred. (Convergence of the tunnel causes the extensometers to record an elongation of the borehole in which they are installed.)

The method of computing the modulus is based on the assumption that displacements within the rock surrounding a tunnel cross-section are analogous to those in an elastic and isotropic material surrounding a hole in an infinite plate when subjected to biaxial stress. The *in situ* stresses in the rock were measured by flat jacks and confirmed by the borehole deformation method. *In situ* normal stresses of approximately 500 lb/in² were found to exist vertically and horizontally, both parallel and perpendicular to the river channel. With uniform stresses in all directions, the analysis was somewhat simplified. The modulus was obtained from the following equation:

$$E = (1 + \nu)(r - a)\frac{a}{r}\frac{p}{\delta},$$

in which p is the uniform primary stress in rock mass (existing prior to excavation of the tunnel), a is the tunnel radius, r is the tunnel radius plus length of rock-bolt extensometer and δ is the elongation recorded by extensometer over the length of the bolt. The value of the Poisson's ratio determined from core samples was assumed to be a reasonable approximation for the rock mass and the value of 0·25 was used in the formula. (It is suggested that in even moderately jointed rock with a ratio of vertical and horizontal *in situ* stresses equal to one, a higher Poisson ratio should have been assumed.)

Main results.   Rock at the dam site is hard, dense, generally fine-grained with specific gravity about 2·94. Unconfined compression tests on core samples indicated a Young's modulus averaging 12 900 000 lb/in² (910 000 kg/cm²) and a compressive strength of 40 000 lb/in² (2820 kg/cm²). Tensile and cohesive strength determined from the same cores were 3300 lb/in² (233 kg/cm²) and 5400 lb/in² (381 kg/cm²) respectively. The rock is metamorphic with high percentages of amphibole and is best described as an amphibolite. It appears massive although a rough schistose structure is not uncommon. Narrow bands of schists and a random system of well-healed joints and occasional shear zones are found throughout the test area.

Results from tunnel convergence measurements show that the modulus for each array of extensometers averaged from 1 200 000 to 3 310 000 lb/in². The five arrays together averaged 2 600 000 lb/in². The flat jacks yielded much higher values, ranging from $5·15 \times 10^6$ to $12·5 \times 10^6$ lb/in² for ten sites and an average of $7·5 \times 10^6$ lb/in² (530 000 kg/cm²). Modulus determined from the plate-bearing tests had to be corrected because the plate tilted. Values are given as follows: minimum $E = 995\,000$ lb/in² corrected to 1 272 000 lb/in², maximum $E = 1\,555\,000$ lb/in² corrected to 1 834 000 lb/in², average (5 sites) 1 500 000 lb/in².

13 Incidents, accidents, dam disasters

A paper entitled 'Dam disasters' was read before the Institution of Civil Engineers, London, in January 1963 by E. Gruner. He mentioned a Spanish publication (*Revista de Obras Publicas*, 1961) which listed 308 dams where serious accidents had happened. The causes of failure were attributed as follows:

foundation failure	40%
inadequate spillway	23
poor construction	12
uneven settlement	10
other causes	15

Most of the foundation failures concern earth-fill and rock-fill dams built on pervious alluvium. The number of failures in concrete dam foundations is relatively small. A few are listed as follows.

(*a*) The designers of the Bouzey dam (France) ignored uplift forces and the dam collapsed on 27 April 1895. A committee, headed by Maurice Lévy, was set up in the same year to investigate the disaster.

(*b*) The Gleno buttress dam failed in the same way on 1 December 1923. Poor design (excessive shear stresses) and bad construction were blamed but Mary (1968), attributed the failure to uplift forces under the foundations.

(*c*) On 13 August 1935, one of the two dams at the Molare development in the Italian Appenines was destroyed by floods. The maximum flood expected on the small catchment area (141 km^2) was about 850 m^3/s. But during an unexpectedly heavy storm the floods rose to 2500 m^3/s, over-topping the dam and deeply eroding the dam toe, which collapsed.

(*d*) In April 1959, the catchment area of the Rio Negro (Uruguay) severely flooded for a period of four weeks. The inflow into the Rincon del Bonete reservoir reached a peak of 605 000 cusecs. Of this, a total of 340 000 cusecs were discharged over and around the dam which flooded the power-house to generator top level, but did not destroy the river barrage.

(*e*) On 13 March 1928, the St Francis dam belonging to the Los Angeles Water Board failed through defective foundations. The rupture of the 16-m-high Moyie thin-arch dam (Idaho, USA) and the 19-m Lanier thin-arch dam (North Carolina) were also caused by flooding and rupture of the dam lateral rock abutment.

An unsuccessful attempt was made during the Spanish Civil War by the troops of General Franco to blow up the Ordunte Dam near Bilbao. In

September 1941 the retreating Soviet Army destroyed the Dnjeprogues dam. On the night of 16 March 1943 the Möhne dam and the Eder dam in the Ruhr (Germany) were successfully bombed by the Royal Air Force.

Three more recent cases of dam accidents or dam disasters will now be analysed in detail. All three were caused by rock weakness.

13.1   The Idbar experimental thin-arch dam (Yugoslavia)

The Idbar dam, on the river of the same name in Yugoslavia was built to accumulate alluvium from the catchment area of the large storage reservoir of Jablanica (fig. 13.1). As the dam is not a water store it can be used for *in situ* tests on the behaviour of very thin dam shells.

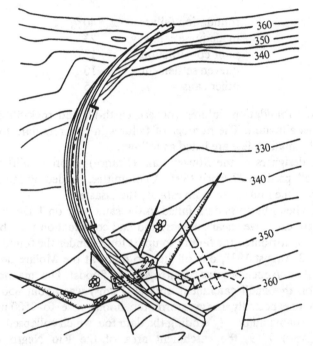

Fig. 13.1   The Idbar dam in Yugoslavia (after Milovarovic, 1958).

The height of the dam is 39 m; the thickness at crest level is only 1·11 m and at the base 2·65 m, where it widens to 3·48 or 4·20 m at foundation level. The arch dam itself required only 4100 m³ of concrete and the right bank concrete abutment 2900 m³. Two 700-mm conduits on the right bank can discharge the whole storage within two days into the large Jablanica reservoir if required.

The main characteristic of this thin structure is the lack of homogeneity of the foundation rock. On the left bank, triassic limestones, with an *E* modulus of more than 200 000 kg/cm², are very hard. Similar rocks on the river bed show a modulus of 100 000 to 150 000 kg/cm², but at the bottom of the right side of the gorge the E modulus is only 1000 to 5000 kg/cm². Higher up the slope (widely fractured triassic limestone) rocks again harden to 40 000 to 60 000 kg/cm² in the direction of the arch thrust and 30 000 kg/cm² in a direction normal to the thrust measured.

Leakages occurred during the first filling, when the water reached the level of the contact zone between the schists and the fractured limestones on the right bank. The small reservoir was emptied and an impervious blanket built on the upstream side of the right bank. A sudden flood (December 1959) filled the reservoir to top level and water spilled over the dam crest before the blanket was finished. It leaked through the rock fractures, washing away filling material (clay) and causing several large limestone rocks to fall into the gorge. The right concrete abutment, left without proper support, was displaced by 10 cm in the vertical direction. A fissure opened in the concrete shell of the dam at 45° from the river bottom to half the height of the dam. But the dam, still overtopped by flood water, withstood the strain perfectly.

When the water level in the reservoir again dropped to three-quarters of the height a 2-m-diameter cylindrical hole (6·6 m³ of volume) was blasted through the concrete shell at about one-quarter height. It took a shaking from the explosion, but again withstood, which demonstrates the extreme toughness of thin dam-shells. After subsiding 10 cm in the right abutment the Idbar dam could adapt to its new foundation. It is of great importance to keep these facts in mind when discussing the case of the Malpasset dam.

13.2 The Frayle arch dam (Peru)

The Frayle dam is a thin double-curvature arch dam, built in 1958 at 4010 m above sea level, upstream of Arequipa in the south of Peru. It was designed to store 200 million m³ of irrigation water for the Arequipa area.

The dam is 70 m high with crest at level 4012 m. The thickness at the crest is 1·58 m and 6·20 m at the base. The bold double curvature closes a narrow gorge, which widens for the upper 10 m of the dam. This meant that substantial concrete abutments had to be built, especially on the right bank of the gorge where a curved 'wing wall' protects the abutment and forms a second smaller arch dam (fig. 13.2). On the left side a scour gallery (diameter 2·90 m) crosses the rock under the abutment at level 3964 m. On the right bank at level 3942 m a small conduit 0·75 m in diameter, could be used to completely empty the reservoir.

The whole area of crystalline rocks and sediments is volcanic and still active. The gorge is cut deeply in hard andesite, with crushing strength varying between 500 kg/cm² and 1800 kg/cm². Above level 4005 the andesite is covered with lavas and slightly cemented ashes of lower strength. The concrete abutments rest directly on the solid andesite.

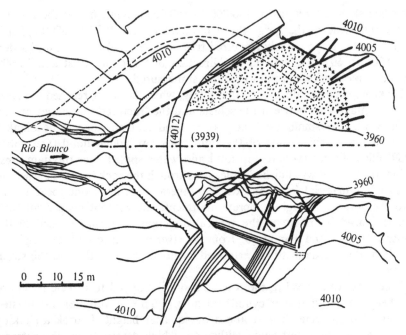

Fig. 13.2   The Frayle dam: rock joints. Stippling shows area of rock slide (after Mary, 1968).

The left dam abutment is strongly anchored in jointed andesite. The nearly vertical joints were more or less parallel to the abutment, one of them being very near to the concrete abutment. A secondary joint system, perpendicular to the first is less important. A large grouting screen penetrates to reasonable depth in the rock mass.

An active volcano, the Misti, is located near Arequipa at about 40 km from the dam. Earthquakes of intensity 4 (Mercali scale) shook the area in 1959 and 1960, causing fissures in buildings near the dam, but none was detected in the dam itself or in the rock masses near it. However, it is possible that the impervious grouting screen mass may have been damaged by the tremors.

The reservoir was slow to fill. On 12 April 1961 the water level reached 4000·70 for the first time, which corresponds to a storage of 10^8 m^3. At 2 a.m. the next day a very large rock slide occurred on the left side of the gorge very close to the dam abutment, on a length of 40 m. Over 15 000 m^3 of rock slid. The main scour gallery was severely damaged. Turbulent water, about 60 m^3/s, escaped through the mass of rocks and accumulated at the bottom of the gorge. The dam, with its weakened abutment, was left in a most dangerous position. Nothing could be done to accelerate the emptying of the reservoir and the inhabitants of Arequipa, downstream from the dam, lived in fear for about a month.

Inspection of the scour gallery showed shear fracture cutting through it, and some damage to the steel and concrete linings. A longitudinal displacement of 30 cm and a lateral displacement of 80 cm of the lower part of the gallery were measured. Several of the grout holes starting from the gallery were found to have lost all the cement filling. One of them was creating a free passage to a main rock joint. The dam itself had not been damaged and the left abutment was only very slightly displaced. Wetness on the face of the rock masses had been observed and in March 1961 water seepage had turned into real jets. It was established that the water was coming from the lake through the joints in the rock mass with only small loss of head (about 10 to 15 m) but the dam foundations still remained watertight (fig. 10.12).

The cause of the rock slide can be clearly established. Water seeping from the reservoir filled the joints in the rock and put them under pressure which was not relieved by any natural drainage. On 3 April 1961, within a few minutes, slides of rock had fallen one after the other into the gorge, pushed by the hydrostatic water thrust. The slide stopped very near to the dam abutment. A large spring opened about 10 m downstream of the dam, indicating that the slide had made some joints open to relieve the interstitial water pressure. This might have saved the abutment from further sliding. It is likely that the grout curtain was either damaged or bypassed, as high water pressure in the joints downstream of the abutment has been observed. An empty borehole, where the cement grout had been washed away, might have linked the scour gallery to one of the main joints.

Mary (1968), whose publication on 'Arch Dams' is the only available source of information about Frayle, says that proper rock drainage is an absolute necessity. He concludes that a grout curtain without drainage may not be safe.

Reinforcement work. A large wall was built to support the steep rock face on the left bank. The width of the wall is such that the thrust from the arch abutment can spread out at an angle of at least 20° (figs. 13.3, 4). The wall is attached to the rock by steel anchors. After making a new grouting curtain, upstream from the old one, the wall and the rock have been drained by drilling deep holes in the rock. The reinforcement wall itself has been consolidated with three heavy buttresses.

A new grout curtain and drainage system equally reinforce the right-hand side of the gorge. A preliminary design suggested to buttress the left bank direct to the right was abandoned because it was not possible to estimate stress and strain conditions should an earthquake occur.

13.3   The Malpasset dam disaster

The rupture of the Malpasset dam, near Fréjus, France occurred at 9.10 p.m. on 2 December 1959. Well over 400 people lost their lives, and part of the town of Fréjus was destroyed. It was a national disaster for France.

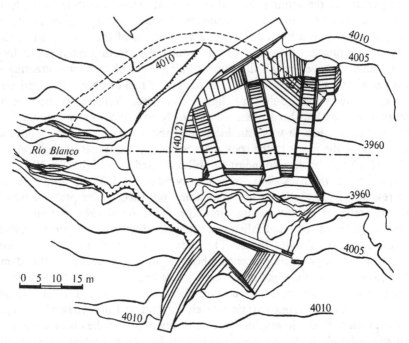

Fig. 13.3   The Frayle dam: reinforcement of the left abutment (after Mary, 1968).

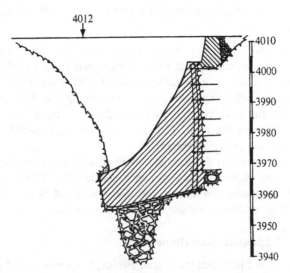

Fig. 13.4   The Frayle dam: cross-section through the buttress (after Mary, 1968).

Technical information as to the cause of the disaster was slow to come. An official report was published (Jaeger, 1963c) which did not clearly explain the cause of the rupture and dam collapse. For several years it was impossible to understand why this particular thin-arch dam, among so many others, had collapsed. As long as the Malpasset failure remained a mystery then there would be doubts about the safety of other structures. Several experts volunteered explanations which can be compared with those suggested in the official report. None was convincing as all of them could be used to prove that other dams, still standing with no sign of weakness, were equally unsafe.

In the meantime the research laboratory of the École Polytechnique (Paris), was trying to determine why fissured gneiss had proved to be a dangerous unreliable dam foundation. Some previously suspected facts became clear when Mary (1968) published his own version of the disaster. With these in hand, it is now possible to understand the cause of failure.

We begin with an analysis of the French experts' opinions. Then we will look at some explanations put forward by other experts which are most interesting as they reveal many shortcomings. Finally, we will study the most recent research work of French engineers.

13.3.1 Official report

An enquiry commission was set up on 5 December by the French Ministry of Agriculture who owned the dam, and a provisional report was submitted by the Commission a few months after the disaster. The final Official Report is dated August 1960 and became available to experts from other countries in the summer of 1962 (see Jaeger, 1963c).

The report itself is only fifty-five pages long, but it is substantiated by forty annexes bound in three thick volumes, and includes two geological reports.

(1) *Main data* (figs. 13.5 and 13.6).   The main data concerning the Malpasset dam on the Reyran river are:

top of the dam	102·55 m above sea level
bottom of river bed about	42 m above sea level
highest water level	100·40 m above sea level
dam radius (upstream crest)	105 m
total subtended angle	121°
dam thickness at the top	1·50 m
dam thickness at the base	6·76 m

Contractors for the dam were Entreprises Leon Ballot and Gianotti, who began work on site on 1 April 1952. The water levels obtained were:

May 1954	level 60·60 m
September 1955	level 79·75 m
July 1959	level 94·10 m

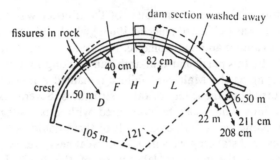

Fig. 13.5   Plan of Malpasset dam showing displacements.

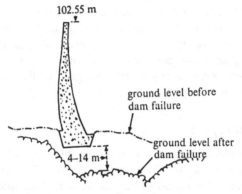

Fig. 13.6   Typical cross-section of dam near left-hand abutment after failure.

The dam collapsed when the water level had reached 100·12 m. A few hours before rupture the dam had been inspected as usual by the warden.

(2) *Analysis of possible causes of rupture; discussion of facts and reports by the Commission.* The Commission examined the following as possible causes of the rupture:

Quality of concrete. The Report concludes: 'The construction work was very good, particularly as regards the quality of the concrete and the bondage of the concrete with the foundation rock.'

The crushing strength of the concrete was between 333 and 533 kg/cm² (4720 and 7850 lb/in²). The modulus of elasticity of the concrete was measured between 218 000 and 300 000 kg/cm² (3 100 000 and 4 260 000 lb/in²) a figure to be remembered for future reference.

Grouting. Contact grouting was carried out as usual, at 2·50-m centres on the upstream side and at 5-m spacing on the downstream side. Injections for a proper grouting curtain were started in the summer of 1953 but they were unsuccessful. It was concluded that the tight rock did not absorb large quantities of grout, and so the process was stopped. The Report agrees with this decision: 'To sum up, the detailed examination of the conditions prevailing during the construction do not reveal any element which could explain the catastrophe.'

Check on the dam analysis and further trial-load analysis. The dam designers (A. Coyne and J. Bellier) adjusted the crown displacements in a radial direction. The calculations were carried out for the following cases (E_c = modulus of concrete; E_r = modulus of rock):

$$E_c/E_r = 0, 1, 5 \text{ and } 10.$$

These calculations were checked and confirmed by Electricité de France (EDF). Calculated stresses are all within acceptable limits, equal to or lower than 61 kg/cm² (compression), showing the safety factor for concrete to be 5·3 (compression). Some tensile stresses (10 kg/cm²) were detected in the crown cantilever. Temperature stresses and concrete shrinkage were neglected in these calculations. (They were estimated at less than 15% of the total stresses.) EDF also checked the dam assuming the arches to be independent.

General trial-load adjustment. This was carried out for a ratio $E_c/E_r = 10$, for five arches and eleven cantilevers. A value $E_r = 10\ 000$ kg/cm² was assumed by the Commission. (Displacements of the dam foot measured in 1955 and 1959 for water levels 79·75 m and 96·10 m showed a probable ratio E_c/E_r to be at least 6·5, possibly 15 to 25.) This additional analysis does not reveal any unusual stress concentration. On the whole the conditions are shown to be rather more satisfactory than with the simplified approach.

Check on the left-hand concrete abutment. The dimensions of this abutment in a horizontal section were 20 × 6 m on the first design, but ultimately they were increased to 22 × 6·50 m. It is protected on the upstream side by a vertical 'wing wall', the space between wall and abutment being correctly drained in order to avoid hydrostatic pressures and uplift pressures on the concrete abutment.

The abutment was calculated for a total arch thrust on the 16-m-high concrete block (as for an isolated elastic arch, not adjusted to cantilevers) of 13 100 tonnes. The maximum compression stress, σ, on the rock foundation was calculated at 6 kg/cm², the shear stress, τ, was 9 kg/cm², with $\tan \phi = 1·45$ which is rather high. But the shear force was in fact absorbed by the pure shear strength of the bondage concrete to rock (cohesion strength), which EDF estimated to be 36 kg/cm². (The formula used, $\tau = 28 + 1·3\sigma$, is based on tests made by EDF at the Grandval dam.)

These values, said the Report, are all right if there is a substantial compression component (larger than 5 kg/cm²). Owing to the way the foundation was designed by almost horizontal steps, this may not be the case. Additional calculations showed, nevertheless, that the concrete abutment was safely anchored in the rock.

Enough concrete samples with rock attached were found after the disaster to confirm the fact that the rupture did not occur along the boundary between concrete and rock, but inside the rock itself (fig. 13.5).

The rock. Preparing laboratory samples for crushing tests proved to be extremely difficult because the densely fissured rock fractured along many planes. Only one cube of 5-cm side length was obtained, and this was used in favour of a 7-cm diameter cylinder. Average crushing strength reached 324 kg/cm² for the blocks and 425 kg/cm² for cubes. These figures were confirmed in similar tests carried out by the consultants. Permeability tests were also done.

Rock tests on site were carried out by Electricité de France in two galleries which were supplemented by seismic tests. The commission's view is that the dam shell and abutment form a whole, but nevertheless it decided to discuss their failures separately.

The geological reports.   Of a first series of geological reports by Professor Corroy, University of Marseilles, the more important ones were dated 15 November 1946 and 11 May 1949. In his 1946 report he describes the geology of the whole area in detail; mainly carboniferous (Houiller du Reyran) with gneiss underlying and then merging in the area where the dam was built (Gneiss du Tanneron). The report predicted accurately that the reservoir would be watertight because of the nature and shape of the under-lying geological formation. It also estimated correctly the thickness of the alluvium on the future site.

In his various reports Professor Corroy mentions the following points: the presence of pegmatite intrusions in the gneiss (which proved less dangerous than his forecast); the downstream plunging strata on the dam site about which he was worried; some fine fissuration of the gneiss in the foundation area.

He stressed the necessity for extensive grouting of the foundation area. The consultants thought that such extensive grouting was not required – an opinion to which the Commission agreed after having considered all the facts. Professor Corroy did not hesitate to recommend construction of the Malpasset dam on the site and for the load finally chosen by the consultants.

Report by Jean Goguel.   The thirty-page-long report dated 17 April 1960, goes into great detail about the geological approach of such a major project. He writes that the first investigations by Corroy were for a site for a gravity dam. It is not certain whether this geologist was ever told that the design had been changed to that of an arch dam. Corroy was concentrating on problems like thickness of alluvium (important for a gravity dam) and paying little attention to rock quality on the top of the banks where an arch dam requires stiff rock resistance. Goguel's report deals extensively with the following: (*a*) the gneiss and the pegmatite (which proved to be less dangerous than expected); (*b*) the petrography of the gneiss, its macroscopic description and microscopic analysis and crystallography, and the history of gneiss formation; (*c*) the presence of sericite in the gneiss on the left bank which is reason to expect the rock to be lower in mechanical strength – a point which he stresses several times; (*d*) action of pore water in the gneiss and descriptions of results obtained from samples put under water in the laboratory; (*e*) schistosity, diaclases and rock fissures and fractures are analysed.

He insists that rupture of the left rock abutment occurred by sliding along well-defined lines of fissuration. It happened at a shallow depth under the concrete and the deep erosion which can now be seen is partly due to water sucking and entraining huge slabs of rock. Nowhere does Goguel say that these fissures could have been detected before the dam's construction. Nor does he state that detailed microscopic analysis of the rock or inspection of the site would have given warning of what was to happen. In fact it gives rather the opposite impression.

Neither the report by Corroy nor that signed by Goguel after the dam's collapse gives any real clue to its cause.

(3) *Description of the dam failure by the French Commission.* The Commission expressed the opinion that the dam design was correct and the construction showed no trace of fault. So the cause had to be in the rock.

A careful survey of the concrete masses still standing on the right river bank and in the river bed, of the rock masses on both river banks and of the many concrete blocks entrained by the water waves and deposited along the river bed downstream of the dam site provided many clues.

The French experts have published a plan view of the dam as it stands now (fig. 13.5) which shows the displacements of the concrete foundations. It can be clearly seen that the left abutment consisted of a V-shaped buttress, with a 'wing wall' which protected the main concrete abutment from direct water pressure. The dam shell was normally pushing against this powerful structure.

The report describes the failure of the buttress and of the dam shell separately as follows.

Failure of the concrete buttress on the left river bank. Figure 13.5 shows that there was a tangential displacement of the buttress of about 2·10 m. The present position of the abutment and the distortion of one vertical reinforcement bar still anchored vertically in the rock and concrete show that this total displacement was a combination of: shearing inside the rock and displacement of about 1·30 m, which occurred first, and a fracture and failure between rock and concrete with further displacement of 0·80 m, which happened after. The thinner 'wing wall' on the upstream side of the abutment and at an angle to it, was partly caught up by the abutment until it was sheared. It came to a standstill, but the main abutment continued to slide.

The force which caused the abutment and some of the underlying rock to move was extremely high. The movement occurred in a direction tangential to the curved dam and parallel to a rock fault not detected by the geologists but now clearly visible.

The Commission estimates that the force which ruptured the root of the lateral 'wing wall' was at least 18 000 tonnes. To this force was added the resistance of the abutment itself and of the entrained rock. 18 000 tonnes is far greater than the total arch thrust between levels 86 m and 100 m above sea level, as calculated by the trial-load method. It may be that the concrete block started to move relative to the rock when the lateral wall was ruptured.

Rupture of the arch.   Figure 13.5 shows that the right half of the arch has rotated round the right abutment. The displacement at the base of the central cantilever is about 0·80 m. (The left half of the arch and the under-lying rock have been completely washed away and only the abutment remains.)

If the dam shell is assumed to be a rigid body the displacement of the left-hand abutment would have been 1·10 m. It was actually over 2 m, which may be explained by the overloading of the upper arches and the deformability of the whole shell. The opinion of the Commission was that near the higher upstream rock fissure the foundation appears to have been more deformable than elsewhere. Because of its rigidity, the arch could not be deformed as much as was required to transmit the calculated thrust to the rock in this area. The part of the thrust not taken by the foundations was transmitted by the arch to the two ends of the zone, which substantially increased the stresses on the points, extending the zone towards the top of the abutment, which became progressively overloaded, and towards the bottom. The Commission estimated that these redistributions of stresses in the dam shell (first phase of the rupture) were slow and they might have lasted days or even weeks. The second phase was a nearly instantaneous rupture of the whole dam.

The Commission analysed the traces of overstraining left on some concrete blocks and their displacements as far as they could from the available information. On the whole the Commission thought that just before the final collapse the arches worked as 'falling arches' (similar to those of the well-known Roselend dam). An 'active arch' was formed inside the shell which transmitted forces from E (where concrete was crushed) to E^* where the concrete abutment slid (fig. 13.7). The 'cantilevers' were unable to contribute to the stability of the overstressed shell, and when the arches finally gave way they collapsed too.

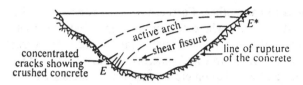

Fig. 13.7 Schematic representation of an active arch inside the dam shell.

According to the Commission, pressures under the dam caused by water seepage did initiate the first phase. The warden who had inspected the foot of the dam shortly before the rupture saw no leakage.

The absence of a proper grouting screen cannot be incriminated. Such a screen would not have stopped the dam failure.

13.3.2 Comments and discussions by other experts

A close analysis of the official French report shows that the experts emphasize two phases in the collapse: one of slow progressive deformations of the rock foundations and straining of the dam shell, the second of spectacular, nearly instantaneous failure of dam and foundations. The report does not say why the disaster occurred and does not point to any cause which makes the site and rock of Malpasset different from that of any other.

Shortly after the disaster, several well-known experts delved into the problem and they came up with several physical explanations. All these tentative explanations have to be examined carefully from two angles: do they explain the failure of the rock and of the dam? do they explain why no failures similar to Malpasset have occurred anywhere else?

(1) *K. Terzaghi* (1962*a*) commented as follows:

The left abutment of this dam appears to have failed by sliding along a continuous seam of weak material covering a large area. A conventional site exploration, including careful examination of the rock outcrops and the recovery of cores from 2 in boreholes by a competent driller, would show – and very likely has shown – that the rock contained numerous joints, some of which are open or filled with clay.

From this data an experienced geologist could have drawn the conclusion that the site is a potentially dangerous one, but he could not have made any positive statement concerning the location of the surface of least resistance against sliding along such a surface. . . . All foundation failures that have occurred, in spite of competent subsurface exploration and strict adherence to the specifications during construction, have one feature in common. The seat of the failure was located in thin weak layers, or in 'weak spots' with very limited dimensions.

None of the methods of exploration, including those used by mining and petroleum engineers, provides adequate information concerning such minor geological details.

Figure 13.6 clearly shows how the rock foundation was ruptured along two inclined faults, one upstream the other downstream of the dam, nearly at right angles to each other. Furthermore, the rock on the left-hand bank of the river at the level where the rupture occurred is densely fissured. A detailed study of the rock made after the failure indicated a treble system of microfissures, microfractures and macrofractures. This confirms Terzaghi's opinion to some extent. But on closer analysis the facts are not in agreement.

The left-hand, V-shaped concrete abutment of the dam was extremely strong. It failed at the very last moment when the dam shell had already suffered severe strains, and when the thrust on it had already altered its direction and increased far more than normally expected.

The two upstream and downstream faults and the clayish material filling them has been carefully analysed in the Research Laboratory of the École Polytechnique in Paris. The friction factor of this material was found to be not larger than $\tan^{-1} \phi_{\max} = 60°$ and not lower than $\tan^{-1} \phi_{\text{ult}} = 45°$ which is too high to justify Terzaghi's hypothesis of rock sliding along a weak fault. Even when this material is in a wet state, its friction factor remains too high for possible rupture by shearing along the faults. It is likely that cohesion of the material was very low; the rupture of the dam was not caused by this particular type of rock weakness. The left bank finally ruptured, as described, after something else – still to be discovered – had occurred.

(2) *Laginha Serafim* at the Thirteenth International Meeting of the Austrian Society of Rock Mechanics in Salzburg (5 October 1962) gave the results of model tests he had made to find local rock weakness. In fig. 13.8 a local rock

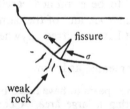

Fig. 13.8   Assumed dam failure by weak rock and tensile fissure of the shell (Serafim).

mass with low modulus of elasticity is shown along the dam foundation. According to Serafim, under such circumstances high tensile stresses may develop in a direction parallel to the foundation, near the area of rock weakness.

The French examined the problem of rock elasticity and rock plasticity with great care. Electricité de France carried out tests inside two galleries near the dam site on the left abutment. The most characteristic data are reproduced in table 13.1.

These figures are exceedingly low for gneiss and the ratios of highest to lowest values about 3 to 1 or 2·5 to 1. The French concluded that the ratio

Table 13.1. *Malpasset gneiss, left bank, Modulus of elasticity E_r and modulus of total deformation E_t in kg/cm².*

	E_r		E_t total
	1st cycle	3rd cycle	3rd cycle
1st gallery	3200 to 17 000	4000 to 18 000	2300 to 7300
2nd gallery	5000 to 18 000	6000 to 25 000	1200 to 8000

$n = E_c/E_r$ was larger than 6·5 and probably less than 25. The Commission accepted a value $E_r = 10\,000$ kg/cm² as probable but did not exclude a value $E_r = 5000$ kg/cm². A trial-load calculation has shown the dam to be safe even for this low value $E_r = 10\,000$ kg/cm².

The conclusions of the Commission are indirectly confirmed by many model tests and other mathematical dam analysis which have proved that thin and very thin arch dams, designed to rule, built in sound concrete and anchored in homogeneous rock (even rock with a low modulus of elasticity), have a great reserve of strength and a high safety factor for normal hydrostatic loads. Rupture of such dam models occurred only when they were heavily overloaded (Semenza, Oberti, Rocha, Serafim *et al.*). The most striking example of how tough these dam shells are is the disaster of the Vajont Gorge rock slide. This very thin arch dam, designed by Semenza, withstood a tremendous overload; it suffered only very minor damage even when overtopped by a deep wave.

The tests for the Tang-e-Soleyman dam in Iraq were made in Portugal (Lisbon). A weak layer of rock E on the left abutment was represented on the model. When overloading the model, failure occurred along the left abutment. The lines of failures were as shown in fig. 13.9. The line of rupture

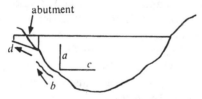

Fig. 13.9 Assumed dam failure by weak rock and shear fracture of the rock. (*a*) Fissure caused by bending; (*b*) shear fracture in weak rock; (*c*) shear fracture; (*d*) gliding of the abutment.

'*a*' by compression or bending, which in the case of homogeneous rock occurs vertically along the crown of the dam, is now displaced towards the left abutment. There is a clear inclined shear fracture *b* in the rock, parallel to the inclined left-hand abutment and another shear fracture *c* to the right, in a nearly horizontal direction (in one of the tests this shear fracture was inclined upwards to the right). There is a marked similarity between the rupture of the Tang-e-Soleyman dam model and the Malpasset failure which definitely confirms the theory that final failure at Malpasset occurred on the left abutment, the inclined shear fracture going not through the concrete but through the rock as the weakest element of the combined system.

But there are still important unexplained points. If the Malpasset design had been examined by experienced dam designers who had had an idea that there was a local weakness in the rock on the left abutment, their reaction would have been to point out the unusual strength of the left-hand V-shaped concrete abutment (which was in fact partly sheared through before the rock finally gave way). A local weakness of the rock, like a low modulus of elasticity suggested by Serafim or a low shear strength of a thin fault as suspected by Terzaghi, would be compensated by the flexibility of the thin shell-shaped dam. Over 500 arch dams and thin-arch dams still stand undamaged, proving the basic soundness of similar designs, even when built on indifferent rock. (Two Italian thin-arch dams withstood earthquakes without damage (Glover, 1957).)

The lines of rupture observed at Malpasset compared to those on the model of the Tang-e-Soleyman dam again confirm the assumption that failure occurred on the left side but do not explain why it happened. As time went on, more experts began to suspect a possible detrimental action of water percolating through the macrofractures of the rock and tried to estimate the danger of uplift forces.

(3) *Serafim, Jaeger, Pacher and Londe.* Pacher (1963) and his Japanese colleagues mathematically analysed water percolation round Kurobe IV dam and checked it on models. Jaeger (1961*b*, 1964*a*, *b*) in several papers, mentioned the problems arising from differences in rock permeability due to rock grouting under dams. Londe (1965) developed a new analytical method for calculating the resulting uplift under a dam (first submitted to the Eighth Congress on Large Dams, Edinburgh, 1964, then to the first Congress on Rock Mechanics, Lisbon, 1966).

In a paper published shortly before his death Terzaghi (1962*b*) made a second suggestion regarding a possible cause of failure of the Malpasset dam. In an introductory remark he writes:

If water leaks out of a reservoir formed by a concrete dam, the greatest cleft water pressure develops in the joints of the rock at the foot of the slope downstream from the toe of the dam. If a slope failure should occur as a result of cleft water pressures it would start at the foot of the slope.

Terzaghi's point could be made clearer if instead of cleft water pressure, he had introduced the idea of a pressure gradient inside the rock foundation (fig. 13.10).

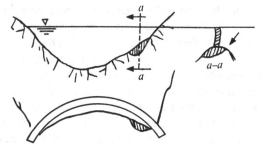

Fig. 13.10   Blow-out of rock under dam foundation as suggested by Jaeger, 1961*b*, 1964*a*; Terzaghi, 1962*b*; Pacher, 1963; and Londe.

After discussing conditions at the toe of a gravity dam, Terzaghi continues:

If the dam is a thin-arch dam, the cleft-water pressures are also greatest in the proximity of the toe of the dam at the foot of the slope. However, they are very much greater than the corresponding pressures near the toe of a concrete gravity dam of equal height, because the base of an arch dam is much narrower. Furthermore the downstream toe of an arch dam rises along the slopes of the valley in a downstream direction and enters the area occupied by a potential slide scar. Finally, also in the proximity of the foot of the slope the effects of the cleft-water pressures combine with those produced by the thrust of the arch which has a component in the downstream direction and thus tends to push the rock out of the slope.

Terzaghi refers then to the Malpasset failure. He comments on the accompanying sketch (fig. 13.10) as follows:

It can be seen that a slide initiated by a blow-out at the foot of the slope would deprive the upper portion of the base of the dam of its support. Hence if a blow-out occurs in the rock supporting an arch dam the consequences are likely to be catastrophic. The failure of Malpasset Dam was probably started by such a blow-out. Fortunately the development of cleft water pressures within the rock downstream from arch dams can be avoided by adequate drainage.

Jaeger (1961*b*, 1966*b*), analysing similar conditions, warned about the danger of tensile stresses which forcibly occur in the rock at shallow depth under the upstream dam heel. These are too often neglected; only compression and shear stresses along the dam base are calculated by most experts. He mentions that tensile strains have a tendency to spread beyond the area obtained by theoretical analysis of a continuous foundation, making conditions worse than foreseen (1966).

13.3.3   Fundamental research on rock permeability and instability

(1) *Stability of the foundation rock masses.*   The most decisive attack on this vital problem of the Malpasset failure was made by a team of French engineers who worked consistently over many years to discover what detrimental effects water percolation has on dam foundations.

Habib (1968) and Bernaix (1966) have systematically investigated the correlations between the Darcy coefficient of permeability k, the rock stresses and the behaviour of different types of rocks. They examined different limestones, sandstones, marls and gneisses and showed that the coefficient k for most sandstones and similar rocks does not vary with the state of stress, k being about the same for compressed rock and rock under tensile stresses. The same will occur with most rocks where the voids are spherical. On the contrary, in microfissured and specially in microfractured rocks, k values depend on the state of stress. When such rock samples are put under tensile stress, they are far more pervious than when the same sample is compressed. The k value can vary from 1000 to 1 depending on the stresses.

This is particularly relevant for schists and gneiss. The Malpasset gneiss proved to be the worst of all rocks tested by Bernaix in Paris. A new percolation test, introduced by him, is used to classify rocks which would be dangerous types for use as dam foundations (see section 4.12).

Krsmanović and others have studied the stress distribution in models of rock masses built up by parallelepipedal blocks of rock, which cannot transmit tensile stresses in a direction perpendicular to the block faces. These have shown that the distribution of compression stresses in such a 'fissured half-space' is very different from that calculated in a continuous half-space (Boussinesq, 1885). There is a high concentration of compression stresses under the thrust load acting on the surface of the half-space formed by parallelepipedal blocks, whereas the classical theories assume a lateral spreading of the compression stresses.

Let us now assume that the foundations of a thin-arch dam are anchored on a highly fissured and fractured gneiss. The gneiss samples may have a relatively high crushing strength. Because of the fissuration of the rock masses, and depending on the direction of the fissures, there will be a high stress concentration under the dam in the direction of the main thrust. If the rock mass is not already fissured, tensile stresses may develop near the dam heel and progressively penetrate deeper and deeper into the rock. This is what must have happened at Malpasset, remembering how the rock on the left bank, where the foundation finally gave way, was markedly worse than the rock on the right bank where a deep fissure, running along the heel of the dam, can still be seen as proof of what happened all along the dam. The concentration of compression stresses compacted the macrofissured gneiss under the dam, which became impervious. Rock under the dam heel and upstream of it, being under tensile stresses or already fissured, became very pervious. Dam designers very often fail to investigate conditions deep within the rock mass. But a new concept is now being developed (Pacher, Jaeger, Londe *et al.*) whereby the true forces acting on the rock foundations due to water seepage in the rock faults are analysed. When calculating these forces, the real values of the coefficient of water seepage k in the different parts of the rock masses must be introduced.

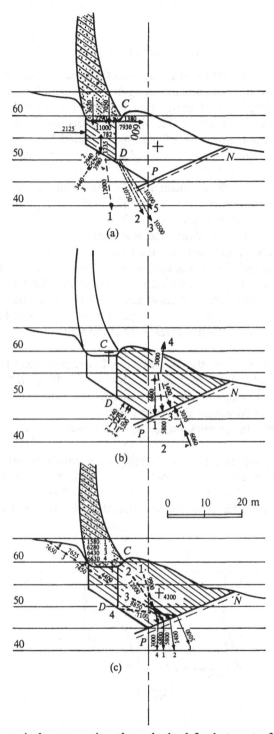

Fig. 13.11 A typical cross-section through the left abutment of the Malpasset dam: (a) upstream block of rock alone; (b) downstream block of rock alone; (c) the two blocks combined. (1) Resultant thrust, uplift neglected; (2) triangular uplift on foundations down to point P where pressure is nil; (3) triangular pressure down to P, half pressure at D, nil at N; (4) full hydrostatic pressure down to P, triangular pressure from P to N (after Mary, 1968).

Londe and Bernaix (1966, 1967) applied these basic ideas to the stability analysis of the Malpasset dam, assuming first a uniform k value for the rock masses, secondly a k value depending on the stresses in the rock. To determine these stresses the conventional Boussinseq analysis had been abandoned in favour of a stress distribution of the type obtained for fissured rock. The ideas of Bernaix and Londe are backed by Mary in his authoritative book on arch dam failures (1968).

The basic idea of these calculations is shown in fig. 13.11 where two typical stability conditions for 'no uplift' and 'full uplift' are analysed. Assuming conventional conditions for homogeneous rock, the Malpasset dam would be stable. Applying the new basic concepts to the same dam there is a possibility for the foundation rock under the dam to slide along one of the major rock faults. The rupture of the rock occurred in a way similar to the one described by Terzaghi but for reasons which he did not suspect.

Summarizing many years of effort and research, it can now be said that the geologists who mentioned that the gneiss was microfissured had, without knowing it, pointed to the main weakness of the foundation rock.

(2) *The rock fissure along the dam heel.* A few experts have pointed to the rock fissure which follows the upstream heel of the dam, on the right bank, and wondered why it was not mentioned in the Commission's report. A series of recent publications and reports to Congresses emphasize the danger of tensile stresses in the rock on the upstream side of concrete dams of all types.

Mary (1968) supplied some interesting information about the displacements of the dam foundation, mainly on the right side of the dam centre. The displacement of the base of the dam measured between July 1958 and July 1959 had been far greater than calculated (fig. 13.12). The dam had a

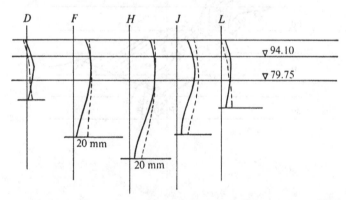

Fig. 13.12 Deformations of the Malpasset dam before rupture. Broken line: measured deformations. Solid line: calculated deformations. See fig. 13.5 for reference points *D–L* (after Mary, 1968).

double movement of rotation; one around its right abutment and another
simultaneously around the dam crest on a line slightly above the crest.
This rotation caused an unforeseen displacement of the foundation and it
could have been noticed as early as July 1959, but it is even more apparent

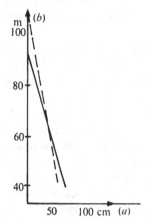

Fig. 13.13   Displacements of fix points. (*a*) Displacements; (*b*) levels above the sea.
Broken line: average displacements; solid line: joint *H* (after Mary, 1968).

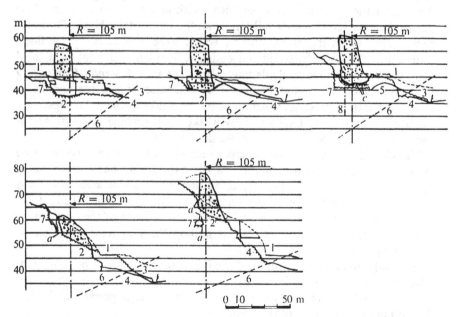

Fig. 13.14   The Malpasset dam: rock fracture along the dam heel, right bank.
(1) Natural rock surface; (2–3) probable depth of excavation; (4) rock surface after
failure; (5) blocks of concrete; (6) rock fracture; (7) trench or gallery; (8) borehole;
(*a*) rock fracture; (*b*) mud; (*c*) fissure (after Mary, 1968).

on the concrete plots measured after the disaster (figs. 13.13, 14). It is obvious that this displacement must have caused, at an early stage and long before rupture, the opening of a fissure 10 mm to 20 mm wide on the upstream side of the dam under varying hydraulic pressure. The water levels were 87·30 m in 1958 and 94·10 m in 1959, the top level of 98·50 m was reached on the day of the disaster when the displacements shown on figs. 13.5 and 13.13 caused failure.

Such conditions were obviously dangerous, and it can be said that by July 1959 the rock foundations had already been weakened all along the periphery of the dam on the right wing and probably on the left wing too. This weakening is confirmed by the fact that, during the rupture, the whole right wing rotated around the right abutment, the crown of the dam, in river axis, showing at its base a final displacement of 820 mm. According to Mary, this caused a widening of the upstream fissure in the rock. The downstream lip of the gap is now 10 cm lower than the upstream lip, because of the movement of the dam shell around its crest. It is vital to stress here that this double rotation was geometrically possible only if the left abutment was still in place. The rupture of the left abutment and the final collapse of the dam occurred after the foundation had been badly damaged along the whole periphery of the shell.

Thin-arch dams have an astonishing reserve of strength against rupture: the case of the Idbar dam has been dealt with in a previous chapter. Two thin-arch dams, the Moyie dam (Idaho) and the Lanier dam (North Carolina) completely lost their left abutments, which were washed away by flood waters, but the dam shells did not collapse. The left abutment of a third dam, the Frayle, was severely damaged but the dam still stands. Comparing this information with what happened at Malpasset it can reasonably be assumed that there must have been some special weakness there. The upstream fracture of the rock had substantially weakened the whole structure and it was probably an important cause of the final rupture.

The Commission of French experts assumed the failure to have occurred in two phases: A slow deformation of the dam during one or two weeks before the final rupture. Then the formation of an active arch within the dam with a concentration of the thrust at its two ends which caused the left V-shaped abutment to slide and the sudden collapse of the dam. Mary largely agrees with this interpretation putting the weight on the Londe–Bernaix hypothesis.

A slightly different description of the dam rupture can now be suggested, which must have occurred in several phases:

(1) A slow build-up of water pressures in the mass of gneiss and fissuration of the rock upstream of the dam.

(2) The slow progressive opening of a 10-mm to 20-mm gap along the dam heel, and from July 1958 to July 1959 a progressive displacement of the dam foot and rotation of the dam shell around its crest.

(3) Dangerous conditions all along the dam foundations; mainly on the left side.

(4) Then in rapid succession: the displacement of the dam foot increased, reaching in plan view 820 mm in river axis (fig. 11.5); at the same time the whole dam shell was rotating round the dam crest and around a point on the extreme right of the shell, this double rotation is proof that at this moment the dam was still a 'shell' and that the left abutment was still there as a fixed point, it seems that only such a double rotation describes the movement of the shell correctly. Then an active arch was formed within the shell. Because the concrete shell was then more or less loose from the solid rock foundation on all its periphery, a tremendous thrust was transferred to the still-standing, left abutment.

(5) A blow-out occurred on the rock mass on the left bank and the left concrete abutment slid causing collapse of the shell. This blow-out occurred in a way similar to the one described by Terzaghi but for reasons and under conditions not suspected by him.

13.4 Final comments on dam foundations and dam failures

The dam incidents and dam failures described in this chapter illustrate the points developed in chapters 11 and 12.

No two cases can be identical. Rock conditions vary from one site to the other. The weak rock investigated by Talobre for the Dez dam in no way compares with the compact foundations of Morrow Point dam. Talobre was concerned with the rock crushing strength, whereas Müller, for Kurobe IV dam, investigated mainly the shear strength along rock faults and fractures. In many cases, the path of the percolating water and the pressure gradient of the interstitial water is of great importance to dam foundation stability. Model tests similar to those initiated by the ISMES research laboratory are a great help to designers.

Chapters 11, 12 and 13 give enough reasons why new rules for the design of dam foundations should be established.

14 The Vajont rock slide

14.1 General information

The Vajont reservoir is located in the lower reaches of the river bearing the same name, not far from its confluence with the River Piave. The Vajont dam, a dome-shaped shell, built by the Società Adriatica de Elettricità (Chief Engineer Carlo Semenza), is the highest arch dam in the world, being 265 m high with a 160 m chord, which spans a very deep and narrow gorge. Several plaster models were built for testing the dam and its deformations under full load. The rock on both sides of the gorge had been skilfully reinforced (sections 6.6.1(3) and 11.4.1(1)).

During the night of 9 October 1963, a mass of rock estimated at 200–300 million m³, broke loose from the sides of Mount Toc and crashed into the reservoir below in which the water level was about 700 m above sea level. (The dam was designed for a water storage level of 722·5 m a.s.l.) Almost 2500 lives were lost when a wave overtopped the dam and flooded the Piave Valley, destroying the small town of Longarone. The Vajont dam was not damaged and it still stands as the highest arch dam in the world, proof of the remarkable work done by the Italian engineers when reinforcing the fractured rock abutments on both sides of the gorge.

The slide, mainly rocky formations which, though strongly fractured, have maintained many of their original morphological features, extended from the close vicinity to the dam to as far as 1800 m upstream. Evidence on the slopes of Mount Toc reached an elevation of 1400 m and extended as far as 1600 m away from the shores of the lake. The rock slide swept 300 m to 400 m across the deep gorge and rose more than 100 m to 150 m on the opposite bank. According to a private report by Electroconsult (Milan, November 1963):

> The collapsed material consists of lower Cretaceous and Malm limestone formations, covered by a thin layer of detritus which slid over other underlying Malm and perhaps Dogger limestone formation, the contact planes between adjacent formations being all more or less pronouncedly inclined toward the reservoir. It has not yet been possible to determine the exact location of the sliding plane.

When the disaster occurred the water line in the reservoir was about 23 m below maximum. The volume of the reservoir above water level was about 55 million m³. It occurred very suddenly, with extreme violence, and the wave which spilled over the dam crest, overtopped it by 250 m at one point on the right bank and 150 m on the left bank. It is estimated that about 30 million m³ of water poured into the narrow gorge which has a depth of over 240 m and a width of only 150 m at the crest and 20 m at the bottom. The water gushed through this narrow canyon at a tremendous speed until it reached the River Piave, where after a right-angle turn it entirely destroyed the small town of Longarone and nearby villages. The collapsed material is now heaped up in what used to be the lake and its total height exceeds the dam crest elevation by some 100 m.

A considerable amount has been published on the Vajont disaster, and most of the more important papers are listed in the reference list.

14.2 The period before 9 October 1963

Most information available on the Vajont rock slide is obtained from two detailed reports published by Müller (1964, 1968). It seems that from the very beginning, the chief engineer Carlo Semenza had some doubts about the stability of the left bank of the gorge. Several geological reports were submitted to him in 1958 and 1959. In 1959 he entrusted Müller with a detailed study of local conditions, which was carried out in collaboration with the geologists Eduardo Semenza and D. F. Giudici. Between February and November 1960, the water level in the reservoir rose from 580 m a.s.l. to 650 m a.s.l. Movements of the rock were detected in September and November and when they reached about 3·5 cm a day it was decided to empty the reservoir down to 600 m a.s.l. Rock displacements rapidly stopped with the falling water level.

A first rock slide occurred on 4 November 1960 when 700 000 m³ of debris slid into the lake within about ten minutes, just upstream of the dam on the left bank. At the same time a long M-shaped fissure was detected high up on the slopes of Mount Toc clearly defining the area which was to slide. Müller visited the area again in November 1960 and then after the major rock slide of October 1963.

According to Müller (1964), geological reports dating from 1958, confirmed by seismic tests carried out a year later, concluded that the area was formed by *in situ* rock showing no sign of earlier movements. Dal Piaz described the rock near the gorge as being consolidated through cementation and he expected only minor superficial slides. In the Spring of 1959 Müller and E. Semenza reached a different conclusion, confirmed by detailed geological surveys. They felt that the rock mass on the left-hand side of the river was part of an old prehistoric slide which had come from Monte Toc. Remains of this mass were found on the right bank of the Vajont, discordantly stratified. The geology of the area is well known. At greater depth a Dogger formation is shaped like a concave saddle, the upper part of which is inclined at about 30° to 40°, the lower part being nearly horizontal. In this saddle, Malm and Cretaceous formations were resting in apparent equilibrium (fig. 14.1).

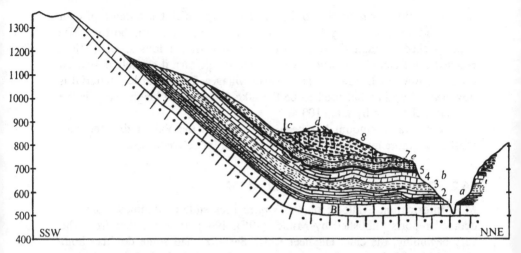

Fig. 14.1   Idealized geological section through the left bank of the Vajont gorge.
(*a*) Vajont gorge; (*b*) north face of Mount Toc; (*c*) Pozza; (*d*) antithetic fractures;
(*e*) fracturing due to external rotation; (*B*) Dogger-malm formation (Oolitic limestone); (1) malm; (2) upper malm; (3–5) Lower Cretaceous; (6–8) Upper Cretaceous
(after Broili, 1967).

Additional studies were carried out after the 1960 slide. Borings and seismic
investigations followed. Some of the experts maintained their favourable
diagnosis of the local rock conditions. Müller and Semenza disagreed with
them.

The rock masses were continuing to slide very slowly. Müller's report
analyses in detail the vertical and the horizontal displacements of the creeping
rock masses and correlates water table levels in the rock (measured in
boreholes) to the varying water levels in the reservoir basin. During 1962
the water in the reservoir rose steadily to level 700. In 1963, it slowly oscillated
between levels 700 and 650 and back to 713, never reaching the dam crest
(at 722·5). In the final days before the catastrophe, the velocity of the slide
reached 20 to 30 cm a day.

Müller, in his report says:

Surprisingly, no appreciable instability was observed throughout the period between
1961 and 1963, although during that time the water level twice reached an elevation
of 100 m above the initial level of 1961 and 50 m above the level at the start of the
creeping in 1960. During this time the effect of the buoyancy must rather have increased and reached its theoretical maximum value (one tonne per cubic metre) since
in the meantime the degree of separation of the joint net and the permeability of the
rock mass had increased, as can be seen from the gradual flattening of the mountain
ground water level.

According to some remarks by Müller, the water levels in the vertical
holes (period from 1962 to 1963) varied with the reservoir levels, but '. . . no
correlation could be established with the precipitation during the final

months'. He thinks that pore pressure, water thrust and weakening of the shear strength in rock joints soaked with water were vital factors in this whole process. The final sentence in Müller's report reads:

Such behaviour [last phase of the slide], connected with properties hardly investigated, and which can in a sense be regarded as thixotrophy, was completely unexpected, and it is the conviction of the author that the sliding could not possibly be foreseen by anybody in the form in which it actually took place and, in fact, nobody had foreseen or predicted it.

The engineers in charge of the Vajont dam had ordered hydraulic tests to be carried out on a very large model of the reservoir basin. They were based on geological findings. It was assumed that a large mass of rock would slide, being progressively accelerated so that the rockslide would occur in a few minutes. Tests carried out on these assumptions showed that there was no danger of the dam crest being overtopped, provided the reservoir level stayed below a determined safety level. When, at the beginning of October 1963, there was danger of an imminent rock slide, the water level in the reservoir was hastily lowered to this safety level.

The prevailing theory on the probable rock slide can be summarized as follows. All the information available to the Italian experts indicate that there is a progressive increase in velocity of the masses which are being accelerated from zero to the final speed, Such an acceleration would normally take several minutes. A time of about ten minutes was acceptable. Furthermore, observation of the creeping movement on the Monte Toc slopes indicated that the pressure of the rock caused rock masses on the edge of the gorge to be overthrown. It was thought that this type of movement would eventually fill the gorge and therefore bring the sliding movement to a stop.

The Italian engineers accepted the fact that the reservoir would be partially filled and maybe cut in two parts. They had already built a bypassing gallery on the right bank of the gorge which should have connected the upper part of the reservoir to the lower part, in case the falling rocks cut the reservoir completely in two. It is clearly established that everyone was fully aware that there was an immediate danger of a major rock slide. The actual rock slide was far larger in volume than expected by most of them, but its volume had been correctly predicted by Müller and Semenza. Furthermore, it occurred with a violence and a suddenness which nobody had expected. Müller writes in his 1964 report:

... Only a spontaneous decrease in the interior resistances to movement would allow one to explain the fact that practically the entire potential energy of the slide mass was transferred without internal absorption into kinetic energy, and that the front of the sliding mass was pushed 400 m and 140 m up on the opposite slope, while moving over an 80 m wide gorge without falling down into it. Such behaviour of the sliding mass was beyond any possible expectation: nobody predicted it and the author believes that such behaviour was in no way predictable.

14.3   Geophysical investigations after the rock slide

Immediately after the catastrophe additional geophysical and technical investigations were started in order to find a scientific explanation. Müller's factual 1964 report caused many reactions. He published a second report in 1968, summarizing and commenting on all the work done by himself and others from 1964 to 1968.

14.3.1   The sliding surface

Müller and E. Semenza, investigating the possibility of a rock slide in the years 1960–3 noted the shape of the geological strata. At that time the contact surface between Dogger and Malm was suspected as a possible slip surface. Detailed investigations by Broili (1967) indicated that the slip surface was located slightly higher, cutting inside the Malm formations and not at the contact with the Dogger.

Figure 14.1 shows the geological section of the area, as established after the slide by Broili (1967). This section differs somewhat from similar documents published previously by other geologists (Giudici & Semenza, 1960; Müller, 1964; Kiersch, 1964). Figures 14.2, 14.3*a, b* reproduce the findings

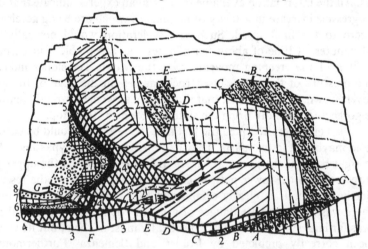

Fig. 14.2   Geological map of the Vajont gorge; *A–A* to *F–F* are cross-sections of the slip mass; (1 to 8) are geological strata with reference to fig. 14.1 (after Broili, 1967).

of Broili concerning the location of the slip surface and the position of the rock masses after the slide. According to him, the sliding surface can be subdivided into two parts. The limit coincides roughly with the old Massalezza river bed or ditch in the central part of the slide. Westwards of this line the slip surface became a very accentuated chairlike shape, determined by the attitude of the stratifications. The 'seat' of this chair has a length of 500 to 700 m, and the 'back' inclined by about 35° to 40° has a length of about

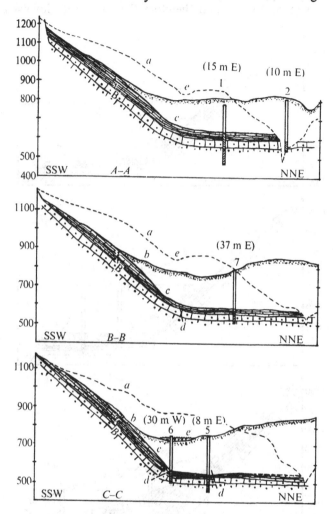

Fig. 14.3*a*   Geological sections *A–A* to *C–C* through the slip surface; reconstruction on the basis of borehole results. (*a*) Topographical surface before the slide; (*b*) topographical surface after the slide; (*c*) reconstructed failure surface and slip surface; (*d*) assumed faults; (*e*) Pozza depression. (1–17) boreholes (after Broili, 1967).

900 m. The bend connecting both is rather accentuated in many parts. The curvature radius was actually between 30 m and 320 m.

East of the 'Massalezza ditch', the chair-like shape is less accentuated, the 'seat' reduces progressively to about 200 or 300 m. The connecting bend is no longer sharp, the curvature widens to radii between 350 m and 1100 m.

In the lower western zones, the slip surface is mainly developed in layers of Malm, mostly upper Malm and lower Cretaceous. The Cretaceous layers are very thin (5 to 15 cm), and therefore the rock mass involved here is

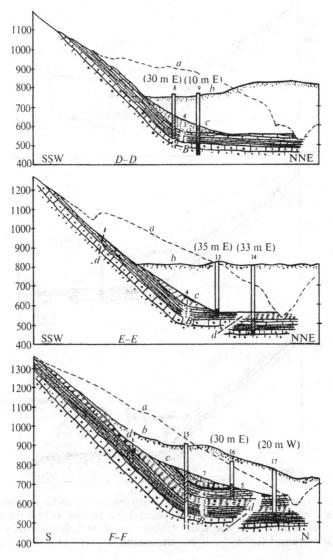

Fig. 14.3*b* Geological sections *D–D* to *F–F* through the slip surface (after Broili, 1967).

rather deformable. The upper Malm and lower Cretaceous layers are thicker, the average thickness ranging from 20 to 100 cm. The rock complex is therefore very rigid. In the middle and lower eastern zones the slip surface involves layers from the lower Cretaceous to the upper Cretaceous. The slip surface must not be regarded as a mere surface but as a zone probably several metres thick. Mylonites of changeable thickness produced by the slide movements have been encountered in the boreholes. According to Broili (1967) and Müller (1968) the mylonites consist mostly of sand, silt and clay materials. Their thickness probably reaches a few metres in some places, and indicates very clearly the dimensions of the slide movement and the conditions under which it developed.

According to Broili's reconstruction, in the central and western zones and in the upper parts of the eastern zone of the slide the slip surface coincides with the bedding planes. Some large and very accentuated 'folding' observed must have affected some zones, resulting in a higher shear resistance. Sometimes the slip surface went through the folds which were then sharply cut. In these zones, one cannot speak of frictional but of true 'shearing' resistances, of far greater magnitude, which had to be overcome in order to accomplish movement.

In the intermediate and lower parts of the eastern zone, owing to the divergencies between stratifications and the slip surface, the slip surface itself could not develop along the planes of the bedding joints but had to cut through the layers shearing completely new joints or forming steps of pre-existing joints and rupture planes.

14.3.2　The creeping movement

The slow 'creeping movement' of the mass of rock has been very systematically observed since 1960 up to October 1963. Figure 14.4 reproduces, after Müller (1968), the main results of these observations.

14.4　Discussion of the observed facts: *a posteriori* calculation

14.4.1　Shear and friction factors

The main result of the geological investigations concerns the shearing along the slide surface. The former assumption of thin layers of clayey material lubricating a polished slip surface is no longer valid. There were no such layers and any clayey seam was found to be highly compact, as expected under such a high overburden. The friction factor of sound rock on rock *in situ* was found to be reasonably high and so was the shear factor of the same rock. No special rock weakness along the slip surface has been detected which can explain the rock failure.

14.4.2　*A posteriori* calculations

Several authors (Mencl, 1966*a, b*; Kenney, 1967; Nonveiller, 1965, 1967*a, b*; Müller, 1968) have published *a posteriori* calculations in order to determine

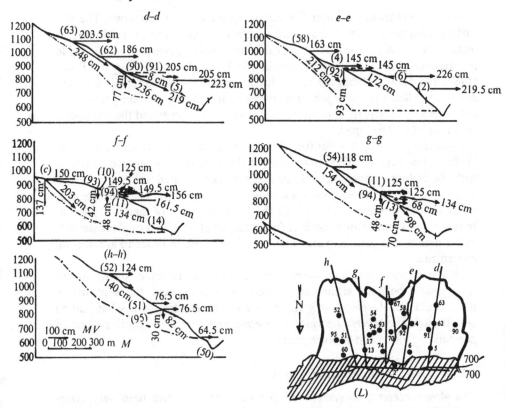

Fig. 14.4 Displacement vectors computed from control point measurements. Solid line, measured; broken line, reconstructed; *d-d* to *h-h* sections; numbers in parentheses, control points; (*M*) scale of the section; (*MV*) scale of the displacements; (*L*) plan.

the required $\tan \phi_{0_1}$ value for which the slip could start moving. These calculations are based on 'static' conditions and it should be stressed that they correspond to a slide velocity $v \equiv 0$. Table 14.1 summarizes the main results.

Table 14.1 ϕ_{0_1} values

Water level above sea level	Lowest values			Highest values		
	ϕ_{0_1}	$\tan \phi_{0_1}$	Author	ϕ_{0_1}	$\tan \phi_{0_1}$	Author
600	18·8°	0·340	Muller	22·1°	0·406	Nonveiller
650	20·1°	0·366	Kenney	22·1°	0·406	Nonveiller
700	17·5°	0·316	Mencl	28·5°	0·542	Nonveiller

These calculations can be carried out with the following formulae (Müller, 1969). They refer to static conditions before the start of the slide:

$$\lambda = \frac{\tan\phi \sum F_i \cos \zeta_i + c \sum l_i}{\sum F_i \sin \zeta} > 1,$$

and for $\lambda = 1$:

$$\phi_{req} = \arc\tan \left[\frac{\sum F_i \sin \zeta_i - c \sum l_i}{\sum F_i \cos \zeta_i} \right] = \phi_{0_1},$$

λ = total safety factor for the sum of all elements i. As will be explained in section 14.8 λ is a scalar factor not very different from the vectorial factor to be introduced in the dynamic equation. ζ_i = angle between external force F_i and normal to slip surface of segment i of length l_i; c = cohesion factor. This 'static' approach does not explain how the masses can accelerate, as it is assumed that positive and negative forces are balanced ($P - R = 0$ and $v = 0$ and $\lambda = 1$).

Commenting on these results Skempton (1966) states that values of about 20° were 'ridiculously small' even for clays and could be explained only if they were not compared with the usual peak strength values but with the reduced values of the 'residual strength' of clays. Nonveiller also relies upon the difference between peak and residual values, but he does not consider them unusually small; by assuming too unfavourable a slip surface, he obtained values that were much higher (22·1° to 28·5°) (see Müller, 1968).

14.5 The velocity of the slide: the 'dynamic' conditions for an accelerated rock slide

On page 82 of Müller's 1968 paper a speed of 25 m/s (90 km/h) is indicated as the most likely maximum velocity of the sliding masses. How can this figure be correlated to the suggested friction constants $\tan\phi_{0_1}$ and c and how is this figure at all acceptable?

The previously calculated ϕ_{req} value corresponds to the highest required friction for which the positive and negative forces are just out of balance which may start a slow slide. Heim, in his 1932 paper, had already defined 'slow' and 'rapid' slides. All geologists expected the Vajont slide to be of the slow type because of the 'chair-like' shape of the rock formations. Nobody expected it to be one of the most rapid and most explosive ever recorded.

Some scientists, mainly in Italy, have worked on this problem from the other end and their thinking can be summarized as follows: the whole problem is a dynamic one which is amenable to Newton's law and the momentum theories. The law of Newton applied to the centre of gravity of a sliding volume V, mass $\rho_m V$ (fig. 14.5) can be written:

$$\rho_m V \frac{d^2 s}{dt^2} = \gamma_m V(\sin\alpha - f\cos\alpha),$$

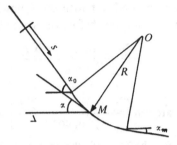

Fig. 14.5   Accelerated slide along a curved surface. s = curved abscissa; α = variable angle; R = radius of curvature at point M.

where $\rho_m = (\gamma_m/g)$ is the density of the material, s the distance, t the time and f the friction factor. Furthermore, $\mathrm{d}^2s/\mathrm{d}t^2 = R\,\mathrm{d}^2\alpha/\mathrm{d}t^2$ with R = radius of curvature. This basic equation can be integrated and yields the velocities $v = \mathrm{d}s/\mathrm{d}t$, and the time t_1 for the maximum velocity to be obtained.

On the other hand, the well-known momentum theorems of fluid mechanics (Jaeger, 1956) allow the calculation of the maximum wave height in the reservoir when a mass $\rho_m V$ penetrates at a velocity $v = \mathrm{d}s/\mathrm{d}t$ (fig. 14.6).

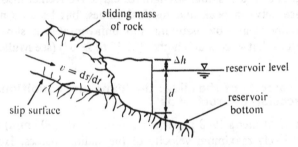

Fig. 14.6   Schematic representation of rock-sliding mass causing a wave, Δh on top of reservoir level.

The wave height is known to have reached about 220 m above water level. A rather lengthy tentative calculation shows that the maximum velocity must have been about 32 m/s, requiring a $\tan\phi_{0_2}$ value of not more than 0·176 ($\phi_{0_2} \leqslant 10°$) after an unpublished report by Professors Supino, Evangelisti and Datei. Another private report by Professor Stragiotti suggests a maximum velocity of 25 m/s and a 20-s duration of the fall. A possible alternative estimate of velocities is based on the lifting of some rock masses at the front of the 'rock wave' by 150 m. It yields an estimated 50 m/s for some points of the front wave (Jaeger, 1965a).

Skempton and Müller expressed their astonishment at the low values of ϕ_{0_1} obtained by most of the *a posteriori* 'static' calculations. The results obtained when calculating the 'dynamics' of the system were far more disconcerting. Mencl (1966b) was concerned with the additional losses which occurred at the sharp curvature of the sliding surface and found them to be quite significant. They add to the negative forces. Assuming the dynamic ϕ_{0_2} value to be about 10° (Supino, Stragiotti and others) a $\Delta\phi_0$ value of about 3° should be subtracted for the bend losses. This would leave for ϕ_{0_2} (average value): $\phi_{0_2} = 10° - 3° = 7°$ only, a remarkably low value, to be taken as an average over all the sliding surface and during the whole slide. Some other authors discussed the slide as a 'creep process' (Haefeli, 1967*b*) or as a 'progressive failure' (Müller, 1964; Haefeli, 1967*b*; Kenney, 1967; Nonveiller, 1967*b*). But none of these efforts brings us nearer to an explanation of the dynamics of the slide.

14.6 The basic contradictions between 'static' and 'dynamic' approaches

The movement of the sliding rock masses is, at any time, given by the equation of Newton (Jaeger, 1968*a*, *b*):

$$M\frac{\mathrm{d}^2s}{\mathrm{d}t^2} = P_s - R_s,$$

where M is the mass of the rocks, ds the small movement of the centre of gravity along the paths, P the positive forces and R the negative (resisting) forces acting on the mass M. This equation of Newton is a vectorial one and the components P_s and R_s in the direction of ds have to be considered. Alternatively, the components P_x and R_x along an axis x could be considered and the equation becomes:

$$M\frac{\mathrm{d}^2x}{\mathrm{d}t^2} = P_x - R_x.$$

At the time $t - \epsilon$, there is no movement and $P_s \leqslant R_s$.

When the movement starts at the time $t + \epsilon$, the positive forces P are just a little larger than the negative forces R and the movement is slow. When the sliding surface is 'chair shaped', as shown in fig. 14.7, the slide normally

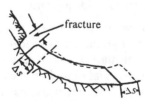

Fig. 14.7   Rock slide on a 'chair-shaped' sliding surface.

comes to a stop: when it progresses by a length Δs the positive forces along the steeper part of the slide decrease, whereas the negative forces along the less inclined section increase. This is the reason why so many masses of rocks which show clear signs of sliding at the surface really never move.

The geologists in charge of the survey of Mount Toc before the slide started its final burst had implicitly accepted a similar interpretation (fig. 14.8). The lower edge of the slow creep had a movement of slow rotation;

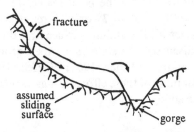

Fig. 14.8   First tentative explanation of the rock creep along Mount Toc.

rocks were falling in the gorge and it was expected that the progressive filling of the gorge would bring the slide to a natural stop.

The passage from rest, at the time $t - \epsilon$ to a sudden tremendous acceleration of 250 000 000 m³ of rock at the time $t + \epsilon$ requires tremendous forces. The probable decrease of the $\tan \phi_{0_1}$ value due to the 'self polishing' of the sliding surface during the slide cannot explain this fantastic acceleration of the masses.

14.7   Explanation based on the dynamics of a discontinuous flow of masses

There is no reason for the creep of 250 000 000 m³ of rock masses to have been continuous in time and space over a period of three and a half years. As it is not possible to explain the Vajont slide with the dynamics of a continuous progressive flow of masses, an attempt will be made to develop a theory of discontinuous flow of rock masses.

14.7.1   The progressive failure of the rock masses; the formation of a thick sliding zone

It has been amply explained in previous chapters that progressive small displacements of crystals and grains during shear tests cause a drop of the shear strength in most sedimentary rocks. It is acceptable to compare this test result obtained on rock material to rock masses.

Borings have clearly established that there was no real 'gliding surface', but a 'gliding zone probably several metres thick' (Müller, 1968). The friction factor within such a zone must have been very low. Such an assumption is physically acceptable. It is also required for mathematical reasons. If it is accepted that the $\tan (\phi_0 + \Delta\phi_0)$ was less than $0 \cdot 176$ ($\phi_0 + \Delta\phi_0 \leqslant 10°$) as average value during the slide, the $\tan \phi_{0_2}$ must have been locally and temporarily even well below this very low value.

It is very tempting to think that percolating water penetrating in the rock fissures has been instrumental in lowering the friction factor. It can hardly be so, because as will be seen, the first rupture to occur – on the steeper region of the slopes – was located well above the water level in the reservoir. Furthermore, the correlations between water-level variations and rock slides were so immediate that slow percolation of interstitial water cannot be an explanation. Finally, it has been proved (Müller, 1964) that there was no correlation between rainfalls and rock displacements.

14.7.2 Deep-reaching weakening of the rock masses; seismic wave measurements

General worsening of rock conditions was proved by seismic wave measurements by Professor Caloi (1966). According to him, over a short period, the longitudinal wave velocity in the rock masses had decreased from about 5000 or 6000 m/s to only 2500 or 3000 m/s. Caloi reported that already between 1959 and 1960 the fracturing process had penetrated into deeper-lying rock levels owing to increasing pressure and yielding of the rock. The depth of broken rock layers at the surface, which in 1959 was about 10 m to 20 m, was as thick as 50 m to 70 m or even 150 m in 1960.

These facts are clues and proof of the far-reaching deterioration of rock masses years before the catastrophe.

14.7.3 Measured creep velocities

This deterioration and physical weakening of the rock was not uniform. There is one diagram published by Müller (1964) which is most informative. It has been reproduced in fig. 14.9. Rock displacements were measured in boreholes at different depths. On the boreholes located in the upper range of the creeping rock mass, the diagram shows that the rock displacements are parallel and all identical whereas on the borehole located near the lower end of the creeping mass of rock the velocities are highest at the top and nil at the bottom. In fluid mechanics, and in rheology, parallel uniform velocities represent a flow where there is no internal viscosity or turbulence and with low wall friction. On the other hand, the unequal velocity distribution, as measured in the borehole at the downstream end of the creeping rock masses, represents a flow where some energy is being destroyed. There is no doubt that these most general theories of rheology are also valid for the creeping displacement or flow of large rock masses (C. Jaeger, 1956).

Figure 14.10 represents Müller's impression of the disruption of rock masses at the lower edge of the rock-creeping movement. Rocks were continuously falling in the gorge by 'extreme rotation'. This confirms the previous figure.

14.7.4 The discontinuous flow or creep of rock masses

The theory of 'progressive failure' should be developed a step further by introducing the idea of 'discontinuous flow'. Discontinuous flow is well

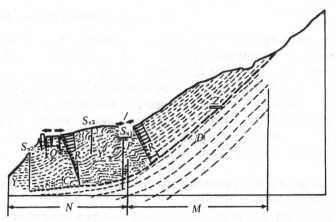

Fig. 14.9 Outlined section through the west zone of the slide. The small arrows indicate the velocity distribution. (Y) outline of rock layers; (l) La Pozza; (M) zone of predominant driving forces; (N) zone of predominant resisting forces; (C) zone of progressive failure; (D) probable slide surface; (P) compression zone; (Q) tension zone; (R) distribution of velocities; (S_n) exploratory boreholes; (Z) exploration tunnel (after Müller, 1964).

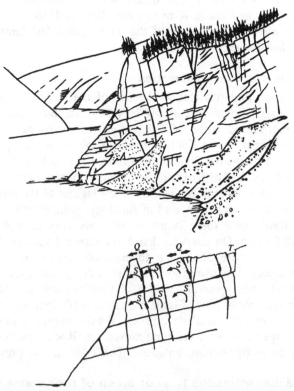

Fig. 14.10 Vertical section through the north front of the rock slide at the 'Pinnacolo'. Above, side view; below, scheme of movement; Q, tension; S, external rotation; T, relative movements (after Müller, 1964).

known in fluid mechanics (C. Jaeger, 1956). An extremely interesting example of a continuous flow becoming progressively discontinuous can be observed on rivers carrying floating ice (Jaeger, 1968a, b).

A river like the Saint Lawrence, Canada, has been observed over a very long period; ice and water levels and ice thickness being measured at regular intervals. The whole thawing process can be reproduced on a model. Such a model – 500 ft long – has been built in the Lasalle Research Laboratory, Montreal, the horizontal scale being 1:600, the vertical scale 1:150. Water levels and velocities can be correctly reproduced on this model and checked in nature at the equivalent locations. This same model has been used to represent the breaking-up of ice in the Spring. The floating ice was represented by pieces of plastic material, about the correct density and average size of floating blocks of ice. The plastic material was chosen because it has no surface tension when floating on water. At the upstream end of the model the discharge of water and of plastic material is maintained constant over a certain period of time. A varying regime is rapidly established downstream with floating pieces of plastic accumulating in those water channels where ice would normally build up. Some areas of the river where the flow is slow and where the water is covered with a thin sheet of ice, are reproduced on the model by a thin layer of plastic over the water. The depth of the floating plastic cover on the model can be checked against the real depth of ice. The accumulation of floating material at narrow passages causes the water level to rise in the upstream area. Potential energy is being accumulated, and the level rises until the shear strength along the river bed is exceeded. The barrier breaks down and a wave of water and ice progresses rapidly downstream on the model as in nature. This unsteady discontinuous flow is typical of the accumulation of potential energy and momentum behind an obstacle and the explosive character of the flow when the forces are relieved.

Figures 14.9 and 14.10 clearly show that a similar explanation is acceptable for the first phase of the slow sliding masses along the slopes of Mount Toc. Sometime in 1960 an M-shaped perimetral crack occurred high up along the slopes of Mount Toc (up to elevation 1400 m); the slow-sliding masses in the steep upper reaches, where the angle α of the slopes was probably steeper than the $\tan \phi_{0_1}$ of the friction factor ($\tan \alpha > \tan \phi_{0_1}$) started moving along a 'sliding zone' which acted as a lubricated smooth surface. According to fig. 14.9 (zone *l*) high pressures were being developed by the upper zone of the slide on the lower zone, pressures which also explain the rotation of the masses nearest to the gorge. Geologists observing the movements at the surface of the slide could not have recognized what had happened deeper inside the mass.

14.7.5 The two-phase slide

It was suggested in 1965 (Jaeger) that the sliding movement consisted of two phases: a first phase was similar to a 'visco–plastic discontinuous rock creep, or flow' which was the one observed by the geologists from 1960 up to the night of 9 October 1963. The second phase was a short-lasting brittle fracture in the lower part of the slide; along the less inclined branch of the chair-like sliding surface. This type of flow was suggested by fig. 14.9 but also later confirmed by some geological diagrams published by Broili (1967) and Müller (1968). On these diagrams it can clearly be seen that the slide, which in the upper reaches occurred parallel to the strata, cut right through the strata, suggesting a brittle fracture in the lower part rather than a sliding movement.

This short-lasting brittle fracture is now confirmed by Caloi, who commented in a private report about seismic wave recordings that a brittle fracture occurred first, which lasted 60 to 70 seconds, immediately followed by the slide which for the whole slide area lasted 20 seconds.

The basic assumption concerning the two phases (Jaeger, 1965a, 1968b, 1969b), one visco-plastic during which forces and momentum were concentrated, the second, a nearly instantaneous brittle fracture, is now positively confirmed by Caloi's remark.

14.8   The correlation between rock displacements and the water variations in the reservoir: action of the uplift forces

The correlation between the rock displacements and the water-level variations in the reservoir from 1960 to 1963 proves that the description of the slide given in section 14.7 is correct. Müller published in 1964 some rather astonishing diagrams concerning this correlation, (fig. 14.11). They show

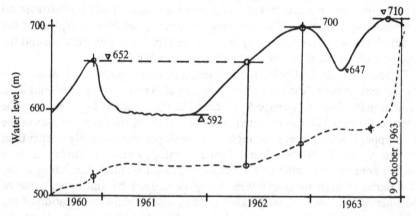

Fig. 14.11 Correlation between water levels (solid line) and rock displacements (broken line) (after Müller, 1964).

that the rock sliding was always accelerated by rising water-levels and stopped by falling water-levels. This contradicts the experience gained with slips occurring in earth dams containing water reservoirs: the earth slides always occur when water levels fall in these reservoirs.

The whole problem of the stability of rock slopes on the bank of reservoirs with varying water levels had to be re-examined as it probably gave warning of the Vajont rock slide (Jaeger, 1969b).

It is always possible to define a scalar 'safety factor'

$$\lambda_1 = R/P$$

as a ratio of the negative forces over the positive forces acting on the rock mass M. The vectorial equation of Newton can be written:

$$M \frac{d^2 s}{dt^2} = P_s - R_s = P_s(1 - R_s/P_s) = (1 - \lambda^*).$$

The ratio $\lambda^* = (R_s/P_s)$ is not identical to λ_1, but for all practical purposes the difference between the absolute value of λ^* and λ_1 is small and it is acceptable, in first approximation, to write that $1 - \lambda_1 \cong 1 - \lambda^*$ is a measure of the acceleration of the masses. The slope is safe if $\lambda_1 > 1$; it becomes unsafe for $\lambda_1 < 1$.

In more general terms (as represented by fig. 14.12), the law of Coulomb shows that the negative retaining force R can be written:

$$R = C + N \tan \phi,$$

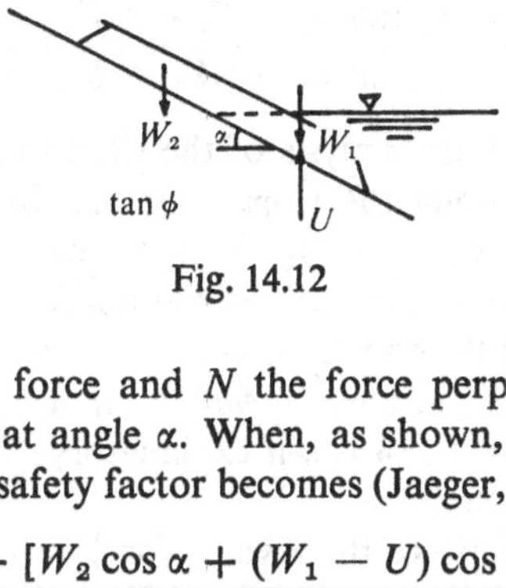

Fig. 14.12

where C is a cohesion force and N the force perpendicular to the plane sliding surface inclined at angle α. When, as shown, the fissured rock mass is partly immerged, the safety factor becomes (Jaeger, 1969b):

$$\lambda = \frac{C + [W_2 \cos \alpha + (W_1 - U) \cos \alpha] \tan \phi}{W_2 \sin \alpha + (W_1 - U) \sin \alpha}, \tag{1}$$

where W_2 and W_1 are the weights of the rock masses above and under the horizontal water line; U is the uplift force on the immerged fissured rock mass. Assuming that, for rupture and sliding conditions, the force C is negligible, then:

$$\lambda = \frac{(W_2 + W_1 - U) \cos \alpha \tan \phi}{(W_2 + W_1 - U) \sin \alpha} = \frac{\tan \phi}{\tan \alpha}. \tag{2}$$

Whatever the uplift force U, its effect on the stability of the slope near rupture is nil, provided the conditions of the rock mass can be represented by fig. 14.12.

When the water recedes rapidly in the reservoir, there is a supplementary dynamic force to be considered, due to the flow of the water in the rock mass towards the reservoir. This secondary force explains why earth slips occur by receding water levels. This force would have been too small to suddenly accelerate the Vajont rock masses to velocities of 25 m/s. In any case, it is incapable of explaining the correlation diagram of fig. 14.11, where sliding occurs by rising water levels.

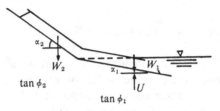

Fig. 14.13

Figure 14.13 represents a more general case, where rock sliding occurs along a 'chair-like' gliding line similar to the Mount Toc banks. The safety factor now becomes (Jaeger, 1969*b*):

$$\lambda = \frac{C_2 + W_2 \cos \alpha_2 \tan \phi_2 + C_1 + (W_1 - U) \cos \alpha_1 \tan \phi_1}{W_2 \sin \alpha_2 + (W_1 - U) \sin \alpha_1} \quad (3)$$

If we accept the previous analysis of the situation on Mount Toc, where rupture occurred first along the slopes high up in the mountain, we can write $C_2 = 0$ and $\tan \phi_2 < \tan \phi_1$. It then becomes obvious that, for $\phi_2 < \phi_1$, the λ factor really depends on the uplift. The ratio ϕ_2/ϕ_1 becomes the principal variable of the equation, as *for low values of ϕ_2/ϕ_1 the dangerous effect of the uplift force, U, becomes predominant*. To prove this two diagrams have been drawn. Figure 14.14 is self explanatory, showing a sharp drop of λ with ϕ_2/ϕ_1 and also with U.

In fig. 14.15 r is the ratio of the λ value for $U = 0.4W_1$ to the λ value for $U = 0$; r is a measure of the damaging effect of the uplift forces on the stability of the rock masses which also depends on ϕ_2/ϕ_1.

Any time the water rises in the reservoir the powerful uplift forces increase almost immediately. The hydrostatic effect of U additionally weakens the rock mass and favours internal rupturing and fissuration of the rock. When the water recedes, there is a general stabilization of the slide due to the reduction of U and to the 'chair-like shape' of the sliding surface: a phase of 'auto-stabilization' of the slide begins.

During all this period from 1960 to October 1963 the λ factor must have been very near to unity ($\lambda \cong 1$) with C_1 still positive. The sudden brittle fracture occurring during the night of 9 October 1963 caused C_1 to drop to zero within a few seconds and λ to drop well below the limit $\lambda \cong 1$.

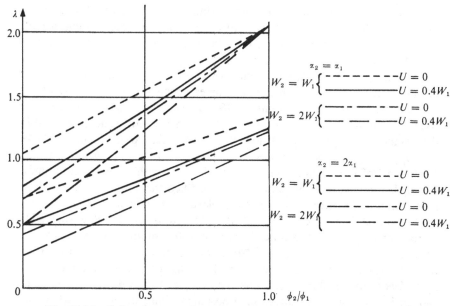

Fig. 14.14 Safety factor against sliding, λ; see equation (3) section 14.8.

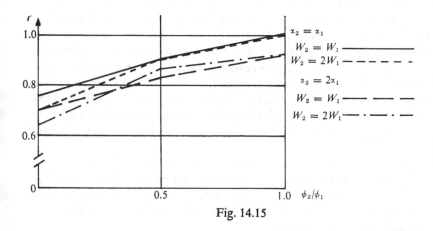

Fig. 14.15

14.9 Summary of the different phases of the Mount Toc rock slide

(1) Carlo Semenza, one of the most careful and experienced European dam designers, became suspicious of possible rock instability along the slopes of Mount Toc at an early stage. Favourable geological reports were submitted to him in 1958, when it was declared that the slide was only very superficial.

(2) The reports by Eduardo Semenza, F. Giudici and L. Müller, based

on their 1958–9 survey, were far more pessimistic. They indicated that a mass of about 250 000 000 m³ of rock was slowly creeping on a deep sliding surface, probably the contact zone between Malm and Dogger. The movement was described as a slow creep of rock masses.

(3) Filling of the reservoir started early in 1960, before the dam was finished. This is the time when rock masses began to move because of the rising water levels. On 4 November 1960, the water having in the meantime risen to level 645 m, a slide of about 300 m in breadth came down within the cataclastic malm layers, below the 'Plan della Pozza' (Müller, 1964). The upper edge of the slide was at an elevation of 850 m. Approximately 700 000 m³ of rock debris then slid into the lake within about ten minutes. It was determined that the falling mass had filled the gorge over a length of several hundred metres, in part up to an elevation of 600 m.

This first slide apparently had all the characteristics of a slow rock slide or rock creep. After it the water level rose to 652 m and was then progressively lowered to level 600. On 6 November 1960, the great M-shaped rock fracture at the upper edge of the rock slide was formed. During the period up to December 1960 most of the measuring points had moved about 47 to 140 cm in the horizontal direction. Loosening along the upper part of the sliding surface and lowering of the tan ϕ probably occurred before the formation of the M-shaped rock fracture. With levels at 600 m the displacements were reduced to about 6 cm to 13 cm during the period from December 1960 to October 1961 (Müller, 1964).

During this period the geologists observing the surface of the slide thought they were measuring a slow continuous rock creep, of a conventional type (fig. 14.11). Seismic wave measurements by Caloi proved the progressive fracturing of the rock masses (measurements were made in November 1959 and November 1960). The seismic measuring sections were at the foot of the slide and did not cover the more interesting upper regions during the period after 1961.

Water levels rose from October 1961 to November 1962 and dropped until March 1963. Displacements measured in this period indicated that the whole rock mass, in the west and east zones, was moving more or less homogeneously (Müller, 1964). Correlation between rising water levels and rock displacements were observed too.

It is interesting to note that during the period 1960 to the beginning of 1963, rock displacements slowed with falling water levels and eventually stopped. The displacements restarted when the water levels reached or passed the previous maximum levels. This confirms that there was an intermediary period of 'auto-stabilization' (Jaeger, 1969*b*).

These alternate movements of the rock masses are explained by equation (3) of section 14.8 and the 'chair-like shape' of the sliding surface. During the last phase from July to October 1963, when the water rose above level 700 to level 710, the rock displacements became suddenly most alarming and the disruption of the rock masses irreversible. Rupture by brittle fracture of the base of the slide occurred suddenly.

From 1960 to 1963 the visco-plastic creep was continuous apparently only on the surface. Deeper down it must have been discontinuous, with accumulation of pressures and momentum until the rock at the lower edge of the slide was weakened, cracked, fissured and finally gave way by brittle fracture on the night of 9 October 1963.

In the particular case of the Vajont slide, the shape of the geological strata, their inclination, the relative stiffness of the strata and the relative values of the friction factors tan ϕ at various levels, determined the physical type of the slide. The action of uplift forces (often negligible at other sites) became the prime factor of the rupture at Vajont.

Recordings of the rock displacements at the surface of the slide were completely inadequate to reveal the discontinuous nature of the bulk sliding movement and its most dangerous implications.

15 Two examples of rock slopes supported with cables

15.1 Consolidating a rock spur on the Simplon Pass road

15.1.1 Introduction

The Simplon Pass (2005 m above sea level) was used by the Romans. In the Middle Ages it became a vital route for commercial exchanges between Germany and Northern Italy; German Emperors crossed the pass. The present road was built on Napoleon's orders (1801–5) for military reasons, after he won several battles in north Italy. Since 1968, the road has been systematically improved and enlarged on the Swiss and on the Italian sides, to cope with increasing traffic.

On the south side, near Gondo (Switzerland) the dangerous 50 m high rock spur of Baji–Krachen (at 990 m above sea level) had to be partially cut for a length of about 60 m to achieve straightening and the widening of the road from about 7·70 m to 12 m.

The geologists reported that the rock mass forming the spur is very hard gneiss, cut by three sets of joints, some of them being potentially dangerous.

15.1.2 The structure of the spur

Figure 15.1 shows a typical cross-section of the rock with the three main sets of joints. The joints of set I, inclined at 38° over most of their length, with a steeper dip to 45° at the lower limit, are the most dangerous. The vertical joints belonging to set II could also cause trouble, but mainly where they occur as discontinuous structure underlying more gently dipping joints at the base, and near the face, of the rock mass. The subhorizontal set of joints III are not seen as a source of trouble in themselves, although in relationship with those of set II, some instability could result.

The diagram of fig. 15.2 refers to a typical shear test on a plane (σ, τ) carried out in two opposite directions, after reversing the force. The test is on a core sample from a boring, along a joint of set I. The test shows clearly that a first shear level occurs for $c = 0$ and $\phi' = 32°$, in the two directions. After increasing the τ-value, a second shear level is reached for $\phi' = 44°$ in one direction and $\phi' = 36°$ in the other direction. These higher ϕ' values are reached after the joint faces have moved in a relative displacement.

On the right hand side of sketch fig. 15.1 it is shown how the stability of the upper mass of the rock spur was obtained by friction along fissures I,

[424]

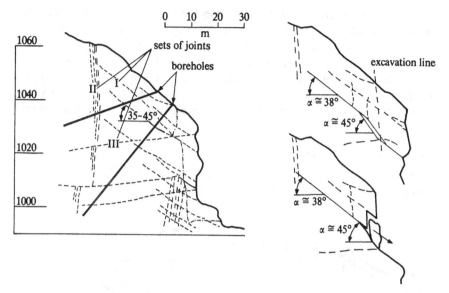

Fig. 15.1 The Baji-Krachen rock spur (Simplon Road). Typical cross-section. I, II, III sets of joints (after Lombardi, 1975).

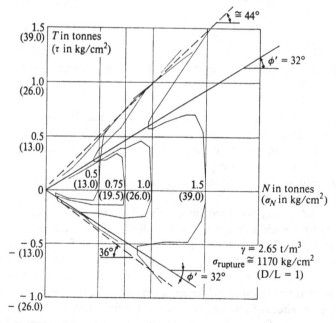

Fig. 15.2 Baji-Krachen rock spur. Rock shear tests on bore hole core (after Lombardi).

inclined at $\alpha = 38°$, after the rock has been displaced, so that condition $\phi' = 44°$ is obtained. The fissures remain tightly closed. This displacement of the upper block causes an opening of the lower end of the joints I, where $\alpha = 45° > \phi'$ and also of the set of vertical joints II. Designers expected that when excavating the rock spur as required by the design of the new Simplon road section, the lower block of rock might slide. Inspection of the rock spur *in situ* after the deformations had taken place proved that it was still stable.

15.1.3 The stabilization scheme

Lombardi (1975) in charge of the consolidation work analysed two possible reinforcement schemes. In the first (fig. 15.3*a*) anchors (1) were to secure the

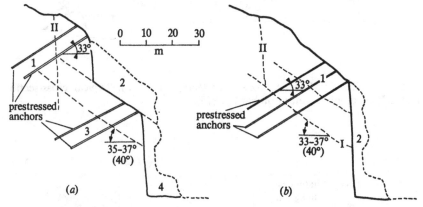

Fig. 15.3 Baji–Krachen rock spur. Two alternative proposals for rock support (after Lombardi).

upper part of the inclined rock before the rock mass (2) was excavated. A second set of anchors (3) were to be installed before finalizing the excavation (4). This reinforcement scheme was rather expensive as it required the cable anchoring to be done in two phases.

The alternative scheme (*b*) was adopted whereby all anchors (1) were placed before excavation of the rock mass (2).

Figure 15.4 shows the final design for the rock mass support. Stability calculations for such a design are straightforward. Interesting is the table of calculated 'safety factors' $\eta = $ (retaining forces/active forces) calculated for different values of the angle ϕ', varying angle α and a series of anchor forces V (in t/m). For $V = 0$, η_1 is obviously less than 1, when $\phi' < \alpha$. But the most important result of these calculations is the fact that, even with very powerful anchorages (anchors of 140 tonnes were used at Baji–Krachen) it is not possible to raise the 'safety factor', η, above 1·3 to 1·5. Designers concerned with the rock supports for large underground excavations using cable anchorages

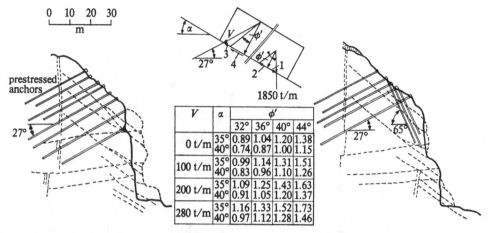

Fig. 15.4   Baji-Krachen rock spur. Final design for rock support (after Lombardi). (1) Active force; (2) natural friction force; (3) negative force due to prestressing; (4) Friction force due to prestressing. Safety factor η = (resisting force/active force). V = prestressing force, α = angle of dipping joint, ϕ' = angle of friction.

arrive at similar conclusions, as explained in the chapters dealing with underground works. Forces in cables remain small compared with the weight of rock masses and other natural forces to be dealt with. Lombardi remarks that these factors refer to the safety against small rock deformations along the fissures. Large deformations for which the safety factor would be higher (according to the shear tests) are no more likely when stiff anchorages with cables give rigidity to the reinforced mass.

   The sketch on the right of fig. 15.4, shows, in addition to the main system of cables which are inclined at 27° to the horizontal, a second system of cables at α = 65°, nearly at right angles to the first system of cables. These cables are supposed to pin blocks of rock in danger of sliding down the rock face after excavation. It was considered that drilling work on the nearly vertical unstable rock face would entail unacceptable dangers to the workers. This explains the reason for this second family of anchorages.

15.2   Stabilization of a very high rock face for Tachien Dam foundations (fig. 15.5)

15.2.1   Introduction

The Tachien Dam is part of a scheme to harness the hydro-electric potential of the Tachia River in central Taiwan (Formosa). The project developed by Electroconsult, Milan, has a rated installed capacity of 234 MW and operates under a maximum gross head of 163 m. The 180-m high dam is a double-curvature arch dam; the crest at elevation 1411 m above mean sea level has a

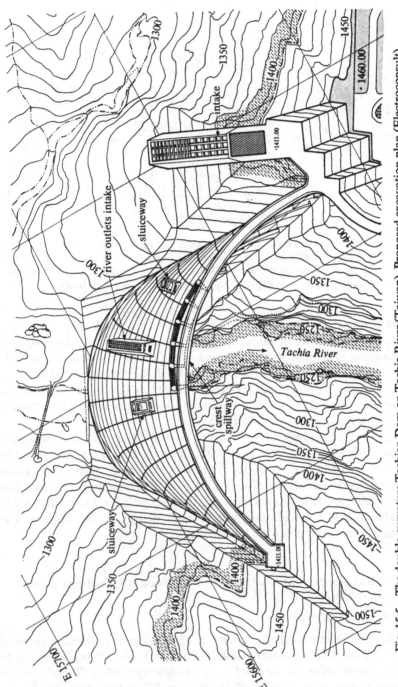

Fig. 15.5   The double-curvature Tachien Dam on Tachia River (Taiwan). Proposed excavation plan (Electroconsult).

length of 290 m. The dam is located at the entrance of a narrow precipitous gorge. At level 1236 m, the gorge is only 20 m wide.

The geologists describing the project area mention two major geologic formations: a quartzite formation, total thickness in the order of 1000 m, and a slate formation, 2000 m thick. The rock can be divided into five classes, varying from pure quartzite, through impure quartzite, slaty quartzite or quartzite slate to pure slate and impure slate. The dam is located in the transition area between slate and quartzite, where massive layers of quartzite (about 35–45 cm thick) are separated by thinner beds of slate (about 3–15 cm thick). The slates are sometimes associated with clay and graphitic material usually 3 to 5 cm thick, rarely up to 20 cm. The strata strike approximately normal to the axis of the gorge and dip upstream at an average angle of about 60°, which is favourable both to the control of seepage and the stability of the rock walls. Four main families or sets of joints cross the rock mass.

The surface geology of the dam site itself has been explored by 45 adits and 16 core borings. Sound rock in the river channel is found a few metres below the alluvium material. The geologists described the walls of the gorge as 'rather fresh and compact' for elevations below 1320 m. Above this elevation, they have been affected by weathered joints and stress relief cracks which tend to extend more deeply into the abutments with increasing distance above riverbed.

15.2.2 Specifications for the support

The basic decision by the designers was to support the dam abutments on the right and the left banks direct on the steeply dipping quartzite stratification. This was possible because the rock masses dip in the upstream direction. Major features to be considered are a large fault on the right bank and stress relief cracks. Due to the geological features mentioned above, the dam excavation was controlled by two definite specifications:

(a) *Stratification.* No undercutting of the stratification planes should be allowed. Undercutting would produce unstable conditions and could trigger off a considerable rock slide along the stratification. The excavation was to be limited to the scaling of the surface weathered rock for a depth of about 5–6 m, 'peeling off' the quartzite stratification. The main job of the geologists was to select a convenient quartzite stratum, continuous from top to bottom and sound enough to provide a rock abutment to the arch dam.

(b) *Joint set 'A'*, a very characteristic joint family, which has an average inclination of about 70°–80°, was recommended to be adopted as a lateral facing for the excavations, in order to minimize the amount of excavation.

Conditions on both banks of the gorge are similar, but were far more difficult on the right bank, which is the only one to be discussed in detail in this section. Similar geotechnical problems had to be solved for the power intake structure and the spillway tunnel.

15.2.3 Rock characteristics

The general geologic survey was supplemented with extensive geotechnical investigations. The figures in table 15.1 illustrate some of the rock characteristics.

Table 15.1

		Quartzite	Quartzite slate	Slate
Compressive strength	kg/cm²	1140	1450	840
Tensile strength	kg/cm²	170	220	130
Static modulus	10³ kg/cm²	340	524	280
Dynamic modulus	10³ kg/cm²	590	890	400
Poisson's ratio		0·18	0·10	0·15
Specific gravity		2·7	2·7	2·7

Core recovery from eight boreholes varied from 55% to nearly 100%, depending on the location of the borehole and on the depth from the ground surface. Water absorption and cement absorption, important for the design of the lateral grout curtain, were determined. They too indicated that the rock characteristics vary rapidly, the rock showing signs of weathering and weakening near the surface and in the upper reaches of the gorge walls. Gouge and seam material was analysed. A thorough exploration of faults and relief cracks was carried and detailed plans were developed for rock treatment, rock stabilization, grouting and plugging of cracks. Adits and connecting shafts were excavated for this effect. In particular the stabilization of the unloaded faces of the gorge (parallel to the gorge), called faces 'A' (just downstream of the dam abutments), was considered to be most important, in view of the very high reaction load from the dam abutments. The need for heavy anchoring was foreseen. The two sides of the gorge had to be reinforced.

These figures and facts confirm the geologists' prediction that the rock, conveniently reinforced, drained and grouted, is suitable to the stability of the dam abutment below dam crest level 1411 (fig. 15.6).

15.2.4 The stabilization scheme

Figure 15.5 shows the dam and the excavation plan, as it was used for a searching stress–strain analysis of dam and abutments. A three-dimensional

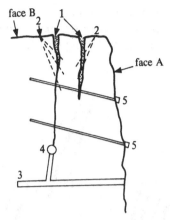

Fig. 15.6   Reinforcing of face A (horizontal section). (1) Open cracks filled with pumped concrete. (2) Boreholes for low-pressure grouting. (3) Adit. (4) Control shaft. (5) 120 tonne anchors.

finite element method programme was developed and published by Zienkiewicz, Taylor (Swansea University) & Gallico (Electroconsult, Milan) (1970).

The geologists had located on the right river bank a quartzite layer starting higher than level 1550, which could be followed down to the bottom of the gorge to level 1230. The contractor (Torno, Italy and Kumagai, Japan) had been instructed to follow this steep rock layer, also called face B, to 'peel' it without undercutting it, down to the lowest level. Laterally the excavations were to follow, whenever possible, the joint set A, to form the face C of the excavations, almost parallel to the gorge (figs 15.6 and 15.8).

Face A, the unloaded face, had to be sustained with 120 tonne anchor cables, forming a dense pattern. For the 300-m-high face B, small rock bolts, sometimes steel nets or shotcrete were thought to be sufficient. When, early in June 1970, the contractor entered face B at levels 1500 to 1550, about 140 m higher than the future dam crest, for excavation, the real structure of the steep rock bed became apparent. The quartzite was cut by more joints than expected, the dispersion of the sets was also larger than expected. Razor-thin fissures were observed. There was a general tendency for the quartzite bed to split in a direction parallel to the rock face, forming plates. The joints, often forming acute angles, cut the bedding into plates or blocks which could become loose. Cohesion between plates of quartzite and between these plates and the slates was probably low. Water pressure was building up behind the plates and water seeping through the drains. It became obvious that the general stability of the face was not secure. A large rock slide was possible, which would have jeopardized the whole dam construction.

Consulting engineer and contractor agreed that action was imperative. All possible situations which could cause a rupture of the bedding were considered. The stability of the inclined bedding under its own weight, under normal circumstances, was analysed. It was found that this case could not cause a rupture; but danger was possible when the plates were cut by oblique joints or weak seams, or undercut when blasting (see fig. 15.7 *a, b*). In fact, during

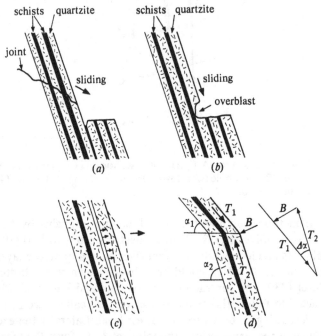

Fig. 15.7   Possible failure of the steep quartzite slope. (*a*) Along an inclined joint. (*b*) Due to overblast. (*c*) Buckling due to hydrostatic pressure. (*d*) Buckling due to change of dip, $\Delta\alpha = \alpha_2 - \alpha_1$.

excavation, in spite of all the care taken, two minor rock slides occurred when plates became loose.

It was found that the most dangerous situation occurred when the dip of the bedding suddenly changed. Figure 15.7*d* shows how a sudden increase of the dip causes a knee in the rock surface. Forces may cause the buckling of the face and possible overturning of the rock mass or a rupture of the rock surface. A typical rupture of the rock bedding caused by a sudden change of the dip of the rock face could be seen on the natural rock surface near the future power intake. There the rock face had to be supported with deep 120-tonne cables anchored in the rock mass. When, on the contrary, the dip of the face suddenly decreases, there is a danger of the quartzite bedding being damaged when excavated and undercut.

The case of cleft-water pressure in rock fissures was also analysed, fig. 15.7*c*. The weight component of a steeply inclined rock slab is not capable of neutralizing hydrostatic pressures of the magnitude to be expected after a rain storm of the high intensity common in that area of Taiwan.

This analysis led to many decisions concerning the construction progress and the required rock support:

Rock support already planned for the unloaded face A was reinforced. Drainage of the area, plugging of all the open faults and relief cracks was carried out, with great care, from adits and connecting shafts (fig. 15.6).

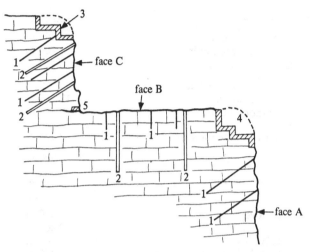

Fig. 15.8   Tachien Dam. Schematic representation of the consolidation of faces A, B, and C. (1) Rock bolting (10- to 120-tonne tendons); (2) Drainage ($l = 10$ m, $d = 7.5$ cm); (3) Shotcrete or concrete facing of unstable rock; (4) Excavated rock; (5) Plugged open joint.

Face B, the dipping quartzite bedding (see fig. 15.8) was systematically supported with monostrand tendons (short cables), rather than with mechanically anchored steel rods, not convenient in fissured rock masses. Any potentially dangerous rock mass was anchored with deep cables, some of 30 or 60 tonnes, many 120 tonnes. Very dangerous rock masses were additionally supported with reinforced concrete beams, anchored with 120 tonne cables. Use of steel mesh and/or shotcrete was recommended whenever necessary. A similar reinforcement programme was developed for face C. It was found that many rocks along this face had a tendency to overturn.

Drainage of all the rock faces was carried out systematically. Figure 15.8 represents schematically the work to be done. A careful excavation technique was worked out. After any blasting, the most probable dip of the quartzite face was estimated and the next blasting undertaken so as to avoid undercutting of the face B.

This difficult excavation of a 300-m-high rock face was finally carried out successfully. But the dip of the face was different from what could be anticipated from the contract drawings. As it is not possible to modify the shape of a double curvature arch dam, the whole dam had to be shifted by 8 m in the upstream direction, along the dam main axis.

Table 15.2 gives some figures concerning the number and the length of the main anchorages which were achieved, not only on the right bank of the Tachien Gorge, but also on the left bank and intake area.

Table 15.2

Steel ribs (spillway tunnel)	1 300 m
Expansion rock-bolts (excavations)	4 100 m
Grouted anchors	6 450 m
10-tonne monostrands:	
Power house	12 245 m
Outdoor excavation	10 255 m
Total 10-tonne monostrands	22 500 m
30-tonne tendons	900 m
60-tonne tendons	350 m
120-tonne tendons:	
Spillway	2 000 m
L.B. face A	9 800 m
R.B. face A	5 100 m
R.B. faces B + C	2 300 m
Total 120-tonne tendons	19 200 m

The number of anchors and tendons used on the right bank is given in table 15.3.

Table 15.3

	120 tonne	Anchor bars	10 tonne	Total
Face A	114	42	15	
Face B	50	318	367	
Face C	23	12	546	
Total	187	372	928	1496

16 Three examples of large underground hydro-power stations

General information on tunnelling and underground work is given in the preceding chapters, mainly in part three. The case histories developed in this chapter concern the Kariba South Bank and the Kariba North Bank underground works, excavated in igneous rock and the Waldeck II machine house in sedimentary rock. In all three cases the rock required support. The methods used were very different in each case. It is interesting to compare designs, excavation techniques and rock support.

16.1 The Kariba South Bank power scheme (fig. 16.1)

16.1.1 The geology of the site (figs. 16.2 and 16.3)

The most impressive element of the Kariba power scheme is the high double-curvature arch dam impounding water in the Kariba Reservoir. Most of the geological investigation on the site was concentrated on the dam foundations and on the rock abutments on both sides (see section 11.4.3) and on the reinforcement work on the South Bank.

The geology of the Kariba area was recorded and interpreted by the late Dr Francis Jones and Dr Dubertret (Paris). Dr J. L. Knill and Dr K. S. Jones have published an extensive geological description of the area, including the areas where the South Bank and the North Bank underground power houses had to be located. They reported that:

The foundations of the dam are mainly composed of biotite, with occasional layers of amphibolite which are more abundant on the south bank. A thick band of quartzite, folded into a syncline, outcrops on the upper part of the south bank. The quartzite is inter-stratified with the gneissic succession. Frequent dykes of pegmatite are present and the gneiss is so ramified by granitic veining that it locally grades into migmatitic rocks. The main trend of the foliation and the individual lithological horizons is north-east to south-west, although there is both regional and small-scale evidence for a later folding on axes normal to the dominant trend. The structural phase which has affected the bedrock at the site most significantly, from an engineering viewpoint, is the minor thrusting which resulted in the formation of the fractured synclinal flexure on the upper part of the south bank. This fold, which is associated with deep weathering in and below the quartzite, has created special engineering problems. A fault trending parallel to the gorge, as well as several on echelon splinter faults, has been located near the river bed and on the north bank.

The geology of the north bank is straightforward in that the sound gneisses are overlain by a weathered layer of fairly uniform thickness (10 m approximately).

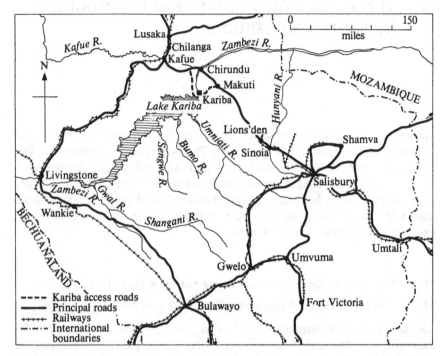

Fig. 16.1   Map showing Kariba and Kafue power-stations and access facilities (after Olivier, 1961).

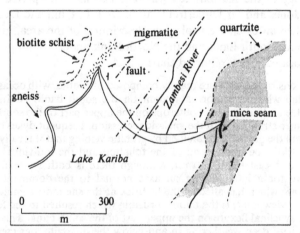

Fig. 16.2   Geological map of Kariba Dam (after Knill & Jones, 1965).

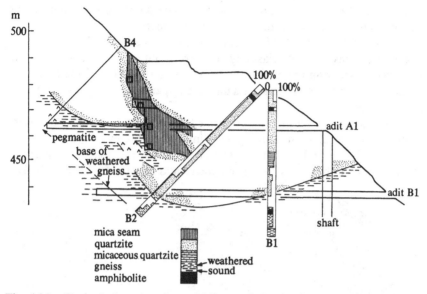

Fig. 16.3 Geological section through the south bank of Kariba, illustrating the correlation of information obtained from boreholes and adits (after Knill & Jones, 1965).

The exploratory boreholes and adits, carried out before excavation of the dam foundation, provided an accurate assessment of the geological conditions. The investigations, carried out in connection with the Second Stage power station, further confirmed this geological picture.

The geology of the south bank is more complex in that the gneisses are overlain by quartzites which frequently dip steeply towards the river in a downstream direction. The quartzites are highly fractured and sheared, the individual joint surfaces being grooved and slickensided. The joint blocks of quartzite are generally only moderately weathered owing to a loss of cementation, but the intervening fractures are infilled by clayey and silty material. Intercalated bands of gneiss and micaceous quartzite occur within the main quartzite mass. A zone of soft micaceous gneiss, composed largely of coarse-grained biotite and muscovite, referred to as the 'mica seam' forms a curved zone within the quartzite. The mica seam appears to be inter-bedded with the quartzites and the boundaries of the seam are approximately parallel to the quartzite-gneiss contact. The outcrop of the mica seam above the dam was strongly sheared to a narrow strip about 3 m in width, yet in depth the breadth increased considerably. The contact between the gneiss and the quartzite has a curved profile, with the axis of curvature lying parallel to the valley.

The original investigations on the south bank (1956) were as usual by boreholes, pits and adits. It was realized that the bank in the fractured quartzite was structurally suspect as an abutment and that an extensive programme of jetting and grouting was required (Lane, 1964). On the surface, there was no sign of the mica seam, which was detected during the excavation of an access road. A new series of adits, at two levels, and boreholes were used to determine the real structure of the rock. The quartzite was found to be deeply fractured and permeable, the gneiss weathered.

The consolidation work of the south bank has been described in section 11.4.3. Problems of rock elasticity, compressibility of the mass, its bulk resistance to shearing forces and the general watertightness of the abutment on the right-hand side of the arch dam were of paramount importance to the dam designers.

The survey of the south bank confirmed that the underground halls were located almost entirely in biotite gneiss, injected in places by some dykes of granite pegmatite and would in general lie below the zone of superficial alteration where the gneiss has disintegrated and become friable (fig. 16.4).

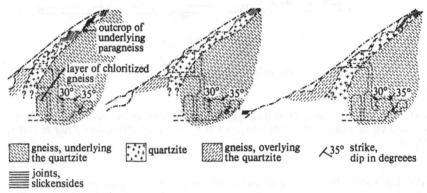

gneiss, underlying the quartzite   quartzite   gneiss, overlying the quartzite   ⟨35° strike, dip in degreees

joints, slickensides

Fig. 16.4 Kariba South Bank underground power-station. Typical cross-sections showing geology of site and locating the excavations in gneiss (after Olivier, 1961).

16.1.2 The design of the underground works (fig. 16.5)

The 128-m-high double-curvature arch-dam at Kariba stores the flow of the Zambesi River in the immense Kariba Reservoir. According to the records available in 1961, the flow at Kariba varies from 8000 cusecs (227 m³/s) to possibly 570 000 cusecs (16 150 m³/s). The firm flow is estimated to be 42 000 cusecs (1190 m³/s). The maximum water level was designed to be at 1590 ft with the dam crest at 1616 ft. The six 100 MW generating sets will generate an estimated 8500 million kWh a year.

Figure 16.5 shows the underground works with six vertical intake shafts, four permanent intakes at level 1510 and 2 temporary intakes at level 1370, the great machine hall, and the associated transformer hall. The discharge from the six turbines is through three unlined, horseshoe-shaped tailrace tunnels, 34 ft (10·40 m) in diameter, each provided with a concrete-lined surge chamber, 63 ft (19·25 m) in diameter.

The machine hall has a length of 468 ft (142·75 m), a width of 75 ft (22·85 m) and a height of 132 ft (40·25 m), the top of the excavation being at level 1326. The concrete arch of the underground cavern has a radius of 36 ft (10·98 m)

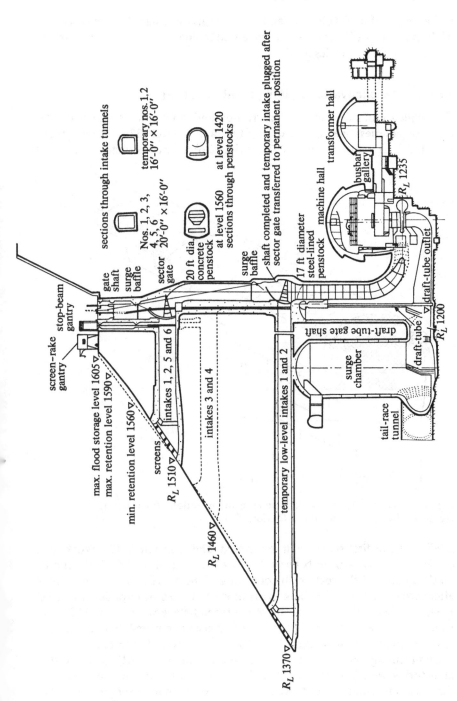

Fig. 16.5 Kariba South Bank. Underground works, typical cross-section (after Olivier, 1961).

and a height of 45 ft (13·73 m) and is thus nearly circular. The general shape
of the machine hall is ovalized, a shape well in line with the theoretical thinking at the time it was designed.

16.1.3 The excavations, linings and grouting

The sequence of operations for excavation and concreting is significant: (1)
the pilot galleries, one at the crown of the vault at level 1326 and two others,
40 ft (12·2 m) apart, at level 1307 (see fig. 16.6) were constructed and a 68-ft

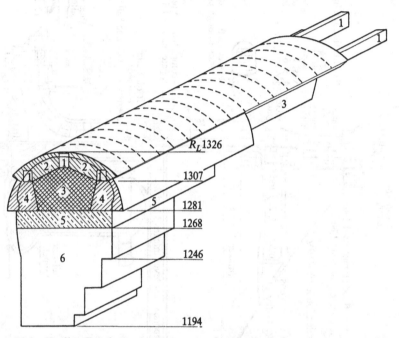

Fig. 16.6   Kariba South Bank. Sequence of operations for excavation and concreting of the machine hall (after Olivier, 1961).

(20·7 m) wide first-stage vault excavated and concreted. (2) Work was
carried out in bays from between 8 to 24 ft wide (2·4 to 7·3 m), depending
on the quality of the rock, with concreting of the bays following immediately
afterwards. (3) Thereafter bulk excavation continued in bays down to level
1281, together with (4) concreting of the second-stage vault, 75 ft (22·9 m) in
width. Once the concreting of the vault had been completed, bulk excavation
in the machine hall was taken down by steps to level 1194 (5 & 6).

Commenting on the underground works, Olivier (1961) remarks that
arch forms were generally used to form vaults of permanent excavations,
the most important being the vault of the machine hall. Near-circular forms
were adopted wherever possible for shafts and adits. Initially, it was intended

to use a complete eggshell lining for the machine hall, but this was rejected on account of the greater cost and the difficulty of construction.

For the much smaller transformer hall, a similar programme was adopted. The volume of excavation for the machine and transformer halls, adits, shafts and tunnels was 680 000 yd³ (519 000 m³) against 460 000 yd³ (351 000 m³) for surface excavations.

The average head of water to the centre line of the power-house was taken as 200 ft (91·5 m) and the linings were designed to take full water pressure. In addition, the rock behind the linings was grouted to reduce any free flow from the reservoir to the structure. Most rock joints appeared to be too fine to be grouted but there is no doubt that pressure will build up behind the linings, hence a comprehensive system of drains was provided to assist in relieving pressures against the linings. It was argued by the designers that grouting also assisted in consolidating rock which had been disturbed by blasting. It is known that redistribution of stresses in the rock occurs mainly during excavation, prior to concreting. Grouting of the power house and transformer hall was therefore purposely delayed to allow internal movements to take place. Further strains will of course take place during the filling of the reservoir.

The diversion tunnels were left unlined, but rock bolting was resorted to during construction as a safety measure.

Accoustic strain meters were embedded in the concrete during construction of the machine-hall vault and in the transformer-hall vault, in the end walls of the machine hall and in two draught-tube gate shafts. It was found that a general state of compression exists along the intrados and extrados of the machine and transformer-hall vaults (see Olivier). In some cases there was evidence of minor tensile stresses in the extrados. It may be that there has been a slight initial inward deflection of the springing of the vaults. The maximum stress in the concrete is of the order of 1000 lb/in² (70 kg/cm²), and it is believed that the concrete is working monolithically with the rock.

Olivier reports that, in the machine hall, one of the upstream transverse buttresses opposite a faulted zone (concreted between the machines to act as temporary struts until incorporated in the mass concrete foundations) has been subjected to sufficient compression to cause shear cracks in the concrete. The indications are that the other buttresses were also subjected to high compression.

Extensive use was made of grouting to consolidate the rock. Radial holes were drilled in a 15-ft grid in the two major vaults, and grouted in three stages:

(*i*) 3 ft (0·92 m) into rock, grouted at 20 lb/in² (1·4 kg/cm²) (contact grouting between concrete and rock).

(*ii*) 10 ft (3·05 m) into rock, grouted at 50 lb/in² (3·5 kg/cm²).

(*iii*) 30 ft (9·15 m) into rock, grouted at 150 lb/in² (10·5 kg/cm²).

Unlined rock faces and skin walls were equally grouted. A lake-side curtain was provided to protect the shafts until the linings of the shafts were sufficiently high to counteract the effects of the rising reservoir. The main object was to consolidate the jointed quartzite in continuity with the dam curtain in 32 months. Approximately 131 614 ft (40 142 m) were drilled, in 7003 holes, taking 8389 tonnes of cement, and 2663 tonnes of sand.

16.1.4   New rock support techniques. Comments on Kafue underground machine hall

The tensile stresses measured along the extrados of the concrete arch vault linked to possible inward movements of the side walls of the excavation. A more detailed analysis of such a problem would have indicated that the wall movements depend not only on the shape of the vault, but also on the quality of the rock on each side of the cavern and on possible reactions from other nearby excavations, such as transformer hall or surge chambers. The prediction of the wall movements, on which the stresses in the concrete arch depend, is most difficult. In fact, the concreted vault acts more as a lining to the rock vault, and not as an independent loaded arch. The concrete vault deforms as rock deforms. This is also the general assumption most mathematicians make when using the finite element method for calculating strains and stresses in the concrete lining of the rock vaults.

Shortly after Kariba South Bank power system had been put into service L. von Rabcewicz outlined the new Austrian tunnelling method (NATM) which he used mainly in galleries and tunnels. Very keen designers used the same method for large excavations like underground power houses (Rescher (1968) on Veytaux, Mantovani (1970) on Lago Delio, etc.).

The Kafue underground power station on the Kafue River, upstream of Kariba, is a very good example of designs adopted at that time. Sten Rosenström describes it (1972), as shown on fig. 16.7.

Photoelastic tests were carried out to get the stress distribution around the machine hall and the transformer Hall. The tests were carried out for a horizontal principal stress $p_h = 200$ kg/cm^2 and a vertical stress $p_v = 135$ kg/cm^2. The tests showed that tensile stresses might have occurred in the walls which could be avoided by proper shaping of the cavern. During the final design of the profiles of the cavern it was specified that the walls were to be as flat as possible, and that transition radii were to be as large as possible. (Similar results were obtained when testing other caverns like Waldeck II, Germany, and Ruacana on the Cunene River.)

Rosenström mentions that, in order to ensure the stability of the rock vault, the roof surfaces were rock-bolted and netted. The high walls of the machine hall were also bolted, the bolting and netting being carried out consecutively during the blasting operation. Figure 16.7 shows the importance of the anchorages used. The longitudinal crane beams, and the wall slabs above them in the machinery hall, were concreted against the rock walls from the first bench. They are anchored into the rock. In both caverns, spaces of 30 cm

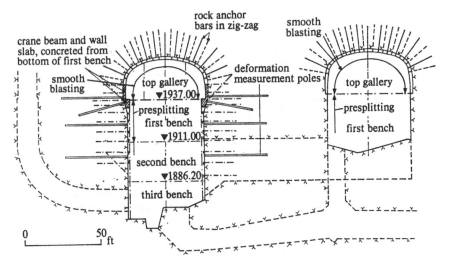

Fig. 16.7   Kafue underground power-station (after Rosenström, 1972). A vertical section through the underground power and transformer caverns showing the routine bolting of roof and wall surfaces with 1 in (2·5 cm) deformed bars, having a yield point of 40 kg/mm². The 4·5 m long bolts placed in the roof were at 2 m centres in a zig-zag pattern, and the 6 m bolts in the walls were at 2 m centres. The roof surface was also secured by mesh (after Rosenström, 1972).

against the rock had been planned for roof arches, but these arches were not constructed.

The photoelastic model test had shown heavy rock stresses. Disturbances due to such stresses were not observed *in situ*, although in the transformer hall a fissure was observed on part of the roof in a direction approximately parallel to the cavern axis.

The example of Kafue is an interesting link in the sequence of events to be related in the next paragraphs of this chapter.

16.2   The Kariba North Bank power scheme (Kariba Second Stage)

16.2.1   The design (figs. 16.8 and 16.9)

Kariba Second Stage, on the north bank of the river, consists of a large underground machine hall, designed for four vertical-axis Francis turbines and generators. The concreted sill of the intakes is at level 460·25 m, the water level in the reservoir is assumed to vary between levels 475 m minimum and 489·20 m maximum. Four 6·75-m-diameter penstocks feed the turbines, of which the horizontal axes are at level 374 m. The draft tubes are 28 m long and the tail-race pressure tunnels are 126·50 m long. The tail water varies between 381·61 m and 403·86 m. A steeply inclined busbar shaft connects

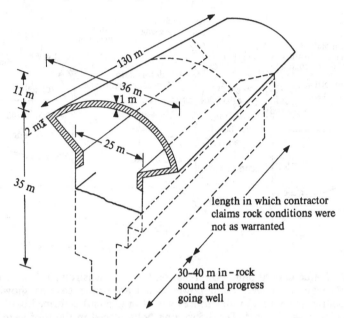

length in which contractor
claims rock conditions were
not as warranted

30–40 m in – rock
sound and progress
going well

Fig. 16.8   Kariba North Bank Machine Hall (from Jaeger, 1973). Diagrammatic
view of the machine hall showing progress of work and rock-fall area.

the generators to the transformers, which are located outside, well above dam
crest level. There are no surge tanks; this obliged the designers to keep the
tail races as short as possible, placing the machine hall as near as possible to
the river gorge.

The excavation of the large machine hall, with adits, access tunnels and
tail races was the main object of the contract with a well-known English
contractor. Figure 16.8 shows a diagramatic view of the machine hall, 130 m
long, 35 m high from bottom to haunch level, and 25 m wide. The lateral
walls are vertical and the rock vault above the excavation is 36 m wide and
only 11 m high, forming a flat arch with 5·50-m-deep haunches. The design
is therefore very different from the Kariba First Stage machine hall, and also
in contrast to sketches to be found in the preliminary *Report on exploratory
work for Second Stage Power Station, November 1961, of the Federal Power
Board, Federation of Rhodesia and Nyasaland,* where the cavern is shown to be
ovalized. The rock vault of the contract design is shown supported by a 1-m
to 2-m-thick concrete vault.

The whole design with flat vault, deep haunches and vertical side walls
shows that the designer had full confidence in the exceptionally high quality
of the rock on the North Bank, as he knew it from the construction of the
left-hand dam abutment. The rock was coherent, watertight and showed high
crushing strength and high modulus of elasticity.

16.2.2 The geology of the North Bank

(*i*) *The Report on exploratory work* (dated November 1961) The Report on exploratory work, based mainly on information obtained by Dr Louis Dubertret mentions:

The north bank is almost exclusively composed of biotite gneiss with pegmatite dykes and migmatite forming the rest of the massif . . . (the biotite gneiss) is extremely hard and gives an excellent proportion of core recovery from borings, nearly 100 percent . . . The presence of these dykes (pegmatite) encourages the stability of the surrounding rock formations . . . (migmatite) is a very hard and compact rock formed as a result from the injection of quartz and felspar in a gaseous state in the gneiss . . . The area of the main works is remarkably clear of faulting and slips . . . In the region of the exploratory works the gneiss is well folded and nowhere in the adits is the bedding horizontal.

The geologist's opinion was based on detailed examination of adits and boreholes. The machine hall having been shifted, only one borehole ('borehole number 4' mentioned in a paper by Lenssen (1973)) passed vertically through the future excavation, indicating solid gneiss on the whole borehole depth.

Consultant and contractor agreed on a programme of work whereby the large rock vault was to be excavated nearly full-face down to the level of the haunches. The concreting of the concrete vault was to follow progressively.

There is no doubt that the consultants who had worked previously on the foundation of the dam and on Kariba First Stage were fully confident on the success of the excavation work, which started in May 1971.

(*ii*) *The geological report by Dr G. D. Matheson* (1971). On the 28/29th June 1971 Dr G. D. Matheson, senior petrologist, Geological Survey, Zambia, and Mr Rowbottom, Mines Inspector, Mines Department, Zambia, visited the site. Dr Matheson produced a report mentioning that:

The entire North Bank Project is located in the complexly folded and highly metamorphosed rocks of the basement system. The commonest rock type is a migmatitic gneiss containing numerous quartzo-feldspathic bands, veins and other bodies of segregation origin. Biotite schists bands and amphibolite horizons are common and nods, lenses and more regular dyke-like bodies of migmatite are sporadically distributed throughout . . . (he mentions) biotite-rich bands with a distinct schistose texture . . . there is no evidence of faulting . . . but jointing and fracturing is very prominent . . .

The most important passage in this report reads:

In underground workings biotite schist bands, sometimes containing feldspar porphyroblasts, are common and vary in size from a few centimetres to several metres in width. From a combined weathering, water seepage, and stability point of view these rock types are potentially the most dangerous in the project area, especially when feldspar porphyroblasts are present . . .

In the conclusions, Dr Matheson again mentions fracture zones, jointing, biotite schists and weathered feldspar as potentially dangerous rock types. A second report dated 1 May 1972 (after the rock falls had occurred), signed by J. D. Keppie, G. D. Matheson and S. Vrana of the Geological Survey reads '. . . the biotite schist bands are potentially the weakest rocks in the area, and split very readily parallel to the schistosity . . .' In the same report the authors presented a map of the area of the large excavation, which is

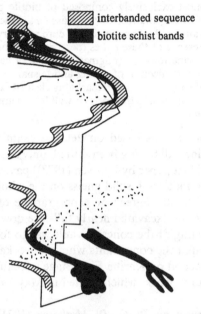

Fig. 16.9   Rock profile of the dangerous half of the Machine Hall, as drawn by Dr D. G. Mathseon of the Zambian Geological Survey. This shows how the black bands of biotite schists fold round other rocks, leaving dangerous 'noses'. (From Jaeger, 1973.)

partly reproduced on fig. 16.9. This figure clearly shows the very dangerous biotite schist bands and interbanded sequence along the rock roof and the equally dangerous folding of the bands near the haunches.

16.2.3   The progress of excavations and concreting of the great vault

The excavation was successful for about 30 m to 40 m from the west face of the excavation. Then difficulties were encountered in mid-July 1971 with major rock falls from the rock vault which over a period of about a year, up to August 1972, varied in weight from about a tonne to 25 tonnes. Most of the falls were located between chainage 32 m and chainage 90 m. The buttressing effect of the two vertical end walls probably contributed to this distribution of rock falls. The falls were located along a line nearly parallel to the axis of the vault and slightly offset on the southern (river) side of the axis. Consecutive rock falls occurred as excavation progressed. At the beginning of 1972, the contractor was ordered to proceed to anchor 3-m-long bolts in the rock. A systematic pattern was adopted for the bolting of the vault. Most bolts, also called 'pins' by the contractor, were epoxy grouted. A steel mesh was fixed to the bolts.

16.2.4 Comments by visiting geologists on the situation created by the rock falls

Rock falls occurred both in the bolted and in the unbolted areas. The heaviest rock fall occured in the bolted area and bolts could be seen still adhering to the fallen blocks.

The second report by geologists from the Geological Survey, Zambia, (dated 1 May 1972) confirmed the pessimistic interpretation of rock conditions given in Dr Matheson's early report (June 1971). The rock falls in the large machine hall excavation were considered to be of such gravity that the excavation work was halted. A panel of three experts, chaired by Brigadier John Edney and including Dr John Knill and Dr Charles Jaeger visited the site in August 1972. On August 18 a small rock fall which occurred during the previous night along the bolted and meshed South Wall could be seen lying on the floor.

In the meantime, as the situation on the site became tense, the Kariba North Bank cavern was visited by several geologists. According to reports by Lenssen in *New Civil Engineer* (8, 15 and 22 February 1973) the conclusions of the reports by Prof. J. G. C. Anderson (30 June 1972) and Prof. R. N. Shackleton (July 1972) confirm the views expressed by Dr Matheson and his colleagues from the Geological Service, Zambia. 'The igneous rock is a "gneiss complex" or "migmatite".' Two geological factors were blamed: the sub-horizontal foliation providing a potential structural defect and continuous layers of biotite schist parallel to the foliation. According to Prof. Shackleton, the falls are due 'to biotite schist, thickened in the hinges of folds tending to fall out, fractures associated with faulting, and some weathering along joints.' The visiting panel could confirm these views during their visit on site and *in situ* inspection.

16.2.5 The significance of jointing and of 'tension gashes'

In addition to the remarks in the previous paragraph, the geologists mentioned the very obvious main vertical jointing system, sub-parallel to the gorge, and therefore sub-parallel to the cavern main axis, and cutting through the gneiss folds. This system of joints does not compare with the jointing observed in sedimentary rock masses and it can be assumed that the joints are of tectonic origin. This seems confirmed by some open cracks, about 0·30 m wide, which could be observed at different levels, from the top rock surface, above the cavity, down to the lower levels of the pressure shafts and on the north face (mountain side) of the excavated cavity. Dr Matheson declared that he had observed similar cracks or 'tension gashes' at different locations along the Zambesi River valley. They can be explained by the structural geology of the river: the river valley was formed by subsidence, not by erosion, and the area next to it was under tension. He concluded that the whole rock mass where the large cavity has been excavated is not under compression in a horizontal direction perpendicular to the gorge.

From the point of view of rock mechanics, this decompression of the rock mass causes very unfavourable stress–strain patterns around the cavity. Furthermore, the stress pattern varies during the different stages of the excavation: tensile stresses can develop at the soffit of the rock vault and become a cause for rock falls, even in bolted areas. Depending on the extension of the tensile areas, short rock-bolts will not hold in areas damaged by tensile stresses. The 3-m-deep rock-bolts used at Kariba were obviously too short. Tensile stresses which would not be dangerous in solid rock may be very much so in biotite schists (see Jaeger, 1973).

16.2.6 The concrete vault

Inspection of the concrete vault, up to chainage 30 m, showed the concrete arch to be fissured. The pattern of fissures was consistent with stresses in an overstressed, deforming rock vault. The thickness of the concrete is 1 m at the arch crown, increasing to 2 m at the springings. It can be suggested that the 1 to 2-m-thick concrete is a facing for the 50 to 100-m-deep mass of rock deforming by creep around the excavation. Fissures on the concrete surface passing through the concrete mass are the projected image of strains inside the rock mass. (This is also the interpretation given by mathematicians analysing on a computer program the deformation of a concrete lined rock mass.) Further fissures, found to be developing in August, showed that the rock mass had not stabilized, although rock vaults of this size usually stabilize after one or two months. The small rock fall which occurred on 18 August 1972 confirmed that strains were still developing in the rock mass at that date.

Engineers expect strains and deformations to occur in sedimentary rock masses. What about igneous rocks? Lecturing in Zurich on the design and construction of the Innertkirchen underground power-station, Dr A. Kaech, the designer, mentioned large deformations of the rock mass inside the cavity of up to 20 cm. The excavation was in gneiss of good quality. At that time (in the middle of 1940) this remark was not well understood. Engineers were mainly concerned with the poorer quality of jointed, sometimes creeping, sedimentary rock masses. Today many designers are more suspicious of migmatic rock masses, some gneisses or gneiss complexes than of stratified jointed sedimentary rock masses, where dangerous rock situations are obvious and can more easily be put on a computer program.

16.2.7 Comments on the situation at Kariba North Bank machine hall excavation

(*i*) There was enough evidence available on the geology of the site to accept the explanation of the geologists that the presence of biotite schist had been a major factor in causing the rock falls, in spite of the efforts in using short rock bolts.

(*ii*) The fissures in the concrete vault were considered to be proof of the rock strains and deformations in the rock vault. Deformations of some amplitude should have been expected in a most complex igneous rock mass, where folded bands of gneiss were predominant. No measurements of such deformations had been made and there was no means on the site to do so. But there were other signs to suggest that, in August 1972, the rock vault had not completely settled and that the utmost care should be taken when resuming the downwards excavations. The decision of the consulting engineer was to finalize the work on bolting and netting the rock vault and to spray shotcrete on the whole rock surface not already protected with concrete. Excavation work in the downward direction was delayed by about nine months, which gave the rock mass time to attain equilibrium.

(*iii*) It was reasonable to assume that the rock mass to be excavated was decompressed. Finite element analysis of other underground cavities – assuming decompressed rock – had shown the danger of tensile stresses developing along the vertical walls. On the other hand, the very flat arch formed by the rock vault was quite unusual in such a very large underground excavation. For these two reasons the horizontal deformations to be expected along the very high flat vertical walls were unpredictable. High stress concentrations would also occur near the deep haunches at the base of the concrete arches.

(*iv*) From a far more general point of view, the rock falls at Kariba North Bank show the difficulty of correct prediction in rock mechanics. The consulting engineer had a first-hand knowledge of the rock on the North Bank. When building the left-hand abutment of the large Kariba arch dam he found that the rock was an excellent foundation rock, hard and watertight. The RQD – if it had been calculated – would have shown very high values, proving the good quality of the rock. The modulus of elasticity of the gneiss is high and so is its crushing strength. A first glance at these figures would have satisfied an expert on rock mechanics that the gneisses would be suitable for easy excavation. The tender documents were based on this assumption.

On the contrary more detailed geological inspection of the rock masses showed them to be dangerous. The gneisses contained folded schists. Some were to be found in a subhorizontal position at roof level. Other schist folds, near the haunches, caused a rock fall which had fatal results: a staff member was killed by this fall. It is most important for the complete analysis of this case, to note that Dr Matheson's report, signed before the first rock fall occured, clearly predicted the dangers.

The Kariba North Bank case shows the difficulty of establishing general rules in rock mechanics. An 'Engineering classification of rocks' as developed in sections 6.7 and 10.9 would have emphazised the high qualities of the rock mass. It was most suitable as a dam abutment. A few features, which expert geological insepection had detected, showed some weaknesses as a mass to be excavated. Design and excavation methods should have been better adapted to real rock conditions.

The next example on Waldeck II underground power-station shows how collaboration between geologists, designers and contractors can result in difficult situations being handled successfully.

16.3   The Waldeck II pumped storage power-station

16.3.1   General remarks on new rock consolidation methods

The 1957 and 1964/65 papers by L. von Rabcewicz on the 'new Austrian tunnelling method' were mainly concerned with the construction of tunnels and galleries. Very soon designers of large underground cavities investigated the possibility of similarly using cables for supporting wide rock vaults. Long before L. v. Rabcewicz's first paper, A. Kaech had used cables for reinforcing the walls of the Verbano underground machine hall and Talobre discusses the use of deep rock-bolts for supporting arched roofs excavated in weak rock masses, but the papers by the Austrian engineer were most important to the evolution of techniques. In 1968 Rescher published an extensive description of Veytaux underground machine hall, 136·50 m long, 30·50 m wide and 26·85 m high which had just been completed. He successfully used deep cables for supporting the rock vault. Several papers submitted to the Second International Congress on Rock Mechanics (1970) showed similar progress in the design of wide cavities excavated in Italy, France, Japan and elsewhere.

16.3.2   Waldeck II, principal data

The choice of Waldeck II (fig. 16.10) pumped storage power-station as a typical example of modern techniques is justified by the care taken by the designers to investigate all details.

In 1929, the Preussische Elektrizitäts-Aktiengesellschaft, Hanover (Germany), made the decision to build Waldeck I pumped storage station designed for an output of 140 MW. The Waldeck II pumped storage power-station built in 1969–73, at the same time as the Würgassen nuclear station, has a generating capacity of 440 MW (two sets of 220 MW each). The main data concerning the two stations are given in table 16.1.

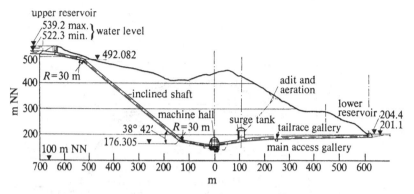

Fig. 16.10 Waldeck II underground pumped storage scheme (after Barth, 1972 and Jaeger, 1973).

Table 16.1

		Waldeck I	Waldeck II
Generating capacity	(MW)	140	440 (880)*
Pumping input	(MW)	96	460 (920)*
Number of machine sets		4	2 (4)*
Water reserve	(m³)	760 000	4 400 000
Mean head	(m)	296	329
Storage capacity	(MWh)	500	3 520

(* final design)

The upper reservoir is at level 522·30/539·20, the lower reservoir at level 201·10/204·40, but, as usual with vertical-axis pumping-generating sytems, the very high machine hall is located below that level, at 129·00/183·00, as can be seen on fig. 16.11. Because of the length of the tail-race pressure tunnel, a very large costly surge tank was required to absorb surges and water hammer waves.

16.3.3 Geological and geophysical exploration; finite element method analysis

In order to carry out geological exploration of the site of the underground works, exploratory galleries were driven (1300 m length) and 1091 m of structural core holes were drilled in the area of the machine hall cavern and 565 m in the area of the surge chamber, partly from the surface and partly from within the galleries. Borings drilled for testing cables (312 m) were also used for gathering geological information. Photoprobes were used to determine the directions of core drilling and television probes were used to carry out tectonic analysis of such factors as the degree of jointing and the fault orientation data.

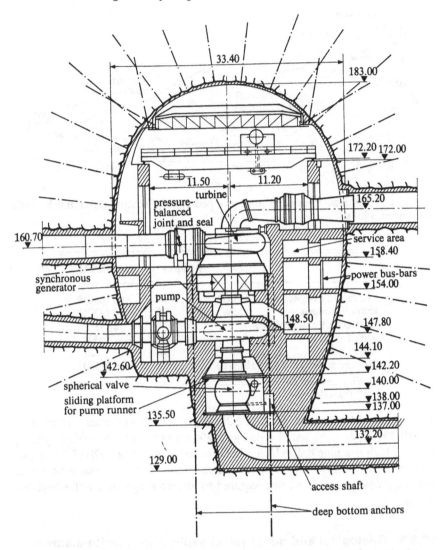

Fig. 16.11 Waldeck II underground pumped storage scheme (after Abraham, 1974).

The formation containing the machine hall cavern and the surge tank consists of lower Carboniferous schist–shales and greywake sandstones. More specifically, these materials occur mostly in the form of an alternating sequence of sand-banded dark schist–shale and fine to coarse greywake. The rock structure is crossed by number of joints. Tectonically, the strata belong to a large trough tending to the north-west and dipping north-east.

16.3.4 Geomechanical investigations

For the purpose of obtaining precise *in situ* information on the rock characteristics, a test cavern 11 m high, 9 m wide and 15 m long was excavated in
the area of the future machine hall excavation. Additionally a gallery was
specially equipped for measurements to be made during excavation. Rock
strains measured in the test cavern with extensometers were of the order of
2·15 mm whereas the rock convergence measured in the gallery attained 15
mm. Blasting caused a loosening of the rock on the test cavern walls which
went about 1·50 m deep. Deformations ceased within a few days of the
extensometers being installed.

Plate loading tests were carried out in the cavern. Because of the relatively
thin alternating bands of greywake and schists and the number of joints, only
limited importance could be given to local rock microstructure.

Table 16.2 summarizes the rock characteristics as obtained from rock-
mechanical investigations.

Table 16.2

	$E_{elastic}$	E_{total}
Schists	30 000	17 000 kg/cm²
Schistose sandstone	50 000	40 000 kg/cm²
Greywake	80 000	70 000 kg/cm²

Uniaxial crushing strength	
Schists	500 kg/cm²
Greywake	800 kg/cm²

k-value = 0·4
Locally measured maximum k-value 2·5

Depth of zone damaged by blasting 1·5 m

Friction factor parallel to joints $\phi = 20°$
    perpendicular to joints $\phi = 37°$

Cohesion, c, parallel to joints 1·5 kg/cm²
    perpendicular to joints 15 kg/cm²

Specific weight of rock 2·65 t/m³

Inclination of principal stress to vertical 20°

In situ tests have shown that ϕ and c depend essentially on the relative
direction of shear stresses to the direction of the joints. The thickness of the
weathered and weakened rock mass near rock faults caused especially low
values of ϕ and c. These facts had to be considered when carrying out the
finite element method calculations and also for the photoelastic model tests.

16.3.5   Testing the prestressing anchors

Five firms were invited to test their anchors at Waldeck II. The design of the anchors were checked for durability, protection against rusting, and against possible damage during transport and erection. Two anchors of each type were fixed inside a 3-m wide and 2·8-m high test gallery, one in the soffit, the other in a side wall. Two other anchors were introduced in the wall of the test cavern at different levels. The relative ease of introducing the anchors in the boreholes was estimated, the possible damage to the anchor protection checked. One anchor was tested for adherence. This anchor was cemented on only 1 m length to the rock and tested to rupture in order to get the maximum shear strength with rock.

Finally three types of anchors were selected:

The VSL anchor from Losinger (Switzerland),
The DW anchors from Dyckerhoff and Widmann (Germany) and
The PZ anchors from Polensky and Zollner (Austria).

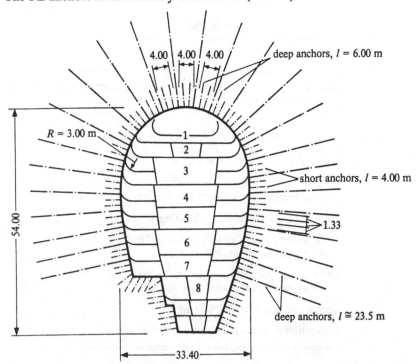

Fig. 16.12   Waldeck II; excavation phases and prestressing anchors.

The shear tests for DW 176 tonne anchors reached $\tau = 55\cdot5$ kg/cm². They were $\tau = 44\cdot7$ kg/cm² for the 170 tonne VSL anchors and $\tau = 15\cdot7$ kg/cm² for the 70 tonne PZ anchors. Table 16.3 gives some details about the types and numbers of anchors used for the machine hall and for the surge tank cavern.

Table 16.3

Type	VSL	VSL	DW	PZ	DW	PZ
Rupture	170	125	175	170	12	125 tonne
Permanent load	133	97	135	134	10	98 tonne
Number of wires	33 × 8 mm	23 × 8 mm	9 × 16 mm	37 × 7 mm	1 × 15 mm	24 × 7 mm
Anchorage length	4.5	4.5	4.5	3	0.5	3.0 m
Number in machine hall	688	64	202		4000	
Number in surge tank				45	1018	151

16.3.6 The stress–strain pattern (fig. 16.13)

Extensive research was carried out in order to determine the stress–strain pattern around the future excavation of the machine hall and the surge tank shaft. The finite element method was used for this analysis. Results were checked on photoelastic models. It was possible to introduce in the computer analysis the different moduli of elasticity, E, for the rock mass: 70 000 kg/cm² for the greywake, 20 000 kg/cm² for the schists and 50 000 kg/cm² for alternating layers of schists and greywake. Rock strains were calculated along the rock surface of the cavity and at depth of 18 m and 30 m. Inhomogeneity of the rock mass could also be considered when testing photoelastic models. Tests were carried out for k-values varying from 0·35 to 2·5. *In situ* tests in the test cavern had proved that small-scale jointing could be neglected in the calculations.

The finite element calculations showed that tensile stresses were developing along the vertical walls of the machine hall. (The same result has also been obtained when analysing other large excavations on the computer.) This led the designer to a complete change in the basic cavern design. The finite element method was also used for the analysis of the progress of the excavations. The calculations proved that extremely high and dangerous stress concentrations could be obtained during excavations unless any sharp corner was avoided. This led to an excavation method to be discussed in the next paragraph.

16.3.7 The design of the underground works (fig. 16.11)

The first designs for the Waldeck II machine hall were on conventional lines, with rigid concrete arch and sub-vertical walls (fig. 16.14). Conventional excavation methods were considered. Then the possibility of a design based on

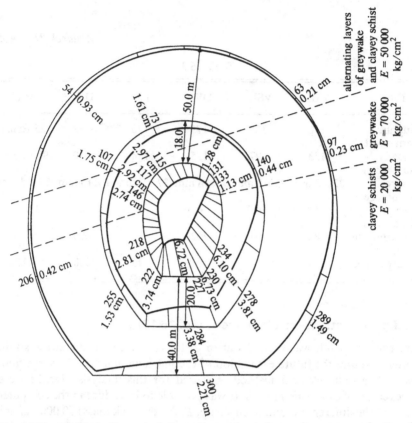

Fig. 16.13 Waldeck II; rock deformations about the cavity, measured at rock surface, at 18 m and at 30 m from the excavation boundary, as calculated by the finite element method (according to Zienkiewicz).

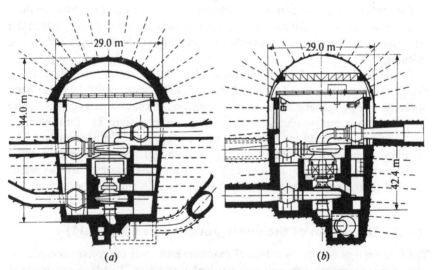

Fig. 16.14 Waldeck II; successive designs (after Abraham, 1974 and Siemens). (a) Conventional design with concrete arch and nearly vertical walls; (b) anchored cables supporting rock vault. (Figure 16.11 shows the final design as adopted by designer.) Cavern cross-sections of 33.40 m width and 46 m height.

the new Austrian tunnelling method (NATM) was investigated. Such a design consisted in supporting the rock vault and walls with deep anchors. It was then decided to use the 'observational method' advocated by R. B. Peck (1969). The design would be based on 'average' rock quality. Alternative designs would be worked out in detail for more severe rock conditions. Systematic measurements would check rock stresses and strains, rock deformations and, if there were signs of excessive strains, the design would be altered. Agreements on this method were reached by all concerned: the Preussische Elektrizität, the designers and the main contractor.

Few examples of very large excavations supported with cables and anchorages (NATM) were available, Veytaux (Switzerland) was a recent one and information from this site was very useful. For Waldeck II the concrete arch over the machine hall was abandoned, the rock vault had to be given a shorter radius and a nearly circular shape to decrease stresses. Another impotant problem concerned the tensile stresses along the vertical side walls of the excavation, which were apparent on the finite element method analysis. They could have been balanced by increasing the horizontal pressure on the walls with a higher density of anchorage cables. The designers found it more convenient to give an ovalized shape to the whole excavation (fig. 16.11).

The shape of the cavern had to be adapted to the mechanical equipment of a medium-head pumped storage plant, located in the centre of a large electric distribution net, and having to face sudden and extremely rapid change-overs from generating to pumping conditions and vice-versa. Such conditions called for two independent runners for the turbine and for the pump, located on the same very high vertical shaft. (Veytaux, with a much higher head has the runners arranged on a horizontal axis, which basically modifies the design of the cavern.) The problem was the location of the pump valves. A flat bottom of the machine hall causes high tensile stresses to develop under the machine sets. Some designers suggest a balancing of these stresses with the weight of the machines and the foundation concrete. For Waldeck II the pump valve was located under the pump, which allowed a better shaping of the cavern, as shown on fig. 16.11.

A possible arrangement for the transformers consists in locating them in an excavation parallel to the machine hall (Ruacana, on the Cunene River, South West Africa). Electrical connections between generating set and transformer are reduced in length. But the finite element method shows how the two large excavations react one on the other. For Waldeck II it was decided to locate the transformers in a cavern in line with the machine hall.

The whole design required a careful balance between hydraulic and mechanical requirements and the conditions imposed by the rock mass structure.

Similarly the whole excavation programme was checked on the computer. It was found that sharp corners caused very dangerous stress concentrations in excess of the rock crushing strength. Figure 16.12 shows the excavation programme which was finally worked out to satisfy stress–strain conditions and the possibility of anchoring cables in deep boreholes.

The pilot gallery, at the top of the cavern excavation was 6·50 m high and 17 m wide. Blasting occurred every 3 m. Immediately after blasting 2-m-deep small anchors were grouted using epoxy resins in small boreholes to support locally unstable blocks of rock. A 3 to 8-cm-thick shotcrete layer was consolidated with a wire mesh. Then 12-tonne, 6-m-deep anchors were grouted in a regular 1·33 × 1·50 m pattern. These anchors were part of the final rock support scheme and had to be protected against rusting in the same manner as the heavy anchors. These 170-tonne anchors, 19 m to 28·50 m deep could be cemented in the 116 mm boreholes and tensioned only after a length of 40 to 50 m pilot gallery had been excavated. The boreholes were drilled on a regular pattern 4·0 m × 3·0 m, but the directions of the boreholes were decided on the spot, according to the jointing and structure of the rock. Figure 16.15

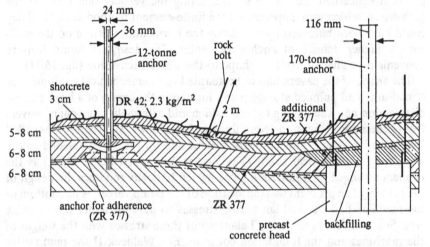

Fig. 16.15   Detail of prestressing anchor for Waldeck II (Abraham, 1974).

shows details of an anchorage. The possibility of requiring intermediate anchors to be fixed had been foreseen from the beginning, but there was no need for any local reinforcement. Table 16.3 gives the number of heavy 170 tonne and 130-tonne anchors which were required for supporting the main machine hall.

16.3.8   Checking rock strains and deformations. Measurements (figs. 16.16 and 16.17)

The observational method adopted for Waldeck II requires a constant checking of all rock deformations. In any case systematic measurements of rock strains and deformation is an obvious requirement in an excavation of such magnitude.

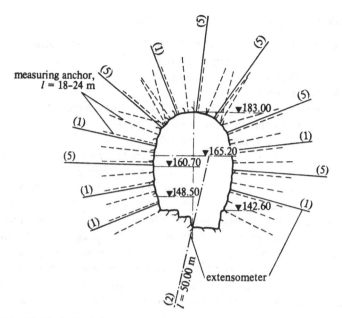

Fig. 16.16 Waldeck II: Measurement equipment in section 3–3 (after Abraham *et al.*, 1974). The whole equipment consists of 48 extensometers of 35–40 m length, 2 inductive deformation measuring cascades of 60 m and 50 m length and 96 anchor load cells, numbers in parenthesis = number of measuring points.

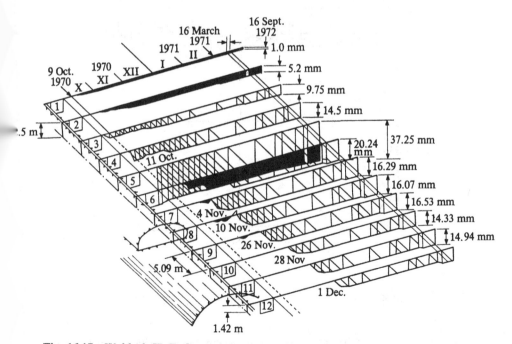

Fig. 16.17 Waldeck II. Deformations measured with the IDI cascade before and during excavation of the cavern crown and during subsequent enlargement (Abraham, 1974).

Strains and deformations were checked by a system consisting of 48 extensometers, 35 to 40 m long, 2 inductive deformation measuring cascades of 50 m and 60 m length and 96 anchor load cells. Extensometers were located in the crown of the cavern, some others were located in the walls. They measured rock deformations at the rock surface and also at 10 m and 25 m distance from the rock surface. Rock strains were negligible at 25 m distance inside the rock mass (fig. 16.16).

The inductive deformation measuring cascades were located in horizontal boreholes and were in position before the beginning of the excavations, permitting a detailed survey of strains from the start of excavation. The deformations measured with this method above the crown of the rock vault were less than 20 mm. It was possible to follow deformation as a function of time and as a function of the distance from the progressing heading (fig. 16.17). The stabilizing action of the anchors could equally be checked. Deformations of rock faults or of disturbed rock areas were checked with small 7-m-long extensometers. The forces in the anchors were controlled with the anchor load cells (as it had been done before in the Veytaux excavation).

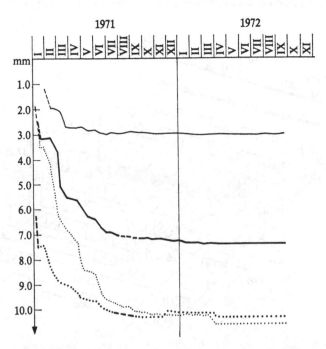

Fig. 16.18 Waldeck II. Deformations measured by four extensometers in the cavern crown (Abraham, 1974).

16.3.9 The surge tank

As a minimum suction head of 50 m was required for the pump impellers, there was practically no alternative to an underground power system, with large surge tank. The maximum flow rate was 154·4 m³/s. The surge tank, located on the downstream side of the turbines, is a vertical cylinder 37·40 m high (the overall height including bifurcation collar is 52·0 m) and 23·80 m in diameter. Such a large excavation had to be analysed very carefully. The NATM design was also adopted for it, with heavy anchors supporting the rock. The whole system was analysed on a computer using the finite element analysis. 196 PZ anchors (permanent load 125 tonne, test load 170 tonne) were required. The average support pressure developed by the anchors was 1·8 kg/cm² on the hemispherical roof and 1·1 to 1·3 kg/cm² on the sides of the vertical shaft.

Deformations of the rock mass were recorded with 30 to 35-m-long extensometers. One extensometer was fitted at the roof apex, before excavations started. It was a multiple instrument with fixes at 1 m, 5 m, 10 m, 25 m, and 35 m. The total deformation up to total rock consolidation was 7 mm at rock surface and 4·85 mm at a distance of 10 m from the rock surface. Tension in the PZ anchors showed an increase of 2 to 5%, except for a few anchors in which there was a decrease of up to 3%.

References

ABEL, J. F. (1970) Tunnel support. Where does geology come in? *Civil Engineering, A.S.C.E.*, February, 69–71.

ABRAHAM, K. H. (1972) Stand der Kavernenbautechnik für Pumpenspeicherwerke in Mitteleuropa. *Wasserwirtschaft*, **62**, No. 4, 103–10.

ABRAHAM, K. H. (1973) The surge chamber and tailrace tunnel for the Waldeck II plant. *Water Power*, **25**, No. 10, October, 395–92.

ABRAHAM, K. H. & PAHL, A. (1974) Planung und Berechnung grosser Felsbauten unter Berücksichtigung felsmechanischer Kontrollmöglichkeiten. *3rd Congress International Society of Rock Mechanics, Denver, Colorado*, 1–7 September.

ABRAHAM, K. H. & PAHL, A. (1976) Bauwerksbeobachtung der grossen Untertagsräume des Pumpspeicherwerkes Waldeck II. *Die Bautechnik*, **53**, No. 5, May, 145–55.

ABRAHAM, K. H. & PORZIG, R. (1973) Die Felsanker des Pumpspeicherwerkes Waldeck II. *Die Baumaschine und Bautechnik*, **20**, Nos. 6/7, June/July, 209–20; 273–85.

ADDINALL, E. & HACKETT, P. (1964) Tensile failure in rock-like materials. *Proceedings of the 6th Sympysium on Rock Mechanics, University of Missouri*, pp. 515–38.

AGTERBERG, F. P. (1974) *Geomathematics, Mathematical Background and Geo-science Applications*. Elsevier, Amsterdam.

ALEXANDER, L. G. (1960) Field and laboratory tests in rock mechanics. *3rd Australia–New Zealand Conference on Soil Mechanics and Foundation Engineering*.

AMBRASEYS, N. N. (1968) Dynamic sources and effects in rocks. In K. G. Stagg & O. C. Zienkiewicz (eds): *Rock Mechanics in Engineering Practice*. New York.

American Society for Testing and Materials (1966) *Testing Techniques for Rock Mechanics. Special Technical Publication No. 402*. Philadelphia.

AMPFERER, O. (1939) Über einige Formen der Bergzerreissung. *Sitzgsb. Akad. Wiss., Math. Nat. Klasse*. Vienna.

AMSTUTZ, E. (1950/53) Das Einbeulen von Schacht – und Stollenpanzerungen. *Schweiz. Bauzeitung*, **68**, 4 March 1950, 102 and **71**, 18 April 1953.

ANDERSON, D., PATON, T. A. L. & BLACKBURN, C. L. (1960) Zambesi hydro-electric development at Kariba, first stage. *Proc. Inst. Civil Eng.*, **17**, 39–60.

ANDERSON, H. W. & DODD, J. S. (1966) Finite element method applied to rock mechanics. *1st Congress International Society for Rock Mechanics*, **2**, 317–22.

ANDREA, C. (1926) *Der Bau langer tiefliegender Gebirgstunnel*. Berlin, 1926, (French translation, Lehman, Zurich, 1948).

ANDREA, C. (1961) Tunnels and tunnelling in Switzerland. In J. Comrie (ed.): *Civil Engineering Reference Book*, Vol. 2, pp. 383–95. London.

ANONYMOUS (1952) Santa Giustina II. *Water Power*, **4**, 324.

ANONYMOUS (1953) The Nechako–Kemano hydro-electric scheme in British Columbia. *Engineer*, **195**, 777, 811.

ANONYMOUS (1958) Static and dynamic test results, Oroville foundation. *U.S. Bureau of Reclamation: Concrete Laboratory Report* no. 6.

ANONYMOUS (1962) Progress report by the Task Committee on cement grouting of the Soil Mechanics and Foundations Division. *Proc. A.S.C.E.*, **88**, SM2, 49–98.

ANONYMOUS (1968) Boulons d'ancrages pour sols et roches. *L'Entreprise*, **64**, 47.

ANONYMOUS (1970) The Hongrin–Léman development. *Water Power*, May/June–July/August, 1970.

ANONYMOUS (1971) The Waldeck II station. *Water Power*, August, 275–85.

ANONYMOUS (1975) Earth curvature dictates route of Orange–Fish tunnel. *International Construction*, **14**, December.

ARGYRIS, J. H. (1965) Continua and discontinua. *Proceedings of Conference on Matrix Methods in Structural Mechanics, Air Force Institute of Technology.*

AVRAMESCO, A. (1963) Étude de certaines ondes élastiques dans les milieux stratifiés. *Annales Ponts et Chaussées*, **8**, 159.

BÄCHTOLD, J. (1952) Erfahrungen beim Bau des Kraftwerkes Handeck II. *Schweiz. Bauztg.* **70**, 573, 587, 612.

BALMER, G. B. (1952) A general analytical solution of Mohr's envelope. *Proc. Am. Soc. Testing and Materials*, **52**, 1260–71.

BANKS, J. A. (1955) The employment of prestressed techniques on Allt Na Lairige dam. *5th Congress on Large Dams, Paris*, **2**, 341–57.

BANKS, J. A. (1957) *Proc. Inst. Civil Eng.*, **6**, March.

BARBIER, R. (1974) La géologie au service des grands travaux. In Calembert, L. (ed.) La géologie de l'ingénieur. *Annales Soc. Geol. Belgique, Liège*, September, 9–13, 77–112.

BARRON, K. & TOEWS, N. A. (1963) Deformation around a mine shaft in salt. *Rock Mechanics Symposium, Queens University, Kingston, Ontario.*

BARTH, S. (1972) Felsmechanische Probleme beim Entwurf der Kaverne des Pumpspeicherwerkes Waldeck II. *Bautechnik*, **49**, No. 3, March, 73–83.

BARTON, N. R. (1971) A relationship between joint roughness and joint shear strength. *International Symposium on Rock Mechanics, Nancy, France*, Rep. 1/8.

BARTON, N., LIEN, R. & LUNDE, J. (1974) *Engineering classification of rock masses for the design of tunnel support.* Norwegian Geotechnical Institute, No. 106.

BAUDENDISTEL, M., MALINA, H. & MÜLLER, L. (1970) The effect of the geologic structure on the stability of an underground power house. *2nd Congress International Society of Rock Mechanics, Belgrade*, paper 4/56.

BELLIER, BERNÈDE, BOLLO, DUFFAUT *et al.* (1964) La déformabilité des massifs rocheux. Analyse et comparaison des résultats. *8th Congress on Large Dams, Edinburgh*, **1**, 287.

BELLIER, LONDE, DUFFAUT, HABIB *et al.* (1964) Les effets physico-chimiques de l'eau dans les appuis de barrage. *8th Congress on Large Dams, Edinburgh*, **1**, 329.

BELLIER, PUYO, LANGLOIS *et al.* (1964) Résultats des mesures d'auscultation effectuées sur les barrages de Lanoux et de Grandval. *8th Congress on Large Dams, Edinburgh*, **2**, 219–37.

BENSON, R. P., MURPHY, D. K. & McCREATH, D. R. (1969) Modulus testing of rock at the Churchill Falls underground powerhouse, Labrador. *ASME, Techn. public.* 477 (1969).

BENSON, R. P., CONLON, R. J., MERITT, J., JOLI-COEUR, J. & DEERE, D. U. (1971) Rock Mechanics at Churchill Falls. *Proceedings of the Symposium of the American Society of Civil Engineers, Phoenix, Arizona*, 407–86.

BENSON, R. P., KIERANS, T. W. & SIGVALDSON, O. T. (1970) *In situ* and induced stresses at the Churchill Falls underground power house, Labrador. *2nd Congress International Society for Rock Mechanics, Belgrade*, **2**, report 4/60.

BERGHINZ, C. (1965) Role of drawdown in the Vajont reservoir disaster. *Civil Engineering, A.S.C.E.*, March.

BÉRIGNY, C. (1832) *Mémoire sur un Procédé d'Injection.* Paris.

BERNAIX, J. (1966) Contribution à l'étude de la stabilité des appuis de barrages, Étude géotechnique de la roche de Malpasset. Ph.D. thesis, Paris.

BERNAIX, J. (1967) *Étude Géotechnique de la Roche de Malpasset*. Paris.

BERNAIX, J. (1975) Propriêtés des roches et massifs rocheux. *Industrie minérale*, special issue, 15 December, 45–69.

BERNOLD, J. (1970) 'Concrete lining in tunnel construction according to Bernold System.' Private publication, 1970.

BERRY, D. S. & SALES, T. W. (1960) An elastic treatment of ground movement due to mining. *J. Mech. Phys. Solids*, **8**, 280.

BERRY, D. S. & SALES, T. W. (1961) An elastic treatment of ground movement due to mining. *J. Mech. Phys. Solids*, **9**, 52.

BERRY, D. S. & SALES, T. W. (1962) An elastic treatment of ground movement due to mining. *J. Mech. Phys. Solids*, **10**, 73.

BEUSCH, E. & GYSEL, M. (1974) Felsmechanische Untersuchungen für den Sonnenberg Tunnel. *Schweiz. Bauzeitung*, **92**, No. 18, 2 May.

BEUSCH, E. & STUDER, W. (1972) Ausbruch und Verkleidung des Sonnenberg Tunnels. *Schweiz. Bauzeitung*, 7 September.

BIENIAWSKI, Z. T. (1969) Deformational behaviour of fractured rock under multi-axial compression. *International Conference on Structure, Solid Mechanics and Engineering Design in Civil Engineering Materials, Southampton.*

BIENIAWSKI, Z. T. (1970) Time-dependent behaviour of fractured rock. *Rock Mechanics, J. Int. Soc. Rock. Mech.*, **2**, no. 3.

BIENIAWSKI, Z. T. (1973) Engineering classification of jointed rock masses. *Trans. South Afr. Inst. Civil Engineers*, **15**, No. 12, December, 335–44.

BINNIE, G. M., CAMPBELL, J. G., EDDINGTON, R. H., FOGDEN, C. A. & JUMSON, N. H. (1958) The Dokan project: the dam. *J. Inst. Civil Engrs.*, **17**, 157.

BINNIE, G. M., GERRARD, R. T., ELRIDGE, J. C. *et al.* (1967). Mangla, part I: Engineering of Mangla; Mangla, part II: Construction of Mangla. *Proc. Inst. Civil Engrs.*, **38**, 343–575.

BISCHOF, R. (1975) Geotechnische Probleme beim Bau der Staumauer Ferden. *Mitteil. Schweiz. Gesell. Boden-und Fels Mechanik*, No. 91, 2/3 May.

BISHOP, A. W. (1966) The strength of soils as engineering materials. *Géotechnique*, **16**, 91.

BJERRUM, L. & JORSTAD, F. (1957) *Rockfalls in Norway*. Oslo (Norwegian Geotechnical Institute).

BLEIFUSS, D. J. (1949) Diversion tunnel and power conduit of Nantahala hydroelectric development. *Proc. A.S.C.E.* **75**, 1409.

BLYTH, F. G. H. (1967) *A Geology for Engineers*, 5th edition. London.

BOLLO, M. F. (1964) Résultats de recherches géotechniques sur une série de sites de barrages. *8th Congress on Large Dams, Edinburgh*, **1**, 405–23.

BONNET, G. & JOUANNA, P. (1975) Vers une approche énergétique des mécanismes de ruine des massifs poreux fissurés. *Industrie minérale*, special issue, 15 December, 21–7.

BOROWICKA, H. (1962) *Soil Mechanics – Rock Mechanics*. Vienna (Vienna University, Soil Mechanics Division).

BOURGUIGNON (1954) Contrôle de l'aération de la galerie de l'Isère-Arc. *Revue de l'Industrie Minérale*, **35**, no. 612.

BOUSSINESQ, V. J. (1885) *Application des potentiels à l'étude de l'équilibre et du mouvement des solides élastiques avec des notes étendues sur divers points de physique mathématique et d'analyse*. Paris.

BOWEN, R. (1975) *Grouting in Engineering Practice*. Applied Science Publishers, Barking, England.

BOZETTO, P. (1974) Premières constatations sur les mesures de déformation de l'usine de la Saussaz. *Industrie minérale*, special issue, 15 April.

466   References

BRACE, W. F. (1964) In W. R. Judd (ed.): *State of Stress in the Earth's Crust*, pp. 111–74. New York.

BRAY, J. W. (1967) A study of jointed and fractured rock. Parts I and II. *Rock Mechanics and Engineering Geology*, **5**, nos. 2, 3, 4.

British Standards Institution (1957) *Site Investigation*. Brit. Standards Institution 1957.

BROILI, L. (1967) New knowledge on the geomorphology of the Vajont slide slip surface. *Rock Mechanics and Engineering Geology*, **5**, no. 1.

BROWN, P. D. & SHAW, J. R. (1953) The *in situ* measurement of Young's modulus for rock. A dynamic source. *Géotechnique*, **3**, 283.

BUREAU OF RECLAMATION (1953) Physical properties of some typical foundation rocks. *Concrete Laboratory Report No. SP-39*. Denver.

BURGER, A. (1969) Influence de l'hétérogénéité des roches cohérentes sur leur perméabilité. *Bull. Techn. Suisse Romande*, **95**, 193–8.

BURMISTER, D. M. (1938) Graphical distribution of vertical pressure beneath foundations. *Trans. A.S.C.E.* **103**, 303–43.

BURMISTER, D. M. (1945) The general theory of stresses and displacements in layered systems. *J. Appl. Phys.* **16**, 16, 89, 126.

BÚRO, M. (1970) Prestressed rock anchors and shotcrete for large underground power-house. *Civil Engineering, A.S.C.E.*, pp. 60–4.

CALDERAÙ, A. (1968) Application des tirants précontraints. *Bull. techn. Suisse Rom.*, **94**, no. 18, 267–8.

CALEMBERT, L. (1974) (ed.) *La Géologie de l'Ingénieur*. Société géologique de Belgique, Liège.

CALOI, P. (1966) L'evento del Vajont nei suoi aspetti geodinamici. *Public. dell'Istit. Naz. di Geofisica, Roma, Annali di Geofisica*, **19**, no. 1.

CALOI, P. & SPADEA, C. (1960–63) *Serie di esperience geosismiche esseguite in sponda sinistra a monte della diga del Vajont*. Reports submitted to the Law Court at Aquila for the Vajont law suit, 4 February 1960, 3 May 1960, 10 February 1961, 5 October 1963. Not published.

CAMBEFORT, H. (1964) *Injection des Sols*. Paris.

CANDIANI, G. & GAVAZZI, P. (1964) Influence des déformations de la roche de fondation d'un barrage sur l'écran d'imperméabilisation. *8th Congress on Large Dams, Edinburgh*, **1**, 571.

CASAGRANDE, A. (1961) 1st Rankine Lecture: Control of seepage through foundations and abutments of dams. *Géotechnique*, **11**, 161–81.

CECIL, O. S. (1970*a*) Correlation of rock bolts, shotcrete support and rock quality parameters in Scandinavian tunnels. Ph.D. thesis, University of Illinois, Urbana, Ill.

CECIL, O. S. (1970*b*) Shotcrete support in rock tunnels in Scandinavia. *Civil Engineering, A.S.C.E.* pp. 74–9.

CERRUTI (1887) *Rendiconti della Reale Accademia dei Lincei*. Roma.

CHOLET, H. (1968) Propagation d'une fissure sollicitée normalement à son plan. *Revue de l'Industrie Minérale*, special issue 15 May, pp. 48–57.

CHUNNETT, E. R. P. (1976) Ruacana Hydro-Power Scheme – rock engineering studies. *Proceedings of the Symposium on Exploration for Rock Engineering, Johannesburg*, November 1976.

CLARK, G. B. (1966) Deformation of moduli in rocks. In *Testing Techniques for Rock Mechanics*. Am. Soc. Testing and Materials Publ. No. 402, pp. 133–74. Philadelphia.

CLOUGH, R. W. (1960) The finite element in plane stress analysis. *Proceedings of 2nd A.S.C.E. Conference on Electronic Computation, Pittsburgh*.

COATES, D. F. (1964) Classification of rocks for mechanics. *Int. J. Rock Mech., Min. Sci.*, 421–9.

COATES, D. F. & GYENCE, M. (1966) Plate-load testing on rock for deformation and strength properties. In *Testing Techniques for Rock Mechanics*. Am. Soc. Testing and Materials Publ. No. 402. Philadelphia.

COATES, D. F. & PARSONS, R. D. (1966) Experimental criteria for classification of rock substances. *Int. J. Rock Mech., Min. Sci.*, **3**, 181–9.

COLBACK, P. S. B. & WILD, B. L. (1965) *The influence of moisture content on the compressive strength of rock*. South African C.S.I.R. Publication.

COLLIN, A. (1846) *Landslides in Clay*. Trans. W. R. Schriever, 1956. Toronto.

COMMISSIONE PARLAMENTARE D'INCHIESTA SUL DISASTRO DEL VAJONT (1965) *Relazione finale. Atti Parlamentari, Senato della Republica.*

COMRIE, J. (ED.) (1961) *Civil Engineering Reference Book*, 2nd edition, 4 vol. London.

COOKE, J. B. (1959) Haas hydro-electric power project. *Trans. A.S.C.E.* **124**, 989.

CORDING, E. J. & DEERE, D. U. (1972) Rock tunnel support and field measurements. *North American Rapid Excavation Conference, Chicago*, vol. 1, pp. 567–600.

CORDING, E. J., HENDRON, A. J. & DEERE, D. U. (1971) Rock engineering for underground caverns. *Proceedings of the Symposium of the American Society of Civil Engineers, Phoenix, Arizona.*

COYNE, A. (1936) Le barrage de Molare. *Annales des Ponts et Chaussées*, No. 1.

CUNDALL, P. A. (1971) A complete model for simulating progressive, large scale movements on blocky rock systems. *Symposium of the International Society for Rock Mechanics, Nancy, France.*

CUNDALL, P. A. (1974) *Rational design of tunnel supports. A computer model for rock mass behavior using interactive graphics for input and output of geometrical data.* Tech. Report MRD, 2–74. Missouri River Division, Corps of Engineers, Omaha, Nebraska.

DACHLER, R. (1936) *Grundwasserströmung*. Vienna.

DAEMEN, J. J. K. (1975) *Tunnel support loading caused by rock failure*. Tech. Report MRD, 3–75. Missouri River Division, Corps of Engineers, Omaha, Nebraska.

DAEMEN, J. J. K., FAIRHURST, C. & STARFIELD, A. M. (1969) Rational design of tunnel supports. *2nd Symposium on rapid excavation, Sacramento, California*, 16/17 October.

DAEMEN, J. J. K. & FAIRHURST, C. (1972) Communication at the *Rock Mechanics Symposium, Lucerne, Switzerland*, September.

DAGNAUX, J. P., LAKSHANAN, J. & GARNIER, J. C. (1970) Seismic vibration testing of chalk subjected to laboratory and *in situ* stresses. *Proceedings of the 2nd Congress of the International Society for Rock Mechanics, Belgrade*, report 4/57.

D'ALBISSIN, M. (1968) Données de la luminescence dans l'analyse de la déformation des roches. *Revue del'Industrie Minérale*, special issue 15 May, pp. 29–37.

D'ALBISSIN, M. (1969) Essai de synthèse sur la réaction des cristaux et de leurs agrégats aux contraintes mécaniques. *Revue de Géographie physique*, **11**, 533–78. (Extensive bibliography.)

DAL PIAZ, G. (1948) *Sulla struttura geologica della valle del Vajont agli affeci smottamenti dei fianchi che possono derivare del projettato invaso e delle oscillazioni del livello del lago.* Report submitted to the Law Court at Aquila for the Vajont law suit, 21 December; appendix 18 November 1953. Not published.

DARCY, H. (1856) *Les foutaines publiques de la Ville de Dijon.* Paris.

DAVIN, M. (1956) Études statistiques sur la résistance des corps primatiques soumis à des champs de contraintes uniformes. *Annales des Ponts et Chaussées*, no. 6.

DAVIN, M. (1957) Études statistiques sur la résistance des corps primatiques soumis à des champs de contraintes uniformes. *Annales des Ponts et Chaussées*, nos. 1, 2.

468   *References*

DEARMAN, w. R. (1974) The characterisation of rock for Civil Engineering practice in Britain. In Calembert, L. (ed.) *La Géologie de l'Ingénieur.* Liège, Belgium.

DEERE, D. U. (1961) Subsidence due to mining – a case history from the Gulf Coast region of Texas. *Proceedings of the 4th Rock Mechanics Symposium, Pennsylvania State University.*

DEERE, D. U. (1963) Technical description of rock cores for engineering purposes. *Rock Mech. Eng. Geology,* **1,** 18–22.

DEERE, D. U. (1968) Geologic considerations. In K. G. Stagg & O. C. Zienkiewicz (eds.): *Rock Mechanics in Engineering Practice.* New York.

DEERE, D. U. & MILLER, R. P. (1966) Engineering classification and index properties for intact rock. *Air Force Weapons Laboratory Technical Report No. A FWL-TR-65-116.* Kirtland, New Mexico.

DEERE, D. U., HENDRON, A. J. JR, DALTON, F. D. & CORDING, E. J. (1966) Design of surface and near-surface construction in rock. *8th Symposium on Rock Mechanics, Minnesota.*

DEL CAMPO, A. & PIQUER, J. S. (1962) Cimentación de presas en terrenos homogeneos e isotropos. *Laboratorio de reologia y geotecnia, Publication no. 11.* Madrid.

DELGADO, RODRIQUES J. (1975) *Alterabilité des roches schisteuses.* Lab. Nac. de Engenharia Civil, No. 464, Lisbon.

DENKHAUS, H. G. (1958) The application of the mathematical theory of elasticity to problems of stress in hard rock at great depth. *Assoc. Mine Managers of South Africa. Papers and Disc.* 271–310.

DENKHAUS, H. G. (1964) Critical review of strata movement, theories and their application to practical mining problems. *J. South Africa Mining & Metallurgy,* **64.**

DENKHAUS, H. G. (1970) Discussion on Theme I. Engineering geological considerations. *Proceedings of the International Symposium on large underground openings, Oslo,* pp. 125–6.

DESCOEUDRES, F. & RECHSTEINER, G. (1973) Etude de corrélations entre géologie, les propriétés mécaniques et la forabilité des roches de Crespera Gemmo. *Schweiz. Bauzeitg.,* **91,** No. 12, March.

DIETHELM, W. (1974) Geologie und Felsmechanik im Untertagbau. *Mitteil. Schweiz. Gesellschaft für Boden und Felsmechanik,* No. 90, 9 November.

DODDS, R. K. (1966) Measurement and analysis of rock physical properties on the Dez project, Iran. In *Testing Techniques for Rock Mechanics.* Am. Soc. Testing and Materials Publ. No. 402. Philadelphia.

DOMINY, F. E. & BELLPORT, B. P. (1965) *Morrow Point Dam and Power Plant Foundation Investigations.* Denver (U.S. Department of the Interior, Bureau of Reclamation).

DONATH, F. A. (1964) Strength variation and deformational behaviour in anisotropic rock. *Conference State of Stress, Santa Monica,* pp. 281–97.

DONATH, F. A. (1966) A triaxial pressure apparatus for testing of consolidated or unconsolidated materials subjected to pore pressure. In *Testing Techniques for Rock Mechanics.* Am. Soc. Testing and Materials Publ. No. 402, pp. 41–51.

DREYER, W. (1975) *The Science of Rock Mechanics,* Part I. *Strength properties of rocks.* Trans Tech Publications, Cleveland, USA.

DUFFAUT, P. (1968) Effet d'échelle dans l'écrasement de blocs de forme irrégulière. *Revue de l'Industrie Minérale,* special issue 15 May, pp. 62–7.

DUFFAUT, P. & PIRAUD, J. (1975) Soutènement des tunnels profonds, autrefois et aujourd'hui. *Industrie Minérale,* special issue, 15 December, pp. 28–44.

DUNCAN, N. (1965) Geology and rock mechanics in civil engineering practice. *Water Power,* **17,** 25–32, 63–8, 99–102, 145–52, 192–4, 225–9.

References 469

DUNCAN, N. (1966) Rock mechanics and earthwork engineering. *Muck Shifter*, June–November.
DUNCAN, N. (1967) Rock mechanics and earthwork engineering. *Muck Shifter*, January, February.
DUNCAN, N. (1969) *Engineering Geology and Rock Mechanics*, London.
DUNCAN, N. & SHEERMAN-CHASE, A. (1965) Rock mechanics in civil engineering works. *Civil Engineering, London*, December.
DUNCAN, N. & SHEERMAN-CHASE, A. (1966) Rock mechanics in civil engineering works. *Civil Engineering, London*, January–June.
DUNCAN, N., DUNNE, M. H. & PETTY, S. (1968) Swelling characteristics of rock. *Water Power*, 20, 185–92.
ECKENFELDER, G. V. (1952) Spray hydro-electric power development. *J. of Engineering, Inst. of Canada*, 35, 288.
EGGER, P. (1973) *Einfluss des post-failure Verhaltens von Fels auf den Tunnelbau.* Institut für Bodenmechanik, University of Karlsruhe, No. 57.
EINSTEIN, H. H., BAECHER, G. B. & HIRSCHFELD, R. C. (1970) The effect of size on strength of a brittle rock. *Proceedings of 2nd Congress, International Society of Rock Mechanics, Belgrade.* Report 3/2.
ELEKTRO-WATT (1959) Erfahrungen beim Betrieb der Kraftwerke Mauvoisin. *Schweiz. Bauztg.*, 77, no. 39.
EMERY, C. L. (1964) Strain energy in rocks. *Proceedings of the International Conference on State of Stress in the Earth's Crust*, pp. 235–69. New York.
ENERGIEVERSORGUNG OSTBAYERN A. G. (1957) *Das Pumpspeicherwerk Reisach-Rabenleite*, Regensburg.
ENERGIEVERSORGUNG OSTBAYERN A. G. (1961) *Das Pumpspeicherwerk Tanzmühle.* Regensburg.
ENTE NAZIONALE PER ENERGIA ELETTRICA (1964) *Relazione sulle cause che hanno determinato la frana nel serbatorio del Vajont.* Rome.
EPSTEIN, B. (1948) Statistical aspects of fracture problems. *J. Appl. Phys.*, 19, 140–7.
ESTÈVES, J. M. (1970) Application of apparent electrical resitivity maps to the study of dam sites. *1st International Congress, International Association of Engineering Geology, Paris.*
EVISON, F. F. (1953) An improved electromechanical seismic source tested in shattered rock. *New Zealand Sci. Techn. Bull.*, 35, no. 4.
EVISON, F. F. (1956) The seismic determination of Young's modulus and Poisson's ratio for rocks *in situ*. *Géotechnique*, 6, September.
FAIRHURST, C. (1964) On the validity of the Brazilian test for brittle materials, *Int. J. Mech. Mining Sci.*, 1.
FANELLI, M. & RICCIONI, R. (1970/1973) Calcoli svolti per l'interpretazione delle misure di spostamento durante l'excavazione della Centrale in caverna del Lago Delio. *X Convengo di geotecnica, Bari, Italy, 1970* (also *ISMES, Bergamo*, No. 58, 1973).
FARMER, I. W. (1968) *Engineering Properties of Rocks.* London.
FARRAN, J. & THENOZ, B. (1965) L'altérabilité des roches, ses fractures, sa prévision. *Annales del'I.T.B.T.P.*, no. 215.
FENNER, R. (1938) *Untersuchungen zur Erkenntnis des Gebirgsdruckes.* Santiago and Essen.
FERRANDON, J. (1948) Les lois de l'écoulement de filtration. *Génie Civil*, 125.
FÖPPL, L. (1944) *Drang und Zwang*, vol. 2. Berlin.
FÖPPL, L. (1957) Elastische Spannungszustände in Körpern mit ebenen Schnitten. *Geologie und Bauwesen*, 23, no. 1.
FOSTER, C. R. & AHLVIN, R. J. (1954) Stresses and deflections induced by a uniform circular load. *Proc. Highway Research Board*, 33.

470   References

FRANKLIN, J. A. & CHANDRA, R. (1972) The slake durability test. *Int. J. Rock Mech. Min. Sci.*, **9**.

FRENCH COMMITTEE OF EXPERTS (1960) *Rapport Définitif*. (Final report on the failure of the Malpasset arch dam, four volumes, signed by Gosselin, Olivier-Martin, Calvet, Duffaut, Diserens & Talureau.) Not published.

FREY BAER, O. (1944) Die Berechnung der Pfeiler aufgelöster Staumauern. *Schweiz. Bauztg.*, **123**, no. 9.

FUMAGALLI, E. (1966a) Equilibrio geomeccanico del banco di sottofondazione alla digha del Pertusillo. *ISMES Publication No. 31*. Bergamo.

FUMAGALLI, E. (1966b) Stability of arch dam rock abutment. *1st Congress International Society for Rock Mechanics, Lisbon*, **2**, 503.

FUMAGALLI, E. (1967) Stability of arch dam rock abutments. *ISMES Publication No. 32*. Bergamo.

FUMAGALLI, E. (1968) Model simulation of rock mechanics problems. In K. G. Stagg & O. C. Zienkiewicz (eds.): *Rock Mechanics in Engineering Practice*. New York.

FUMAGALLI, E. (1973) *Verification par modèles des revêtements des tunnels*. ISMES, Bergamo, public. No. 55.

FUMAGALLI, E. & CAMPONUOVO, H. F. (1975) *Nota on alcune esperienze di modellazione di frane di roccia eseguite al 'ISMES'*. ISMES, Bergamo, public. No. 71, 1975.

GAYE, F. (1972) Efficient excavation, cutting head design of hard rock tunnelling machine. *Tunnels and Tunnelling*, January/March.

GIBB, COYNE, & SOGEI (1962) The Kariba dam abutment investigations. *Rhodesian Engineer*, January.

GICOT, H. (1955) Influence du sol de fondation sur les déformations du barrage de Rossens. *5th Congress on Large Dams, Paris*.

GICOT, H. (1961) Conceptions et techniques de quelques barrages-voûtes suisses. *Cours d'Eau et Énergie*, **6**, 194–205.

GICOT, H. (1964) The deformation of the Rossens arch dam during fourteen years service. *8th Congress on Large Dams, Edinburgh*, **2**, 419–29.

GIGNOUX, M. & BARBIER, R. (1955) *Géologie des Barrages*. Paris.

GILG, B. & DIETLICHER, E. (1965) Felsmechanische Untersuchungen an der Sperrstelle Punt dal Gall. *Schweiz. Bauztg.*, **83**, no. 43.

GIUDICI, F. & SEMENZA, E. (1960) *Studio geologico sul serbatoio del Vajont*. Report submitted to the Law Court of Aquila for the Vajont law suit, June. Not published.

GLOSSOP, R. (1960) The invention and development of injection processes. Part I: 1802–1850. *Géotechnique*, **10**, 91–100.

GLOSSOP, R. (1961) The invention and development of injection processes. Part II: 1850–1860. *Géotechnique*, **11**, 225–74.

GLÖTZL, F. (1957) Statische Betrachtungen zur Injektionsvorspannung. *Das Pumpspeicherwerk Reisach-Rabenleite*. Regensburg.

GLOVER, R. E. (1957) Arch dams: review of experience. *Proc. A.S.C.E. (Power Div.)*, PO 2, paper 1217.

GOFFI & OBERTI (1964) Communication. *15th Symposium on Rock Mechanics, Salzburg*.

GOLSER, J., HÄCKL, E. & JÖSTL, J. (1977) Tunnelling in soft Ground. *International Construction*, December.

GRAMBERG, J. Rock mechanics – the axial cleavage fracture. *Report No. 1.07, Laboratory of Mining, Technological University of Delft*.

GRANT, L. F. (1964) Grouting evaluation. *Proc. A.S.C.E. (Soil Mech. & Found. Div.)*, **90**, SM 1, 63–92.

GRIFFITH, A. A. (1921) The phenomenon of rupture and flow in solids. *Phil. Trans. Roy. Soc.*, A **228**, 163–97.

GRIFFITH, A. A. (1924) Theory of rupture. *Proceedings of the 1st International Congress on Applied Mechanics, Delft*, pp. 55–63.

GRIGGS, D. T. (1936) Deformation of rocks under high confining pressures. *J. Geology*, **44**, 541–7.

GRUNDY, C. F. (1955) The treatment by grouting of permeable foundations of dams. *5th Congress on Large Dams, Paris*.

GRUNER, E. (1963) Dam disasters. *Proc. Inst. Civil Engineers*, **24**, 47–60.

GSTALDER, S. & MARTY, G. (1968) Relation entre résistance à la compression simple et la dureté en poinçennage des roches. *Revue de l'Industrie Minérale*, special issue 15 May, pp. 74–81.

GUTHRIE BROWN, J. (1970) ICOLD: its history and activities. *Water Power*, **22**, nos. 5/6, May/June.

HABIB, P. (1950) Determination of the elastic modulus of rocks *in situ*. *Annales Inst. Techn. du Bâtiment, Paris*, September.

HABIB, P. (1968) La recherche de mécanique des roches au Laboratoire de Mécanique des Solides de l'École Polytechnique. *Revue de l'Industrie Minérale*, 15 December, pp. 31–6. Introduction à la fissuration des roches. *Revue de l'Industrie Minérale*, special issue 15 May, pp. 5–15.

HABIB, P. & BERNAIX, J. (1966) The fissuration of rocks. *1st Congress International Society for Rock Mechanics, Lisbon*, **1**, 185.

HABIB, P., BERNÈDE, J. & CARPENTIER, L. (1965) Résultats des mesures de contraintes effectives dans divers souterrains en France. *Annals Inst. Techn. Bât. et Travaux Publics*, **18**, no. 210.

HABIB, P. & VOUILLE, G. (1966) Sur la disparition de l'échelle aux hautes pressions. *C.R. Acad. Sci., Paris*, **262**, B, 715–17.

HABIB, P., VOUILLE, J. & AUDIBERT, P. (1965) Variation de la vitesse du son dans les roches et les sables soumis à des contraintes élevées. *C.R. Acad. Sci., Paris*, **260**, 4909–11.

HAEFELI, R. (1967a) Kriechen und progressiver Bruch in Schnee, Boden, Fels und Eis. *Schweiz. Bauztg.*, nos. 1, 2.

HAEFELI, R. (1967b) Zum progressiven Bruch in Schnee, Boden, Fels und Eis. *Rock. Mech. Eng. Geology*, **5**, no. 1.

HAMROL, A. (1961) A quantitative classification of the weathering and weatherability of rocks. *Proceedings of the 5th International Conference of Soil Mechanics and Foundation Engineering, Paris*, **2**, 771–4.

HAMROL, A. (1962) A quantitative classification of the weathering and weatherability of rocks. *Lab. Nacional de Engenharia Civ. Publ. No. 142*. Lisbon.

HANDIN, J. & HAGER, R. (1957) Experimental deformation of sedimentary rocks under confining pressure. Tests at room temperature on dry samples. *Am. Soc. Petroleum Geologists Bull.*, **41**, 1.

HANDIN, J., HIGGS, D. V., LEWIS, D. R. & WEYL, P. K. (1957) Effects of gamma radiation on the experimental deformation of calcite and certain rocks. *Geol. Soc. Am.*, **68**, 1203–24.

HANNA, T. H. (1973) *Foundation Instrumentation*. Trans Tech Publications, Cleveland, USA.

HARDY, H. R. (1959) Time-dependent deformation and failure of geologic materials. *Quarterly J. of the Colorado School of Mines*, **54**, 135–77.

HARDY, H. R. JR (1966) A loading system for the investigation of the inelastic properties of geologic materials. In *Testing Techniques for Rock Mechanics*. Am. Soc. Testing & Materials Publ. No. 402. Philadelphia.

472   References

HARRISON, J. V. & FALCON, N. L. (1937) The Saidmarreh landslip, south-west Iran. *Geographical Journal*, **89**.

HARTMANN, B. E. (1966) *Rock Mechanics Instrumentation for Tunnel Construction*. Wheatridge, Colorado.

HARZA, R. D. & EDBROOKE, R. F. (1960) Design of Karadj hydro-electric project. *Proc. A.S.C.E. (Power Div.)*, **86**, PO 4.

HAST, N. (1958) The measurement of rock pressure in mines. *Sveriges geol. Undersokn. Arsbok. Ser. C*, **52**, no. 3.

HAST, N. (1969) The state of stress in the upper part of the earth crust. *Tectonophysics*, 169–211.

HAST, N. & NILSSON, T. (1964) Recent pressure measurements and their implications for dam building. *8th Congress on Large Dams, Edinburgh*, **1**, 601.

HAYASHI, M. (1966a) Strength and dilatency of brittle jointed mass. The extreme value stochastics and anisotropic failure mechanism. *1st Congress International Society of Rock Mechanics, Lisbon*, **1**, paper nos. 3–12.

HAYASHI, M. (1966b) A mechanism of stress distribution in the fissured foundation. *1st Congress International Society of Rock Mechanics, Lisbon*, **2**, 509.

HAYASHI, M. & HIBINO, S. (1968) Progressive relaxation of rock masses during excavation of underground cavity. *International Symposium on Rock Mechanics, Madrid*.

HAYASHI, M. & HIBINO, S. (1970) Visco-plastic analysis on progressive relaxation of underground excavation works. *2nd Congress International Society for Rock Mechanics. Belgrade*, report 4/25.

HAYASHI, M., KITAHARA, Y. & HIBINO, S. (1969) Time-dependent stress analysis in under-ground structure in visco-plastic rock masses. *International Symposium on the Determination of Stresses in Rock Masses, Lisbon*.

HEILAND, C. A. (1940) *Geophysical Explorations*. New York.

HEIM, A. (1878) *Mechanismus der Gebirgsbildung*. Basle.

HEIM, A. (1905) Geologische Nachlese, Tunnelbau und Gebirgsdruck. *Vierteljahrschrift der naturforsch. Gesellschaft, Zürich*, **50**.

HEIM, A. (1912) Zur Frage der Gebirgs- und Gesteinsfestigkeit. *Schweiz. Bauztg.* **50**, February.

HEIM, A. (1932) Bergsturz und Menschenleben. *Vierteljahrschrift der naturforsch, Gesellschaft. Zürich*, **77**.

HELLER, S. R. JR, BROCK, J. S. & BART, R. (1958) The stress around a rectangular opening with rounded corners in a uniformly loaded plate. *Transactions of the 3rd U.S. Congress on Applied Mechanics*.

HENDRON, A. J. (1968) Mechanical properties of rock. In K. G. Stagg & O. C. Zienkiewicz (eds.): *Rock Mechanics in Engineering Practice*. New York.

HEUZÉ, F. E., GOODMAN, R. E. & BORNSTEIN, A. (1970) Numerical analysis of deformability tests in jointed rock—finite element analysis. *Rock Mechanics, J. Int. Soc. Rock Mech.*, **2**, no. 3.

HILL, R. (1950) *The Mathematics Theory of Plasticity*. Oxford.

HILTSCHER, R., FLORIN, G. & STRINDELL, L. (1966) Arbeitsanleitung zur spannungs-optischen Messung ebener Spannungszustände mit Polariskop und lateral Extensometer. *Bautechnik*, **43**, 41–5; 91–5.

HOBBS, D. W. (1970) The behaviour of broken rock under triaxial compression. *Int. J. Rock. Mech. Min. Sci.*, **7**, no. 2.

HOEK, E. (1968) Brittle failure of rock. In K. G. Stagg & O. C. Zienkiewicz (eds.): *Rock Mechanics in Engineering Practice*. New York.

HOEK, E. (1970) Estimating the stability of excavated slopes in open-cast mines. *Trans. Inst. Min. Metall., London*, **79**, A.109–20.

HOEK, E. (1972) *Rock slope engineering.* Imperial College, London, Rock Mech. Progress Report No. 8, July.

HOEK, E. (1973) Methods for the rapid assessment of the stability of three-dimensional rock slopes. *Quarterly J. of Engineering Geology,* **6**, no. 2.

HOEK, E. & BIENIAWSKI, Z. T. (1965) Brittle fracture propagation in rock under compression. *Int. J. Fracture Mech.,* **1**, no. 3, 137–55.

HOEK, E. & BRAY, J. W. (1973) *Rock Slope Engineering.* Institute of Mining and Metallurgy, London.

HOEK, E. & LONDE, P. (1975) Travaux de surface au rocher. *Industrie Minérale,* special issue, 15 December, 70–110.

HOLMES, A. (1965) *Principles of Physical Geology.* 2nd edition. Nelson, London.

HOSKINS, G. (1931–2) The construction, testing and strengthening of a pressure tunnel for the water supply of Sydney, N.S.W. *Proc. Inst. Civil Engineers,* **234**, 25.

HOUPERT, R. (1968) La résistance à la rupture des granites. *Revue de l'Industrie Minérale,* special issue 15 May, pp. 21–3.

HOWLAND, R. C. J. (1934) Stresses in a plate containing an infinite row of holes. *Proc. Camb. Phil Soc.,* pp. 471–91.

HUCKA, V. (1965) A rapid method for determining the strength of rocks *in situ.* *Int. J. Rock. Mech. Min. Sci.* July.

IKODA, K. (1970) A classification of rock conditions for tunnelling. *Proceedings of the Congress, International Society of Engineering Geology, Paris.*

IRMAY, S. (1964) *Theoretical Models of Flow Through Porous Media.* Paris.

JAECKLIN, F. P. (1965*a*) Felsmechanische Grossversuche. *Schweiz. Bauztg.* **83**, no. 15.

JAECKLIN, F. P. (1965*b*) Felsmechanik im Tunnelbau. *Schweiz. Bauztg.* **83**, no. 27.

JAECKLIN, F. P. (1966) Rock tunnelling and optimum shape. *1st Congress on Rock Mechanics, Lisbon,* **2**, 397–403.

JAEGER, C. (1933) *Théorie Générale du Coup de Bélier.* Paris.

JAEGER, C. (1948*a*) Water hammer effects in power conduits. *Civil Engineering & Public Works Review,* **43**, nos. 500–3.

JAEGER, C. (1948*b*) Underground hydro-electric power stations. *Civil Engineering, London,* December.

JAEGER, C. (1949*a*) Underground hydro-electric power stations. *Civil Engineering, London,* January.

JAEGER, C. (1949*b*) *Technische Hydraulik.* Basle.

JAEGER, C. (1950) Modern trends in arch dam construction and design. *Civil Engineering, London,* **45**, 526–31.

JAEGER, C. (1951) Modern trends in arch dam construction and design. *English Electric Journal,* **12**, no. 4.

JAEGER, C. (1955*a*) Present trends in the design of pressure tunnels and shafts for underground hydroelectric power stations. *Proc. Inst. Civil Engineers,* March, 1953, pp. 116–200. Correspondence on the paper: *Proc. Inst. Civil Engineers,* July 1955, pp. 545–96. See also *Water Power,* February–May 1955.

JAEGER, C. (1955*b*) The new techniques of underground hydro-electric power stations. *English Electric Journal,* **14**, 3–29.

JAEGER, C. (1956) *Engineering Fluid Mechanics.* Glasgow.

JAEGER, C. (1958/64) Arch dams. In J. Guthrie Brown (ed.): *Hydro-electric Engineering Practice.* Glasgow.

JAEGER, C. (1961*a*) Hydraulic power plants. In J. Comrie (ed.): *Civil Engineering Reference Book,* 2nd edition, vol. 2, pp. 183–334. London.

JAEGER, C. (1961*b*) Rock mechanics for hydro-power engineering. *Water Power,* **13**, nos. 9, 10.

JAEGER, C. (1961c) Recent British experience on underground work and rock mechanics. *7th Congress on Large Dams, Rome,* question 25, report 6.

JAEGER, C. (1963a) Tunnels for hydro-electric power. In C. A. Péquignot (ed.): *Tunnels and Tunnelling,* pp. 365–417. London.

JAEGER, C. (1963b) Vibrations and resonance in large hydro-power systems. *10th Congress Intern. Assoc. Hydr. Research, London.*

JAEGER, C. (1963c) The Malpasset Report. *Water Power,* **15,** 55–61.

JAEGER, C. (1963d) The theory of resonance in hydro-power systems. Discussion of incidents and accidents occurring in pressure sysrems. *Trans. A.S.M.E., J. Basic Engineering,* pp. 631–40.

JAEGER, C. (1964a) Rock mechanics for dam foundations. *Civil Engineering, London,* **59,** May.

JAEGER, C. (1964b) Rock mechanics and dam design. *Water Power,* **16,** 210–17.

JAEGER, C. (1964c) Underground power stations. In J. Guthrie Brown (ed.): *Hydro-electric Engineering Practice,* Vol. 1, 2nd edition. Glasgow.

JAEGER, C. (1964d) Rock mechanics for dam foundations. *8th International Congress on Large Dams, Edinburgh,* supplement, pp. 3–19.

JAEGER, C. (1964e) Bemerkungen zum Problem Felsmechanik und Wasserkraftwerkbau. *Die Wasserwirtschaft,* **54,** 149–57.

JAEGER, C. (1965a) The Vajont rock slide. *Water Power,* **17,** 110–11, 142–4.

JAEGER, C. (1965b) Reflections on the 8th Congress on Large Dams. *Civil Engineering, London,* **61,** 1031–5, 1177–9.

JAEGER, C. (1965c) Tendances actuelles et difficultés en mécanique des roches. *Schweiz. Bauztg.* **83,** 789–93.

JAEGER, C. (1966a) Some problems regarding the mechanics of rock masses. *Water Power,* **18,** 403–6.

JAEGER, C. (1966b) Der Stand der Felsmechanik nach dem 8. Kongress grosser Staumauern (Edinburgh, 1964). *Die Wasserwirtschaft,* **56,** 155–60.

JAEGER, C. (1966c) Discussion. *1st Congress International Society for Rock Mechanics, Lisbon,* **3,** p. 533.

JAEGER, C. (1968a) Discontinuous creep of masses. *Water Power,* **20,** 197–8.

JAEGER, C. (1968b) Discussion of the paper by L. Müller, on 'New considerations of the Vajont slide – The dynamics of the slide'. *Rock Mechanics & Engineering Geology,* **6,** 243–7.

JAEGER, C. (1969a) Felsmechanik und Gründung von Staumauern. In H. Press (ed.): *Wasser-Jahrbuch.* Berlin.

JAEGER, C. (1969b) The stability of partly immersed fissured rock masses, and the Vajont rock slide. *Civil Engineering, London,* **64,** 1204–7.

JAEGER, C. (1970) Engineering and rock mechanics, problems, practical examples and recent trends. *Water Power,* **22,** 203–9, 253–9.

JAEGER, C. (1971) Measurements in rock mechanics. Methods, techniques, examples and results. *Water Power,* **23,** 10–14.

JAEGER, C. (1973) Engineering problems and rock mechanics; some examples. *Engineering Geology,* **7,** 333–58.

JAEGER, C. (1975/76) Assessing problems in underground structures. *Water Power,* December 1975, January 1976.

JAEGER, C. (1977) *Fluid Transients in Hydro-Electric Engineering Practice.* Blackie, Glasgow.

JAEGER, J. C. (1956) *Elasticity, Fracture and Flow.* Methuen, London.

JAEGER, J. C. (1959) The frictional properties of joints in rock. *Geofisica pura e applicata,* **43,** 148.

JAEGER, J. C. (1962) *Elasticity, Fracture and Flow,* 2nd edition, Methuen London.

JAEGER, J. C. (1963) Extension failures in rocks subject to fluid pressure. *J. Geophys. Res.* **68**, 6066–7.

JAEGER, J. C. (1966) Brittle fracture of rock. *8th Symposium on Rock Mechanics, Minneapolis.*

JAEGER, J. C. (1967) Brittle fracture of rocks. In C. Fairhurst (ed.): *Failure and Breakage of Rock*, pp. 3–58.

JAEGER, J. C. & COOK, N. G. W. (1969) *Fundamentals of Rock Mechanics.* London.

JIMENES-SALAS, J. A. & URIEL, S. (1964) Some recent rock mechanics testing in Spain. *8th Congress on Large Dams, Edinburgh*, **1**, 995.

JOHN, K. W. (1961) Die Praxis der Felsgrossversuche, beschrieben am Beispiel der Arbeiten an der Kurobe IV Staumauer in Japan. *Geologie und Bauwesen*, **27**, no. 1.

JOHN, K. W. (1962) An approach to rock mechanics. *Proc. A.S.C.E. (Soil Mech. & Found. Div.)*, SM4, paper no. 3223.

JOHN, K. W. (1968) Graphical stability analyses of slopes in jointed rock. *Proc. A.S.C.E. (Soil Mech. & Found. Div.)*, SM2, paper no. 5865.

JOHN, K. W. (1970) Engineering analysis of three-dimensional stability problems utilising the reference hemi-sphere. *2nd Congress, International Society of Rock Mechanics, Belgrade.*

JONES, P. F. F., LANCASTER & GILLOTT, C.A. (1958) The Dokan project: the grouted cut-off curtain. *J. Inst. Civil Engineers*, **14**, 193.

JUDD, W. R. (ed.) (1964) *State of Stress in the Earth's Crust.* New York.

KARMAN, TH. V. (1911) Festigkeitsversuche unter allseitigem Druck. *Zeits. Ver. dt. Ingenieure*, **55**, 1749–57.

KASTNER, H. (1962) *Statik des Tunnel- und Stollenbaues.* Berlin.

KEITH, R. E. & GILMAN, J. J. (1960) Dislocation etch-pits and plastic deformation in calcite. *Acta Metallurg. U.S.A.*, **8**, no. 1.

KENNEDY, B. A. (1970) Discussion on Theme 7: Stability of natural and excavation slopes. *2nd Congress, International Society of Rock Mechanics, Belgrade*, **4.**

KENNEDY, B. A. & NIERMEYER, K. E. (1970) Slope monitoring system used in the prediction of major slope failure at the Chuquicamata mine, Chile. *Proceedings of the Symposium on planning open pit mines, Johannesburg.*

KENNEY, T. C. (1965) Causes of the Vajont reservoir disaster. *Civil Engineering, A.S.C.E.* September.

KENNEY, T. C. (1967) Stability of the Vajont Valley slope. *Rock Mech. & Eng. Geology*, **5**, no. 1.

KENT, P. E. (1965) The transport mechanism in catastrophic rock falls. *J. Geol.* **74**, 79.

KEPPIE, J. D., MATHESON, G. D. & VRANA, S. (1972) *Interim Report of the geology of the Kariba North Bank project.* Geological Survey, Zambia, 1 May 1972 (manuscript).

KIEMANS, H. (1972) Pumped storage plant at Raccoon Mountain in U.S.A. *Tunnels and Tunnelling*, March.

KIERSCH, G. A. (1963) Trends in engineering geology in the United States. *14th Congress of the International Society of Rock Mechanics, Salzburg.*

KIERSCH, G. A. (1964) Vajont Reservoir disaster. *Civil Engineering, A.S.C.E.*, pp. 32–9.

KIESER, A. (1960) *Druckstollenbau.* Vienna.

KIESLINGER, A. (1958) Restspannungen und Entspannungen im Gestein. *Geologie und Bauwesen*, **24**, no. 2.

KIESLINGER, A. (1960) Residual stress and relaxation in rocks. *International Geol. Congress, 21st Session, North.*

KIMISHIMA, H. *et al.* (1970) Analysis of strain energy of jointed rock mass during direct shear test *in situ. 2nd Congress, Inetrnational Society of Rock Mechanics, Belgrade*, Report 3/28.

KNILL, J. L. & JONES, K. S. (1965) The recording and interpretation of geological conditions in the foundations of the Roseires, Kariba and Latiyan dams. *Géotechnique*, 15, no. 1.

KOBOLD, F. (1968) Méthodes géodésiques pour la détermination des mouvements de roches ou de terrains dans les zones de glissement. *L'Entrepise, Zürich*, 67, 1007–16.

KOMMEREL, O. (1940) *Statische Berechnungen von Tunnelmauerwerk.* 2nd edition, Ernst, Berlin.

KOVARI, K. (1969) Ein Beitrag Zur Bemessung von Untertagbauten. *Schweiz. Bauzeitung*, 87, no. 9.

KRYNINE, D. P. & JUDD, W. R. (1957) *Principles of Engineering Geology and Geotechnics.* New York.

KRSMANOVIĆ, D. (1967a) Initial and residual shear strength of hard rock. *Géotechnique*, 17, no. 2.

KRSMANOVIĆ, D. (1967b) Contribution to the study of the failure problem in rock mass. *Proceedings of the Geotechnical Conference, Oslo*, 1.

KRSMANOVIĆ, D. & LANGOF, Z. (1963) Large scale laboratory tests on shear strength of rocky materials. *14th Congress of the International Society of Rock Mechanics, Salzburg.* See also: *Rock Mechanics & Engineering Geology* (1964), Suppl. 1, pp. 20–30.

KRSMANOVIĆ, D. & MILIC, S. (1963) Model experiments on pressure distribution in some cases of discontinuum. *14th Congress of the International Society of Rock Mechanics, Salzburg.* See also: *Rock Mechanics & Engineering Geology* (1964), Suppl. 1, pp. 72–87.

KUJUNDZIĆ, B. (1964) Idbar dam. *8th International Congress on Large Dams, Edinburgh*, question 28, discussion.

KUJUNDZIĆ, B. (1965) Experimental research into mechanical rock masses in Yugoslavia. *Int. Rock. Mech. Mining Sci.*, 2.

KUJUNDZIĆ, B. (1966) Behaviour of rock masses as structural foundations. General Report of Theme 8. *1st Congress, International Society of Rock Mechanics, Lisbon.*

KUJUNDZIĆ, B. (1969) A contribution to the investigation of the pressure grouting effect on consolidation of rock masses. *1st Congress, International Society of Rock Mechanics, Lisbon*, 2, 633–8.

KUJUNDZIĆ, B. (1974) Anwendung kombinierter statisch-dynamischer Methoden zur Ermittlung der Verformbarkeit des Gebirges. *Neue Bergbautechnik*, 4, No. 11, November.

KUJUNDZIĆ, B. & GRUJIC, N. (1966) Correlation between static and dynamic investigations of rock mass *in situ. 1st Congress, International Society of Rock Mechanics, Lisbon*, 1, 565–70. Rep. 3/56.

KUJUNDZIĆ, B., JOVANOVIC, L. & RADOSAVLJEVIC, Z. (1970) A pressure tunnel lining using high pressure grouting. *2nd Congress, International Society for Rock Mechanics, Belgrade*, report 4/66.

LANE, K. S. & HOCK, W. J. (1964) Triaxial testing for strength of rock joints. *6th Symposium on Rock Mechanics, University of Missouri*, pp. 98–108.

LANE, R. G. T. (1964a) The jetting and grouting of fissured quartzite at Kariba. *Proceedings of the Conference on Grouts.* Butterworth, London.

LANE, R. G. T. (1964b) Rock foundations; diagnosis of mechanical properties and treatment. *8th ICOLD Congress*, 1, 141–65.

LANE, R. G. T. & ROFF, J. W. (1961) Kariba underground works, design and construction methods. *7th Congress on Large Dams, Rome*, paper R.16.

LANE, R. G. T. & SERAFIM, J. L. (1962) The structural design of Tang-e-Soleman dam. *Proc. Inst. Civil Engineers*, 22, 257–90.

LANG. T. A. (1957) Rock behaviour and rock bolt support in large excavations. *Symposium on Underground Power Stations, New York.*

LANG, T. A. (1961) Theory and practice of rock bolting. *Trans. S.M.E., A.I.M.E.,* p. 229.

LANG, T. A. (1971) Underground rock structures challenge the engineer. *Proceedings of the Symposium on Underground Rock Chambers, American Society of Civil Engineers, Phoenix, Arizona.*

LANGE, E. *Sonnenburgerhof Tunnel, Insbruck.* Published privately by Bernold, A. G., Wallenstadt, Switzerland (no date).

LAUFFER, H. (1958) Gebirgsklassifizierung für Stollenbau. *Geologie und Bauwesen,* 24, 46–51.

LAUFFER, H., NEUHAUSER, E. & SCHOBER, W. (1967) Der Auttrieb als Ursache von Hangbewegungen bei der Füllung des Gepatschspeichers. *9th International Congress on Large Dams, Istanbul.*

LAUFFER, H. & SEEBER, G. (1962) Die Bemessung von Druckschachtauskleidungen für Innendruck auf Grund von Felsdehnungsmessungen. *Österr. Ingenieurzeitschrift,* no. 2.

LAUFFER, H. & SEEBER, G. (1966) Measurement of the rock deformability with the radial jack. *1st Congress, International Society of Rock Mechanics, Lisbon,* 2, 347–56.

LAURENT, D. & VOUILLE, G. (1968) Recherche des directions de moindre résistance dans le plan des roches à structure plainaire. *Revue de l'Industrie Minérale,* special issue 15 May, pp. 24–8.

LEE, E. H. (1962) Visco-elasticity. In W. Flügge (ed.): *Handbook of Engineering Mechanics.* New York.

LEEMAN, E. R. (1969) The 'Doorstopper' and triaxial rock stress measuring instruments developed by the C.S.I.R. *J. South African Inst. Mining and Metallurgy,* 69, 305–39.

LEEMAN, E. R. (1970) Experience throughout the world with the C.S.I.R. 'doorstopper' rock stress measuring equipment. *2nd Congress, International Society for Rock Mechanics, Belgrade.*

LEGGET, R. F. (1962) *Geology and Engineering,* 2nd edition. New York.

LELIAVSKY, S. (1958) *Uplift in Gravity Dams.* London.

LENSSEN, S. (1973) Kariba rock sinks Mitchell. The story of Borehole 4, Kariba North. *New Civil Engineer,* 8, February.

LÉVY, M. (1895) In *Mémoires de l'Académie des Sciences,* 5 August.

LIBBY, J., COOK, B. & MADILL, J. T. (1962) Kemano tunnel repair. *Eng. News Record,* October.

LING CHIH-BING (1948) On stresses in a plate containing two circular holes. *J. Appl. Phys.,* 19, 77–82.

LINK, H. (1962) Zur Querdehnungszahl von Gestein und Gebirge. *Geologie und Bauwesen,* 27, 89, 100.

LINK, H. (1964) Evaluation of elasticity moduli of dam foundation rock determined seismically in comparison with those arrived at statically. *8th Congress on Large Dams, Edinburgh,* 1, 833–58.

LINK, H. (1966) On Poisson's ratio in rock and rock masses at stresses near fracture strength. *1st Congress, International Society for Rock Mechanics, Lisbon,* 1, 425–31.

LINK, H. (1967) Zur Beurteilung und Bestimmung der Gleitsicherheit von Gewicht- und Pfeilerstaumauern. *Die Wasserwirtschaft,* pp. 35–46.

LINK, H. (1968) Zum Verhältnis seismischer und statisch ermittelter Elastizitätsmodeln von Fels. *Rock Mech. & Engineering Geology,* Suppl. 4, pp. 90–110.

478   *References*

LOCHER, H. G. (1968) Some results of direct shear tests on rock discontinuities. *International Symposium on Rock Mechanics, Madrid*, Report 11/6.

LOMBARDI, G. (1966) Contribution to discussion. *1st Congress, International Society of Rock Mechanics, Lisbon.*

LOMBARDI, G. (1969) Der Einfluss der Felseigenschaften auf die Stabilität von Hohlräumen. *Schweiz. Bauzeitung*, **87**, no. 3, 16 January.

LOMBARDI, G. (1970) The influence of rock characteristics on the stability of rock cavities. *Tunnels and Tunnelling*, **2**, nos. 1, 2.

LOMBARDI, G. (1971) Zur Bemessung der Tunnelauskleidung mit Berücksichtigung des Bauvorganges. *Schweiz. Bauzeitung*, **89**, no. 32, 12 August, 793–801.

LOMBARDI, G. (1972) *Contribution to the Symposium on Rock Mechanics, Lucerne, Switzerland.*

LOMBARDI, G. (1974) The problem of tunnel supports. *3rd Congress, International Society of Rock Mechanics, Denver.*

LOMBARDI, G. (1974) La prévision dans la construction des tunnels; géologie et mécanique des roches. In Calembert, L. (ed.): *La géologie de l'ingénieur*. Liège.

LOMBARDI, G. (1975) Consolidation de l'éperon rocheux de Baji-Krachen. *Mitteil. der Schweiz. Gesellschaft für Boden und Felsmechanik*, no. 91, 2/3 May.

LOMBARDI, G. & DAL VESCO, E. (1966) Die experimentelle Bestimmung der Reibungs-koeffizienten der Staumauer Contra (Verasca). *1st Congress, International Society of Rock Mechanics, Lisbon.*

LOMBARDI, G., ELECTROWATT ENGINEERING SERVICES & HAERTER, A. (1972) St Gotthard road tunnel project, the original concept and design. *Tunnels and Tunnelling*, September/October and November/December.

LONDE, P. (1965) Une méthode d'analyse à trois dimensions de la stabilité d'une rive rocheuse. *Annales des Ponts et Chaussées*, no. 1, 1–24.

LONDE, P. (1968) Le rupture des roches (d'après la Conférence Géotechnique d'Oslo, September 1969). *Revue de l'Industrie Minérale*, no. 1, pp. 3–18.

LONDE, P. (1973) The mechanics of rock slopes and foundations. *Quarterly J. Engineering Geology*, **6**, no. 1.

LONDE, P. & SABARLY, F. (1966) Permeability distribution in arch dam foundations. *1st Congress, International Society Rock Mechanics, Lisbon*, **2**, 517; **3**, 449.

LONDE, P. et al. (1969) Stability of rock slopes, a three-dimensional study. *Proc. A.S.C.E.*, **95**, S.M.1, Paper 6363.

LONDE, P., VIGIER, G. & VORMERINGER, R. (1966/70) Stability of rock slopes, graphical methods. *Proc. A.S.C.E.*, **96**, S.M.4, Paper 7435.

LOTTES, G. (1970) Waldeck II pumped storage station. *Water Power*, **22**, no. 5/6, May/June.

LOTTI, C. & BEAMONTE, M. (1964) Execution and control of consolidation work carried out in the foundation rock of an arch gravity dam. *8th Congress on Large Dams, Edinburgh*, **1**, 671–95.

LOTTI, C. & PANDOLFI, C. (1966a) L'azione meccanica dell' acqua nell' equilibrio degli ammassi rocciosi. *Geotecnica*, no. 14.

LOTTI, C. & PANDOLFI, C. (1966b) The mechanical action of water on the equilibrium of rock masses, with special regard to the presence of watertight surfaces (natural or artificial). *1st Congress, International Society of Rock Mechanics*, pp. 769–72.

LOUIS, C. (1974) *Rock hydraulics*. Bureau Recherches géologiques et minières, Orléans, France.

LOUIS, C. & MAINI, Y. N. T. (1970) Determination of 'in-situ' hydraulic parameters in jointed rock. *2nd Congress, International Society for Rock Mechanics, Belgrade.* Report 1/32.

LOVE, A. E. H. (1944) *Mathematical Theory of Elasticity*, 4th edition. New York.

LUGEON, M. (1933) *Barrages et Géologie*. Paris.

MCCLINTOCK, F. A. & WALSH, J. B. (1962) Friction on Griffith cracks in rocks under pressure. *Proceedings of the 4th National Congress on Applied Mechanics, Berkeley*, pp. 1015–21.

MCQUEEN, A. W. F., SIMPSON, C. N. & MCCAIG, I. W. (1958) Underground powerplants in Canada. *Proc. A.S.C.E. (Power Div.)* **84**, PO 3, paper 1670.

MCWILLIAMS, J. R. (1966) The role of microstructure in the physical properties of rock. In *Testing Techniques for Rock Mechanics*, Am. Soc. Testing & Materials *Publ. No. 402*, pp. 175–89. Philadelpnia.

MADHOW, M. R. & PRANESH, M. R. (1966) Nomographs for the design of pressure tunnels adjacent to rock slope. *1st Congress, International Society for Rock Mechanics, Lisbon*, **2**, 243–8.

MAILLART, R. (1922/23) De la construction des galeries sous pression intérieure. *Bull. Techn. Suisse Romande*, **68**, 25 October, 11 November and 25 November 1922 and *Schweiz. Bauzeitung*, 1923.

MAMILLAN (1968) Méthode de classification des pierres calcaires. *Inst. Techn. du Bâtiment et Trav.* Publ. 11, 469–526.

MANFREDINI, G., MARTINETTI, S., ROSSI, P. P. & SAMPAOLO (1975) *Observations on the procedures and the interpretation of the plate-bearing test*. ISMES, Bergamo, Rep. No. 69, 1975.

MANTOVANI, E. (1970) Method for supporting very high rock walls in underground power stations. *2nd Congress, International Society for Rock Mechanics, Belgrade*, report 6/5.

MANTOVANI, E., BERTACCHI, P. & SAMPAOLO, A. (1969) Measurements of rock deformation during excavation of a large underground power house. *Intern. Symposium Stresses in Rock Masses, Lisbon*.

MANTOVANI, E., BERTACCHI, P. & SAMPAOLO, A. (1970) Geomechanical survey for the construction of a large underground powerhouse. *2nd Congress, International Society for Rock Mechanics, Belgrade*, report 4/24.

MARY, M. (1968) *Barrages-voûtes, Historique, Accidents et Incidents*. Paris.

MASUR, A. (1970) Efficiency of rock consolidation grouting in mountains around intake pressure galleries of hydroelectric power plants. *2nd Congress, International Society for Rock Mechanics, Belgrade*, report 6/17.

MATHESON, G. D. (1971) *Report on a visit to Kariba North Bank Project*. Geological Survey, Zambia, June 1971 (Manuscript; see also Lenssen, 1973).

MATHESON, G. D. (1972) See Keppie *et al.*, 1972.

MATHIAS, F. & HUBER, W. G. (1954) Kemano underground. *J. Engineering, Inst. of Canada*, **37**, 1398, 1413.

MAURY, V. (1970*a*) *Mécanique des Milieux Stratifiés*. Dunod, Paris.

MAURY, V. (1970*b*) Distribution of stresses in layered systems. *Water Power*, **22**, 195–202.

MAURY, V. & DUFFAUT, P. (1970) Stress distribution model analysis in a two families discontinuity medium. *2nd Congress, International Society for Rock Mechanics, Belgrade*, report 8/19.

MAURY, V. & HABIB, P. (1967) Étude de distribution de contraintes en milieux discontinus. *Revue de l'Industrie Minérale*, **49**, no. 8.

MAYER, A. (1953) Les propriétés mécaniques des roches. *Géotechnique*, **3**, 329.

MAYER, A. (1963) Recent work in rock mechanics. (Third Rankine Lecture) *Géotechnique*, June.

MAYER, A. (1966) A lecture given in Montreal.

MAYER, A., BOLLO, DUFFAUT, *et al.* (French National Committee) (1964) Mesure des modules de déformation des massifs rocheux dans les sondages. *8th Congress on Large Dams, Edinburgh*, **1**, 313–27.

480 References

MAZANTI, B. B. & SOWERS, G. F. (1965) Laboratory testing of rock strength. In *Testing Techniques for Rock Mechanics. Am. Soc. Testing & Materials Publ. No. 402*, pp. 207–31. Philadelphia.

MAZENOT, P. (1965) Interpretation de nombreuses mesures de déformations exécutées sur massifs rocheux par E.d.F. *Annales de l'Institut Technique du Bâtiment et des Travaux Publics*, **18**, no. 206.

MENCL, V. (1966a) Mechanics of landslide with non-circular slip surfaces. *Géotechnique*, **16**.

MENCL, V. (1966b) The influence of the stiffness of a sliding mass on the stability of slopes. *Rock Mech. & Engineering Geology*, **4**, no. 2.

MERRILL, R. H. & PETERSON, J. R. (1961) Deformation of a borehole in rock. *U.S. Bureau of Mines Report R.I.5881*.

MILLER, D. V. Giant waves in Lituya Bay, Alsaka. *Bull. U.S. Geological Survey*, Paper 354C.

MILLER, R. P. (1965) Engineering classification and index properties for intact rock. Thesis, University of Illinois.

MILLI, E. (1967) *Sondaggi geognostici sulla frana del Monte Toc*. Report submitted to the Law Court of Aquila for the Vajont law suit. Not published.

MILOVANOVIC, D. (1958) Idbar Dam. *6th International Congress on Large Dams, Paris*, **4**, 571.

MINTCHEV, I. T. (1966) On the distribution of stresses in regions near workings with elliptical and circular cross sections driven in laminated rocks with steeply inclined strata. *1st Congress, International Society for Rock Mechanics, Lisbon*, **2**, 465–9.

MITCHELL, J. H. (1900) The stress in an aeotropic elastic solid with an infinite plane boundary. *Proc. London Math. Soc.*, **247**, June.

MOGI, K. The influence of the dimensions of specimens on the fracture strength of rocks. *Bulletin of the Earthquake Research Institute, Tokyo University*, **40**, 107–24, 125–73, 815–29, 831–53.

MONTARGES, P. (1968) Microfissuration des granites. Liaison avec la résistance à la traction. *Revue de l'Industrie Minérale*, special issue 15 May, pp. 9–15.

MOORE, B. W. (1961) Geophysics efficient in exploring the sub-surface. *Proc. A.S.C.E. (Soil Mech. & Found. Div.)* **87**, SM 3, paper 2838.

MORE, G. & SIRIEYS, P. H. (1965) Rupture des matériaux à deux directions de joints *C.R. Acad. Sci. Paris*, **260**, April.

MORGAN, H. D. (1961) Tunnelling practice. In J. Comrie (ed.): *Civil Engineering Reference Book*. London, Vol. 2, pp. 366–82.

MORGAN, T. A. & PANEK, L. A. (1963) A method for determining stresses in rock. *U.S. Bureau of Mines Report R.I. 6312*.

MORLIER, P. (1968) Relation quantitative entre la fissuration et la célérité des ondes dans les roches fissurées. *Revue de l'Industrie Minérale*, special issue 15 May, pp. 16–20.

MORRELL, R. J., BRUCE, W. E. & LARSEN, D. A. (1970) *Tunnelling, boring technology. Disc cutter experiments in sedimentary and metamorphic rocks*. U.S. Dept. Interior, Bureau of Mines, Rep. 7410, July.

MÜLLER, L. (1958) Geomechanische Auswertung gefügekundlicher Details. *Geologie und Bauwesen*, **24**, no. 1. p. 4.

MÜLLER, L. (1959) The European approach to slope stability problems in open pit mines. *Symposium on Rock Mechanics, Colorado School of Mines*.

MÜLLER, L. (1961) *Talsperre Vajont*. Report submitted to the Law Court of Aquila for the Vajont law suit, 3 February. Not published.

MÜLLER, L. (1963a) *Der Felsbau* (2 vols.). Stuttgart.

MÜLLER, L. (1963*b*) Die technischen Eigenschaften des Gebirges. *Schweiz. Bauztg.* **81**, 125–33.

MÜLLER, L. (1964) The rock slide in the Vajont valley. *Rock Mechanics & Engineering Geology*, **2**, 148–212.

MÜLLER, L. (1966) Der progressive Bruch in geklüfteten Medien. *1st Congress, International Society of Rock Mechanics, Lisbon*, **1**, 679–86.

MÜLLER, L. (1968) New considerations on the Vajont slide. *Rock Mechanics & Engineering Geology*, **6**, 1–91.

MÜLLER, L. & JOHN, K. W. (1963) Recent development of stability studies of steep rock slopes in Europe. *Trans. Soc. Mining Engineers*, **226**, 326–31.

MÜLLER, L., MALINA, H. & BAUDENDISTEL, M. (1970) The effect of the geologic structure on the stability of an underground powerhouse. *2nd Congress, International Society for Rock Mechanics, Belgrade*, report 4/56.

MÜLLER, L. & PACHER, F. (1962) Baugeologie und Gefügemechanik. *Zeitschrift der deutschen geologischen Gesellschaft*, **114**, 337–43.

MÜLLER, L. & PACHER, F. (1964) Bearing capacity of abutments of dams, especially arch dams in rock. *8th Congress on Large Dams, Edinburgh*, **1**, 935–54.

MÜLLER, L. & PACHER, F. (1965) Modellversuche zur Klärung der Bruchgefahr geklüfteter Medien. *Rock Mechanics & Engineering Geology*, Suppl. 2, pp. 7–24.

MURRELL, S. F. (1962) A criterion for brittle fracture of rocks and concrete under triaxial stress and the effect of pore pressure on the criterion. *Proceedings of the 5th Symposium on Rock Mechanics*, p. 563, Oxford.

MUSKAT, M. (1937/56) *The Flow in Homogeneous Fluids through Porous Media.* New York.

NADAI, A. (1956) *The Theory of Flow and Fracture of Solids.* New York.

NAEF, E. (1969) Quelques considérations concernant le calcul de la stabilité d'un versant rocheux sous l'influence de forces extérieures. *Bull. Techn. Suisse Romande*, **95**, 213–19.

NATIONAL ACADEMY OF SCIENCES (1974) *Advances in rock mechanics. 3rd Congress, International Society of Rock Mechanics, Denver.*

NAYLOR, D. J. Pressure tunnels in which the pressure is carried by the rock. Dissertation, Imperial College London, not dated.

NEFF, T. L. (1965) Equipment for measuring pore pressure in rock specimens under triaxial load. In *Testing Techniques for Rock Mechanics, Am. Soc. Testing & Materials, Publ. No. 402*, pp. 3–18. Philadelphia.

NONVEILLER, E. (1954) The determination of the deformation of loaded rock in tunnels. *Proc. Yugoslav. Soc. Soil Mech.*, no. 8, p. 43.

NONVEILLER, E. (1965) The stability analysis of slopes with a slip surface of general shape. *Proceedings of the 6th International Conference of Soil Mechanics and Foundation Engineers, Montreal*, **2**, div. 6.

NONVEILLER, E. (1967*a*) Zur Frage der Felsrutschung im Vajont-Tal. *Rock Mechanics & Engineering Geology*, **5**, no. 1.

NONVEILLER, E. (1967*b*) Mechanics of land-slides with noncircular surfaces with special reference to the Vajont slide. *Corr. Géotechnique*, no. 2.

NONVEILLER, E. (1967*c*) Shear strength of bedded and jointed rock as determined from the Zalesina and Vajont slides. *Proc. Geot. Conf. Norw, Geot. Inst.*, Oslo.

NORSE, P. M. & FESHBACH, H. (1953) *Methods of Theoretical Physics.* New York.

OBERT, L. (1963) *An Inexpensive Triaxial Apparatus for Testing Mine Rock.* U.S. Bureau of Mines publication.

OBERT, L. & DUVALL, W. I. (1967) *Rock Mechanics and the Design of Structures in Rock.* New York.

OBERT, L., WINDES, S. L. & DUVALL, W. I. (1946) Standardized tests for determining the physical properties of mine rock. *U.S. Bureau of Mines Report No. R.I.3891.*

482 References

OBERTI, G. (1948) Ricerche sperimentali sulla deformabilita della roccia di fondazione della diga del Piave. *Giornale del Genio Civile*, no. 11, 607.

OBERTI, G. (1960) Experimentelle Untersuchungen über die Charakteristik der Verformbarkeit des Felsen. *Geologie und Bauwesen*, 25, 95–113.

OBERTI, G. & REBAUDI, A. (1967) Bedrock stability behaviour with time at the Place Moulin arch-gravity dam. *9th International Congress on Large Dams, Istanbul: ISMES Publications No. 36*, pp. 3–21.

OLIVEIRA, R. (1974/75) *Underground constructions. Engineering geological investigations in 'in-situ' testing.* Int. Assoc. Engineering Geology, Sao Paulo, Brazil (1974) and Laboratorio Nacional de Engenharia Civil, Lisbon, Rep. No. 467, 1975.

OLIVEIRA, R. et al. (1975) *Geotechnical studies of the foundation rock mass of Valhelhas Dam, Portugal.* Laboratorio Nacional de Engenharia Civil Report No. 463, Lisbon.

OLIVIER, H. (1961) Some aspects relating to the civil engineering constructions of the Kariba Hydro-electric scheme. *Die Siviele Ingenieur in Suid Africa*, April.

OLIVIER, H. (1973) Swelling properties and other related geomechanical parameters of Karoo Strata as encountered in the Orange–Fish Tunnel. *Proceedings of the 15th Congress, Geological Society of South Africa.*

OLIVIER-MARTIN, D. & KOBILINSKY (1955) L'exécution d'un grand souterrain pour l'aménagement hydro-éléctrique d'Isère-Arc. *Techn. des Trav.*, pp. 145–56. See also *Schweiz. Bauztg.* **70**, nos. 17, 18, 19.

ONODERA, T. F. (1963) Dynamic investigation of foundation rocks *in situ. Proceedings of the 5th Symposium on Rock Mechanics*, pp. 517–33. Oxford and New York.

PACHER, F. (1958) Kennziffern des Flächengefüges. *Geologie und Bauwesen*, 24, no. 3/4.

PACHER, F. (1961) Zur Auswertung von Grossversuchen. *Geologie und Bauwesen*, 27, no. 1.

PACHER, F. (1963) Die Lage des Dichtungsschirmes von Bogenstaumauern und der Einfluss auf die Sicherheit der Felswiderlager. *Geologie und Bauwesen*, 28, no. 3.

PACHER, F. & YOKOTA, J. (1962) Experimental studies on the design of the grouting curtain and the drainage system of the Kurobe IV Dam. *12th Congress of the International Society of Rock Mechanics, Salzburg.*

PAHL, A. & ALBRECHT, H. (1970) Felsmechanische Untersuchungen zur Beurteilung der Standfestigheit grosser Felskavernen. *2nd Congress, International Society for Rock Mechanics, Belgrade.*

PANCINI, M. (1961a) Observations and surveys on the abutments of the Vajont dam. *Geologie und Bauwesen*, 26, no. 3.

PANCINI, M. (1961b) Results of the first series of tests performed on a model of the Vajont dam. *Colloquium on Rock Mechanics, Salzburg*. See also: *Geologie und Bauwesen* (1962), 27, 105–19.

PANEK, L. A. (1964) Design for bolting stratified roof. *Trans. S.M.E., A.I.M.E.* **229.**

PANET, M. (1967) Étude de la structure d'un massif rocheux. *Bulletin Liaison Labo, Routiers P. et Ch.*, no. 28, ref. 391.

PANET, M. (1969) Quelques problèmes de mécanique des roches posés par le tunnel du Mont Blanc. *Bulletin Liaison Labo. Routiers P. et Ch.*, no. 42, 115–45.

PATTERSON, F. W., CLINCH, R. L. & MCCAIG, I. W. (1957) Design of large pressure conduits in rock. *Proc. A.S.C.E. (Power Div.)*, PO 6, paper 1457.

PAULDING, B. W. JR. (1966) Techniques used in studying the fracture mechanics of rock. In *Testing Techniques for Rock Mechanics*, Am. Soc. Testing & Materials Publ. No. 402, pp. 73–86. Philadelphia.

PAULMANN, H. G. (1966) Measurements of strength anisotropy of tectonic origin on rock specimens. *1st Congress of the International Society for Rock Mechanics,*

Lisbon, **1**, 125–31. (This paper can also be consulted for statistical analysis of test results.)

PECK, R. B. (1969) Advantages and limitations of the observational method, applied soil mechanics. *Géotechnique*, **19**, no. 2, 171.

PELLETIER, J. (1953) The construction of Tignes Dam and Malgovert Tunnel. *Proc. Inst. Civil Engineers*, **2**, 480.

PÉQUIGNOT, C. A. (ed.) (1963) *Tunnels and Tunnelling*. London.

PERAMI, R. & THENOZ, B. (1968) Perméabilité et porosité des roches. *Revue de l'Industrie Minérale*, special issue 15 May, pp. 38–47.

PFISTER, R. (1974) Untertagbau im Valanginien Mergel. In *Ingenieurrgeologie*, publication of Schweiz. Gesellschaft für Boden-und Felsmechanik, No. 90.

PIERSON, L. N. (1975) *Rock Dynamics and Geophysical Exploration*. Elsevier, Amsterdam.

PINCUS, H. J. (1966) Capabilities of photoelastic coatings for the study of strain in rocks. In *Testing Techniques for Rock Mechanics*. Am. Soc. Testing & Materials Publ. No. 402. Philadelphia.

PIRRIE, N. D. (1970) Design and development of a rock tunnelling machine. *Tunnels and Tunnelling*, March/April.

PÖCHHACKER, H. (1976) Rock classification and contracts for large highway tunnels. *International Construction*, **15**, July.

Proceedings of the 1st Congress of the International Society of Rock Mechanics, Lisbon (1966) 3 vols.

Proceedings of the 6th Symposium on Rock Mechanics, University of Missouri (1964).

Proceedings of the 8th Symposium on Rock Mechanics, University of Minnesota (1967).

PROCTOR, R. V. & WHITE, T. (eds.) (1946) *Rock Tunnelling with Steel Supports*, pp. 15–99. Commercial Shearing and Stamping Co., Youngstown. (See also Terzaghi, K., *Rock defects and loads on tunnel supports*.)

RABCEWICZ, L. V. (1953) The Forçacava hydro-electric scheme. *Water Power*, **5**, 333, 370.

RABCEWICZ, L. V. (1957) Die Ankerung im Tunnelbau ersetzt bisher gebräuchliche Einbaumethoden. *Schweiz. Bauztg.*, **75**, 123–31.

RABCEWICZ, L. V. (1963a) Bemessungen von Hohlraumbauten. *Geologie und Bauwesen*, pp. 224–44.

RABCEWICZ, L. V. (1963b) Bemessung von Hohlraumbauten, die neue Österreichesche Bauweise und ihr Einfluss auf Gebirgsdruckwirkungen und Dimensionierung. *Felsmechanik und Ingenieurgeologie*, **1**, nos. 3–4.

RABCEWICZ, L. V. (1964) The new Austrian tunnelling method. *Water Power*, **16**, November, December.

RABCEWICZ, L. V. (1965) The new Austrian tunnelling method, *Water Power*, **17**, January.

RABCEWICZ, L. V. & GOLSER (1974) Application of the N.A.T.M. to the underground works at Tarbela. *Water Power*, September.

RABCEWICZ, L. V., JOLSER, J. & HACKL. E. (1972) Die Bedentung der Messung im Hohlraumbau. *Bauingenieur*, **47**, no. 7.

REED, J. J. (1966) Putting rock to work. *Testing World*, no. 17.

RELLENSMANN, O. (1957) Rock mechanics in regard to static loading caused by mining excavation. *2nd Symposium on Rock Mechanics, Quart. Colorado School of Mines*, **52**.

RESCHER, O. J. (1968) Aménagement Hongrin-Léman. Soutènement de la centrale en caverne de Veytaux par tirants en rocher et béton projeté. *Bull. Techn. Suisse Romande*, **94**, no. 18, pp. 249–68.

RESCHER, O. J. (1972) Ein Kavernenbau mit Ankerung und Spritzbeton unter Berückseichtigung der geomechanischen Bedingungen. *Rock Mech.*, no.1.

RESCHER, O. J., BRÄUTIGAM, F., PAHL, A. & ABRAHAM, K. H. (1973) Ein Kavernenbau mit Ankerungen und Spritzbeton unter Berücksichtigung geomechanischer Bedingungen. *Rock Mech.*, Suppl. 2, 313–54.

REYES, S. F. & DEERE, D. U. (1966) Elastic-plastic analysis of underground openings by the finite element method. *1st Congress, International Society for Rock Mechanics, Lisbon*, **2**, 477–86.

REYNOLDS, H. (1961) *Rock Mechanics*. London.

RIBES, G. & BUTIN, P. (1958) Surélévation du barrage des Beni-Bahdel. *6th Congress on Large Dams, Paris*, **4**, communication no. 15.

RICCIONI, R. (1973) *Interpretazione delle misure di spostamente durante l'excavatione di una grande centrale in caverna*. ISMES, Bergamo, no. 57.

RICHEY, J. E. (1964) *Elements of Engineering Geology*. London.

RICHTER, R. (1964) Grundlegende Betrachtungen zum Ankerbau. *Bergbauwissenschaften*, **11**, 393–402.

RIPLEY, C. F. & LEE, K. L. (1961) Sliding friction tests on sedimentary rock specimens. *7th Congress on Large Dams, Rome*.

ROBERTS, A., EMERY, C. L. *et al*. (1962) Photoelastic coating technique applied to research on rock mechanics. Parts I and II. *Bulletin. Inst. Mining and Metallurgy, London*. See also: Measurement of strain and stress in rock masses; in K. G. Stagg & O. C. Zienkiewicz (eds.): *Rock Mechanics in Engineering Practice*, pp. 157–202. London.

ROBERTS, C. M. (1958) Underground power plants in Scotland. *Proc. A.S.C.E.*, paper 1675.

ROBINSON, L. H. (1959) The mechanics of rock failure. *Quart. Colorado School of Mines*, **54**, 136–77.

ROBINSON, L. H. JR (1959) The effect of pore and confining pressure on the failure process in sedimentary rock. *3rd Symposium on Rock Mechanics, Colorado School of Mines*.

ROCHA, M. (1963) Some problems on failure of rock masses. *14th Congress of the International Society of Rock Mechanics, Salzburg*.

ROCHA, M. (1964) Mechanical behaviour of rock foundations in concrete dams. *8th Congress on Large Dams, Edinburgh*, **1**, 785.

ROCHA, M. (1969) *New technique for the determination of the deformability and state of stress in rock masses*. Laboratorio Nacional de Engenharia Civil, Lisbon, Publ. No. 328, 16 pp.

ROCHA, M. (1971a) Assessing the deformability of rock masses. *Water Power*, August.

ROCHA, M. (1971b) *A method of integral sampling of rock masses*. Lab. Nacional de Engenharia Civil, Publ. No. 382, Lisbon, 1971, and *Rock Mechanics*, **3**, No. 1 (1967).

ROCHA, M., BRITO, S. & NIEBLE, C. (1975) *Application of advanced techniques to the study of the foundations of São Simão Dam*. Laboratorio Nacional de Engenharia Civil, Publ. No. 458, Lisbon, 1975.

ROCHA, M., SERAFIM, J. L. & DA SILVEIRA, A. F. (1955) Deformability of foundation rocks. *5th Congress on Large Dams, Paris*, **3**, 531.

ROCHA, M. & SILVERIO, A. (1969) *A new method for the complete determination of the state of stress in rock masses*. Laboratorio Nacional de Engenharia Civil, Lisbon, Publ. No. 329, 17 pp. See also: *Géotechnique*, **19**, 116–32.

ROS, M. & EICHINGER, A. (1930) *Die statische Bruchgefahr fester Bau- und Werkstoffe*. Zürich.

ROS, M. & EICHINGER, A. (1949) *Die Bruchgefahr fester Körper*. E.M.P.A. Publication No. 172. Zürich.

ROSA, S. A., FEDORENKO, A. N., ERISTOV, V. S. & TOKACHIROV, V. A. (1969) Studies of deformation properties of rock foundations of high arch and gravity dams in the USSR. *8th Congress on Large Dams, Edinburgh*, 1, 1023–40.

ROSENSTRÖM, S. (1972) Kafue Gorge hydraulic power project. *Water Power*, June/July.

ROUSSEL, J. M. (1968) Etude théorique et expérimentale du module dynamique des massifs rocheux. *Revue de l'Industrie Minérale*, 50, 1–28.

RUCKLI, R. *et al.* (1970) *Der Gotthard Strassentunnel No. 2.* Instituto Editoriale Ticinese, Bellinzona, Switzerland.

RUESCH, H. (1960) Research towards a general flexural theory for structural concrete. *J. Am. Concrete Inst.* 32, no. 1. See also: *Proc. Am. Concrete Inst.* (1966) 57, 1–28.

RUTSCHMANN, W. (1974) Vorschlag für ein System der Gebirgsklassifikation für mechanischen Vortrieb. *Schweiz. Bauzeitung*, 92, no. 18, 2 May.

RZIHA, F. (1874) *Lehrbuch der gesammten Tunnelbaukunst*, 2nd edition. Berlin.

SABARLY, F. (1968) Les injections et les drainages de fondations de barrages en roches peu perméables. *Géotechnique*, pp. 229–49.

SAVIN, G. N. (1961) *Stress Concentrations around Holes*. Oxford.

SCHEIBLAUER, J. (1962) Modellversuche zur Klärung des Spannungszustandes in steilen Böschungen. *13th Symposium on Geomechanics, Salzburg*.

SCHMIDT, A. (1975) Les travaux d'amélioration de la route du Simplon. *Mitteil. Schweiz. Gesellschaft für Boden- und Felsmechanik*, No. 91, 2/3 May.

SCHMIDT, J. (1926a) *Statische Grenzprobleme in kreisformig durchörterten Gebirge*. Zürich and Berlin.

SCHMIDT, J. (1926b) *Statische Probleme des Tunnel und Druckstollenbaues*. Berlin.

SCHNEEBELI, J. (1966) *Hydraulique Souterraine*. Paris.

SCHNEIDER, B. (1966) Contribution à l'étude des massifs de fondation de barrages. Unpublished thesis, mentioned in J. M. Roussel: Étude théorique et expérimentale du module dynamique des massifs rocheux, *Revue de l'Industrie Minérale* (1966), 50, 1–28.

SCHNEIDER, B. (1967) Moyens nouveaux de reconnaissance des massifs rocheux. *Annales Inst. Techn. Bât. Travaux Publics*, July, August.

SCHNITTER, G. & WEBER, E. (1964) Die Katastrophe von Vajont in Oberitalien. *Wasser- und Energiewirtschaft*, 56, 61–9.

SCHOELLER, H. (1962) *Les Eaux Souterraines*. Paris.

SCHRAFL, A. (1924) Kurzer Bericht über die Druckstollen versuche der Schweizerischen Bundesbahnen. *Schweiz. Bauztg.* 83, January. See also: *Schweiz. Bauztg.* (1920), 76.

SCHWARTZ, A. E. (1964) Failure of rock in the triaxial shear test. *6th Symposium on Rock Mechanics, University of Missouri*, pp. 109–51.

Schweizerischer Ingenieur und Architekten Verein (S.I.A.) (1976) *Etude du massif rocheux pour les travaux souterrains*. Recommandation SIA 199, Zürich, SIA, 1976.

SCOTT, P. A. (1952) A 75-inch diameter water main in tunnel: a new method of tunnelling in London clay. *Proc. Inst. Civil Engineering*, 1, 302.

SEEBER, G. (1961) Auswertung von statischen Felsdehnungsmessungen. *Geologie und Bauwesen*, 26, 152–76.

SEEBER, G. (1964) Einige felsmechanische Messergebnisse aus dem Druckschacht des Kaunertal Kraftwerkes. *Rock Mechanics & Engineering Geology*, 1, Suppl. 1, 130–48.

SELDENRATH, T. R. (1951) Can coal measures be considered as masses of loose structure to which the laws of soil mechanics may be applied? *Proceedings of the International Conference on Rock Pressure, Liège, Belgium.*

SEMENZA, E. (1966) Sintesi degli studi geologici sulla frana del Vajont dal 1959 al al 1964. *Memorie del Museo Tridentino di Scienze Naturali A. XXIX-XXX.* **16,** 5–51.

SERAFIM, J. L. (1963) Rock mechanics considerations in the design of concrete dams. *Conference on State of Stress in the Earth's Crust, Santa Monica, California,* p. 611.

SERAFIM, J. L. (1964) The uplift area in plain concrete in the elastic range. *8th Congress on Large Dams, Edinburgh,* **5,** communication C.17.

SERAFIM, J. L. (1968) Influence of interstitial water on the behaviour of rock masses. In K. G. Stagg & O. C. Zienkiewicz (eds.): *Rock Mechanics in Engineering Practice.* New York.

SERAFIM, J. L. & DEL CAMPO, A. (1965) Interstitial pressures on rock foundation of dams. *Proc. A.S.C.E. (Soil Mech. & Found. Div.),* SM 5, 65.

SERAFIM, J. L. & LOPEZ, J. J. B. (1961) *In situ* shear tests and triaxial tests of foundation rocks of concrete dams. *Proceedings of the 5th International Conference on Soil Mechanics and Foundation Engineering, Paris,* **1.**

SERDENGECTI, S. & BOOZER, G. D. (1961) The effects of strain rate and temperature on the behaviour of rocks subject to triaxial compression. *Proceedings of the 4th Symposium on Rock Mechanics, Bull. Mineral. Indus. Exper. Sta. Perm. St. Univ. No. 76.*

SERIEYS, P. M. (1966) Porosité, degré de saturation et lois de comportement des roches. *1st Congress, International Society for Rock Mechanics, Lisbon,* **1,** 471.

SHUK, TH. (1964a) Caracteristicas de deformación de una roca compuesta. *Bull. Venezuelan Soc. Soil Mech.,* no. 16, 12–22.

SHUK, TH. (1964b) Discussion on paper 'Tests for Karadj arch dam'. *Proc. A.S.C.E. (Soil Mech. & Found. Div.),* **90,** SM 2, 157–61. See also: *1st Congress, International Society for Rock Mechanics,* **1,** 319–27.

SKEMPTON, A. W. (1964) Long term stability of clay slopes. (4th Rankine Lecture.) *Géotechnique,* June.

SKEMPTON, A. W. (1966) Bedding-plane-slip, residual strength and Vajont landslide. *Inst. Civil Engineers,* pp. 82–4.

SLOT, T. (1960) A study of the photoelastic coating technique. *General Electric Report 60 G.L. 72.*

SONDEREGGER, A. (1956) Spritzbeton im Stollenbau. *Schweiz. Bauztg.* **74,** 211.

SPRADO, K. H., GLÖGGLER, W. & PAHL, A. (1976) Die Einführung der Triebwasserleitung in die Maschinenkaverne des Pumpspeicherwerkes Waldeck II. *Die Bautechnik,* **53,** no. 4, April.

STAGG, K. G. (1968) *In situ* tests on the rock mass. In K. G. Stagg & O. C. Zienkiewicz (eds.): *Rock Mechanics in Engineering Practice.* New York.

STAGG, K. G. & ZIENKIEWICZ, O. C. (1968) *Rock Mechanics in Engineering Practice.* New York.

STINI, J. (1941) Unsere Täler wachsen zu. *Geologie und Bauwesen,* **13,** no. 3.

STINI, J. (1942) Talzuschub und Bauwesen. *Die Bautechnik,* **20.**

STINI, J. (1950) Neuere und ältere Vorschläge zur Einschätzung des Gebirgsdruckes. *Gebirgsdrucktagung, Loeben & Vienna.*

STINI, J. (1952a) Der Gebirgsdruck und seine Berechnung, *Geologie und Bauwesen,* **19,** no. 3.

STINI, J. (1952b) Neuere Ansichten über Bodenbewegungen und ihre Beherrschung durch den Ingenieur. *Geologie und Bauwesen,* **19,** 31–54.

STROPPINI, E. W. & KRUSE, J. H. (1963) Discussion of paper 3576 on 'Foundation modulus test for Karadj arch dam.' *Proc. A.S.C.E.* July.

STROPPINI, E. W. & KRUSE, J. H. (1964) Oroville dam. *Proc. A.S.C.E. (Soil Mech. & Found. Div.)*, **90**, SM 2, 191–204.

STUART, W. H. (1961) Influence of geological conditions on uplift. *Proc. A.S.C.E. (Soil Mech. & Found. Div.)*, **87**, SM 6, 1–18.

STUCKY, A. (1924) Der Talsperrenbruch im Val Gleno. *Schweiz. Bauztg.*, **83**, no. 6/7; *Bull. Techn. Suisse Romande*, no. 6/7.

STUCKY, A. (1953) Le centre de recherches pour l'étude des barrages. *Centenaire de l'École Polytechnique de Lausanne, Switzerland*, p. 106.

STUCKY, A. (1955) Le barrage d'accumulation de Ben-Metir en Tunisie. *Bull. Techn. Suisse Romande*, **81**, no. 21/22.

SUTCLIFFE, H. & MCCLURE, C. R. (1969) Large aggregate shotcrete challenges steel ribs as a tunnel support. *Civil Engineering, A.S.C.E.*, pp. 51–5.

SUZUKI, K. & ISHIJIMA, Y. (1970) The simulation technique to analyse the rock pressure applied to tunnel supports. *2nd Congress, International Society for Rock Mechanics, Belgrade*, report 4/27.

Swiss National Committee on Large Dams (1964) *Behaviour of Large Swiss Dams.* Zürich.

SZECHY, K. (1966/70) *Traité de la Construction des Tunnels.* Dunod, Paris, 1970. (First published by Hungarian Academy, Budapest, 1966.)

TALOBRE, J. (1957) *La Mécanique des Roches.* Paris.

TALOBRE, J. (1962) *Tests. General Report.* New York (Development & Resources Corporation).

TATTERSHALL, WAKELING & WARD (1955) Investigations into the design of pressure tunnels in London clay. *Proc. Inst. Civil Engineers*, **4**, 400.

TAYLOR, I. G. (1966) The influence of discontinuities on the stability of an underground opening. *1st Congress, International Society for Rock Mechanics*, **2**, 329–33.

TELLER, E., TALLEY, W. K., HIGGINS, G. H. & JOHNSON, G. W. (1968) *The Constructive Uses of Nuclear Explosions.* New York.

TERZAGHI, K. (1943) *Theoretical Soil Mechanics.* New York.

TERZAGHI, K. (1945) Stress conditions for the failure of saturated concrete and rock. *Am. Soc. Testing & Materials*, **45**, 777–801.

TERZAGHI, K. (1946) Rock defects and loads on tunnels. In *Rock Tunnelling with Steel Support.* Ohio (The Comm. Shearing and Stamping Co.).

TERZAGHI, K. (1962a) Does foundation technology really lag? *Eng. News Record*, p. 263.

TERZAGHI, K. (1962b) Stability of steep slopes on hard unweathered rock. *Géotechnique*, **12**, 251–70. See also: *Géotechnique*, **12**, 67–71.

TERZAGHI, K. & FRÖHLICH, O. K. (1939) *Théorie du tassement des couches argileuses.* Paris.

TERZAGHI, K. & RICHART, F. E. (1952) Stresses in rock around cavities. *Géotechnique*, **3**, 57–99.

THOMA, E. (1906) *Über das Wärmeleitungsproblem bei welliger Oberfläche und dessen Anwendung auf Tunnelbauten.* Karksruhe.

THOMA, E., KOENIGSBERGER, J. & GOETZ, H. (1908) Versuche über primäre und sekundäre Beeinflussung der normalen geothermischen Tiefenstufe. *Ecl. Geol. Helvetica*, **10**.

THOMANN, G. (1969) The Rönkhausen pumped-storage project. *Water Power*, **21**.

TIMOSHENKO, S. (1940) *Theory of Plates and Shells*, p. 228. New York.

TIMOSHENKO, S. (1947) *Strength of Materials*, parts I and II. New York.

TIMOSHENKO, S. & GOODIER, J. N. (1951) *Theory of Elasticity*, 2nd edition. New York.

TINCELIN, M. E. (1952) Mesures des pressions de terrains dans les mines de fer de

488   References

l'Est. *Annales de l'Institut Technique du Bâtiment et des Traveaux Publics*, no. 58, 972–90.

TORROJA, E. (1951) *Elasticidad*. Madrid.

TOURENQ, C. & DENIS, A. (1970) La résistance à la traction des roches. *Laboratoire Central des Ponts et Chaussées Publ. No. 4*. Paris.

TREFETHEN, J. M. (1949) *Geology for Engineers*. New York.

TROLLOPE, D. H. (1957) The systematic arching theory applied to the stability analysis of embankments. *Proceedings of the 4th International Conference on Soil Mechanics and Foundation Engineering*, 2, 382.

TROLLOPE, D. H. (1961) The mechanics of rock slopes. *Trans. A.I.M.E. (Mining Div.)*, **222.**

TROLLOPE, D. H. (1962) The mechanics of rock slopes. *Trans. A.I.M.E. (Mining Div.)*, **223.**

TROLLOPE, D. H. (1968) The mechanics of discontinua or clastic mechanics in rock problems. In K. G. Stagg & O. C. Zienkiewicz (eds.): *Rock Mechanics in Engineering Practice*, pp. 275–320. New York.

TROLLOPE, D. H. & BROWN, E. T. (1965) Pressure distribution in some discontinua. *Water Power*, August.

TROLLOPE, D. H. & BROWN, E. T. (1966) Effective stress criteria of failure of rock masses. *1st Congress, International Society for Rock Mechanics, Lisbon*, **1**, 515.

TRUCCO, G. (1968) Stollenbau durch Triasschichten – einige Beispiele und Erfahrungen. *L'Entreprise, Zürich*, **67**, no. 34.

U.S. Army, W.E.S. (1963) Project Dribble, petrographic examination and physical tests of cores, Tatum Salt Dome, Mississippi. *Technical Report No. 6–614*. Vicksburg, Mississippi (U.S. Army Waterways Experimental Station).

U.S. Bureau of Reclamation (1962) Triaxial compression tests of salt rock cores for the United States Atomic Energy Commission, Project Dribble. *Laboratory Report No. C-1043*. Denver, Colorado.

VAN HEERDEN, W. L. (1974) *Additional rock stress measurements at the Ruacana power station*, S.W.A.-CSIR Contract Report ME1324, Johannesburg.

VOGT, F. (1925) Über die Berechnung der Fundamentdeformation. *Math. Nat. Klasse, Oslo*. See also: Stresses in thick dams, *Trans. Am. Civil Eng.* (1927), **90.**

VOUILLE, G. (1960) Étude monographique de diverses roches. Laboratoire de Mécanique des Solides de l'École Polytechnique, Paris, internal report; mentioned in J. Bernaix, thesis (1966).

VUTUKURI, V. S. et al. (1974) *Handbook of Mechanical Properties of Rocks*, vols. I and II. Trans Tech Publications, Bay Village, Ohio, 1974.

WAHLSTROM, E. (1973) *Tunnelling in rock*. Elsevier Scientific, Amsterdam and New York.

WALDORF, W. A., VELTROP, J. A. & CURTIS, J. J. (1963) Foundation modulus tests for Karadj arch dam. *Proc. A.S.C.E. (Soil Mech. & Found. Div.)*, **89**, SM 4, 91.

WALLACE, G. B. & OLSEN, O. J. (1965) Foundation testing techniques for arch dams and underground power plants. In *Testing Techniques for Rock Mechanics*. Am. Soc. Testing & Materials Publ. No. 402, pp. 272–89. Philadelphia.

WALSH, J. B. (1965) The effects of cracks in the compressibility of rocks. *J. Geophysical Research*, **70**, 391–8.

WALTERS, R. C. S. (1962) *Dam Geology*. London.

WEIRICH, K. W. (1972) Economic design of tunnel linings. *8th Canadian Symposium on Rock Mechanics*, 30 November–1 December.

WESTERBERG, G. & HELLSTROM, B. (1950) Swedish practice in water power development. *4th World Power Conference*, **4**, 2071.

WESTERGUARD, H. N. (1938) A problem of elasticity suggested by a problem in soil mechanics. In *S. Timoshenko's Sixtieth Anniversary*. New York.

WHITE, J. E. (1965) *Seismic Waves*. McGraw-Hill International Series in the Earth Sciences. New York.

WICKHAM, G. E., TIEDEMANN, H. R. & SKINNER, E. H. (1972) Support determinations on geologic predictions. *Proceedings of the 1st North American Rapid Excavation and Tunnelling Conference, AIME, New York*, pp. 43–64.

WILLIAMSON, T. N. (1970) *Tunnelling machines of today and tomorrow*. Highway Research Board, Div. of Eng., NR, Nat. Acad. Sciences and Eng., Washington, HRR. No. 339.

WILSON, S. D. (1957) The application of soil mechanics to the stability of open pit mines. *Symposium on Rock Mechanics, Colorado School of Mines*.

WINDES, S. L. (1953) Physical properties of mine rocks. *Concrete Laboratory Report S.P. 59*. U.S. Bureau of Reclamation.

WISE, L. L. (1952) World's largest underground power plant. *Eng. News Record*, **149**, 31.

WITTKE, W. (1969) Durchströmung von klüftigem Fels, Theorie, Experiment und Anwendung. *Theodor Rehbok Flussbaulaboratorium Universität Karlsruhe Bull. Nr. 155*, pp. 121–56.

WITTKE, W. & LOUIS, C. (1966) Zur Berechnung des Einflusses der Bergwasser-strömung auf die Standsicherheit von Böschungen und Bauwerken in zerklüftetem Fels. *1st Congress, International Society for Rock Mechanics, Lisbon*, **2**, 201.

WÖHLBIER, H. & NATAN (1969) Der Ausbau unterirdischer Hohlräume mit S + A Blech System Bernold. *Bergbauwissenschaft*, **16**, 117–26.

WOLF, K. (1935) Ausbreitung der Kraft in der Halbebene und ins Halbraum bei anisotropen Material. *Zeitschrift angew. Math. und Mech.*, **15**, 249–54.

WOLF, W. H., BROWN, J. L. & MORGAN, E. D. (1965) Morrow Point underground power plant. *A.S.C.E. Power Division Speciality Conference, Denver, Colorado*, pp. 493–534.

YAMAGUCHI, U. (1970) Seismic field study for the state of stress or cracks of rock around mining openings. *2nd Congress, International Society for Rock Mechanics, Belgrade*, report 4/48.

YOKOTA, J. (1962) Experimental studies on the design of the grouting curtain and the drainage system of the Kurobe IV dam. *Colloquium on Rock Mechanics, Salzburg*, paper 12.

YOKOTA, J. (1963) Experimental studies on the design of the grouting curtain and drainage for the Kurobe IV dam. *Rock Mechanics and Engineering Geology*, **1**, 104–19.

YOSHIDA, M. & YOSHIMURA, K. (1970) Deformation of rock mass and stress in concrete lining around the machine hall of Kisenyama underground power plant. *2nd Congress, International Society for Rock Mechanics, Belgrade*, report 4/29.

YOUNG, L. E. & STOCK, H. H. (1916) Subsidence resulting from mining. *Bull. No. 91*. Engineering Experiment Station, University of Illinois.

ZARUBA, Q. (1960) Plastische Verformung von Schichten in Täler und ihre Bedeutung für die Gründung von Bauwerken. *Zeitschrift für angewandte Geologie*, no. 9, 425–8.

ZARUBA, Q. (1962) Wasserdurchlässigkeitsprüfungen und Probeinjektionen für den Talsperrenbau. *Zeitschrift für angewandte Geologie*, **8**, 78–84.

ZARUBA, Q. (1964a) Geologische Erfahrungen beim Bau der Wasserkraftanlagen an der Vah (Waag). *Österreichische Wasserwirtschaft*, **16**, 262–70.

ZARUBA, Q. (1964b) Aufgaben der Ingenieurgeologie. *Österreichische Ingenieur Zeitschrift*, 109, 155–9.

ZARUBA, Q. & MENCL, V. (1961) *Ingenieurgeologie*. East Berlin and Prague.

ZARUBA, Q. & MENCL, V. (1976) *Engineering Geology*. Elsevier, Amsterdam.

ZELLER, E. (1954) Thermoluminescence of carbonate sediments. In H. Faul (ed.): *Nuclear Geology*. New York.

ZIENKIEWICZ, O. C. (1968) Continuum mechanics as an approach to rock mass problems. In K. G. Stagg & O. C. Zienkiewicz (eds.): *Rock Mechanics in Engineering Practice*. New York.

ZIENKIEWICZ, O. C. & CHEUNG, Y. K. (1964) Buttress dams on complex rock foundations. *Water Power*, 16, 193.

ZIENKIEWICZ, O. C. & CHEUNG, Y. K. (1965) Stresses in buttress dams. *Water Power*, 17, 69.

ZIENKIEWICZ, O. C. & CHEUNG, Y. K. (1967) *The Finite Element Method in Structural and Continuum Mechanics*. London.

ZIENKIEWICZ, O. C., CHEUNG, Y. K. & STAGG, K. G. (1966) Stresses in anisotropic media with particular reference to problems of rock mechanics. *J. Strain Analysis*, 1, 172–82.

ZIENKIEWICZ, O. C. & GERSTNER, R. W. (1961) Stress analysis and special problems of prestressed dams. *Proc. Am. Soc. Civil Engineering*, 19, 209.

ZIENKIEWICZ, O. C. & HOLISTER, G. S. (1965) *Stress Analysis*. New York.

ZIENKIEWICZ, O. C., MAYER, P. & CHEUNG, Y. K. (1966) Solution of anisotropic seepage by finite elements. *Proc. A.S.C.E. (Eng. Mech. Div.)*, EM 1, 111–20.

ZIENKIEWICZ, O. C. & STAGG, K. G. (1966) The cable method of *in situ* testing. *1st International Congress, Society for Rock Mechanics, Lisbon*, 1, 667–72.

ZIENKIEWICZ, O. C. & STAGG, K. G. (1967) Cable method of *in situ* rock testing. *Int. J. Rock Mech. Min. Sci.*, 4, 273–300.

ZIENKIEWICZ, O. C., TAYLOR, C. & GALLICO, A. (1970) Three-dimensional finite element analysis of the Tachien Arch Dam. *Water Power*, 22, 5/6, May/June.

Appendix 1
Comments on the bibliography

When Talobre published his treatise on Rock Mechanics in 1957 (*Mécanique des Roches*, Dunod: Paris) the available literature on the subject was still very limited. The list of references at the end of his book and those contained in a few publications covered most of the more important works of Heim, Schmidt, Lugeon, Terzaghi, etc. (see also the extensive lists of references in Jaeger: *Proceedings of the Institute of Civil Engineers*, March 1955 and Jaeger, *Water Power*, September–October 1961).

In the following ten years the number of valuable publications on rock mechanics increased considerably. The Proceedings of the First Congress of the International Society for Rock Mechanics fill three volumes (1966–7). The proceedings of American symposia are also very important.

Textbooks on engineering geology and rock mechanics

Testing Techniques for Rock Mechanics, American Society for Testing and Materials, Philadelphia, 1966.

Blyth, F. G. H. *Geology for Engineers*, Arnold: London, 1945.

Coates, D. F. *Rock Mechanics Principles*, Dept of Mines and Technical Surveys: Ottawa, 1966.

Desio, A. *Geologia applicata all'Ingegneria*, Hoepli: Milan, 1949.

Evans, I. & Pomeroy, C. D. *Strength, Fracture and Workability of Coal*, Pergamon Oxford, 1966.

Farmer, I. W. *Engineering Properties of Rocks*, Spon: London, 1968.

Gignoux, M. & Barbier, R. *Géologie des Barrages et des Aménagements Hydrauliques*, Mason: Paris, 1955.

Jaeger, J. C. & Cook, N. G. W. *Fundamentals of Rock Mechanics*, Methuen: London, 1968.

Krynine, D. P. & Judd, W. R. *Principles of Engineering Geology and Geotechnics*, McGraw-Hill: New York, 1957.

Lancaster-Jones, P. F. F. *Bibliography of Rock Mechanics*, Cementation Company: London, 1966. (89 pp).

Legget, R. *Geology and Engineering*, 2nd edition, McGraw-Hill: New York, 1962,

Lugeon, M. *Barrages et Géologie*, Lausanne: Rouge and Dunod: Paris, 1933.

Mary, M. *Barrages voûtes, historique, accidents et incidents*, Dunod: Paris, 1968.

Müller, L. *Der Felsbau. Theoretischer Teil*, Felsbau über Tag; 1. Teil (Theoretical section. Construction on rock). Vol. I, Enke Verlag: Stuttgart. (624 pp. extensive bibliography, mainly in German, about 800 items) (1963).

Obert, L. & Duvall, W. I. *Rock Mechanics and the Design of Structures in Rock*. Wiley: New York, 1967.

Price, N. J. *Fault and Joint Development in Brittle and Semi-brittle Rock*, Pergamon: Oxford 1965 (176 pp).

Ramsey, J. G. *Folding and Fracturing of Rocks*, McGraw-Hill: New York, 1967 (568 pp).
Richey, J. E. *Elements of Engineering Geology*, Pitman: London, 1964.
Reynolds, H. R. *Rock Mechanics*, Crosby Lockwood: London, 1961 (136 pp).
Schultz, J. R. & Cleaves, A. B. *Geology in Engineering*, Wiley: New York and Chapman & Hall: London, 1955.
Stini, J. *Technische Geologie*, Enke: Stuttgart, 1922.
Talobre, J. *La Mécanique des Roches*, Dunod: Paris, 1957; 2nd edition, 1967. (444 pp, extensive bibliography, about 200 items).
Terzaghi, K. *Theoretical Soil Mechanics*, Wiley: New York, 1943.
Trefethen, J. M. *Geology for Engineers*, 2nd edition, Van Nostrand: New York, 1959.
Walters, R. C. S. *Dam Geology*, Butterworth: London, 1962.
Zaruba, Q. & Mencl, V. *Ingenieurgeologie*, Akademie Verlag: East Berlin, 1961.
Zienkiewicz, O. C. & Stagg, K. G. *Rock Mechanics in Engineering Practice*, Wiley: New York, 1968.

Journals dealing with rock mechanics

Geologie und Bauwesen, Springer: Vienna, was for about twenty years the organ of the so-called 'Austrian School' of Geology and Rock Mechanics, In 1963 it became bilingual and is also known as '*Rock Mechanics and Engineering Geology*', also published by Springer, Vienna. Since 1969, *Rock Mechanics*.
Bulletin Geological Society of America.
Open File Reports, U.S. Geological Survey.
Bulletin American Association of Petroleum Geologists.
Journal of Physics of the Earth.
Journal Geophysical Research.
Transactions American Institute of Mining, Metallurgical and Petroleum Engineers.
Mining Engineering.
Geophysics.
Reports, U.S. Corps of Engineers.
Reports, U.S. Bureau of Reclamation, U.S. Dept. of the Interior, Denver, Colorado.
Proceedings, American Society for Testing and Materials.
Journal of Soil Mechanics and Foundation Division, American Society of Civil Engineers.
Civil Engineering, American Society of Civil Engineers.
Philosophical Transactions, Royal Society, London.
Proceedings, Institution of Civil Engineers, London.
Geotechnique, Institution of Civil Engineers, London.
Civil Engineering and Public Works Review, London.
Water Power, London.
Tunnels and Tunnelling, London.
International Journal of Rock Mechanics and Mining Sciences, Pergamon: Oxford, 1964.
Journal of Engineering Geology, Geological Society: London (1967)
Engineering Geology, Elsevier: Amsterdam, 1965.
Annales de l'Institut du Bâtiment et des Travaux Publics. Paris.
Revue de l'Industrie Minérale, Paris.
Publications of the 'Comité Français de Mécanique des Roches' (French Committee on Rock Mechanics).
Memoirs of the Laboratorio National de Engenharia Civil, Lisbon.

Bulletins of the Istituto Sperimentale Modelli e Strutture (ISMES), Bergamo, Italy.
Publicazioni della Università degli studi di Roma, Italy.
L'Energia Elettrica, Milano, Italy.
Publicazioni dell Istituto di Scienza delle Costruzioni, Politecnico, Milano, Italy.
Zeitschrift für Angewandte Geologie.
Der Bauingenieur, West Berlin.
Die Wasserwirtschaft, West Berlin.
Die Schweizerische Bauzeitung, Zurich, Switzerland.
Publications of the Laboratorio de Reologia y Geotecnica, Madrid, Spain.
Publications of the Technical Laboratory, Central Research Institute of Electric Power Industry, Tokyo, Japan.
Publications of the Yugoslav National Committee on Rock Mechanics.
Publications of the Yugoslav University of Sarajevo.

Symposia and congresses

6th International Congress on Large Dams (ICOLD), New York, 1958.
Quarterly, Colorado School of Mines: 3rd Symposium on Rock Mechanics, vol. 54, no. 3, July, 1959.
3rd International Conference on Strata Control, Revue de l'Industrie Minérale, Paris, 1960.
International Symposium on Mining Research, Missouri School of Mines, Rolla, Mo. 1961 and Pergamon: Oxford, 1962.
7th International Congress on Large Dams, Rome, 1961.
4th Symposium on Rock Mechanics (ed. H. Hartmann), Bull. 76, Min. Inds. Exp. Stn.; Penn, State University, 1961.
5th Symposium on Rock Mechanics (ed. C. Fairhurst), Pergamon: Oxford, 1963.
Rock Mechanics Symposium, Queen's University, Kingston, Ontario, Canada, 1963, (ed. Dept. of Mines and Technical Surveys, Ontario, 1964).
International Conference on the State of Stress in the Earth's Crust, Santa Monica, California, May 1963 (ed. W. R. Judd, Elsevier: New York, 1964).
4th International Conference on Strata Control and Rock Mechanics, Columbia University: New York, 1964.
8th International Congress on Large Dams (Edinburgh), Vol. i, 1964.
6th Symposium on Rock Mechanics, Univ. of Missouri: Rolla, Mo., 1964.
Mechanics and Strata Control in Mines, S. African Inst. Min. Met.: Johannesburg, 1966.
1st Congress International Society for Rock Mechanics. Vols. i, ii and iii, Lisbon, 1966 and 1967; *2nd Congress*, Belgrade, 1970; *3rd Congress*, Denver, 1974.
9th International Congress on Large Dams, Istanbul, 1967.
23rd Colloquium Austrian Society for Geomechanics, Salzburg, 1974.

Appendix 2
Measurement conversion tables

Conversion from f.p.s. to metric units.

Linear measures

1 inch (in) = 25·4 mm

1 foot (ft) = 0·3048 m

1 yard (yd) = 0·9144 m

1 mile = 1·60934 km

1 mm = 0·03937 in

1 m = 3·28084 ft = 39·3701 in

1 m = 1·09361 yd

1 km = 0·62137 miles

Square measures

1 in^2 = 6·4516 cm^2

1 ft^2 = 0·0929 m^2

1 acre = 4046·86 m^2

1 mile^2 = 2·58998 km^2

(= 640 acres)

1 cm^2 = 0·155 in^2

1 m^2 = 10·764 ft^2

= 1·19599 yd^2

1 m^2 = 0·00025 acre

1 km^2 = 0·3861 mile^2

Cubic measures

1 in^3 = 16·387 cm^3

1 ft^3 = 0·02832 m^3

1 yd^3 = 0·76455 m^3

1 pint = 0·56825 litres

1 Imp. gallon = 4·54596 litres

1 U.S. gallon = 3·78533 litres

= 0·834 Imp. gallon

= 231 in^3

1 acre-foot = 43 560 ft^3

= 1233·48 m^3

1 cusec = 6·23 gal/s

= 0·539 million gal/day

= 0·03832 m^3/s

1 cm^3 = 0·06102 in^3

1 m^3 = 35·3148 ft^3

1 m^3 = 1·30795 yd^3

1 litre = 1·7598 pints

1 litre = 0·21998 Imp. gallon

1 m^3/s = 35·3148 cusec

Densities

1 lb/in^3 = 27·68 g/cm^3

1 lb/ft^3 = 16·019 kg/m^3

1 g/cm^3 = 0·03613 lb/in^3

1 kg/m^3 = 0·06243 lb/ft^3

Pressures

1 lb/in^2 = 0·00070 kg/mm^2

= 0·07031 kg/cm^2

1 lb/ft^2 = 0·000488 kg/cm^2

= 4·88243 kg/m^2

1 ton/ft^2 = 10·9366 tonne/m^2

1 kg/cm^2 = 14·223 lb/in^2

= 2048·16 lb/ft^2

1 kg/m^2 = 0·20481 lb/ft^2

1 tonne/m^2 = 0·091436 ton/ft^2

SI units conversion

1 N = 1/9·81 = 0·1019 kg force

1 MN = 101·9 tonnes force

1 kg force $= 9\cdot81$ N
1 Pa $= 1$ N/m$^2 = 0\cdot1019 \times 10^{-4}$ kg/cm^2
1 KPa $= 0\cdot01019$ kg/cm^2
1 MPa $= 10^6$ Pa $= 10\cdot19$ kg/cm^2
1 bar $= 10^6$ dynes/cm$^2 = 1\cdot019$ kg/cm^2
1 kg/cm$^2 = 1/10\cdot19$ MPa $= 0\cdot0981$ MPa
1 t/cm$^2 = 1000$ kg/cm$^2 = 98\cdot1$ MPa

Appendix 3
Table of geological formation and earth history

Eras	Systems and series	Max. thickness of stratum (ft)	Approximate ages of base of system or series (years)
Caenozoic (Recent life) Age of mammals (including man in quaternary) and flowering plants	*Quaternary*	4 000	
	Recent		25 000
	Pleistocene or Glacial		1 000 000
	Tertiary	63 000	70 000 000
	Pliocene		
	Miocene	Rise of Alpine Chain	
	Oligocene		
	Eocene		
	Palaeocene		
Mesozoic (Intermediate life) The age of giant reptiles and of primitive conifers	Cretaceous	64 000	135 000 000
	Jurassic	20 000	180 000 000
	a malm		
	b dogger		
	c lias		
	Triassic	25 000	225 000 000
Palaeozoic (Ancient life) Extinct races of fishes, amphibians, early invertebrates, spore-bearing plants	Permian	13 000	270 000 000
	Rise of Hercynian Chain		
	Carboniferous	40 000	350 000 000
	Devonian or Old Red Sandstone	37 000	400 000 000
	Rise of Caledonian Chain		
	Silurian	15 000	440 000 000
	Ordovician	40 000	1 500 000 000
	Cambrian	40 000	1 600 000 000

Pre Cambrian
The probable age of the earth is about 4·5 thousand million years but the oldest rocks found at the earth's surface are dated about 3·5 thousand million years.

(after A. Holmes, 1960; J. E. Richey, 1964; and Neil Duncan, 1968).

Appendix 4
Some petrographic properties of rocks

1. Common rock-forming minerals

(a) *In eruptive rocks*

Felspars
   Orthoclase: $K_2OAl_2O_36SiO_2$
   Plagioclase: $NaAlSi_3O_8$ or $CaAl_2Si_2O_8$ (variable depending on type of feldspar)
   Quartz: SiO_2
Micas
   Muscovite: $KAl_2(Si_2Al)O_{10}(OH)_2$
   Biotite: $K_2(MgFe)_6(SiAl)_8O_{20}(OH)_4$
Amphiboles
   Hornblende: (Na, Ca, Mg, Fe, Al) silicate
   Augite: (Ca, Mg, Fe, Al) silicate
   Olivine; $(MgFe)_2SiO_4$

Remarks: orthoclase contains 64–68 % SiO_2; plagioclase contains 43–62 % SiO_2; albite 64–69 %; muscovite contains mainly K and Al_2O_2 and is white; biotite contains Mg and Fe and is black or dark green; hornblende contains more or less water and can be found in acid rocks (granite and syenite) or in basic rocks (basalts, trachytes, andesites).

(b) *In sedimentary rocks*
Calcite: $CaCO_3$ (soluble in weak acids)
Dolomite: $MgCa(CO_3)_2$ (not easy to dissolve)
Anhydrite: SO_4Ca
Gypsum: $So_4Ca, 2H_2O$
Kaolinite: $Al_2Si_2O_5(OH)_4$
Iron oxides: Fe_2O_3; $2Fe_2O_3, 3H_2O$

2. Classification of igneous rocks according to Silica content (SiO$_2$)

Texture	Acid (>66%)	Intermediate (52–66%)	Basic (<52%)
Plutonic (coarse) Hypabyssal	granite micrugranites	syenite-diorite microsyenite microdiorites	gabbro microgabbros dolerite peridotite
Volcanic (fine) glassy	rhyolite obsidian	trachyte-andesite	basalt
Major Mineral Constituents	quartz orthoclase mica	orthoclase, plagioclase hornblende, mica	augite plagioclase olivine hornblende

3. Classification of rocks according to hardness and density

	Hardness Mohs' scale	Vickers scale	Specific gravity
	1–10		
Diamond	10	10 000	
Corindon	9	2 000	
Topaze	8	1 400	
Quartz	7	1 120	2·65
Felspar	6	800	2·65
Olivine	6–7		3·5
Hornblende	5–6		3·05
Augite	5–6		3·05
Dolomite	4		
Muscovite	2·5		2·8
Biotite	3		3
Calcite	3		2·7
Salt	2		
Gypsum		36	
Kaolinite	1		

Author Index

Abel, J. F. 263, 463
Abraham, K. H. 452, 458–60, 463
Addinall, E. 463
Adriatica di Elettricita 402
Ahlvin, R. G. 469
Airy 89, 90, 290, 291
Albissin, (d') M. 69, 70, 467
Albrecht, H. 482
Alexander, L. G. 463
Ambraseys, N. N. 463
American Geophysical Society 2
American Society of Engineering
  Geologists 2
American Society for Testing and
  Materials 2
American Task Committee 335
Ampferer 210, 463
Amstutz, E. 237, 277, 322, 463
Anderson, Sir Duncan 463
Anderson, H. W. 463
Anderson, J. G. C. (Prof.) 447
Andrea, C. 11, 13, 263, 463
Anonymous 463–4
Argyris, J. H. 463
Army, Corps of Engineers (U.S.) 47, 48
Army (U.S.) W.E.S. 124, 125
Audibert, P. 145, 147, 471
Austrian School 3, 27, 28, 32; see also
  Müller; Stini
Austrian Society for Geophysics and
  Engineering Geology 3, 5
Austrian Society of Rock Mechanics 3, 5,
  130, 212, 265, 301, 392
Avranesco, A. 463

Ballot, L. 385
Balmer, G. B. 463
Banks, J. A. 120, 121, 335, 464
Barbier, R. 160, 491
Barron, K. 125, 484
Bart, R. 472
Barth, S. 451, 464
Barton, N. 55, 56, 156, 283–5, 287, 464
Baudendistel, M. 259, 318, 481
Beamonte, M. 478
Beavis, F. C. x
Bellier, J. 199, 387, 464
Bellport, B. P. 35, 368, 372, 468

Benson 228, 464
Berghinz, C. 464
Berigny, C. 332, 464
Bernaix, J. 29, 53, 55, 67, 68, 72, 82, 117,
  150, 210, 347, 396, 398, 400, 464, 465,
  471
Bernède, J. 260, 261, 464, 471
Bernold, J. 263, 266–8, 465
Berry, D. S. 465
Bertacchi, P. 264
Beutsch, N. 270, 465
Bieniawski, Z. T. 46, 156–8, 160, 281–4,
  286, 287, 324, 465
Binnie, G. M. 270, 465
Bischof, R. 465
Bishop, A. W. 465
Bjerum, L. 5, 212, 465
Blackburn, C. L. 465
Blasius 197, 198
Bleifuss, D. J. 465
Blyth, F. G. H. 465, 491
Bollo, M. F. 116, 465, 479
Bonnet, G. 465
Boozer, G. D. 75, 486
Borowicka, H. 205, 465
Bourguignon, 465
Boussinesq, J. V. 107, 110, 113, 122, 137,
  138, 161, 168, 169, 171, 173, 216, 396,
  398, 465
Bowen, R. 465
Bozetto, P. 310–12, 320, 324, 465
Brace, W. F. 466
Bray, J. W. 178, 466
Brock, J. S. 472
Broili, L. 404, 406–9, 418, 466
Brown, E. T. 488
Brown, J. D. 368, 369, 371, 489
Brown, P. D. 466
Brunel, M. I. 226
Bureau of Mines (U.S.) 2, 76, 97
Bureau of Public Roads (U.S.) 154
Bureau of Reclamation (U.S.) 2, 35, 47, 56,
  95, 98, 107, 108, 119, 347, 368, 372–5, 466
Burmister, D. M. 466
Buro 258, 466
Butin, P. 357, 484

Calembert, L. 466

Caloi, P. 415, 418, 422, 466
Cambefort, H. 337, 466
Campbell, J. G. 465
Campo (Del), *see* Del Campo
Camponuovo, H. F. 210
Candiani, G. 466
Carpentier, L. 260, 261, 471
Casagrande, A. 5, 337–40, 342, 345, 346, 353, 466
Cecil, O. S. 156, 263, 283, 466
Centre d'Etudes du Bâtiment 111
Cerruti 113, 138, 161, 168, 169, 171, 466
Cheung, Y. K. 25, 187, 188, 490
Cholet, H. 466
Chunnett, E. R. P. 97, 466
Clark, G. B. 41, 108, 116, 145, 466
Cleaves, A. B. 492
Clinch, R. L. 482
Clough, R. W. 466
Coates, D. F. 467, 491
Colback, P. S. 467
Collin, A. 23, 467
Commissione Parlamentaria 467
Compagnie des Chemins de Fer du Nord 151
Compagnie Internationale du Canal de Suez 151
Compagnie Nationale du Rhône 14
Comrie, J. 467, 473
Congress on Large Dams, *see* International Commission
Cook, B. 18, 477
Cook, N. G. W. 475, 491
Cooke, J. B. 467
Cording, E. J. 283, 467
Corroy, 388, 389
Coulomb 5, 45, 78, 119, 163–5, 178, 213, 249, 326
Coyne, A. 126, 345, 357, 467
Cundall, P. A. 191, 467
Curtis, J. J. 172, 174–6, 488

Dachler, R. 338, 467
Daemen, J. J. K. 297, 467
Dagnaux, J. P. 467
Dal Piaz, G. 403, 467
Dalton, F. D. 468
Darcy, H. 64, 193, 197, 333, 340, 396, 467
Da Silva, A. F. 484
Datei (Prof.) 412
Davin, N. 467
Dearman, W. R. 155, 468
Deere, D. U. ix, 5, 8, 9, 10, 34, 82, 142, 143, 155, 250, 283, 284, 468, 484
Del Campo, A. 194, 328, 329, 468, 486
Denkhaus, H. G. 154, 468
Denver Laboratory 372
Department of Mines (U.S.) 76, *see* Bureau of Mines

Descoendses, F. 269, 468
Desio, A. 491
Diethelm, W. 307, 468
Dietlicher, E. 104, 109, 470
Dodd, J. S. 463
Dodds, R. K. 365, 468
Dominy, F. E. 35, 368, 372, 468
Donath, F. A. 54, 468
Dubertet 435, 445
Duffaut, P. 263, 305, 468, 479
Duncan, N. Q. 6, 7, 9, 37, 38, 142, 144–7, 150, 155, 468, 469, 496
Dunne, M. H. 6, 469
Duvall, W. I. 2, 96, 228, 229, 250, 273, 274, 290, 481, 491

Eckenfelder, G. V. 469
Ecole Polytechnique, Paris (Laboratory) 47, 52, 56, 61–8, 83, 84, 182, 347, 385, 392
Edbrooke, R. F. 472
Eddington, R. H. 465
Edney, J. 447
Egger, P. 263, 287, 305, 469
Eichinger, A. 205, 484
Einstein, H. 469
Electricité de France (E.d.F.) 14, 60, 111, 118, 120, 125, 130–2, 387, 388, 392
Electroconsult 402, 427, 431
Electro-Watt 306, 469, 478
Emery, C. L. 469, 484
Energieversorgung Ostbayern 469
Ente Nazionale per l'Energia Elettrica (E.N.E.L.) 469
Epstein, B. 469
Eristov, V. S. 485
Estèves, J. N. 469
Evangelisti (Prof.) 412
Evans, I. 491
Evison, F. F. 139, 142, 469

Fairhurst, C. 469
Falcon, N. L. 210, 472
Falconnier, 20
Fanelli, M. 323, 469
Farmer, I. W. 469, 491
Farran, J. 27, 59, 469
Favre, L. 281, 305
Federal Board of Rhodesia and Nyasaland 444
Fedorenko, A. N. 485
Fenner, R. 2, 89, 257, 258, 297, 310, 469
Ferrandon, J. 469
Feshbach, H. 481
Föppl, L. 469
Fogden, C. A. 465
Foster, C. R. 469
Franklin, J. A. 158, 470

French Committee of Experts 385–91, 394, 400, 470
French Electricity Board: *see* Electricité de France
French ICOLD Committee 125, 126, 133, 134, 143
French Ministry of Agriculture 385
French Railway 151
Frey Baer, O. 470
Freyssinet 95, 109, 121, 357
Fröhlich, O. K. 487
Fumagalli, E. 44, 115, 123, 137, 184, 186, 210, 347, 352, 353, 470

Gallico, A. 431, 490
Gamond 151
Garnier, J. C. 467
Gavazzi, P. 466
Geological Survey, Zambia 445–7
Georgia Institute of Technology 51
Gibb, Sir Alexander, and Partners 351, 470
Gicot, H. 126, 127, 201, 345, 470
Gignoux, M. 470, 491
Gilg, B. 104, 109, 136, 470
Gillott, C. A. 22, 475
Gilman, J. J. 69, 475
Giudici, D. F. 403, 406, 421, 470
Glossop, R. 332, 470
Glötzl, F. 470
Glover, R. E. 394, 470
Goblot, H. 325
Goffi, L. 261, 470
Goguel 388, 389
Golser, J. 265
Gramberg, J. 470
Grant, L. F. 470
Griffith, A. A. 68, 71–3, 76, 78, 98, 102, 196, 245, 471
Griggs, D. T. 124, 471
Grujic, N. 476
Grundy, C. F. 471
Gruner, E. 325, 379, 471
Gstalder, S. 77, 79, 471
Guthrie-Brown, J. 3, 471
Gyence, M. 467

Habib, P. 62, 68, 82, 108, 145, 147, 186, 260, 261, 396, 471, 479
Hackett, P. 463
Haefeli, R. 413, 471
Hager, R. 196, 471
Halcrow, Sir William 238
Hamrol, A. 58, 59, 142–4, 196, 471
Handin, J. 69, 196, 471
Hanna, T. H. 471
Hardy, H. R. 75, 76, 117, 471
Harrison, J. V. 210, 472
Hartman, B. E. 98, 259, 260, 472
Harza, R. D. 472

Hast, N. 97, 98, 472
Hayashi, M. 52, 54, 259, 264, 300, 315, 472
Heiland, C. A. 472
Heim, A. 1, 2, 85, 86, 88, 89, 92, 98, 203, 208–11, 220, 411, 472, 491
Heller, S. R., jr 472
Hellstrom, B. 488
Hendron, A. J., jr 468, 472
Hibino, S. 259, 472
Higgs, D. V. 471
Hill, R. 250, 472
Hock, W. J. 49, 50, 476
Hoek, E. 71, 72, 204, 208, 210, 245, 472, 473
Holister, G. S. 490
Holmes, A. 496
Hook, 290
Hoskin, G. 473
Houpert, R. 77, 78, 473
Howland, R. C. J. 473
Huber, W. G. 479
Hucka, V. 473

Ikoda, K. 473
Institution of Civil Engineers (I.C.E.) 337
International Association for Engineering Geology 156
International Commission on Large Dams, (ICOLD) 3, 112–14, 120, 123, 126, 325, 347, 394
International Conference on Soil Mechanics 3, 5, 118
International Society for Rock Mechanics 3, 5, 156, 394, 491
Irmay, S. 473
Ishijima, Y. 259, 487
Istituto Sperimentale Modelli e Strutture (ISMES) 44, 45, 95, 123, 347, 352, 401

Jaecklin, F. P. 131, 257, 473
Jaeger, C. ix, 2, 11, 19, 23, 25, 88, 96, 98, 105, 113, 114, 140, 172, 193, 195, 197, 201, 210, 219, 220, 222, 230, 237–9, 243, 244, 247, 249–54, 258, 259, 287, 290, 300, 308–12, 317–22, 325, 335, 339, 340, 342, 344, 353, 355, 385, 394–6, 412–15, 417–20, 422, 444, 446–8, 457, 473, 474, 491
Jaeger, J. C. 474, 475, 491
Jimenes Salas, J. A. 83, 123, 325, 475
John, K. W. 27, 33, 179, 180, 209, 210, 340, 349, 433, 475
Jones, F. 435
Jones, K. S. 435–7, 476
Jones, P. F. F. 22, 23, 342, 475
Jorstad, F. 155, 465
Jovanovic, L. 258, 322, 476
Judd, W. R. 466, 475, 491
Jumson, N. H. 465

Kaech, A. 319, 322, 448, 450
Karlsruhe University 305
Karman, T. von 205, 475
Kastner, H. 94, 250, 258, 261, 263, 287, 290–6, 310, 317, 475
Keith, R. E. 69, 475
Kelvin, Lord 145, 148
Kennedy, B. A. 210, 475
Kenney, T. C. 409, 410, 413, 475
Kent, P. E. 475
Keppie, J. D. 446
Kierans, T. W. 228, 464
Kiersch, G. A. 5, 406, 475
Kieser, A. 281, 475
Kieslinger, A. 86, 201, 475
Kitahara, Y. 472
Knill, J. L. 435–7, 447, 476
Kobilinsky, 482
Kobold, F. 221, 476
Kommerel, O. 476
Krsmanović, D. 50, 52, 80, 81, 138, 181–6, 216, 396, 476
Kruse, J. H. 376, 486
Krynine, D. P. 476, 491
Kujundzic, B. 96, 105, 258, 280, 322, 328, 476

Laboratorio Nacional de Engenharia Civil 11, 57, 96, 323
Lakshanan, J. 467
Lamé 148
Lancaster-Jones, P. F. F. 491
Lane, K. S. 49, 50, 476
Lane, R. G. T. 350, 351, 436, 437, 476
Lang, T. A. 477
Lange, E. 477
Langlois 464
Langof, Z. 50, 52, 80, 81, 476
Laplace 195
Lapparent, de 151
Lasalle Research Laboratory (Montreal) 417
Lauffer, H. 281–3, 477
Laurent, D. 477
Lee, E. H. 477
Lee, K. L. 484
Leeman, E. R. 97, 323, 477
Legget, R. 477, 491
Leliavsky, S. 331, 332, 477
Lenssen, S. 445, 447, 477
Lévy, M. 330, 331, 379, 477
Lewis, D. R. 471
Libby, J. B. 18, 477
Lien, R. 464
Ling-Chih-Bing 477
Link, H. 41, 42, 477
Locher, H. G. 478
Lombardi, G. 237, 258, 263–6, 281, 287, 297–9, 301–4, 306, 318, 425–7, 473, 474, 478

Londe, P. 208, 210, 339, 341, 347, 394–6, 398, 400, 478
Lopez, J. J. B. 58, 81, 118–20
Los Angeles Water Board 379
Lottes, G. 214, 478
Lotti, C. 214, 478
Louis, C. 197, 478
Love, A. E. H. 478
Lugeon, M. 22, 143, 144, 333–7, 345, 479, 491
Lunde, J. 464

Madhow, M. R. 479
Madill, J. T. 18, 477
Maillart, R. 263, 479
Malina, H. 318, 481
Mamillan 479
Manfredini, G. 109, 479
Mantovani, E. 264, 323, 442, 479
Marty, G. 77, 79, 471
Mary, M. 240, 325, 379, 383–5, 397–400, 479, 491
Masur, A. 263, 281, 479
Matheson, G. D. 445–9, 479
Mathias, F. 479
Mathieu, A. 150
Maury, V. 182, 479
Mayer, A. 21, 41, 59, 60, 111, 112, 187, 479
Mayer, P. 490
Mazanti, B. B. 50, 480
Mazenot, P. 131, 136, 480
McCaig, I. W. 479, 482
McClintock, F. A. 71, 479
McClure, C. R. 487
McQueen, A. W. F. 313–15, 479
McWilliams, J. R. 39, 479
Mencl, V. 208, 210, 409, 410, 412, 480, 492
Mercali 382
Merrill, R. H. 93, 97, 376, 480
Milic, S. 181, 182, 476
Miller, D. V. 480
Miller, R. P. 5, 8, 9, 74, 480
Milli, E. 480
Milovanovic, D. 380, 480
Mintchev, I. T. 250, 480
Missouri Laboratory 48, 49
Mitchell, J. H. 113, 480
Mogi, K. 480
Mohr, 43, 45, 48, 49, 51, 123, 134, 153, 161–165, 178, 180, 229, 248, 255, 256, 291, 295, 310, 366, 367, 375
Montarges, P. 76, 480
Moore, B. W. 153, 480
More, G. 480
Morgan, E. D. 368, 369, 371, 489
Morgan, H. D. 261, 480
Morgan, T. A. 101, 480
Morlier, P. 76, 77, 142–4, 150, 480

Müller, L. 23, 27, 28, 31–3, 53, 54, 138, 155, 179, 204, 208–24, 259, 281, 282, 297, 318, 340, 352, 368, 401–6, 409–16, 418, 421, 422, 480, 481, 491
Murrell, S. F. 481
Muskat, M. 193, 481

Nadai, A. 481
Naef, E. 178, 216, 217, 481
National Coal Board 96, 97
Naylor, D. J. 247, 248, 481
Neff, T. L. 48, 50, 481
Neuhauser, E. 477
Newton, I. 218, 411, 413, 419
Nikuradse 197
Nilsson, T. 98, 472
Nonweiller, E. 409, 410, 413, 481
Norse, P. M. 481

Obert, L. 2, 96, 228, 229, 250, 273, 274, 290, 481, 491
Oberti, G. 95, 130, 261, 349, 353, 366, 393, 470, 482
Oliveira, R. 142, 150, 482
Olivier, H. 158, 436, 438–41, 482
Olivier-Martin, D. 17, 482
Olsen, O. J. 370, 488
Onodera, T. F. 11, 143, 482

Pacher, F. 53, 54, 138, 179, 195, 340–2, 349, 353, 394, 396, 442, 450, 482
Pahl, A. 482
Pancini, M. 348, 482
Pandolfi, C. 214, 478
Panek, K. A. 101, 480, 482
Panet, M. 27, 31, 264, 482
Paris Laboratory, *see* Ecole Polytechnique
Parsons, R. D. 467
Pascal, B. 86
Paton, T. A. L. 463
Patterson 482
Paulding, B. W. 73, 82, 482
Paulmann, H. G. 39, 482
Peck, R. B. 457, 483
Pelletier, J. 14, 16, 17, 483
Péquignot, E. A. 261, 474, 483
Perami, R. 68, 70, 483
Petersen, J. R. 93, 97, 376, 480
Petty, S. 6
Pfister, R. 483
Piaz, *see* Dal Piaz
Pierson, L. N. 483
Pincus, H. J. 483
Piquer, J. S. 328, 329, 468
Piraud, J. 263, 305
Pirrie, N. D. 270, 483
Poiseuille, 194, 197, 198
Poisson 40, 42, 44, 87, 88, 94, 110, 111, 113, 136, 141, 142, 149, 170, 176, 177, 183, 194, 197, 198, 217, 259, 300, 372, 378
Polensky 454
Pomeroy, C. D. 491
Portuguese National Laboratory 56
Prandtl 109, 138, 168
Pranesh, M. R. 479
Preussische Elektrizitäts Aktiengesellschaft 450, 457
Price, N. Y. 491
Proctor, R. V. 483
Puyo 464

Rabcewicz, L. von 201, 256, 257, 262–5, 293, 294, 483
Radosavljevic, Z. 258, 323, 476
Ramsay, J. G. 492
Rankine 111, 337
Rebaudi, A. 482
Rechsteiner, G. 269
Reed, J. J. 483
Reichmuth 38
Rellensmann, O. 229, 483
Rescher, O. J. 264, 273, 323, 442, 450, 483, 484
Reyes, S. F. 250, 484
Reynolds, H. R. 150–2, 484, 492
Rhône, *see* Compagnie Nationale du Rhône
Ribes, G. 357, 484
Riccioni, R. 323, 484
Richart, F. E., jr 90, 91, 93, 317, 487
Richey, J. E. 484, 492, 496
Richter, R. 484
Ripley, C. F. 484
Roberts, A. 97, 484
Roberts, C. M. 484
Robinson, L. H. 484
Rocha, M. 11, 96, 109, 118, 323, 351, 393, 484
Roff, J. 476
Roš, M. 205, 485
Rosa, S. A. 485
Rosenström, S. 442, 443, 485
Roussel, J. M. 145–50, 485
Rowbottom 445
Ruesh, H. 75, 485
Rutschmann, W. 269, 305, 485
Rziha, F. 1, 85, 485

Sabarly, F. 335, 341–5, 347, 478, 485
Sales, T. W. 465
Sampaolo, A. 479
Savin, G. N. 485
Scheiblauer, J. 485
Schmidt 145
Schmidt, A. 485
Schmidt, J. 1–2, 89, 257, 485, 491
Schmidt, W. 30, 33

Schneebeli, G. 485
Schneider, B. 145, 485
Schnitter, G. 485
Schober, W. 477
Schoeller, H. 485
Schrafl, A. 485
Schultz, J. R. 492
Schwartz, A. E. 50, 485
Scott, P. A. 485
Seeber, G. 105, 117, 118, 230, 259, 477, 485
Seldenrath, T. R. 229, 485
Semenza, C. 344, 345, 393, 402, 403, 421
Semenza, E. 23, 345, 403–6, 421, 470, 486
Serafim, J. L. 6, 37, 58, 59, 70, 71, 81, 118–20, 144, 150, 194, 196, 350, 351, 375, 392–4, 486
Serdengecti, S. 75, 486
Serieys, P. M. 196, 486
Shackleton, R. N. 447
Shaw, J. R. 466
Sheermann-Chase, A. 9
Sheffield University 97
Shuk, T. H. 117, 131, 486
Siemens 456
Sigvaldson 228, 464
Silvera, A. F. 484
Silverio 323, 484
Simpson, C. N. 479
Skempton, A. W. 48, 50, 52, 411, 412, 486, 490
Skinner, E. H. 156
Slot, T. 486
Societa Adriatica 402
Sogei 357
Sonderegger, A. 486
Soviet Army 380
Sowers, G. E. 50, 480
Spadea, C. 466
Stagg, K. G. 25, 112–14, 188, 486, 490
Stini, J. 3, 204, 210, 211, 486, 492
Stock, H. H. 2, 489
Stragiotti 412, 413
Stroppini, E. W. 376, 486, 487
Stuart, W. H. 487
Stucky, A. 109, 357, 358, 487
Supino 412, 413
Sutcliffe, H. 263, 487
Suzuki, K. 259, 487
Swiss Federal Railways 1, 239, 306
Swiss National Committee on Large Dams 487
Szechy, K. 487

Talobre, J. A. 2, 31, 102, 106, 117, 131, 134–6, 166, 194, 201, 255, 257, 262–5, 297, 310, 333, 365–7, 401, 450, 487, 491, 492
Tattershall, 487
Taylor, I. G. 431, 487

Terrametrics Inc. 99, 100, 259
Terzaghi, K. von 2, 27, 32, 60, 85–91, 93–97, 109, 138, 168, 204–8, 212, 228, 238, 257, 283, 284, 317, 391, 394, 395, 398, 401, 487, 491, 492
Thenoz, B. 27, 59, 70, 71, 469
Thoma, E. 11, 13, 487
Tiedmann, H. R. 156
Timoshenko, S. 487
Tincelin, M. E. 366, 487
Toews, N. A. 125, 464
Tokachirov, V. A. 485
Torno-Kumagai 272, 431, 433
Torre 78
Torroja, E. 488
Toulouse Mineralogic Research Laboratory 60
Trefethen, J. M. 488, 492
Tresca 250
Trollope, D. H. 138, 181–4, 186, 488

Uriel, S. 83, 123, 475
U.S. Department of Mines 76
U.S.A., *see* Bureau of Mines, Bureau of Public Roads, Bureau of Reclamation, Army, American Societies, etc.

Van Heerden, W. L. 97
Veltrop, J. A. 172, 174, 175, 359–65, 488
Vienna University 2
Vogt, F. 488
Voight, B. 145, 148
Vouille, G. 44, 62, 145, 147, 471, 477, 488
Vrana, S. 446, 475
Vutukuri, V. S. 488

Wahlström, E. 488
Wakeling 487
Waldorf, W. A. 88, 89, 172, 174, 175, 177, 359–65, 488
Wallace, G. B. 370, 488
Walsh, J. B. 71, 73, 479, 488
Walters, R. C. S. 220, 355, 488, 492
Ward 487
Weber, E. 485
Weirich, K. W. 268, 488
Westerberg, G. 3, 488
Westerguard, H. M. 489
Weyl, P. K. 471
White, J. E. 489
Wickham, G. E. 156, 158
Widmann 454
Wild, B. L. 467
Williamson, T. N. 489
Wilson, S. D. 489
Windes, S. L. 481, 489
Wise, L. L. 489
Wittke, W. 197, 198, 489
Wohlbier, H. 266, 489

Wolf, K. 489
Wolf, W. H. 368, 369, 371, 489

Yokota 195, 340, 349, 353, 482, 489
Yoshida 312, 489
Yoshimura 489
Young, L. E. 2, 37, 41, 129, 170, 345, 378

Zaruba, Q. 337, 489, 490, 492
Zeller, E. 490
Zienkiewicz, O. C. 25, 112–14, 122, 123, 138, 186–8, 218, 277, 297, 330, 431, 456, 490, 492
Zollner 454

Index of geographical names, dam sites, reservoirs, tunnels and caverns

Case histories are indicated by italics

Aiguille du Midi 28
Airolo 306
Alamansa dam 325
Allt-na-Lairige dam 120–2, 335, 336
Alps 207, 209, 220, 242, 264
Amsteg tunnel 1, 103, 104, 239
Andes 19, 221, 239
Appennines 379
Arequipa (Peru) 381, 382
Arlberg tunnel 226
Ashford Common water tunnel 238
Aswan (High Dam) 338

Baji-Krachen rock spur 225, 424–7
Bathie 112
Beni Bahdel dam 357
Ben Metir dam 358
Berea 75, 196
Bergamo 352
Bersimis I power station 281, 317
Bersimis II power station 314
Bhakra dam 24
Bilbao 379
Black Canyon gorge 369
Blackwall tunnel 226
Blenio tunnels 279
Blue Mesa rockfill dam 368
Bort dam 24, 208
Bort tunnel, 106, 107, 135
Bouzet dam 330, 379
Brechbergmueller 308
British Columbia 18
British Isles 155
Brommat 308
Bryant Park tunnel 240

Cabora Bassa 150, 310, 318
Calais 151
Calgary 238
Cá Selva dam 352
Camarasa dam 22, 23, 325, 342
Cannes 355
Castillon arch dam 22, 345, 355, 356
Channel Tunnel 139, 150, 151
Chaudanne arch dam 22, 126, 345, 355, 356
Cheurfas dam 120
Churchill Falls 310, 319
Chute des Passes 281, 313, 315

Colorado River Project 368
Crystal Dam 368
Cunene River 228, 442, 457

Denver (Colorado) 2, 374
Dez dam 109, 165–7, 365–7, 401
Dez River 365
Dieppe 332
Dnjeprogues dam 380
Dokan dam 342, 345
Dover 151

Eder dam 380
Egypt 325
Eichen pressure shaft 104
Elche dam 325
English Channel 151, 199
Erraguene dam 358

Fionnay–Riddes tunnels 278
Flims landslide 220
Frankfurt 265
Frayle (or El Frayle) dam 355, 381–3, 400
Frayle diversion gallery 239, 240
Fréjus (France) 383

Gerlos pressure shaft 242, 243
Gittaz gallery 130
Glärnisch landslide 220
Glen Moriston power development 238
Gleno buttress dam 379
Gondo 241, 424
Gondo pressure shaft 241
Göschenen 306
Gotthard railway tunnel ix, 11–13, 226, 281, 301
Gotthard road tunnel 226, 261, 265, 266, 281, 303, 304, 306, 324
Gotthard railway basis tunnel ix, 303, 305, 307
Gran Carero dam 352
Grandvall dam 126, 387
Gulf of Mexico 152
Gunnison River development 368

Haas pressure tunnel 238
Hanover 450
Hoosac tunnel 226

Idaho (U.S.A.) 379
Idbar arch dam 380, 381, 400
Inguri tunnel 263, 266, 280
Innertkirchen 448
Iran 165, 350, 359, 365
Iraq 342, 393
Isère–Arc tunnel 13, 14, 17
Italy 424, 450

Jogne (La Jogne) 24
Jablanica reservoir 380
Japan 268, 340, 450

Kafue 318, 436, 442, 443
Kander 13
Kandergrund tunnel 19–20, 221, 239
Karadj dam 109, 359–65, 367, 376
Karadj River 359
Kariba dam 337, 355–7, 435, 436
Kariba North Bank ix, 160, 272, 317, 319, 321, 435, 443–50
Kariba South Bank ix, 101, 321, 435–42
Kaunertal power development 106
Kaunertal steel-lined shaft 322
Kaunertal tunnel 106, 132, 263, 264
Kebar dam 325
Kemano pressure shaft 241, 242
Kemano tunnel 18
Kemano underground power-station 316, 321, 322
Khusestan province (Iran) 365
Kisenyama power-house 312, 321
Klosters tunnel 222
Kurobe IV dam 195, 209, 340, 341, 349, 350, 353, 355, 368, 394, 401

La Bathie power-station 112
Lago Delio 264, 320, 321, 323, 442
Lamoux arch dam 126, 354
Lanier dam 379, 400
La Palisse dam 126
La Pozza 404, 416, 422
Lasalle Laboratory 417
Les Brévières 14, 15
Lisbon 3, 4, 393
Livishie power-house 238
Lodano–Mosagno tunnel 263
Loen Lake rockfall 212
Longarone 209, 210, 402, 403 (*see also* Vajont rock slide)
Lötschberg tunnel 14, 207, 263
Lucerne 264

Maggia power development 263, 315, 322
Maggia pressure shaft 322
Malgovert 41, 221, 279
Malgovert tunnel 14–17, 221
Malpasset dam 3, 160, 201, 203, 339, 347, 351–3, 383–401

Malpasset site 24, 29, 56, 67, 68, 84, 115, 145
Mangla power development and tunnel 270
Mantaro power development 19
Mantaro tunnel 19–21
Mauvoisin dam 335
Mauvoisin gallery 14
Mauvoisin tunnel 277, 280
Mazzolezza ditch 406, 407 (*see also* Vajont)
Milan 402, 427
Misti Volcano (Peru) 382
Möhne dam 380
Molare dam 379
Monaco tunnel 260
Mont Blanc 28, 294
Mont Blanc tunnel 9, 28, 257, 264
Mont Cenis tunnel 11, 226
Monte Jaque dam 22, 23, 325
Monteynard dam 356
Montreal 417
Morrow Point dam and underground power-station 107, 119, 120, 367–76, 401
Mosset 69
Mount Toc 402, 403, 414, 417, 420, 421 (*see also* Vajont)
Moyie dam 379, 400
Munich 265

Nantahala tunnel 238
Neguilhes dam 354
New York 3
Norad project 49

Oahé dam 269
Orange–Fish tunnel 226, 265
Ordunte dam 379
Orlu River 354
Oroville dam and power scheme 111, 117, 376, 377, 378
Oued Djen Djen (Algeria) 358

Passan–Wegscheld road 224
Pertusillo dam 115
Piave River (Italy) 402, 403
Place Moulin dam 353
Plan della Pozza 404, 416, 422
Pontesei rockslide 210, 220
Projus 308

Quebec Province 240

Rama pressure tunnel 280, 323
Rapel dam 24
Raviège dam 354
Reyrand River 385
Reza Pahlavi dam (*see* Dez dam) 165, 365
Rhône Valley 14
Rincon del Bonete dam 379
Rio Negro (Uraguay) 379

Ritom tunnell, 239
Rome 3
Roselend dam 353, 354, 390
Roselend gallery 130
Rossens dam 126, 201
Ruacana 97, 228, 265, 442, 457
Ruhr (Germany) 380

Saidmarreh landslide 210
Saint Francis dam 379
Saint Gotthard, *see* Gotthard tunnels
Saint Guèrin arch dam 354
Saint Lawrence River (Canada) 417
Saint Vaast 64, 68
Salzburg (Austria) 3, 5, 130, 222, 265, 340, 352, 392
Santa Giustina 21, 315, 319, 320
Santa Mazzenza downstream gallery 277, 319
Saussaz 310–12, 317, 320, 324
Sautet dam 333
Schiffenen dam 345
Sefid Roud dam 358
Serra Ripoli Highway tunnel 264
Serre Ponçon rockfill dam 338
Severn tunnel 226
Simplon Pan Road 424, 425
Simplon tunnel, 12, 13, 226, 263
Snoqualmie 308
Snowy Mountains scheme 265
Sonnenberg tunnel 269, 270
South Africa 226
South West Africa 228
Spain 200
Spitallamm dam 337
Spray hydroelectric power tunnel 238
Stornorrfors tunnel 321, 322
Straight Creek tunnel 98, 123, 259
Swansea 5
Sweden 98

Swedish granite plateau 82, 94, 243
Sydney water tunnel 238, 239

Tachia River 427
Tachien Dam IX, 427, 434
Tachien dam rockface 225, 272
Taiwan 427, 432
Tang e Soleyman dam 350–52, 393, 394
Tauern Mountains 265
Teheran 325, 359
Thames tunnel 11, 226
Tignes dam 14, 41, 126, 221
Toulouse 59, 60
Tunisia 358

Uraguay 379

Vajont dam 25, 86, 142, 201, 220, 348, 349, 355, 402
Vajont gorge 142, 220, 402, 403, 405
Vajont rockslide 3, 16, 23, 25, 154, 160, 201, 204, 208–11, 218–22, 393, 402–23
Venice 11
Verbano underground station 315, 450
Verdon dam 355
Verdon River 355
Veytaux underground station 264, 273, 276, 277, 310, 318–20, 323, 442, 450, 457, 460

Waldeck I 450, 457
Waldeck II ix, 101, 264, 272, 297, 308, 319, 320, 321, 324, 435, 442, 450–61
Wombeyan 75
Würgassen nuclear station 450

Yugoslavia 110, 111, 380

Zambesi River 436, 442, 448
Zambia 445, 447
Zurich 448

Subject Index

acoustic strain meters 441
active arch theory 391
age of rocks 35, 200, 201, 496
  and void index 6, 7, 58, 81
aging of rock masses 200, 201
air, permeability of rock to 59, 70
alluvium 14, 247, 388, 389
alteration of rock 117
  chemical 60, 70, 200
  by water 199
  index of, *see* void index
American Society of Engineering Geologists 2
American Task Committee 335
amphibolite 7, 378
anchorage of cables, *see* rock bolting
  anchored cables 112–15, 120–2, 276, 277, 425–7, 433, 442, 454–8, 461
  compression tests with cables 108, 120, 273, 262–6
  forces in cables 120–2, 223–5, 320, 427, 434, 454, 455, 458
  reinforcing walls 321
  rock reinforcement 222, 224
  safety factor 426
  shear tests with cables 110, 118
  testing 454
anchorages, examples:
  Baji-Krachen rock spur 424–7
  Castillon dam 355, 356
  Chaudanne dam 355, 356
  Kafue 442
  Kariba North 447
  Lago Delio 320, 321
  Tachien Dam rock face 430–4
  underground power stations 442, 443
  Veytaux 276, 277, 318–20
  Waldeck II 320, 454, 455, 458, 459
anchored dams 120–2, 330, 355–6
anchors, steel 273, *see also* rock bolts
anchor systems 454
andesite 381, 382
anisotropy
  analysis of 186–92, 194
  of rock masses 28–30, 111, 112
  of rock material 7, 39
arch of rock, self-sustaining 275, 276, 290–4

arch dams 3, 325, 380, 381, 383
  abutments of 103, 126, 337, 340
  drainage for 343, 344, 353, 355
  foundations of, *see* dam foundations
  reserve of strength of 381, 390, 400–2
  rotation of base of 343–5
arch effect 351, 390
arch gravity dams 353
arching, degree of, in clastic theory 184–6
argolith 125
Austrian method
  for estimating rock jointing 31, 32
  for *in situ* tunnel tests 102, 105–6, 230, 259, 349
  for tunnel lining 241, 262–6; *see also* shotcrete
Austrian School 2
Austrian Society for Geophysics and Engineering Geology 3, 5
Austrian Society of Rock Mechanics 3, 5, 130, 212, 265, 301, 392
Austrian tunnelling method 262–6; *see also* tunnelling methods of, New Austrian tunnelling method (NATM)
autostabilization of rock slides 218, 420, 422
avalanches, comparison of rock slides to 210
axial tension on tests 42

basalt 8, 9, 38, 74, 146
bedding joints 9, 205
  orientation of 206, 207
  planes 8, 9, 206
bending test, for tensile strength 42
Bernold system 266–9
biaxial tension 42
Bieniawski on rock classification 156–8, 285
biotite 373
biotite gneiss 57, 67, 68
biotite schists 42, 369–74
birefringence method of measuring residual rock stress 97
blasting, stresses caused by 96, 372
blow-out of rock masses under dam 395–7
bolting, *see* rock bolts
bolts, *see* rock bolts

boreholes
  in grouting of dam foundations 333–5,
    336, 337
  measuring deformations of 94, 96–8,
    111, 371
borers, tunnelling machines 269–71
boundary conditions 236, 288, 293, 296
Boussinesq
  Boussinesq-Cerruti equations 161,
    168–75
  Boussinesq half space 168–70
Brazilian tensile test 43, 367
  brittle fracture of rock 44, 78, 79, 136,
    137, 178
  at surface 352, 353
  tensile strength and 44, 45, 81
  under dams 115, 123
  at Vajont 219, 417, 418, 420
buckling
  of rock face 432
  of steel lining 237, 322
buttress dams 326, 330, 357, 358
buttresses, failure of 389

cables, anchored 112–14, 262–6, 272, 273,
    276, 312, 320, 442, 443, 450, 454, 455,
    457
  consolidation by means of: dam abut-
    ments and foundations 348, 355–67;
    rock faces 425–7, 430–4; rock masses
    87, 222–5
  for *in situ* tests of deformation 112–14
calcite 50, 303, 361
Cambrian rocks 6, 17, 369, 496
Carboniferous rocks 6, 7, 15, 277, 388,
    496
caving and subsidence 2, 228, 229
cavities, strains and stresses about 2, 89–94,
    96, 98, 226, 230–7, 262–6, 300
  Hayashi's solution 54, 300; *see also*
    galleries, tunnels
cement, grouting with 16, 17, 22, 23, 87,
    240, 262–7, 272, 276, 332–5
cemented rock 6, 7, 36, 262–6
Centre d'Etudes de Bâtiment, Paris 111
chalk 151, 199
characteristic lines for rock (Lombardi)
    303
characteristics
  of rock masses 374, 497, 498
  of rock material 6–11, 35–84
chemical effects of water on rock 60, 70,
    200
circular loaded plate, strains under 106–9,
    176
circumferential stresses, round cavities 91,
    116, 262
classification of rocks 6, 7, 78–82, 310, 374
  conventional classification 8, 34, 80

crushing strength of rocks 34, 80
engineering classification of intact rocks
    8, 9, 10, 34, 78–9
  geological 5, 7, 34, 495, 497
  of jointed rock masses x, 4, 8, 9; Bieni-
    awski's approach 156–8; Lauffer's
    approach 281–3; Norwegian approach
    283–7
  of joints in 8, 9, 156, 283, 284
  by radial permeability tests 63–8
  of rocks by age 7, 46
  of rocks according to: hardness, density,
    silica content 498
  rock material 35–84
  rock quality designation (R.Q.D.) 10, 11,
    156, 283
  rock slides 219, 220, 415–18
  sandstones 7
  strain–stress curves 128–36
  by void index 81
clastic rock masses, theory of stresses in
    134–5, 138, 181–6
clay-shales 6
clayey schists 21
clays 36, 37, 229
  shear strength of 52
  slopes of 23
  strength 411
cleft water 4, 193, 431, 432
cleft water pressure 211–13, 215, 241–4,
    254, 278
  causing rock slides 212
  on slopes 211–13
  on tunnels 240–2, 254, 277, 278, 321
  under dams 330–2
  see also water pressure, interstitial water
    flow of water
coal 146, 229
cohesion of rock 27, 204, 205
  loss of 53, 102, 137
  low values of 84, 115, 116, 137, 392
  and rock falls 206, 207
  and rock slides 207
  and tensile strength 122, 123
collapse of tunnel roof or wall 14–18, 446,
    447
compacted rock 6, 7, 36
compactness of rock masses 115, 116
Compagnie Nationale du Rhône 14
compression (crushing) strength 33, 34, 81,
    157, 163, 164, 307
  correlations of: with elasticity 8, 9; with
    rate of loading 77; with rebound
    number 146, 147; with spacing of
    fissures 29; with swelling strain 37;
    with void index 7, 58, 59, 82
  of dry and wet rock 199
  tests for, *see* compression tests
  uniaxial 46

compression stress
  in dam 351, 352
  distribution of, in rock masses 396
  and elasticity 259
  and permeability 347
compression tests 45
  compression chamber 102, 105, 106, 230, 259, 366
  dispersion of results of 29, 33, 46, 54, 56, 57, 68, 82
  on rock samples 40–42, 45–50, 58, 73–80, 124, 366, 367, 375
  scale effect in 29, 46, 56, 57, 68, 87
  in situ 105–9, 359, 361, 362
  triaxial 47
  uniaxial 46
computers, use of 4, 186, 191–8, 348
concrete
  boundary between rock and 233, 294–7, 388
  comparison of rock and 37, 38, 45, 46, 125–8, 360
  compression tests on 56, 74
  creep of 128
  elasticity of 74, 142, 387
  shear tests on 53
  stabilization of potential rock slides by 222, 223
  tunnels lined with 233–5, 241
  uplift pressure in 331
concrete vaults
  Kafue 442, 443
  Kariba North 443, 444, 448
  Kariba South 438–42
  Waldeck II 454–8
concreting; early concreting technique 440–2
conglomerates 115, 116, 365
consolidation of
  dam foundations 355–8
  large excavations 320, 440–2, 443, 454–61
  rock masses: by cables, *see under* cables; by grouting, *see under* grouting; under load, 116, 134–6
  rock slopes 425–7, 430–4
contact of concrete and rock 233, 294–7, 388
continuum rock mechanics 87–9, 89–94, 161–70, 186–92, 226–37, 290–300
contour diagrams 33, 112
convergence in tunnels 117, 228, 229, 323, 377, 378
  tests for 2, 112, 117, 377
correlations of rock characteristics 30, 37, 39–40
  age of rocks versus void index 6, 58
crushing strength, tensile strength, shear strength versus void index 7, 8, 58, 116

dispersion of crushing strength versus radial percolation tests 68
E (dynamic) versus E(elastic) 42, 373
E_d/E_s versus wavelength, λ 147
E(plate load tests) versus transversal velocity 148
rebound number R versus v_L 146
seismic velocity versus void index 38
size of sample effect versus radial percolation 68
unit swelling strain versus void index 7, 37, 58
uplift forces and rock slides 421, 422
other important correlations 58, 59, 82, 116, 117, 143, 144
see also compression, elastic modulus, permeability, seismic wave velocity, void index '*i*'
Coulomb (Coulomb–Mohr) law of shear strength 5, 45, 163–6, 213, 248, 326, 419
counterforts 355
cracks, in Griffiths theory of rock rupture 71, 73, 76, 82
creep
  of concrete 128
  continuous, in rock slides 219, 220
  of rock masses 16, 123–8, 132, 413–17
  of rock material 60, 74, 82
  in rock slide at Vajont 409, 414–18, 422
  velocity of 415
creeping rock conditions 125–8, 310, 415–17
Cretaceous rocks 402–4, 408, 409, 496; *see also* chalk
critical slope 204, 206; *see also* slope
cross joints, in sedimentary rocks 205
crushing of rock masses, round tunnels 255–6
crushing strength 255–6
  versus grain diameter 78
  versus rate of load increase 75–8
  see also compression strength
crystalline rocks 35, 131, 132, 152, 205; *see also* igneous rocks
cylindrical jacks 111, 112

dam abutment blocks 356, 357, 360, 394
dam abutments 3, 4, 57, 336, 339, 382, 389, 390
  case histories 347–58
  convergence tests 377
  displacements of 125, 348, 381–3
  elasticity of 103
  failure of 379, 383, 385
  models of 123, 347–9, 352, 387
  reinforcement of 355–8, 382–4, 402
  stability of 23, 137, 138, 346–7, 351, 381 389, 390
  stresses in 169, 350–2, 389, 390, 392

dam abutments—*cont.*
　under dams 325–30
dam foundations 4, 325, 326, 330
　anchorage of dams 120
　blow out of rock masses under a dam
　　394–401
　classical approach 325, 326
　consolidation of, *see* consolidation
　deformation (slow) 398, 399
　design and construction of 347–53
　displacements of 125–8, 174–6, 342, 343,
　　401
　drainage of 337–45, 353–5
　drainage for grouting for 253, 332–45,
　　346
　failures of 23, 380–2, 387, 388, 395–401
　geomechanical model tests 347–53
　grouting of, *see* grouting
　percolating water in 330–2
　reinforcement of 355–8
　rocks suitable for 200, 396
　rotation of 343–5, 399, 401
　stability of 339, 340, 395–401
　stratified foundations 362–5
　stresses in 325–30
　zone of tensile stresses 325
dams
　choice of sites for 22, 24, 211, 240,
　　359–78
　failures of, 3, 23, 379–401
　grout curtains under, *see* grout curtains
　in situ tests 348–50
　model tests 341–53
　profiles of 343, 344
　settlement of large 125–8
　shell fracture 392, 393
　see also dam abutments, dam founda-
　　tions, *and different types of dam*
Darcy's law 333, 340
deflectometers 100, 121, 304, 459, 460
deformability of rock 2, 29, 357
deformation
　irreversible 128–30, 133, 260, 357
　measurement of 2, 96, 100, 103–18, 230,
　　375
　modulus of 102–15, 116–18, 125–8, *see*
　　modulus of deformation
　reversible 125–7, 133
　in settlement of large dams 125–8
　shear strength 80
density of rock 35, 57, 498
dented fissure 55
deviation (standard) 56
diabase 8, 9, 74
diagenetic process 6
diorite 17, 40, 361, 362
dip of strata 30
discontinuities in rock 10, 28, 30, 33, 373
dispersion of test results 56

displacement curves, in shear tests 119
displacement vectors, at Vajont 410, 423
displacements
　in dam abutments 126, 348, 381, 383
　in dam foundations 170–6, 398, 399, 401
　inside tunnels 232–6, 255–9, 321
　under loading plates 106–9, 170–6
　under point loads 106, 108, 113, 136,
　　137
Dogger limestone 402–4
dolomite 22, 74, 211, 279, 365, 402–4
'doorstopper' method of measuring strain
　in rock 97, 323
dragbits 270
drainage
　versus grouting for dams 252, 253,
　　332–45, 346
　of large excavations 321
　of rock masses 199, 241, 242, 383
　of tunnels 241–3, 277–81
　under dams 340, 353–5, 431, 433, 441
　for Kariba South 441
　for Tachien rockface 433
dynamic tests, *see* geophysical techniques
dynamics of slides xii, 411–23

earth dams 338, 339, 418
earthquakes 382, 393, 394
École Polytechnique, Paris 47, 52, 56, 61–8,
　83, 84, 182, 347, 385, 392
effective cohesion 204
effective modulus and Poisson ratio 87–9,
　176, 177, 300
effective stress 45, 196, 204
　in rock masses 195
　in rock material 45
　shear stress 167–8
elastic deformation of rock 1, 78, 79, 88, 89
elasticity modulus E of concrete 74, 142,
　387
　compared with that of rock 125–8, 360
elasticity modulus E (static) of rock 2, 3
　compression strength and 8, 9, 39
　compression stress and 259
　correlation of different determinations of
　　116–18
　at depth 176
　determinations of: in designing dam
　　foundations 348–50; in investigation
　　of dam sites 361–5, 371–3, 375, 376;
　　by seismic wave velocities 139–42; *in
　　situ* 34, 102–16; by sonic wave
　　velocities 40–2; by ultrasonic wave
　　velocities 37
　effective in rock masses 87–8, 176, 177,
　　362–5
　fissures and 29, 33
　grouting and 277, 335
　porosity and 195

residual stress and 85, 86, 115, 116
in strain-stress calculations 93, 94, 128, 129, 171–6, 186–92, 225, 226–36
void index and 7, 57, 58, 81, 117, 142, 143
water content and 212
elasticity modulus E_d (dynamic) of rock 40, 141
ratio of static modulus to 80, 81, 147, 148, 323
elasticity modulus E_t (ratio of values at 50% and 100% ultimate strength) 8, 9, 80
elasticity modulus E_{total} (ratio of stress to deformation: bulk modulus) 1, 102–6, 125, 128–30, 366
electrical resistivity method of testing rock masses 152–4
Électricité de France 14, 60, 111, 118, 120, 125, 130–2, 387, 388, 392
engineering approach to
in situ rock measurements 322, 323
rock deformations 281, 282, 283–7, 290–4, 301
rock support estimate 287–90, 294–300, 301–8
engineering classification of jointed rock 154, 156
Bieniawski's classification 156–8, 285
Barton's classification 283–5
epoxy-grouted bolts 272, 447
epoxy resin
coating or rock samples with 61
strength of 49, 50
equipotentials 198
erosion, factors determining 21
excavations, large 276, 277, 435–40, 442, 444–6, 450, 452
excavations, rock movements caused by 220, 229
extensometers 99, 111, 123, 304, 377
for boreholes 99–101, 377
sonor 126

faces (rock) *see* slides, slopes
failure of rock
brittle failure 31, 44, 71
of dam foundations, *see* dam foundations
progressive failure of rock masses 395–401, 417, 418
of rock masses 102–60, 180, 181, 352, 353, 414, 415
of rock material 31, 71, 80
of rock slopes, *see* slides *and* slopes
of tunnels and caverns 13–18, 238–43
shear failure 31, 44, 71, 118–20, 181
tensile failure 180
visco-plastic failure 31, 71
see also fracture

failure criterion, *see* Coulomb law, Griffith–Hoek theory, intrinsic curve, Mohr circle
failures, case histories
of dams 22, 330, 379–83, 384–401
of rock slopes 219–21, 402–23
of tunnels, galleries, caverns 12–18, 207, 220, 237–43, 315, 444, 450
see also Geographical index
failure zone R_L 298
falling hammer method 148
falls of rock 17, 18, 204–7, 219–21
at Vajont 416
faults in rock 25, 29, 32, 102
bolting of 273–6
and engineering works 17, 315, 352, 392
filling material in 83, 84
rock slides along 23
filling material 17, 28, 32, 33, 82–4
effect of percolating water in 84
tests on 83, 84
finite element method (f.e.m.) of numerica
stress analysis 186–92, 228, 277, 317, 318, 319, 330, 455, 457
around galleries 297–9
under dams 330
fissuration porosity 144
fissured vault (Kariba North) 448
fissure shearing 55
fissures in rock xi, 20, 21, 28
along dam heel 398–401
classification of 27, 35
and compression strength 57
deformation of 88
density 31
and effective elasticity 176–7
estimates of extent of 32, 74, 75 139, 222
families 29
grouting of 21, 22, 251, 341, 345, 346, 355, 356, 383
ice in 212, 219
M-shaped above Vajont slide 403, 417, 420, 422
over pressure tunnels 244, 245
rock slides along 389
samples 49, 53, 54
spacing 31
and stability 178, 179
and stress distribution 137
tensile strength and average length of 76
water flow in 193–6
water pressure in 241
flat jacks 97, 377
floods, dam failures caused by 379–81
flow of rock masses 197, 198, 416
continuous flow 218, 220, 414
discontinuous flow 218, 222, 414, 417
see also slides

flow of water in rock masses 60, 61, 63–7,
    197, 198, 212, 346, 347
  computer solution 197, 198
  convergent or divergent flow 63–7
foliation, plane of 39, 40
foundations, *see* dams, dam foundations
fractures of rock 31, 44, 71–9, 381, 414,
    415
  correlation between RQD and frequency
    of 10
  flow of water in 197, 198
  planes of 39, 40
  see also brittle and shear fractures,
    sheeting, *and* tensile failure
friction
  angle of 49, 50, 204, 205, 210, 253, 409,
    410; for different rocks 229; tests for
    373, 374, 425; coefficient of, for filling
    material 33; internal 52, 163, 181
friction factor 72, 197
  among faults 352
  for rock on rock 220, 409
  and shear strength 5
  at Vajont 218, 412, 417, 423
friction force 209
full-face excavation 15–18, 321

gabbro 75
galleries 226–324
  circular 89–94
  discharge 243
  drainage of 241–3
  parallel 226–7, 228
  pressurized, *see* pressure chamber tests
  scour 383
  see also tunnels
gap 27
gas, underground storage of 243, 248
gauges
  strain gauges 94–6, 375
  rosette of 95
geoisotherms 12
geological mapping 30, 373
geological studies (overall) of sites, *in situ*
    rock-testing programmes
  general remarks 345, 346
  Allt-na-Lairige dam 335, 336
  Baji-Krachen 424, 425
  Dez dam 365–7
  El Frayle 382
  Gotthard tunnels 306, 307
  Idbar dam 380
  Karadj dam 362–5
  Kariba dam 356, 357
  Kariba North Bank 445–8
  Kariba South Bank 435–8
  Kutobe IV dam 349–50
  Malpasset dam 388, 389, 398–401
  Morrow Point dam 367, 370–6

Oroville dam 376
Pertusillo dam 115
Tachim dam 430
Tang e Soleyman dam 350, 352
Vajont dam 348, 349
Vajont rock slide 403–9
Waldeck II 451–3
geological planes 30
  see also dip *and* strike
geology
  of dam sites 21–5, 345–7, 361, 370–6,
    388, 389
  and rock grouting 345–7
  and rock mechanics 5–11, 25
  of underground power stations 308, 310,
    315, 317
  of Vajont slide 404, 406–8, 416
geomechanics classification *see* classifica-
    tion of jointed rock names
geophysical techniques, *see* electrical resis-
    tivity, seismic waves
Georgia Institute of Technology 57
geothermal gradients 11, 12
glaciers 200
gneiss 7, 57, 74, 212, 354, 393, 424, 437,
    445
  for dam foundations 200, 356
  effect of water on 212
  fissured 24, 29, 57–9, 212
  kaolinized 264
  Malpasset 67, 68, 82, 84, 145, 147, 388,
    389, 393, 396, 398
  permeability of 63–8, 396
gneiss complex (migmatite) 445
gouge 18
granite 8, 9, 20, 29, 38, 40, 42, 58, 59, 79,
    120, 212, 321
  compression strength of 45, 49, 50
  creep curve for 125
  for dam foundations 200
  density of 59
  fissured 144
  permeability of 66–8, 70
  power stations in 321
  rebound number of 146
  resistivity tests on 152
  shear strength of 119–21
  slopes in 205
  strain–stress curves for 74, 132
  swelling of 36
  weathered 60, 61, 125, 132, 140, 200
gravity dams 326–32, 389, 395
Griffith–Hoek theory of rupture of rocks 68,
    71–3, 67, 78, 82, 98, 102, 196
groundwater flow 197, 198
grout curtains under dams 21, 22, 330,
    332, 333, 338–44
  shape 341–4
  thickness 338

water seepage through 338
grouting
  with cement 16, 22, 87, 332, 334, 335
  consolidation of rock by 21, 22, 251, 267, 277–81, 332, 333, 345, 346, 355–8
  versus drainage, for dams 332–44, 345–7
  at Malpasset dam 386–7
  pressure 10, 251, 277–81, 335
  Rama tunnel 280, 323
  of rock round tunnels 251–3, 277–81
  with silicates 16, 279, 336, 338
  technique of 15, 16, 277, 332–7, 341
  of tunnels 251–3, 277–81
  water losses 340
  water tests 333–5, 337
grouting (examples)
  Blenio tunnels 279
  Inguri tunnel 280, 281
  Kariba South Bank 441
  Malgovert tunnel 10
  Malpasset dam 386, 387
  Mauvoisin dam 336
  Mauvoisin tunnels 277–9
  Rana tunnel 280, 281
  Sautet dam 333
gunite 277, 278
gypsum 20

half-space (Boussinesq–Cerruti) homogeneous elastic 122, 138, 168–78, 186–92
  clastic half-space 136, 137, 181–6
  fissured and continuous 396
  loaded half-space 168–76
hardness of rocks 498
heading, excavation methods 15–18, 301, 303
  bottom heading 17, 18
  top heading 15, 17
heat flow, equation for 12, 13
Heim's hypothesis 85, 88, 89, 92
  see also residual stresses
Heim's paradox 88
  see also effective Poisson ratio
homogeneous zones 27, 32, 138
horizontal stress
  in dam foundations 325–9
  none from hydrostatic force 213–15
  in rock 21, 85, 116
horizontal thrust from rock walls 312, 319, 320, 441, 442, 448
hydraulic gradients 193
hydraulic potential lines 193, 254
hydrodynamic forces on rock masses 213–15 230, 232, 418–21
hydrogeology 13
hydro power tunnels 14–17, 20, 21, 226, 230–7, 277–80, 311–17

hydrostatic pressure
  on dam foundations 325–9, 330–2, 335, 341, 345–7, 381, 395–401
  distribution of 85, 98
  and residual stress in rock 85, 98
  on rock masses 213–15
  in rock salt 124, 125
  stresses round pressure tunnels caused by 105, 106, 230–7
hydrostatic pressure chamber 103, 104, 266; *see* pressure chamber tests

ice
  discontinuous flow of 417
  inrock fissures 212, 219
ideal stress 44, 137
igneous rocks 7, 9, 35, 362
immersed and partly immersed masses 213–16
inductive deformation measuring cascade 460
indurated rock material 6
inflow of water 15, 197, 198
in situ measurements
  about large excavations 259–61
  Gotthard Road Tunnel 304
  Karadj dam 176, 361–5
  Kariba South Bank 441
  about tunnels 259–61
  Saussaz 310–12, 320, 324
  Waldeck II 458–60
in situ stress measurement 97–8
in situ tests 322, 324, 388
  borehole tests 96, 97
  cable tests 112–14, 426
  compression tests 102, 104, 105–11, 230, 259, 348
  on dams 125–8, 348–50, 398, 399
  permeability tests 270, 271, 370
  plate-bearing tests 106–9, 362–4, 366, 388
  prestressed anchors 454, 455, 460
  resistivity tests 152–4
  on rock slides 221, 222, 410, 418
  seismic tests 139–45, 150–2, 371, 415
  shear tests 118–22, 341–52, 373–5
  stress measurements 94–100, 111–15, 121
  triaxial tests 109
Institution of Civil Engineers, London 337, 319
instrumented test sections 265
integral sampling method 11
International Committee on Large Dams (ICOLD) 3
International Society of Rock Mechanics 3
interstitial water, *see* cleft water
intrinsic curve 45, 50, 53, 78, 123, 134–6, 153, 164, 165, 178–81, 275
  definition of 163
intrinsic zone 53

isotherms 13
isotropic rock masses 72, 377
Istituto Sperimentale Modelli e Strutture
    (ISMES) 44, 45, 95, 347, 352, 353, 401

jacks
    cylindrical 111, 112
    field tests with 106–9, 118–22, 124, 125,
        132, 351, 356, 357, 361, 362, 366, 367,
        371, 373, 376, 377
    flat or Freyssinet 95, 109–11, 117, 121,
        335, 336, 348, 365, 366
    in toe of dam 357, 358
joints
    alteration number 286
    bonded joints 49, 50,
    classification 8, 9, 10, 27, 28, 29, 31, 49,
        156, 205, 283, 284
    continuity 156
    cohesion 55, 205
    cooling joints 9
    critical angle of slopes 204, 206
    differences between laboratory and *in
        situ* tests: of deformation 117; of
        elasticity 362–5
    orientation of 30, 31, 33
    roughness number 286
    separation 31, 156
    set numbers 286
    shearing at 54–6, 71, 165–8
    spacing of 8, 10, 31–3, 176–7
    strength of 49–51
    and strength of rock 178–81
    stress relief joints 9; tectonic joints 9;
        tension joints 9, 447; survey 32;
        systems 33
    water reduction factor 286
joint meter 376
Jurassic rocks 6, 7, 29, 496

kaolinization 264, 361
kaolinized gneiss 264
Kastner–Jaeger theory 290–7
Kelvin–Voight model for rock deformations
    148
kinetic energy of rock masses 409–14

Laboratorio Nacional de Engenharia Civil,
    Lisbon 11, 57, 96, 323
Laboratory tests on rock material or filling
    material
    crushing strength 46–50, 72–5, 81, 82,
        366
    shear strength 3, 51, 52, 80, 83, 84, 375
    tensile strength 42, 72, 76, 80
    water percolation 59–68
    wave velocity 141, 142, 143–8
andslide 18
Laplace equation 195

Lasalle Research Laboratory, Montreal 417
Lauffer on rock classification 281–3
lava 382
leakage, *see* loss of water
Leliavsky test 331–3
Levy theory on uplift 330
lias 22, 496
limestone 21, 22, 29, 38, 84, 276, 402
    compression strength of 57, 78, 80
    for dam foundations 200, 355, 381
    elasticity of 40, 41, 196, 347, 348
    permeability of 62–8, 396
    strain–stress curves of 74, 133
    swelling of 36, 37
line-load tests
    for anisotropy 39
    for tensile strength 42, 43, 367
lines of displacement (Lombardi) 301, 302
lining
    stresses in 230–6, 259–62, 280, 281, 283
    of tunnels 21, 241, 280–3
lithology 5
load cells 259–61
load–strain diagram 40
loading
    consolidation under 71, 115, 116, 124,
        125
    rate of, in tests 40, 74–5, 76, 77, 117,
        127–30
    on a slope 216–18
logging boreholes 10
Lombardi on tunnelling 297–9, 301–5, 306,
    307
loss of water
    under dams 339–44
    from tunnels 278, 279
Lugeon
    grouting techniques 333–5
    unit 143, 144, 333
    water test 333–5, 337

macrofractures 28, 29, 33, 57, 79, 80, 392
    and permeability 59, 60, 66, 67
    see also fractures
Malm limestone 402–4, 409
marble 40, 41, 69, 70, 74, 75, 205
marl–clay sandstone 115
marls 6, 7, 22, 36, 37
    permeability of 396
    power station in 312, 315, 317, 319, 320
measurements (techniques) 90, 96, 102, 103
    borehole deformation 96
    convergence of tunnel walls 228, 229,
        323, 377
    deflectometer 100, 121, 304, 312, 459,
        460
    deformations 2, 96, 100, 103–18, 125–8,
        230, 375
    doorstopper 97, 323

electric resistivity 152–4
examples, *see in situ* measurements
extensometer 323, 377
flat jacks 97, 109–11, 259, 260
hydraulic pressure chamber 322
integral sampling 11
jacking tests 97, 101–11
pendulum 349
plate-bearing test 106–9, 322
residual stress 90–4
seismic waves 14, 115, 116, 139–43, 348,
   356, 357, 371–3, 418
strains 96–9, 101
mechanical properties of rock material
   40–84
Mercali scale 382
metamorphic rocks 7, 9, 28, 35, 74, 132,
   200, 369, 378
mica schists 369, 372
micaceous quartzite 369, 372, 373, 374
microfissures 28, 29, 33, 39, 40, 46, 57, 68,
   77, 81, 392, 396, 398
   induced by compression 59
   and permeability 60, 63, 68, 69, 71
   and wave velocity 77, 139
microfractures 28, 29, 33, 39, 40, 57, 69,
   80, 81, 392, 396
   and permeability 60, 61, 66, 67
migmatite 445
miniborers (full-facets) 270
mining engineering xi, 2, 226, 256, 440–2
model testing
   of dam abutments 123, 348
   of dams 116, 137, 340, 341, 347–50,
   351–3, 360, 361, 393, 402
   of percolation of water 198, 340
   of reservoir basin 405
   of rock masses 396
   of St. Lawrence River (thawing) 417
   of sliding surface 210, 405, 413
   three axial tests for Cabora Bassa 318
   of underground power-station 277
modulus 1, 8, 29, 33, 40, 41, 74, 75, 102–60
   correlation of measurements 116, 117
   dynamic modulus E_d 40, 141
   effective modulus 87–8, 176–8, 362–4
   in situ testing of rock masses 87, 99, 103,
   111, 116, 229, 259, 361–4, 365–7, 370–6
   laboratory tests 40, 41, 363
   ratio E_s/E_d 80, 372
   static modulus E_s 9, 40, 362, 363, 364
   in stratified rock masses 113, 176, 364
modulus of deformation or bulk modulus
   102, 126–30, 365–7
modulus of elasticity (Young's modulus),
   see elasticity
modulus of rigidity G 41, 42
Mohr circles 161–5, 167
   examples of use of 43, 45, 48, 50, 53, 54,

134, 135, 178–80, 229, 248, 255, 367
mole, *see* borers, tunnelling machines
moment, displacements caused by 173, 174
momentum equation 412, 413, 419
moraine 24, 211
mortar, grouting with 22, 23
mudstones 6, 7, 36, 37, 146
mylonite 18, 276, 409

National Coal Board, Britain 96, 97
needle test, for anisotropy 39
new Austrian tunnelling method (NATM)
   early attempts at the theory 263
   general information 261–6, 319, 320,
   442, 443, 454–6
   for Gotthard tunnel 306–8
   for Waldeck II 454, 455, 457–8
Newton's law for movement of rock masses
   218, 411, 413, 419
Norwegian theory on rock classification
   (Barton) 283–7

observational method (Peck) 457
odometer 199
openings in rock masses
   circular opening 89–92
   parallel circular openings 228
   rectangular opening 227
   square opening 226–7
   strains and stresses about openings, due
   to residual stresses 86–92, 228; due to
   hydrostatic pressure in the tunnel
   230–7
ophites 67
orthotropic rock 186
overburden
   grout pressure and 237, 243–5, 334, 335
   minimum for pressure tunnels 237–54,
   see pressure tunnels
overcoring method for measuring residual
   stresses in rock 96, 97, 323
overstrained rock round tunnels 96, 97,
   255–9, 290–7

parallel galleries 228
pegmatite 369, 388, 389
pendulum, measurement with 349
percolation of water through rock 60, 84,
   251–4, 337–46, 381, 394
   Darcy's law for 64, 193, 197, 340, 396
   longitudinal 62, 66
   primary and secondary 60, 194
   radial 63–8, 82
   test for, under dam foundations 396
   see also seepage
perfo-anchors, perfo-bolts 264, 272
perimetral crack (Mount Toc) 403
permeability of rock to air 21, 59, 70
permeability of rock to water 60

permeability of rock to water—*cont.*
  compression stress and 69–70, 346, 347
  correlations of: with mechanical pro-
    perties 60, 68, 71; with wave velocities
    142, 144
  grouting and 277
  Lugeon unit for 143, 144, 333, 334
  and microfissures 60
  primary and secondary 60, 212
  tensile stress and 396
  tests for 60–2, 333–4, 371
permeability factor K 61–3, 66–7, 71, 81–2,
    193–4, 346–7
Permian rocks 7, 15, 496
perviousness of rock, *see* permeability
petrographic properties of rocks 497
  petroleum industry 2
photoelastic methods 228, 277, 318, 455
phyllite 117, 142, 212, 278
phyllite quartzite 42, 117, 130, 131
physical properties of rock 35–84, 199, 200
piezometers 18, 340, 342, 343
piezometric line 338–40, 349
pilot tunnels, pilot galleries 15, 440, 458
pinning rock 223–5, 348, 424–7, 434
pipes, theory of thick elastic 230–2
plastic blocks, transmission of stress
    through 183–4
plastic deformation of rock 74, 130, 135,
    136, 247–51, 290–7
plastic fracture 78, 79
plasticity of rock 3, 29, 247–51, 346, 392
plate-bearing tests for deformations 106–9,
    116, 136, 361, 362, 366, 377
  theory of 170–6
point-load test for anisotropy 38–9
Poiseuille's formula 194, 197, 198
Poisson's ratio 29, 33, 40–2, 44, 87, 110,
    136, 141, 142, 375
  compression stress and 88, 89, 300
  effective value of, in rock masses 87, 378
polyethylene sleeves for cables 273
pore pressure 47, 48, 49, 196, 248, 367, 405
  test for 331, 332
porosity
  of grout curtain 340
  of rock 33, 35, 58, 59, 63, 73, 144, 375;
    and elasticity 196; and erosion 21
porosity index 77; *see also* void index
porous media, flow of water in 193, 194,
    215, 340
porphyroblastic gneiss 7
porphyry 42
Portuguese National Research Laboratory
    57
power-stations, underground 242, 308–22
  different types of 309, 311
  rock mechanics for 308–10, 315, 311–22
prepact concrete 242, 322

pressure, relation of percolation to 63–8,
    70, 71
pressure cells, *see* load cells
pressure chamber tests 1, 102, 103–6, 348,
    365
pressure gradients 61, 338, 340, 394, 401
pressure pipes 21, 25, 230–2, 315
  defective air valve in 19
pressure shafts 235–6, 241–3, 313, 314, 322
pressure support 259, 260, 294–9, 301–5,
    307, 457, 458–60
pressure tunnels 230, 237
  failures of 238–40
  grouting round 237, 277–81, 322
  minimum overburden above 237, 238:
    adjacent to rock slopes 245, 246; for
    different types of rocks and galleries
    240–3; in rock liable to plastic defor-
    mation 247–51; under alluvium 247;
    under horizontal rock surface 243–5
  seepage from 253–5
  stresses in, from hydrostatic pressure
    230–7
  theory of 14–21, 89–94, 230–6, 290–7
  see also tunnels
pressurized galleries 104
prestressed anchored cables 272, 273, 454
  testing 454
pseudo-shear 50
pulvino (pressure distribution slab) 344
punching shear test 51, 78

quality indexes
  radial percolation index 63–6, 81, 82
  rock mass quality Q, (Barton) 283–7
  rock quality designation, RQD (Deere)
    10, 34, 211–13
  see also void index
quarrying method of excavation 321
quartz 6, 7, 41, 374
  permeability of 60–7
quartz-diorite 18, 42
quartz-mica gneiss 7
quartz-mica schists 369, 372–4
quartz-monzonite 49
quartzite 40, 41, 74, 79, 80, 200, 429
  creep of 125
  for dam foundations 200, 356, 430, 437,
    438
  fissured 144, 278, 437, 438
  micaceous 369, 372–4, 437, 438
  powdery 15–17
  strain–stress curves for 130
quartzose phyllite 42
quartzose shale 42

radial percolation tests 63–8, 82
rate of loading, *see* loading
rate of strain 117, 118

ratio vertical/horizontal stresses for
   Cabora Bassa 310, 318
rebound number *R* 145-8
rectangle
   strains round a 2, 226, 227
   strains under a loaded 170-3
relaxation of rock masses xii, 377
   see also sheeting
Repeatable Acoustic Seismic Source 150-2
reservoirs 21-5, 213-16, 369, 379, 380
   382, 402-23, 429, 436, 451
   displacements of embankments of 348
   variation of water level in, and rock
      slides 211, 213-16, 355, 403, 404,
      417-22
residual stresses in rock 85-94, 116, 250,
      310, 318, 376
   and elasticity 85-6, 115, 116
   measurements of 94-6, 115, 116, 323,
      366, 367, 376
   relaxation from, *see* sheeting
   static equilibrium method 101
   tensile strength 49
resins, impregnation of rock with
   coloured 59
resistance quotient, of jointed rock 179,
   180
resistivity, *see* electrical resistivity
resonance waves 19, 41
reversible deformations 126, 127, 128, 133
rigidity, modulus of 41
rock arch action 257, 263, 264, 266, 275,
   276
rock bolt extensometers 117, 123
rock bolting 18, 112, 166, 271, 273-6, 277,
   442, 447
rock bolts, different types of 271-2, 447
rock bursts 1, 256, 297, 446-8
rock characteristic lines (Lombardi) 303
rock characteristics for tunnelling
   machines 269, 270
rock creeping at Vajont 409, 415, 422
rock classification
   see: engineering classification of jointed
      rock
rock deformations 128-31, *see* deformation
   drainage 242, 243
   elastic 134, 233, 294
   galleries 240, 243
   plastic 135, 250, 290; *see* plastic
      deformation of rock; without rock
      support 290; with rock support 295;
      fissured rock masses 235; stratified
      rock 113, 132, 176, 364
rock-fill dams 240, 338, 368
rock load
   on lining 260, 261
   on steel supports 260
rock mass quality *Q* (Barton) 283-7

rock masses 27
   classification of, *see* engineering
      classification of jointed rock
   structure and anisotropy of 28-30
rock material 28, 35
   classifications of 80-4
rock quality designation (RQD) 10, 34,
      142, 143, 156, 211-13, 283, 286
rock slides, *see* slides of rock
   classification by Heim 203, 204, 220
   formation of 204, 208, 416, 420
   shape of sliding surface 208, 209, 210,
      413, 416, 420
rock support estimate
   general remarks, 287
   accepted values for pressures 320, 435
   based on rock deformation ix, 301-6
   case of homogeneous rock mass 294, 295
   computer programme by Cundall 191
   elastic rock deformations 294
   finite element method 297-9, 300-18
   limit, R_L, of fissuring 292, 293
   plastic rock deformations 290, 295-7
rock-support techniques 442, 443
   Kafue 442, 443
   Kariba South Bank 440
   Veytaux 320
   Waldeck II 320, 454, 455, 458, 461
rupture of rocks 163, 165
   Griffith-Hoek theory of 68, 72-3, 76,
      78, 98, 102-96
   point of 78, 79
   Torre's theory of 78
   see also fractures, dam foundations,
      rock slides

safety factor 218
   against sliding 419, 421
   for dams 367, 387
   for tunnels 237, 238, 244, 246, 250, 251
salt mines, creep in 123, 124, 125
sandstones 6, 7, 38, 40, 41, 74, 75, 79, 80,
      147, 229
   compressibility of 240
   permeability of 62, 65, 67, 240, 396
   stresses in 196
   swelling of 36, 37
saturation swelling stress 37
scale effect
   in compression tests 46, 56, 57, 68, 82
   for dam testing 172, 364
   in deformation and displacement tests
      102, 125, 126, 176
   in elasticity tests 364
   scale factor 68
schistosity 389
   planes of 119, 120
schists 17, 42, 73, 74, 241, 242, 278, 369,
      378

schists—*cont.*
  creep of 125, 133
  for dam foundations 200
  permeability of 62, 396
  shear strength of 120
sedimentary rocks 9, 28, 35, 152
  slopes in 205
seepage
  from pressure tunnels 252–4, 277–8
  in rock masses 205, 215, 339, 340, 382, 383, 396
  under dams 338, 347, 391, 396
  see also model tests, percolation
seismic wave tests on rocks 10, 14, 115, 116, 139–43, 146–9
  on dam sites 348, 356, 357, 371–3, 409, 415, 422
  for detecting incipient slides 154; and weakening rocks 415, 422
  inside galleries 373–4
  under water 151, 152
  wave path 140
  wave velocity ratio 10, 142, 143
seismic wave velocity 10, 29, 141–4, 149, 348, 415
  correlations of: with aging of rock 200, 201; with elasticity 141, 144, 146, 147, 148, 371, 372; with frequency 148; with permeability 143, 144; with ratio E_d/E_s 96, 147; with rebound number 145, 146; with rock fissuration 77, 139 222; with void index 7, 37, 38, 84, 118
  for testing tunnel walls 96, 257, 258, 263, 317, 323, 372
  longitudinal and transversal 41, 77, 141, 142, 147, 148, 150
  ratio of, to laboratory sonic wave velocity 10, 33, 34, 142, 143
  under compression 147
separation of joints 157, 159
sericite 389
shafts, steel-lined 235, 237, 241–3, 322
shales 6, 7, 23, 27, 37, 38, 42, 196, 229
  rebound number of 146
  swelling of 36, 37
shape of large excavations 319, 440, 442, 444, 454, 455–7
shear fracture 31, 51, 52, 71, 163, 178, 204
  failure of dams by 123, 393, 394
  in plate tests 116
shear strength 3, 51, 55, 83, 102, 119, 120
  classification of rocks by 80
  of rock joints weakened by water 405
  of rock masses 405, 410
  versus shear deformation 52, 55, 80
  tests for: on rock samples 51, 56, 78, 374, 375, 425; *in situ* 34, 118–22, 349, 350, 375
  and void index 58, 81, 119, 120

shear stress 121, 122, 204
  Coulomb's law of 5, 45, 163–7, 178, 213, 248, 326, 419
  on dam foundations 326–30
  displacement caused by 173, 174
  effective 167, 168
  resistance to 179, 180, 204
  and rock slides 204
shear, tests on
  samples 51
  rock masses 118–22, 373–5, 435
sheeting of rock masses (stress relaxation) 9, 86, 201, 204, 212, 220, 370, 371
shotcrete 241, 262, 263, 264
  reinforced shotcrete 241, 264, 442
silica, dissolved from granites 60
silicates, grouting with 16, 279, 337
siliceous rocks 199, 497
silt 154
siltstones 74, 196
single-pulse method 41
skin resistance 257, 258, 264
sleeves for cables 273, 454
slides of rock 16, 23, 201, 204, 207, 220–3, 381–3, 403, 404, 418–23
  along faults 397, 398; and fissures 389
  autostabilization of 218, 420, 422
  classification of 203, 204, 219, 220
  continuous and discontinuous flow of 219, 220, 222, 413, 414, 417, 418
  dynamics of 218, 219, 411–13, 419, 422, 423
  slow and rapid slides 210, 211, 212, 411, 412
  statics of 208–10, 213–16, 409–11
  supervision and stabilization of potential 154, 221–35, 424–7, 427–34
  two-phase slide 417, 418
  at Vajont 402–23
  velocity of 218–20, 405, 410–13
  volume of 220, 221, 402, 403, 414, 422
sliding friction tests 373–5
sliding surface 207–11, 218–21, 389–91, 406–9, 411–14
  deep seated 207–11, 219, 220, 222, 406–8, 416
  model of 210, 219, 405, 417
  shape of 208, 209, 210, 211, 406, 407–9, 411, 420–6
  sliding zone 414
slip, angle of 50
slip surface 406, 409
  zone 409, 414
slopes
  angle of, in rock slides 204, 206
  classification of 203
  critical angle of 204, 206, 207, 220
  external force or load on 216–18, 223, 224

stability ix, 4, 21–5, 204–11; interstitial water and 211–16, 385
tunnels adjacent to 239, 245, 246
snow, and rock falls 212
Sonar system, seismic reflection method 151
sonic waves, ratio of velocities of seismic waves and 10, 141–3
'Sparker' sound source 151
specific gravity of rock
  apparent, dry and saturated 36
  of solid mineral grain 35, 375
springs, in tunnels 19, 20
stability; conditions of, for rock masses 178, 179, 204, 213, 214
  of dam abutments, dam foundations, and slopes, *see under those headings*
  of tunnel bore 302, 303
  of tunnel heading 302, 303
static equilibrium method for determining residual stress 101
steel
  arches of 260, 264
  reinforcements of cavities 264
  shaft and tunnel, linings of 2, 104, 105, 235, 236, 237, 241, 242, 260, 261, 322; grouting behind 277
  buckling of steel linings 237, 322
storage of gas in caverns 243, 248
storage of oil in caverns 255
strain in rock 2, 4, 29, 36, 37, 47, 69
  and deformation modulus 102, 103
  measurements 102, 103
  permeability under varying 64–7
  round cavities, *see* cavities
  and thermoluminescence 69–70
strain meters 94–8, 304, 323, 324, 376
strain–stress
  curves of 33, 74, 80, 102, 128–38; analysis of xii, 74, 128, 129, 138
  mathematical approach to distribution of, in rock masses 161–92, 299, 300, 456
  measurements of 4, 259–61
  stratified rock 113, 174–6, 183, 185, 186, 364
  Talobre diagrams 134–6
strength
  of jointed rock 138, 179–81
  of materials 5, 42
  of rocks, *see* compression, shear, *and* tensile strengths
stress reduction factor 283, 286
stress tensor gauge 323
stresses 26–9, 161–6, 232–7
  caused by hydrostatic pressure 230–7
  effective, in rock masses 196
  hydrostatic distribution of 85, 88
  relief of 260, 261; at surface of rock

masses 9, 86; *see also* sheeting
round cavities, *see* cavities
see also circumferential, compression, shear, *and* tensile stresses
strike 30
struts 312, 321
subsidence and caving 2, 228, 229
surge tanks
  general 208–10, 322, 439–43
  Kariba North (missing tank) 444
  Kariba South 438, 439
  Ruacana 228
  shape of 228
  Waldeck II 261
surveys (*see also* geological studies)
  geological 32
  geophysical 139–54, 370, 406, 415, 422, 423
  of rock slides 221, 222
sustaining arch 264
swelling of rocks 15, 36, 199, 200
swelling strain
  and compression strength 37
  and void index 7, 37, 57–9, 81
Swiss Federal Railways 1
syenite 45

tectonic forces 9, 85
television, in bore holes 371
temperature
  and aging of rock 200
  in tunnels 11–14
tendons 272
tensile failure of rock 64, 71, 105
tensile strength 42, 49, 72, 76, 102, 164
  and brittle fracture 42–4, 80
  correlations 40
  of rock masses 118, 122, 123
  tests for 42–4, 367, 377, 378
  and void index 58, 81
tensile stress
  about cavities 89, 98, 317, 318
  at heel of a dam 325, 326, 328–30, 346, 347, 392, 395–9
  and permeability 64–7, 396–9
  resistance to 179–81
  in rock masses 72, 73, 122, 123, 208–10
tensile tests 42
  bending test 42
  Brazilian test 43, 367
tension gashes 447, 448
tension joints 9
test sections 265
testing anchors 323, 454
testing methods
  cable testing method 112, 113
  in situ tests for *E*-modulus 104–5
  in situ convergence tests 2, 112, 229, 321, 377

testing methods—*cont.*
  jacking tests 106–9, 118–22, 322, 347, 361,
    362, 366, 370, 373–6
  pressure chamber tests 104, 105, 106,
    348, 366
  results 362–4, 366, 367, 373–5, 378
  see also laboratory tests, percolation,
    resistivity, seismic waves, shear
thermoluminescence tests 69, 70
three axial tests for Cabora Bassa 318
three-dimensional f.e.m. model tests 186,
    188
transformers 319, 322, 439, 457
triassic rocks 279, 496
triaxial compression tests 47–50, 58, 59, 69,
    71, 80, 109, 116, 145, 147, 196, 349, 350,
    375
tuff, 74, 361
tuffaceous rock 361
tunnel borers, tunnelling machines
  see borers
tunnelling, methods of 15–18, 226, 241,
    242, 261–71, 318–22, 372
tunnels
  bolting of roof of 273–96
  caving of 15, 17
  erosion 21
  grouting of *see* grouting
  lining of 21, 233–6, 241; estimation of
    required thickness 103, 261, 281, 289
  overstrained rock round 255–9, 290–300
  prediction of rock formations in 13–21
  repair (Kemano) 18
  strain and stress measurements in 94–6,
    99–101, 111–13, 259–61, 312, 322–4,
    377, 458–60
  stresses round *see* cavities
  supports 14–18, 259–61
  temperature in 11–14, 86
  water inflow 15, 16
  see also pressure tunnels
turbines 310, 319, 451, 457

ultimate strength 8
ultrasonic wave velocity in rock 37, 142
underground works 308–24
  anchors 319–21, 454
  design and construction 318–22
  drainage 321
  examples 306–7, 311–17, 435–42, 443,
    441–61
  hydroelectric power-stations 309, 310,
    322
  measurements 322–4, 458–60
  residual stress 318
  shape of excavation 317, 440, 444,
    452
  stress–strain analysis 290–7, 317, 318
unit block of rock 31, 40

United States Army Corps of Engineers
    47, 48–50
United States Bureau of Mines 2, 76, 97
United States Bureau of Public Roads 154
United States Bureau of Reclamation 35,
    47, 56, 95, 98, 107, 108, 119, 347, 368,
    372–5, 466
uplift pressure
  Levy on uplift 330
  on rock immersed in water 419–21, 423
  on rock round pressure tunnels 253–5
  pressure line 338
  tests by Leliavsky 331–2
  under dams 3, 330–2, 337–9, 341–3, 345,
    346, 379, 394, 397
  in undrained rock 241

valleys, rock slides into 210
valves
  to let cleft water into tunnels 278
  in pressure pipe 19
vaults without concrete (Ruacana) 228,
    265, 457
velocity of waves 10, 142–4, 149, 415; *see*
    seismic waves
ventilation of tunnels 13, 452, 457
vertical stress
  in dam foundations 326–9
  in rock 101
vibrations in rock
  measurement of 348
  levels 150
  visco-elastic 145, 148, 149
visco-plastic deformation of rock (some-
    times leading to failure) 45, 71, 136,
    137, 163, 305, 352, 353
  at Vajont 417, 418, 423
void index 6–8, 34, 55–6, 57, 76
  correlations of: with age of rocks 6, 7,
    58, 81; with compression strength 7,
    57, 58, 81; with elasticity 7, 58, 81,
    117, 142, 143; with rebound number
    147; with shear strength 58, 81, 120;
    with swelling strain 3, 37, 57, 81; with
    tensile strength 58, 81; with wave
    velocities 7, 37, 38, 81, 118, 142
  and creep 74, 75
void ratio 35
void in rock
  density of 81
  shapes of 57, 58, 61, 62, 63, 81, 396
volcanic area 381, 382
volumetric strain–stress curve 73

water
  chemical effects of, on rock 60, 71, 200
  flow of, in fissured rock 93–6; and frac-
    tured rock 197–9

laminar and turbulent flow of 197, 198
percolation of, through rock, *see* percolation
permeability of rock to, *see* permeability
rock masses immersed in 213–16, 418–
  20, 423
in rock 35; and compression strength
  199; and deformation modulus 117;
  and elasticity 212; and resistivity 153,
  154; and wave velocity 37
water hammer 19
  waves from 19, 221, 230, 237, 239
water loading test, for determination of
  deformations 104–6
water pressure
  interstitial 16, 45, 338, 399, 400; *see also*
  cleft water pressure

in reservoir slopes 355; *see also* pressure
  tunnels, reservoirs, slides
in rock fissures and joints 241, 334, 383
in testing permeability 333, 334
see also hydrostatic pressure
water table 19, 20, 215, 222
water-tightness of reservoirs 21–3
waves
  on water surface, caused by rock falls
    212, 412
  see also seismic, sonic, *and* ultrasonic
    waves
weathering 156, 157
wing wall 381, 389
work curve 129

zeolite 361